Petroleum Refining Processes

Petroleum Refining Processes

Edited by **Michael Dedini**

SYRAWOOD
PUBLISHING HOUSE
New York

Published by Syrawood Publishing House,
750 Third Avenue, 9th Floor,
New York, NY 10017, USA
www.syrawoodpublishinghouse.com

Petroleum Refining Processes
Edited by Michael Dedini

International Standard Book Number: 978-1-68286-371-8 (Hardback)

Printed in China.

Contents

Preface

Petroleum engineering has undergone rapid transformation in the last few decades. The refining processes for conversion and extraction of crude oil from petroleum deposits particularly have evolved a lot. This book presents a thorough understanding of petroleum refining processes by discussing some of the significant topics in the field like the extraction procedures, design and maintenance of petroleum refineries, distillation of petroleum and petrochemicals, etc. The various studies that are constantly contributing towards advancing technologies and evolution of this field are examined in detail. This book will serve as a reference to a broad spectrum of readers.

The researches compiled throughout the book are authentic and of high quality, combining several disciplines and from very diverse regions from around the world. Drawing on the contributions of many researchers from diverse countries, the book's objective is to provide the readers with the latest achievements in the area of research. This book will surely be a source of knowledge to all interested and researching the field.

In the end, I would like to express my deep sense of gratitude to all the authors for meeting the set deadlines in completing and submitting their research chapters. I would also like to thank the publisher for the support offered to us throughout the course of the book. Finally, I extend my sincere thanks to my family for being a constant source of inspiration and encouragement.

Editor

1

Oxidation of benzylic alcohols over Cr (MCM-41/ZSM-5) assisted by microwaves

Zoubida Lounis · Naouel Boumesla · A. E. K. Bengueddach

Abstract The oxidation of benzylic alcohols in the machine of microwave (Discover, CEM-SP1245 model, USA) by the composite material type Cr-(MCM41/ZSM5) with high crystallinity, exchanged at low content of chromium has dramatically reduced the reaction times and have high conversion rates and high yields compared to conventional heating. Ketones were obtained for only 10 min of reaction. The composite materials used containing both micropore and mesopore structures are stable for this type of reaction. The method uses composite materials Cr (50-MCM41/ZSM5) as the catalyst system and *t*-BuOOH as the stoichiometric terminal oxidant. The reactions carried out by the microwave and made without the use of an organic solvent, because, the substrates are mixed directly or, in our case, the subtract is mixed with silica which is inactive in the reaction.

Keywords Benzylic oxidation · Microwaves · Composite materials · Catalyse · Chromium exchange

Introduction

The oxidation of primary and secondary alcohols to the corresponding carbonyl compounds is a fundamental reaction in organic synthesis (Sheldon and Korchi 1981; Ley and Martin 1991). Such transformations are most frequently accomplished by the use of Cr(IV)-based reagents in amounts ranging from stoichiometric to large excess (Cainelli and Cardillon 1984). Problems in the use of Cr(IV) complexes as oxidation reagents often crop up due to the lack of selectivity in oxidations, safety hazards associated with the use of large quantities of toxic chromium compounds and the need of aqueous acidic or basic conditions for reactions of chromate salts. In attempt to circumvent these problems, attention has been focused recently on the use of transition metal incorporated zeolites (e.g., TS1, VS1, etc.).

The first example of a redox molecular sieve was titanium silicalite-1(TS-1) developed by Enichem workers (Taramasso et al. 1983; Clerici 1991; Clerici et al. 1991; Romano et al. 1990) which proved to be active catalyst for selective oxidations with 30 % aqueous H_2O_2 in the liquid phase under mild conditions. Examples of oxidative transformations catalysed by TS-1 include the hydroxylation of olefins and aromatics (Thangaray et al. 1990), and the selective oxidation of primary alcohols to aldehydes and secondary alcohols to ketones (Esposito et al. 1984). The success of TS-1 stimulated the investigation of broad range of metal-substituted molecular sieves as oxidation catalysts, e.g., ZrS-1 for the hydroxylation of benzene to phenol with H_2O_2 (Dongare et al. 1991), cobalt-substituted aluminophosphates (CoAPOs) as catalysts for the oxidation of cyclo-hexane with O_2 (Czarnetzki 1989), vanadium-substituted aluminophosphates (VAPOs) as olefin epoxidation catalysts (Haanepen and Van Hooff 1992; Rigutto and van Bekkum 1993), chromium-substituted silicalite 1 (CrS-1) for the oxidation of alkyl aromatics, alcohols, and allylic oxidation of olefins (Chen et al. 1995; Lempers and Sheldon 1996) and exchanged CrZSM5 for oxidation of secondary and primary benzylic alcohols to the corresponding organic components(ketones, aldehydes) (Lounis 2006). More recent developments include the incorporation of redox metals into the framework of mesoporous molecular sieves, such as

Z. Lounis (✉)
Institut de Maintenance et de Sécurité Industrielle, Université d'Oran, 1524, El Manouer, Oran, Algeria
e-mail: lounis_amira@yahoo.fr

Z. Lounis · N. Boumesla · A. E. K. Bengueddach
Laboratoire de Chimie Des, Matériaux Université d'Oran, 1524, El Manouer, Oran, Algeria

MCM-41 (Zhang et al. 1996). One aspect which is generally overlooked in the euphoric rush to develop new catalytic systems is the stability and, hence the true heterogeneity of these materials under oxidising conditions.

On the other hand, the fine chemical industry is confronted with market inquiries of increasing molecular complexity, notably in the area of advanced pharmaceutical intermediates. To respond positively to such inquiries, it needs broadly applicable synthetic methods that are effective under mild conditions and that fulfil economic and environmental constraints. With respect to alcohol oxidation, very few currently available methods meet with the above methods. The use of the microwave technology can reduce the reaction times and energy consumption, and may lead to an increase in yields and selectivity in some cases (Gronnow et al. 2005).

However, the microwave as non-conventional source of energy was largely accepted to carry out organic synthesis; this was supported by the increase in numbers of publications in these last years, with the general availability of the new instruments of microwaves. A great number of examples of reactions were described in the organic synthesis (Strauss and Trainor 1995).

The reactions carried out by the microwave and made without the use of an organic solvent, because, the substrates are mixed directly or, in our case, the subtract is mixed with silica which is inactive in the reaction.

We report here a simple and efficient method for the oxidation of secondary alcohols to ketones under mild reaction conditions using the new instrument microwave discover as source of energy. The method uses composite materials Cr (50-MCM41/ZSM5) as the catalyst system and *t*-BuOOH as the stoichiometric terminal oxidant. The oxidation reaction was carried out free from solvents.

The composite materials used as catalyst were molecular sieves with a bimodal pore size distribution, combining the advantages of both the microporous and the mesoporous materials. This twofold molecular sieve might have potential applications in adsorption and catalysis for bulky molecules (Meynen et al. 2007). For instance, these materials proved to be also active for solid acid catalysts (Ordomsky et al. 2007). From catalytic point of view, the composite materials Cr-(50-MCM41/ZSM5) prepared by our research were investigated on the oxidation reaction of the benzylic alcohols.

Experimental section

The chemicals

The chemicals used for the syntheses were colloidal solution of silica (Ludox HS40, dupont, 40 % SiO_2, 60 % H_2O), tetrapropylammonium bromide (TPABr (97 %, Aldrich) cetyltrimethylammonium bromide (CTMA Br) (98 %, Fluka), sodium aluminate $NaAl_2O_3$ (65 % Al_2O_3, 27 % Na_2O, Aldrich), sodium hydroxide (98 %, Fluka) and deionised water.

The products of exchange are the salts of chromium nitrate (>99 %, Aldrich), the deionised water and the $MeNH_2$ (40 wt % in water, Fluka).

The catalyst preparation

A synthesis procedure described earlier (Karlsson et al. 1999) was modified for mixtures of the $C_6[C_6H_{13}(CH_3)NBr]$ and $C_{14}[C_{14}H_{29}(CH_3)_3NBr]$ templates/surfactants. The concentrations of ammonium were scaled according to: C_6 (wt) + C_{14} (wt) = 100 %, where C_6 (wt) = 50 % and C_{14} (wt) = 50 %.

A synthesis of the composite materials Al MCM41/ZSM5 was obtained by two-step procedure:

Solution A

The mixture of the template agents was dissolved in water under magnetic stirring at room temperature for 15 min.

Solution B

For the preparation of aluminosilicate material, aluminate sodium used as source of Al was added to sodium hydroxide and subsequently dissolved in water at 50 °C for 10 min. Then, a solution of silica ludox (40 % SiO_2, 60 % H_2O) was added to the previous solution under stirring with magnetic stirrer at room temperature for 15 min.

The solution B was poured onto the solution A dropwise under continuous stirring for 24 h at room temperature. The gel of the surfactants/silica mixture obtained has the following composition: SiO_2, 0.01 Al_2O_3, 0.20 Na_2O and 700 H_2O. The (C_6 + C_{14}):SiO_2 molar gel ratio was maintained at 0.4

Finally, the gel mixture was transferred into the tubes of glass Pyrex sealed by Teflon ribbon to avoid the loss of material during heating. The tubes were placed vertically into beakers and were introduced into the drying oven. Then, the gel was crystallised at the temperature of 120 °C under static conditions and atmospheric pressure for 4 days. The resulting solids were recovered by filtration, completely washed with distilled water three to four times and then dried to 100 °C overnight. The calcination of the as-synthesised materials was carried out at 500 °C with heating rate of 3 °C/min in air for 6 h to remove the template molecules. The obtained solid is denoted by 50-(MCM-41/ZSM-5).

The cationic exchange of the 50-(MCM-41/ZSM-5) by chromium ions was obtained by the following procedure:

1 g of the calcined product was ion exchanged in 100 ml of a solution of 0.5 M NaCl for 24 h at ambient temperature and pH 5–6, this initial Na^+ ion exchange was followed by a second (24 h) under gentle agitation in fresh solutions to obtain Na-50(MCM41/ZSM5).

The Na-50(MCM-41/ZSM-5) samples were subsequently exchanged with Cr^{3+} ions, again with 1 g solid in 40 ml solution of 0.1 M $Cr(NO_3)_2$. The chromium ion exchanges were done under stirring at 50 °C for 3 h. The pH control was employed throughout all ion exchanges and the pH value was adjusted to the desired range (5–6) by adding $MeNH_2$ (40 wt % in water) before introducing the solid. The final solid was recovered by filtration, washed several times with de-ionised water and dried (150 °C, 5 h) before being analysed by various characterisations. The chromium-exchanged composite material obtained is denoted by Cr-(50-MCM-41/ZSM-5).

Catalyst characterisation

The Cr-(50-MCM-41/ZSM-5) was characterised by powder X-ray diffraction using a Philips PW 1801/29 diffractometer with CuKα radiation. The similarity of the intensity of the main peaks of Cr-(50-MCM-41/ZSM-5) and (50-MCM-41/ZSM-5) spectra indicated that the alteration of the structure in the course of the ion exchanges is not significant (Figs. 1, 2).

Specific surface area measurements were obtained from volumetric nitrogen adsorption using ASAP 2010 instrument. Nitrogen adsorption measurements were performed at liquid nitrogen temperature. The BET surface area measured was 560 m^2/g.

The analysis of X-ray fluorescence give the composition in percentage of the Cr-(50-MCM-41/ZSM-5), from this composition we deduct the concentration of chromium in 1 g of product, in our case we use only 5 mg of product which correspond to 0.05 mmol of chromium.

Catalytic testing

The oxidation of the secondary benzylic alcohols over the Cr-(50-MCM-41/ZSM5) catalysts was performed using the following procedure: amixture of the secondary benzylic alcohol as substrate (1 mmol), 70 % aqueous *t*-BuOOH (4 mmol) + Cr-(50-MCM-41/ZSM5) (5 mg) was mixed with silica gel (5 mg) and it was introduced into the vessel of 50 ml in the machine of microwave (Discover, CEM-SP1245 model, USA), (300 W), with stirring. The option of agitation consists of an electromagnetic dish located below the floor of the cavity of microwave. Agitation occurs when the magnetic field couples with a bar of agitation in the container.

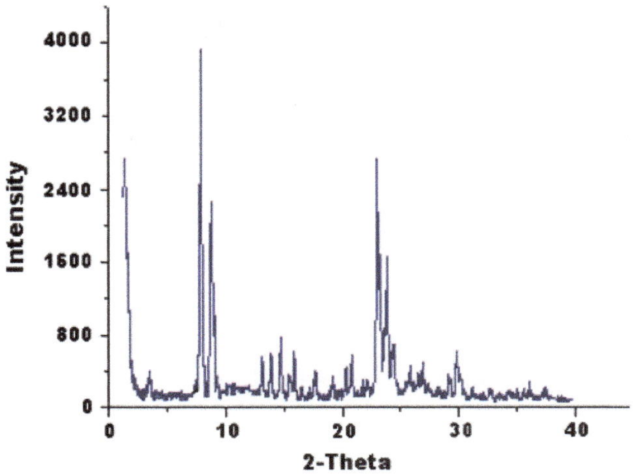

Fig. 1 X-ray diffraction patterns of calcined 50-(MCM-41/ZSM-5)

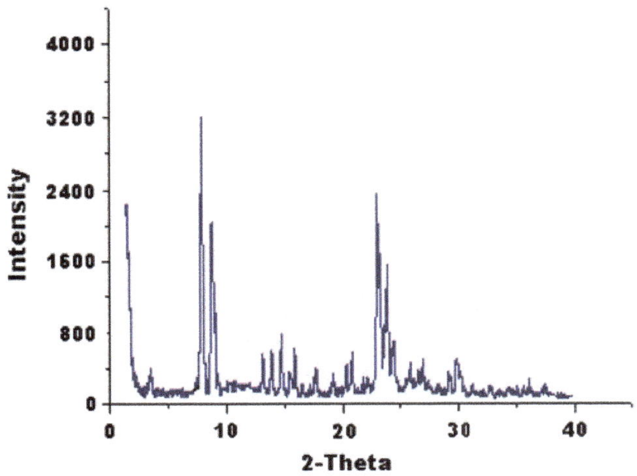

Fig. 2 X-ray diffraction patterns of Cr (50-MCM-41/ZSM-5)

The reaction conditions were recorded on the machine; the temperature of 100 °C is selected under an atmospheric pressure. The oxidation was done under backward flow without organic solvent; the reactants were poured onto silica gel and mixed intimately prior to be introduced in the machine of microwave. The reaction was stopped after 10 min. The product was recuperated from the machine after 20 min until the temperature reached 20 °C. The option of cooling is composed of valves and ports necessary to direct a gas of cooling (nitrogen or clean air) towards the containers in the cavity of the system. This option will decrease the temperature from 150 to 50 °C in less than 120 s.

We suspect that some *t*-BuOOH did not react in the course of reaction. In order to eliminate the nonreacted oxidant, we filtered the obtained product of reaction over a small pad of alumina, because alumina breaks *t*-BuOOH.

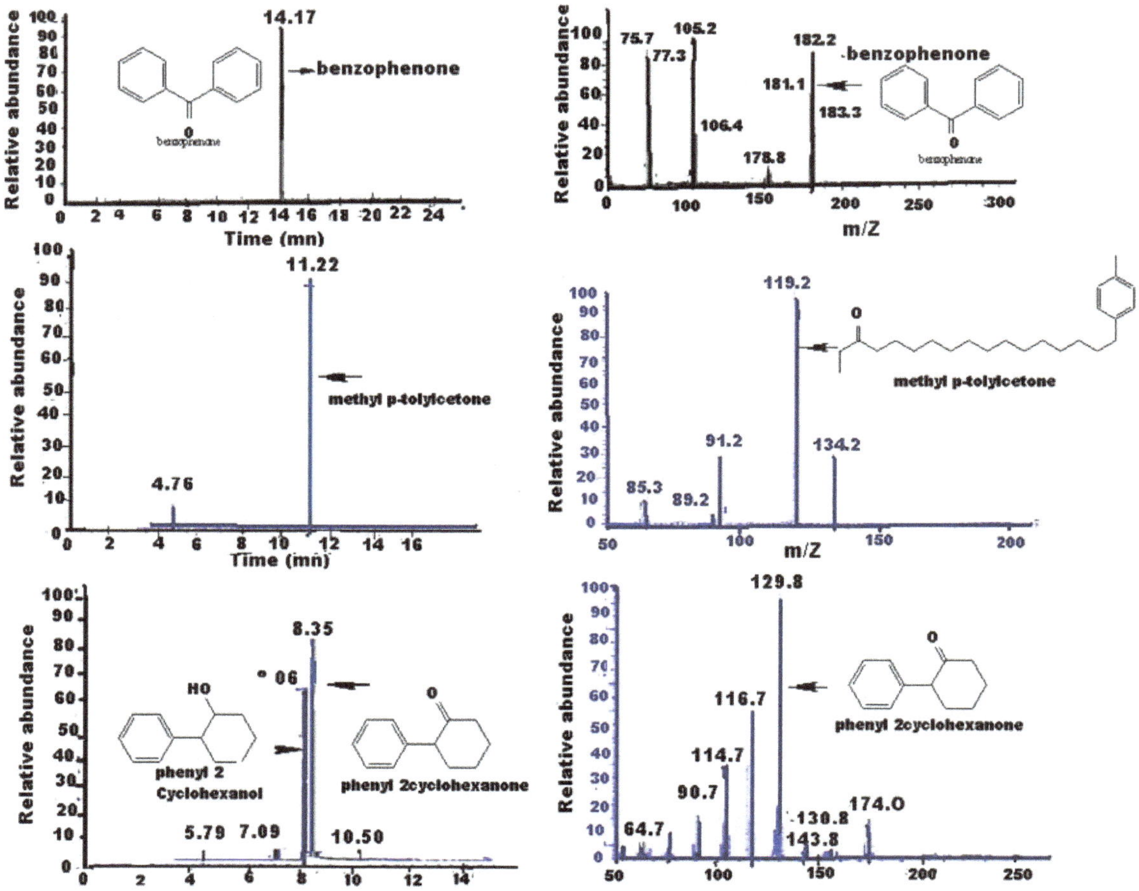

Fig. 3 CPG oxidation analyse of some benzylic alcohols

We used diethyl ether to permit the recuperation of the product from alumina by a vacuum pump.

Results and discussion

Taking into account the analyses carried out by gas chromatography and mass spectrometry (Fig. 3), we noted that the composite material Cr-(50-MCM-41/ZSM5) was selective and efficient catalyst for the oxidation of benzylic alcohols under mild conditions using 70 % *t*-BuOOH as an oxidant, the best yields were obtained to the oxidation of secondary alcohols to the ketones. Interestingly, unactivated alcohols such as 1-phenyl-1,2-ethanediol and *trans* 2-phenyl 1-cyclohexanol have been oxidised to the corresponding ketones in satisfactory yield (Table 1).The conversion of benzylic alcohols to the corresponding ketones was obtained for drastic time, only 10 min using the technology microwaves.

The use of bimodal pores Cr-(50-MCM-41/ZSM5) in oxidation of benzylic alcohols is due to the fact that zeolites with small pores ZSM5 ($\leq$5 Å) cannot undergo the oxidation of bulky molecules, such as (*trans*-2-phenyl 1-cyclohexanol, benzhydrol, etc.).

To determine the recyclability of our material, the oxidation of 1-indanol was chosen as substrate (1 mmol) and was performed under the usual conditions, at the end of reaction, the catalyst was separated from the products by filtration. Before the reuse, the Cr-(50-MCM-41/ZSM5) undergo treatment which consist in putting the zeolite in contact with a solution of NaOH 0.1 M (20 ml) under magnetic agitation at ambient temperature for duration of 15 min to remove the cooked carbonyl compounds deposited on its surface. The composite material was then washed several times with distilled water, dried at 100 °C for 3 h and analysed by powder X-ray diffraction which gave well-defined reproducible patterns (Fig. 4). The reuse of the recycled composite material showed the same activity and selectivity as the new one. The composite material can be exploited several times without any loss of its activity.

Following these results, we try to determine the role of chromium-exchanged composite material in the oxidation of benzylic alcohols. To clarify this issue, two experiments

Table 1 Oxidation of benzylic alcohols on the microwaves (Discover, CEM-SP1245 model)

Substrate	Conversion %	Product (yields %)
Benzhydrol	100	Benzophenone (100)
Trans-2-phenyl-1-cyclohexanol	80	*Trans*-2-phenyl-1-cyclohexanone (72) + d*
Cyclopropyl phenyl carbinol	98	Cyclopropyl phenyl cetone (95) + nd*
Methy-l *p*-tolyl carbinol	100	Methyl -*p*-tolyl cetone (99)
1-Phenyl-1propanol	100	1-Phenyl-1-propanone (95) + nd*
1-Phenyl-1, 2-ethandiol	80	α-Hydroxy-acetophone (71%) + nd*
1-Indanol	100	1-Indanone (99)

+*nd** non-identifiable products

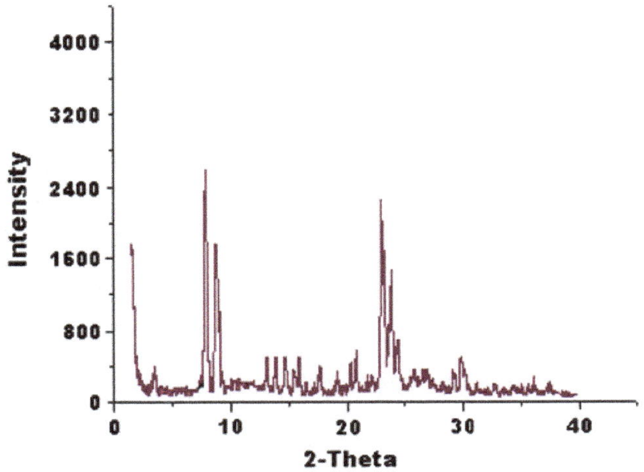

Fig. 4 X-ray diffraction patterns of recyclable Cr-(50-MCM-41/ZSM5)

have been carried out. As checking test reaction, the oxidation of benzhydrol was chosen in the usual conditions without catalyst. The reaction was stopped after 10 min; work up led to benzophenone in 10 % yield. The formation of benzophenone is due probably to the oxidation by 70 % aqueous TBHP.

This test is made to confirm the role played by chromium in the case of oxidation of benzylic alcohols, we have taken composite materials non exchanged with chromium and we have also chosen one alcohols (benzhydrol) in the list of substracts which we have already used. The reaction was stopped after 10 min. Work up led to benzophenone in 23 % yield. The acid activity of the solid catalyst and the presence of 70 % aqueous TBHP as oxidant led to the formation of ketone in poor yields. It is clear from these data that the effect of chromium in this type of reaction is predominant. The team of Muzart and that of Sheldon have worked for more than two decades on the effectiveness and the role of transition cations and more particularly, chromium in the oxidation reactions of benzylic alcohols (Muzart 1987, 1989, 1990, 1992, 1993a, b; N'ait Ajjou et al. 1994). On the basis of literature, the team of Muzart which worked in homogeneous chromium catalyst in liquid phase; they interested on the interactions of the TBHP with transitions metals (Boitsov et al. 2001). This team proposed that the catalysed oxidation of alcohols and activated methylenes implies the formation of complex *tert*-butyl peroxychromic (A) which plays the true role of oxidising species. The reaction of alcohols with (A) leads to the formation of complex alkoxy of chromium (*tert*-butyl peroxychromic). However, an intramolecular isolation of hydrogen by group *tert*-butyloxy releases ketone and the species (R: *t*-Bu).

The interaction of TBHP with transitions metals in the heterogeneous catalyse in liquid phase was undertaken by the team of Sheldon (Chen et al. 1995; Lempers and Sheldon 1996; Sheldon et al. 1994a, 1994b). The mechanism of Sheldon implies firstly the addition of alcohol to oxo-chromic (VI) catalyst to give complexes of alkoxy-oxochromic (B). However, an elimination of hydrogen from B led to the formation of ketone with an occurrence reduction of CrVI to CrIV.

In our case, the oxidation of benzylic alcohols by heterogeneous chromium catalyst occurs in solid phase; the chromium exchanged is within the pores of composite material, in the third state, and counterbalances the negative charge of the structure. However, the interaction of the molecule TBHP with the Cr^{III} into the pores of composite material leads to the oxidation of chromium to the sixth state (Cr^{VI}). Further research on the mechanism of this reaction and the scope of exchanged chromium composite material and related redox molecular sieves as solid catalyst for various solid phase oxidations is in progress.

Conclusions

Generally, these reactions are made in homogeneous medium with chromium oxides of high concentrations (the team of Muzart works with contents of chromium oxide which are 0.5 mmol); these reactions are made in the presence of organic solvents such as toluene or dichloromethane. At the end of the reactions, these oxides are dissolved and can be regenerated. In our case, we worked with solid catalysts. The content of chromium is very low, only 0.05 mmol. The catalyst is regenerated and can be used several times without affecting its activity, and also the reactions are carried out without using the organic solvent. We can show that Cr-(50-MCM-41/ZSM5) is an excellent recyclable solid catalyst for the selective oxidation of benzylic alcohols to the corresponding carbonyls compounds using 70 % aqueous TBHP as oxidant.

The technology of microwaves can reduce the reaction times and the energy consumption, and can lead to an increase in the outputs and selectivity of the oxidative reactions.

References

Boitsov S, Songstad J, Muzart J (2001) Chromium(VI) oxide/*tert*-butyl hydroperoxide interactions: evidence for a *tert*-butylperoxychromium complex and its role in the catalytic oxidation of alcohols. J Chem Soc Perkin Trans 2:2318

Cainelli G, Cardillon G (1984) Chromium oxidations in organic chemistry. Springer, New York

Chen JD, Lempers HEB, Sheldon RA (1995) Ti-Al-β and CrAPO-5 as heterogeneous catalysts for selective oxidations in the liquid phase. Colloid Surf A Physicochem Eng Aspects 101:137

Clerici MG (1991) Oxidation of saturated hydrocarbons with hydrogen peroxide, catalysed by titanium silicalite. Appl Catal 68:249

Clerici MG, Bellussi G, Romano U (1991) Synthesis of propylene oxide from propylene and hydrogen peroxide catalyzed by titanium silicalite. J Catal 129:159

Czarnetzki K (1989) Characterisation and modification of zeolites and related molecular sieves. PhD thesis, Eindhoven University of technology, The Netherlands p 116

Dongare MK, Singh P, Moghe PP, Ratnasamy PP (1991) Synthesis, characterization, and catalytic properties of [Zr]-ZSM-5. Zeolites 11:690

Esposito A, Neri C, Buonomo F (1984) Process for oxidizing alcohols to aldehydes and/or ketones, US Patent No. 4480135

Gronnow MJ, White RJ, Clark JH, Macquarrie DJ (2005) Energy efficiency in chemical reactions: a comparative study of different reaction techniques. Org Proc Res Dev 9:516

Haanepen MJ, Van Hooff JHC (1992) In: Tagungsbericht 9204 proceedings of the DGMK conference selective oxidations in petrochemistry, p 227

Chen JD, Haanepen MJ, VanHooff JHC, Sheldon RA (1995) CrAPO-catalyzed oxidations of alkylaromatics and alcohols with TBHP in the liquid phase (redox molecular sieves, part 8) in Stud.Surf.Sci.Catal 84:973

Karlsson A, Stocker M, Schmidt R (1999) Composites of micro- and mesoporous materials: simultaneous syntheses of MFI/MCM-41 like phases by a mixed template approach. Microporous Mesoporous Mater 27:181

Lempers HEB, Sheldon RA (1996) Allylic oxidation of olefins to the corresponding a,ß-unsaturated ketones catalyzed by chromium aluminophosphate-5. Appl Catal A Gen 143:137

Ley SV, Martin A (1991) Oxidation of activated C-H bonds adjacent to heteroatoms: oxidation adjacent to oxygen of alcohols by chromium reagent. In: Trost BM, Fleming I (eds) Comprehensive organic synthesis, vol 7. Pergamon, Oxford, p 251

Lounis Z et al (2006) Chromium-exchanged zeolite (Cr_E-ZSM-5) as catalyst for alcohol oxidation and benzylic oxidation with *t*-BuOOH. Appl Catal A Gen 309:270

Meynen V, Cool P, Vansant EF (2007) Synthesis of siliceous materials with micro- and mesoporosity. Microporous Mesopor Mater 104:26

Muzart J (1987) Practical chromium VI oxide-catalyzed benzylic oxidations using 70% tert butylhydroperoxide. Tetrahedron Lett 28:2133

Muzart J (1989) Bimetallic oxidation catalysts: oxidations with *tert*-butylhydroperoxide mediated by bis-(tributyltin oxide) dioxochromium (VI). J Synth Commun 19:9

J Muzart (1990) Chromium(VI)-catalyzed oxidation of para-substituted phenols by *t*-butyl hydroperoxide. J Chem Res (S) 96

Muzart J (1992) Chromium-catalyzed oxidations in organic synthesis. J Chem Rev 92:113

Muzart J (1993) Silyl ethers as protective groups for alcohols: oxidative deprotection and stability under alcohol oxidation conditions. Synthesis 1993:11

Muzart J, N'ait Ajjou A (1993) Highly efficient chromium-catalyzed oxidation of secondary benzylic alcohols by aqueous 70 % *tert*-butyl hydroperoxide. Synthesis 8:785

N'ait Ajjou A, Muzart J, Savelon L, Guillaumet G (1994) Chromium-assisted oxidations. A simple and efficient oxidation of oxazolopyridylcarbinols by aqueous *tert*-butyl hydroperoxide. Synthesis: 359

Ordomsky VV, Murzin VY, Monakhova YV, Zubavichus YV, Knyazeva EE, Nesterenko NS, Ivanova II (2007) Nature, strength and accessibility of acid sites in micro/mesoporous catalysts obtained by recrystallization of zeolite BEA. Microporous Mesoporous Mater 105:101

Rigutto MS, van Bekkum H (1993) Vanadium site in VAPO-5: characterization and catalytic properties in liquid-phase alkene epoxidation and benzylic, oxidation. J Mol Catal 81:77

Romano U, Esposito A, Maspero F, Neri C, Clerici MG (1990) Selective oxidation with Ti-silicalite. Stud Surf Sci Catal 55:33

Sheldon RA, Korchi JK (1981) Metal-catalyzed oxidations of organic compounds. Academic, New York, p 152

Sheldon RA, Chen JD, Dakka J, Neeleman E (1994a) Recyclable catalysts for liquid phase oxidations. Stud Surf Sci Catal 83:407

Sheldon RA, Chen JD, Dakka J, Neeleman E (1994b) Redox molecular sieves as heterogeneous catalysts for liquid phase oxidations. Stud Surf Sci Catal 82:515

Taramasso M, Perego G, Notari B (1983) Preparation of porous crystalline synthetic material comprised of silicon and titanium oxide. US Patent No. 4410501

Thangaray A, Kumar R, Ratnasamy P (1990) Studied the oxidation of benzene by hydrogen peroxide to phenol and benzoquinone as catalyzed by various zeolites and molecular sieves. Appl Catal 57:L1

Trainor RW, Strauss CR (1995) L'évolution des micro-ondes assisté en chimie organique. Aust J Chem 48:1665

Zhang W, Wang J, Tanev PT (1996) Catalytic hydroxylation of benzene over transition-metal substituted hexagonal mesoporous silica. J Chem Commun: 979

2

Electron microscopic studies of growth of nanoscale catalysts and soot particles in a candle flame

Wu Zong Zhou · Feng Jiao Yu · Heather F. Greer · Zheng Jiang · Peter P. Edwards

Abstract Scanning electron microscopy and high-resolution transmission electron microscopy have been used to investigate non-classic crystal growth of catalytic nanoparticles, such as zeolites, perovskites, metal and alloy particles. Growth mechanisms of some crystals with novel morphologies, for example, BiOBr flower-like particles and ZnO twin-crystals, have also been studied. A development of sampling method for soot particles inside a candle flame allows us to reveal all four well-known carbon forms, amorphous, graphitic, fullerenic and nanodiamond particles. This article demonstrates that electron microscopy is a powerful tool to study the microstructures of small particles, giving us more freedom to develop new materials.

Keywords Crystal growth · ZnO · BiOBr · $CaCO_3$ · Electron microscopy

Introduction

Development of catalysts is an important field in petrochemical research. Every year, a large number of new catalysts with various morphologies are reported. However, our knowledge of the formation mechanisms of these crystalline materials is still very limited. One of the reasons is that people normally admire the beauty of these final products and their catalytic performance, but pay less attention to the way they form.

Growth of a free crystal can be well described using the classic theory established 100 years ago, i.e. a nucleation step followed by a layer-by-layer deposition of the building units, which can be atoms, or ions, or molecules. Its morphology can be predicted by the Bravais–Friedel–Donnay–Harker (BFDH) law: the final polyhedral morphology is formed by slow-growing faces (with large *d*-spacings), because all the fast-growing faces (with small *d*-spacings) would grow out during the process [1–3]. For example, the largest *d*-spacing in zeolite analcime corresponds to the (211) planes, therefore, both the natural and synthetic crystals of analcime have an icositetrahedral shape consisting of 24 {211} facets. Another obvious fact in this kinetically controlled mechanism is that the particle at any stage during the growth should be a single crystal. But when we examined the intermediate specimens at early stages of a crystal-growth process, we found the classic theory of crystal growth was not always followed. Herein, we demonstrate some examples of non-classic crystal growth and what we learnt from scanning and transmission electron microscopic (TEM) images of particles before their maturity.

Reversed crystal-growth route

The classic theory of crystal growth has been used for over one century and it seems to be very successful in explaining the relation between crystal morphologies and their corresponding crystal structures. This was why when we saw a perfect icositetrahedral particle of analcime synthesised in a system with $C_2H_5NH_2$, we had no reason to doubt its single-crystal state, until its single-crystal

W. Z. Zhou (✉) · F. J. Yu · H. F. Greer
School of Chemistry, University of St Andrews, St Andrews, Fife KY16 9ST, UK
e-mail: wzhou@st-andrews.ac.uk

Z. Jiang · P. P. Edwards
Inorganic Chemistry Laboratory, Department of Chemistry, University of Oxford, Oxford OX1 3QR, UK

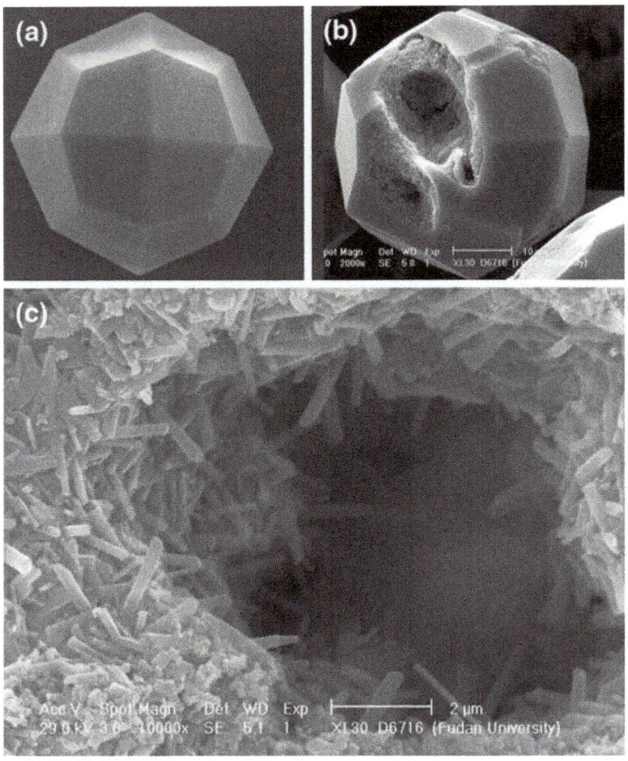

Fig. 1 SEM images of synthetic zeolite analcime. **a** An icositetrahedral particle. **b** An intentionally broken icositetrahedron showing polycrystalline core. **c** Enlarged image of the core nanorods [4]

X-ray diffraction pattern showed a strong polycrystalline ring. Figure 1a shows a scanning electron microscopic (SEM) image of such a crystal with about 50 μm in diameter.

When we made a hole in a crystal which had a perfect icositetrahedral morphology, the core was revealed to be actually polycrystalline, containing randomly oriented nanorods (Fig. 1b, c). This discovery encouraged us to examine the details of the crystal growth step-by-step. At the beginning, it was found that when the analcime crystals grew into very small nanoplates about 20 nm in diameter, they started to aggregate into large spherical particles. $C_2H_5NH_2$ played an important role in aggregation. In this case, most crystallites inside the microspheres lost their original environment for crystal growth. The most active sites were on the surface of the spheres or in the interface between the spheres and the solution. As a result, re-crystallization took place on the surface of the spheres to form many monocrystalline islands. These islands grew up, joined together with adjusted orientations to form a very thin single-crystal shell with perfect polyhedral morphology (Fig. 1a). Re-crystallization then extended from surface to the core and finally achieved a real single-crystal state [4].

This so-called reversed crystal-growth route was later found in other systems, such as zeolite A [5, 6], perovskites $CaTiO_3$ and $BaTiO_3$ [7, 8], metal [9] and alloy particles (F. J. Yu et al. paper under preparation). Recently, it was also found in organic crystals [10].

Zeolite A is an important industry catalyst. When biopolymer chitosan was used as a non-structure directing agent, microcubes of zeolite A were produced without any notable difference to those prepared using the traditional synthetic method. However, we found chitosan induced aggregation of precursor molecules even before nucleation occurred. Consequently, cubic particles had a core–shell structure, an amorphous core covered by a single-crystal shell of only a few nanometers thick as shown in the TEM images in Fig. 2 [6]. After acid treatment, the amorphous cores were removed, leaving very thin zeolite A boxes or hollow cubes. This novel morphology can be easily achieved via the reversed crystal-growth route.

Many hollow crystals have been fabricated in the last 20 years. However, their formation mechanism can only be fully understood when the structure evolution is investigated in steps. We took hollow crystals of perovskite $CaTiO_3$ as an example. In a poly(ethylene glycol) 200 (PEG-200) solution, $CaTiO_3$ nanocubes formed first. The nanocubes underwent oriented self-assembly into spherical particles, enhanced by the surface-adsorbed polymer molecules. These polycrystalline spheres had a relatively low density compared to single crystals. Surface re-crystallization then took place to increase their density and create a high-crystallinity shell. The nanocubes in the core gradually disappeared via an Ostwald ripening process and finally led to hollow crystals (Fig. 3) [7]. Understanding the complete formation mechanism of these hollow $CaTiO_3$ crystals would enable us to control the microstructures of these materials and to explain the formation of many other hollow crystals. Water agate stones are typical natural hollow crystals, which might form with exactly the same mechanism [11].

It is interesting to see such a reversed crystal growth is also a true process in nanomaterials. In fact, several years before the discovery of this non-classic mechanism, we found the formation of cobalt nanocrystallites as a catalyst for Mg_2SiO_4 fishbone-like fractal nanorods followed a non-classic route [9]. When Co nanocrystallites grew up to about 2 nm in size, they aggregated on the surface of Mg_2SiO_4, without the presence of any organic surfactants in this solid–gas reaction system. Only when the aggregates increased their size to over 100 nm, did they recrystallize into single crystals with a cubic morphology. In more recent work, we have obtained clear evidence of a surface to core extension of crystallization in Cu/Pt alloy nanoparticles (F. J. Yu et al. paper under preparation). It seems to be confirmed that the reversed crystal growth is a common phenomenon in the formation of various materials and more examples will be found in future research.

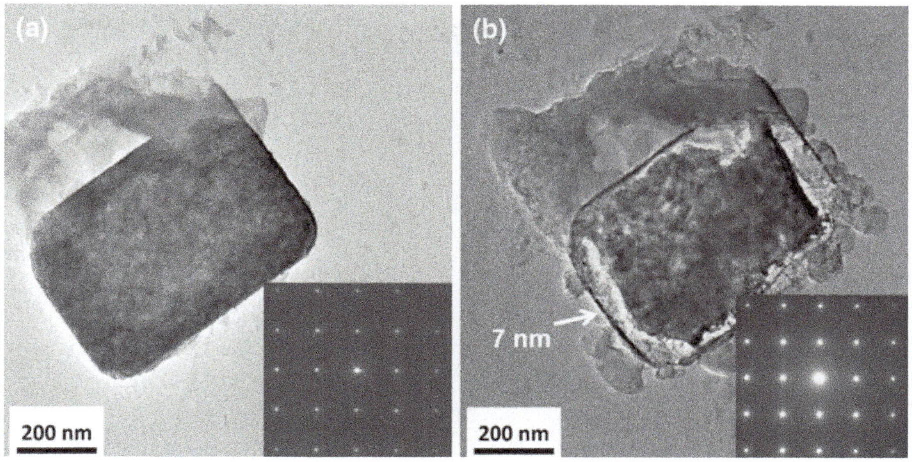

Fig. 2 **a** TEM image of a zeolite A particle after crystal growth for 3 h. **b** The same particle after annealing with the electron beam for a few minutes. *Insets* are the corresponding SAED patterns

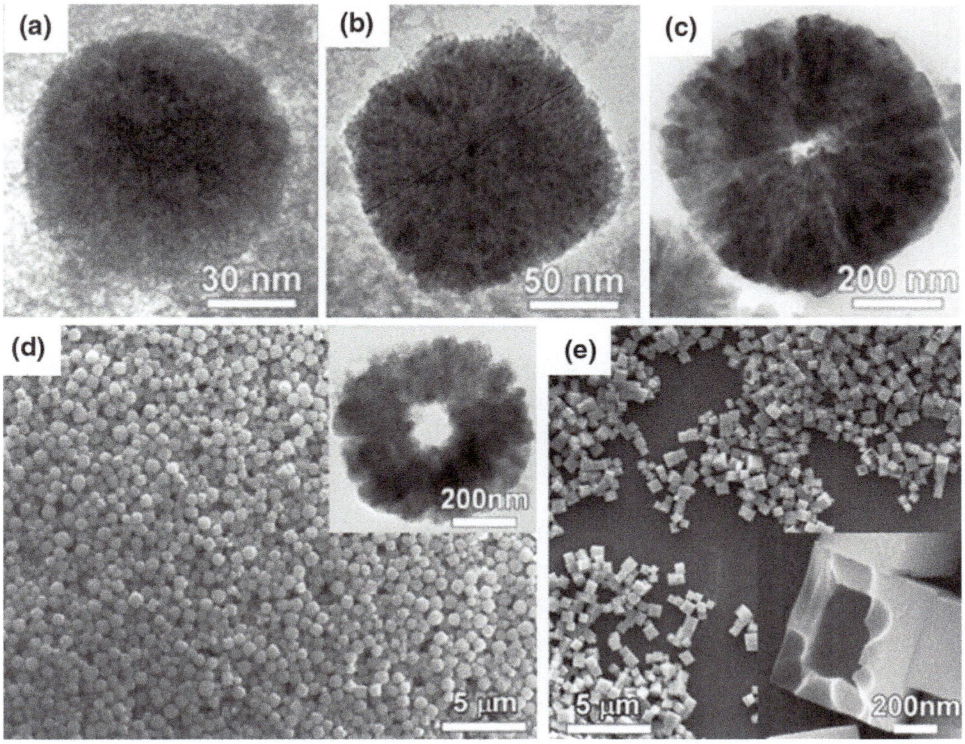

Fig. 3 TEM images of $CaTiO_3$ particles with different crystal-growth times: **a** 1, **b** 2, and **c** 5 h. **d** Low-magnification SEM image of the as-synthesized $CaTiO_3$ product prepared in the water-free poly(ethylene glycol) solution. The inset shows a corresponding cross-section TEM image of a particle showing a hollow structure. **e** SEM image of cubic $CaTiO_3$ particles where the recrystallization and reversed growth was enhanced by adding 5 vol% water. The *inset* is a higher magnification SEM image of a broken cube from the same specimen

Particles with novel hierarchical morphologies

Electron microscopy has been used to search for some hidden particles in solids, which are key particles in the formation of novel hierarchical morphologies. Good examples are twin-crystals of ZnO and flower-like BiOBr.

Wurtzite ZnO is an important semiconductor with a wide band gap of 3.37 eV at room temperature and a large exciton binding energy of 60 meV. These properties make ZnO a good candidate to use in catalysis [12] and many other applications. Twin-crystals of ZnO was synthesized using $Zn(NO_3)_2{\cdot}6H_2O$, $C_6H_{12}N_4$, and gelatin (Fig. 4a) [13]. They formed by oriented stacking of ZnO nanoplates along the [001] zone axis. However, stacking of nanoplates normally leads to larger disk-like particles. An approximate mirror-symmetric manner of stacking implies a special microstructure in the centre of each particle. A similar growth manner was observed by Taubert et al. [14] in 2003. They used high-resolution transmission electron microscopy (HRTEM) to examine the central area of the particles and found a large number of defects in the central grain boundary. The formation mechanism or the origin of the central grain boundary was not revealed.

We then decided to investigate particles at early stages to try to reveal the microstructural evolution [15]. At beginning, the precursor molecules aggregated into some spherical amorphous particles (Fig. 4b). These particles crystallized into nanoplatelets of $Zn_5(NO_3)_2(OH)_8{\cdot}2H_2O$

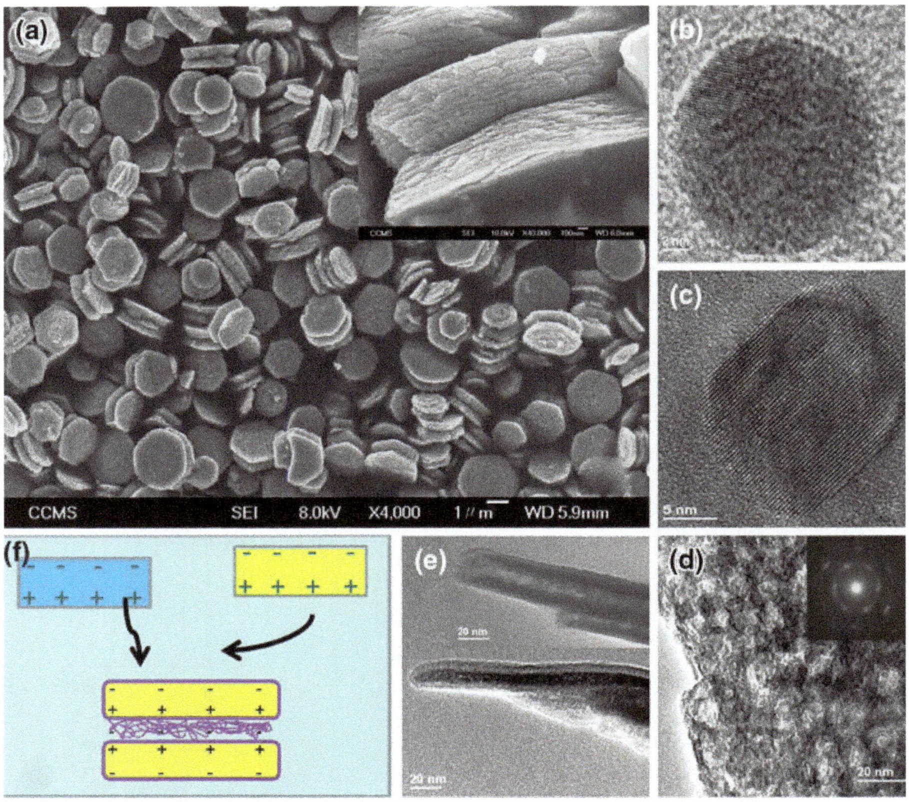

Fig. 4 **a** SEM image of the ZnO twin-crystals after growth for 21 h. The *inset* is an enlarged SEM image with a profile view of a twin-crystal, showing nanoplates as building units. **b** A spherical particle at a very early stage with partial crystallization. **c** HRTEM image of a typical $Zn_5(NO_3)_2(OH)_8{\cdot}2H_2O$ nanoplatelet from the 10-min sample. The *d*-spacing of the fringes is about 0.321 nm. **d** Enlarged TEM image of a single-layer nanoplate showing a porous nature. The *inset* is the corresponding SAED pattern indicating a highly ordered orientation of the nanoplatelets. **e** TEM image of single- (*bottom*) and double- (*top*) layer nanoplates. **f** *Schematic drawing* showing a double-layer ZnO nanoplate attracting single-layer ZnO (*yellow*) or $Zn_5(NO_3)_2(OH)_8{\cdot}2H_2O$ (*blue*) nanoplate

with a size about 15 nm (Fig. 4c), which then further aggregated into large nanoplates with a fairly good crystal orientation, although individual nanoplatelets inside the single-layer nanoplates are still distinguishable (Fig. 4d). The *d*-spacing of 0.321 nm observed in Fig. 4c can be indexed to the (020) planes of $Zn_5(NO_3)_2(OH)_8{\cdot}2H_2O$, but cannot be indexed to ZnO. The most interesting structural evolution is that a phase transformation from $Zn_5(NO_3)_2(OH)_8{\cdot}2H_2O$ to ZnO took place on both surfaces of some single-layer nanoplates, leading to double-layer nanoplates separated by a layer of gelatin molecules (Fig. 4e). This is the first example of the reversed crystal-growth route in nanoplates. Because the gelatin molecules were negatively charged in the synthetic solution with a pH value in a range of 6.2–8, the inner surfaces of the two ZnO layers in the double-layer nanoplates were positively charged and both the outer surfaces were negatively charged. Consequently, this structure of dipolar ZnO made stacking of these double-layer nanoplates very difficult. On the other hand, single-layer nanoplates of $Zn_5(NO_3)_2(OH)_8{\cdot}2H_2O$ can stack on the surface of the double-layer nanoplates much more easily (Fig. 4f). Such self-assembly on both surface of double-layer nanoplates followed by a complete phase transformation resulted in the so-called twin-crystals of ZnO. In the final products, the original double-layer core particles cannot be distinguished from other ZnO nanoplates.

We learnt from this work that to form twin-crystals, two different particles must form independently in the synthetic system, i.e. single-layer and double-layer nanoplates. The former is the building unit and the latter is the core. They can have the same crystal phase, but different microstructures or superstructures with different electrochemical properties. The ratio of the building units and the core particles must be large so that each twin-crystal can form by the stacking of hundreds of building units (single-layer nanoplates) on only one core particle (double-layer nanoplate). Finally, the double-layer core is the key particle for this hierarchical morphology. However, in the final twin-crystals, it grows with other single-layer nanoplates and becomes indistinguishable.

Flower-like BiOBr crystals synthesized by Z. Jiang et al. (paper under preparation) in Oxford shows another novel hierarchical morphology (Fig. 5a), which drew our attention recently. The specimens were prepared via a mixture of KBr and $Bi(NO_3)_3{\cdot}5H_2O$ aqueous solutions followed by ultrasonic treatment. Each particle consists of many small nanoplates, stacking in a non-parallel way, ensuring a large exposed surface, which is beneficial in photocatalysis. To understand its formation mechanism, we also examined microstructures of specimens at different growth stages. We found it is essential to have two types of BiOBr particles with different morphologies so that a flower-like particle can form. One is a large hollow nanoplate

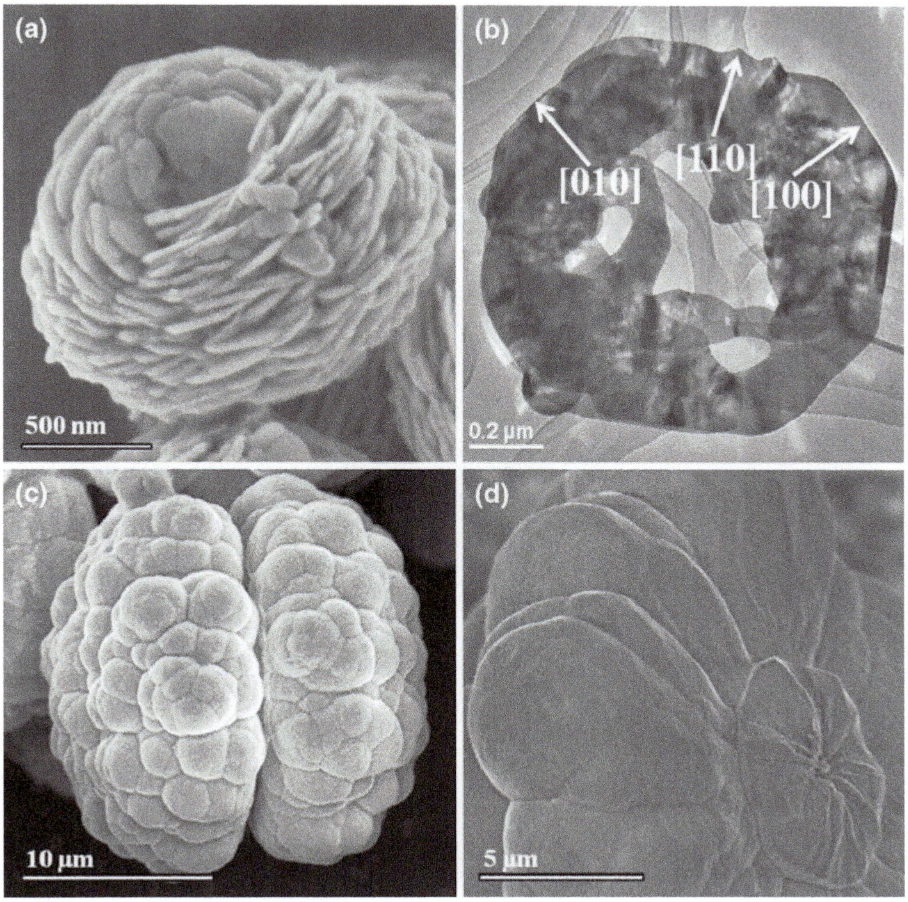

Fig. 5 **a** SEM image of a flower-like BiOBr particle. **b** TEM image of a typical large hollow plate, which acts as a base for a flower-like particle. **c** SEM image of broccoli-like $CaCO_3$ particle. **d** SEM image of a half broccoli-like particle, showing the central nanoplate

servicing as the flower base (Fig. 5b). The other is small solid nanoplate acting as the flower petal. Post-reaction aging and sonochemical treatment were crucial in the formation of large hollow plates.

Figure 5c shows a broccoli-like $CaCO_3$ particle developed from a spherical particle. This morphology change, similar to a cell-division process, is the most important step in a continuous morphology evolution from spherical to a hexagonal prism. The driving force of this change has been found from a central microdisk as shown in Fig. 5d (H. F. Greer, et al. paper in preparation).

As we see from the above examples, formation of some complex hierarchical microstructures requires two types of pre-formed building units. The central hidden particles are crucial, but are often ignored by people. One of the difficulties is that these particles have the same crystal phases as the main building units. It is even more difficult to detect them when they become indistinguishable during the crystal growth (see the case of ZnO). We found HRTEM investigations of the early-stage crystal growth is a very powerful method to discover these hidden particles and to understand the detailed formation mechanisms of some solid materials with novel morphologies.

Soot particles in a candle flame

High-resolution transmission electron microscopy technique has also been used to detect some 'hidden' particles in a candle flame. Candles have a history of over 2,000 years. Candle flame as a research subject has also a long history of several hundred years. However, the soot particles inside a candle flame are not well known. The main problem is that we do not have a good sampling method. If we can collect the soot particles from a candle flame with their original states intact, we may investigate their structures using HRTEM.

When we inserted a thin film of porous anodic aluminium oxide (AAO) with both ends of the pores open into a candle flame for 1 s and removed it quickly, some soot particles smaller than the pore size would enter the pores and further reaction of these particles would be stopped. The particles were then transferred by an ultrasonic treatment into a solution and further transferred to a microscopic specimen grid. We found the structures of the soot particles collected from different positions of a candle flame were different. In particular, all four well-known carbon structural types were observed from the middle position of the candle flame as shown in Fig. 6, i.e.

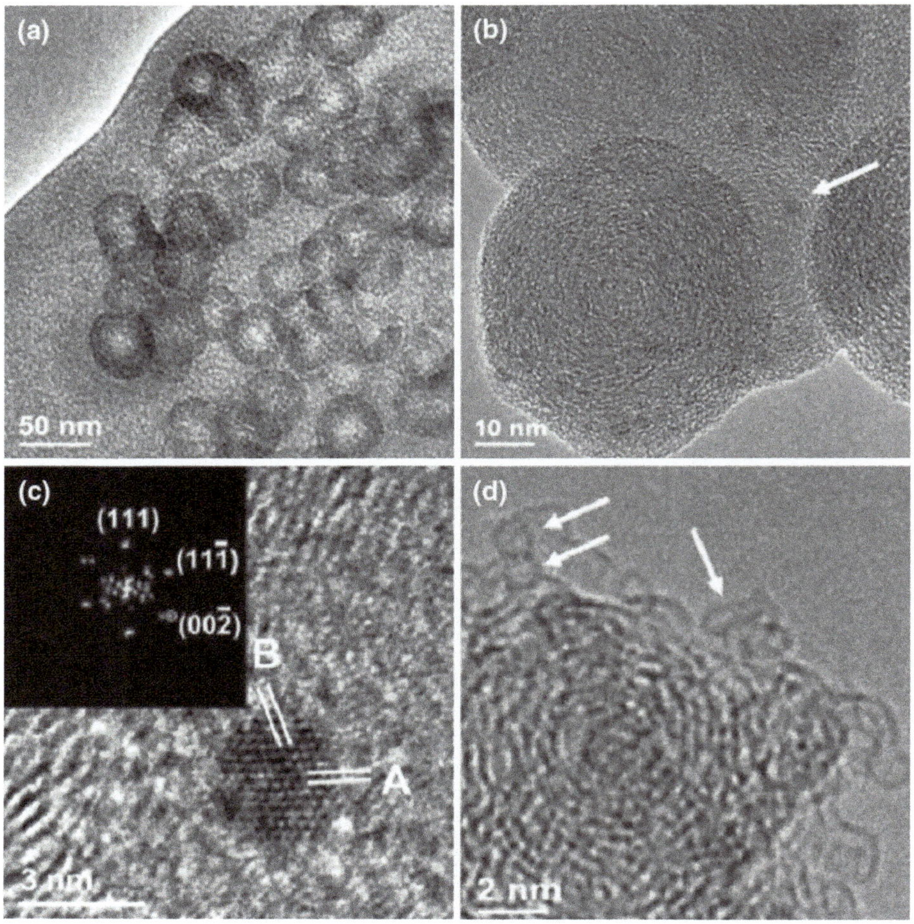

Fig. 6 HRTEM images of some nanoparticles collected from a central position of a candle flame, **a** graphitic spheres, **b** graphitic spheres with nanodiamond particles, **c** a nanodiamond particle surrounded by amorphous carbon compounds, and **d** fullerene particles on the graphitic surface indicated by the *arrows*

amorphous, graphitic, fullerenic and nanodiamond particles [16]. This work may indicate a new formation mechanism of diamond. More importantly, improved knowledge of the soot particles in combustion may help us to better understand the chemical reactions in combustion, including the formation of NO_x particles.

In summary, electron microscopic studies can reveal more details of crystal growth. The discoveries of reversed crystal-growth route and some hidden particles give us more freedom to control the morphologies of catalytic materials. It can be expected the research in this field will be continuously active in the future.

References

1. Bravais A (1866) Études Crystallographic. Gauthier-Villars, Paris
2. Friedel MG (1907) Études sur la loi de Bravais. Bull Soc Fr Mineral Cristallogr 30:326
3. Donnay JDH, Harker D (1937) A new law of crystal morphology extending the law of Bravais. Am Mineral 22:446
4. Chen XY, Qiao MH, Xie SH, Fan KN, Zhou WZ, He HY (2007) Self-construction of core–shell and hollow zeolite analcime icositetrahedra: a reversed crystal growth process via oriented aggregation of nanocrystallites and recrystallization from surface to core. J Am Chem Soc 129:13305
5. Greer H, Wheatley PS, Ashbrook SE, Morris RE, Zhou WZ (2009) Early stage reversed crystal growth of zeolite A and its phase transformation to sodalite. J Am Chem Soc 131:17986
6. Yao JF, Li D, Zhang XY, Kong CH, Yue WB, Zhou WZ, Wang HT (2008) Cubes of zeolite A with an amorphous core. Angew Chem Int Ed 47:8397
7. Yang XF, Fu JX, Jin CJ, Liang CL, Wu MM, Zhou WZ (2010) Formation mechanism of $CaTiO_3$ hollow crystals with different microstructures. J Am Chem Soc 132:14279
8. Zhan HQ, Yang XF, Wang CM, Chen J, Wen YP, Liang CL, Greer HF, Wu MM, Zhou WZ (2012) Multiple nucleation and crystal growth of barium titanate. Cryst Growth Des 12:1247
9. Xie SH, Zhou WZ, Zhu YQ (2004) Formation mechanism of Mg_2SiO_4 fishbone-like fractal nanostructures. J Phys Chem B 108:11561
10. Sander JRG, Bučar D-K, Baltrusaitis J, MacGillivray LR (2012) Organic nanocrystals of the resorcinarene hexamer via sonochemistry: evidence of reversed crystal growth involving hollow morphologies. J Am Chem Soc 134:6900
11. Zhou WZ (2010) Reversed crystal growth: implications for crystal engineering. Adv Mater 22:3086
12. Zhou H, Fan T, Li X, Zhang D, Guo Q, Ogawa H (2009) Biomimetic photocatalyst system derived from the natural prototype in leaves for efficient visible-light-driven catalysis. J Mater Chem 19:2695

13. Tseng Y-H, Lin H-Y, Liu M-H, Chen Y-F, Mou C-Y (2009) Biomimetic synthesis of nacrelike faceted mesocrystals of ZnO–gelatin composite. J Phys Chem C 113:18053
14. Taubert A, Kubel C, Martin DC (2003) Polymer-induced microstructure variation in zinc oxide crystals precipitated from aqueous solution. J. Phys. Chem. B. 107:2660
15. Greer HF, Zhou WZ, Liu MH, Tseng YH, Mou CY (2012) The origin of ZnO twin crystals in bio-inspired synthesis. CrystEngComm 14:1247
16. Su ZX, Zhou WZ, Zhang Y (2011) New insight into the soot nanoparticles in a candle flame. Chem Commun 47:4700

3

Neutron scattering studies of catalyst systems at the ISIS neutron spallation source

Martin O. Jones · Andrew D. Taylor · Stewart F. Parker

Abstract The ISIS neutron spallation facility is a world-leading centre for neutron scattering and has a formidable selection of elastic and inelastic neutron scattering instruments to study the physical properties of solids and liquids by a number of techniques that include diffraction, total scattering and molecular spectroscopy. In addition, complex sample environment apparatus may be utilized with these instruments that allows materials to be studied under controlled gas environments as a function of temperature, pressure and gas flow. Here, we discuss the application of these instruments and various sample environments to materials challenges within the field of catalysis, describe some of the more recent catalysis and catalysis-related experiments and highlight the capabilities of the ISIS facility in tackling catalytic challenges.

Keywords Neutron · Neutron diffraction · Inelastic neutron scattering · Catalysis

Introduction

ISIS [12], (Figs. 1, 2) is a world-leading centre for neutron scattering. The technique is applicable to a great diversity of subjects spanning condensed matter physics, engineering, materials science, chemistry and soft matter. The usefulness and generality of neutron scattering arise from the properties of the neutron, which is an uncharged particle of mass 1.00866 amu (almost the same as the hydrogen atom ^{1}H, 1.00782) with a magnetic moment. The scattering occurs from the atomic nuclei of the sample and since the nucleus of an atom is only ~1/1,000 of the diameter of the atom, neutrons only interact weakly with matter and are consequently highly penetrating; millimeters of steel are essentially transparent.

The nuclear scattering has a number of interesting consequences. The scattering cross section is both atom and isotope dependent and is not a monotonic function of atomic number as it is for X-ray scattering (see Research (London) 7, 257, 1954). Thus for neutron diffraction studies it is straightforward to study light elements in the presence of heavy ones, hydrogen-in-metal systems being the archetype but the precise location of oxygen in heavy metal oxides is also an exemplar. Since X-rays are scattered by electrons, bond distances are more accurate from neutron scattering data, this is particularly the case for bonds involving hydrogen, since the maximum in electron density is not located at the atomic position but is displaced toward the heavier atom. Thus C–H and O–H distances measured by X-ray diffraction are typically 0.1 Å shorter than those measured by neutron diffraction. It is also possible to distinguish elements that are adjacent in the periodic table, thus Al^{3+} and Si^{4+}, as commonly found in zeolites, have identical scattering power for X-rays (since they have the same number of electrons) but different neutron cross sections, 1.503 versus 2.167 barn (1 barn = 1×10^{-28} m^2), respectively, hence providing contrast.

The most striking difference in cross section occurs for ^{1}H and ^{2}H, hydrogen and deuterium: $\sigma_{coh} = 1.76$ (^{1}H), 5.59 (^{2}H) barn, $\sigma_{inc} = 80.27$ (^{1}H), 2.05 (^{2}H) barn (σ_{coh} = coherent scattering cross section, σ_{inc} = incoherent scattering cross section). Hence, hydrogen is overwhelmingly an incoherent scatterer while deuterium is primarily a coherent scatterer. Coherent scattering gives information on long range properties such as structure

M. O. Jones (✉) · A. D. Taylor · S. F. Parker
ISIS Facility, STFC Rutherford Appleton Laboratory,
Chilton, Didcot, Oxon OX11 0QX, UK
e-mail: martin-owen.jones@stfc.ac.uk

Fig. 1 Aerial view of the STFC Rutherford Appleton Laboratory, Chilton, Oxfordshire, UK. The ISIS facility is the group of buildings in the lower half of the picture with the Diamond Light Source (http://www.diamond.ac.uk) and the Research Complex (http://www.rc-harwell.ac.uk) shown in the upper half

while incoherent scattering is a local probe. The neutron may be scattered without a change in energy (elastic scattering) as in diffraction, so this relies on σ_{coh} or they may be exchange of energy between the sample and the neutron (inelastic scattering). The latter may be coherent or incoherent, but for any material containing hydrogen, the incoherent scattering dominates. For catalyst studies this is both a blessing and a curse. Since much of catalysis involves the transfer of hydrogen from reactant to product, for structural studies it is usually necessary to use deuterated materials. For inelastic scattering, it is only the exceptionally large value of σ_{inc} that makes such studies feasible.

In the rest of the paper, we will describe the instruments that have been used for catalyst studies and the infrastructure that has been developed to enable such work. It must be emphasized that for successful investigations both components are essential.

Elastic scattering

Dedicated catalysis and gas handling systems for catalytic reaction have been developed at ISIS [37] to directly probe the structure of catalyst materials in environments that closely approximate those of real industrial processes. This apparatus permits catalysis samples to be studied within the elastic instruments at ISIS while under vacuum, flowing atmospheres of up to 5 bar and at temperatures between 4 and 1,273 K. In 2000, Turner et al. [36] utilized this apparatus to probe the interaction of the active site of a NiNa-Zeolite Y catalyst (NiNaY) with acetylene. Here, two samples of NiNaY (^{Nat}NiNaY and ^{62}NiNaY) were analyzed using the small angle neutron diffractometer for liquids and amorphous samples (SANDALS) at ISIS together with the catalysis apparatus described above. The SANDALS diffractometer at ISIS has been specially constructed to investigate the structure of liquids and amorphous materials over a wide range (0.2–50 $Å^{-1}$) of momentum transfers and the combination of an intense pulsed neutron source and a large number of detectors at low angles make SANDALS particularly useful for measuring structure factors containing light atoms such as hydrogen and deuterium. Total neutron cross scattering was measured for the two NiNaY samples (^{Nat}Ni and ^{62}Ni) with and without the addition of d_2-acetylene (C_2D_2). For acetylene-free samples, a very clear differentiation of the isotopically enriched and natural abundance samples was obtained. When treated with deuterated acetylene the intermolecular C–C and C–D bond lengths could be observed at $r_{CC} = 1.16 \pm 0.01$ Å and $r_{CD} = 2.3 \pm 0.05$ Å, with small changes to the host structure. The complementary use of density functional theory (DFT) modeling with the total scattering studies allowed the investigators to observe a considerable deformation of the acetylene, with the C–C bond lengthening by 0.05 Å, and the C–D bonds bending back from the coordinating metal centre by 11°, with a concomitant reduction in the charge of the nickel ion from +2 to +1.08 due to donation of electron density from the C–C triple bond (Fig. 3). The neutron data were found to be consistent with structural features of the predicted Ni complex associated with the Dewar–Chatt–Duncanson model for π acid coordination. Similarly, Parker et al. [25] have used a combination of inelastic and elastic neutron

Fig. 2 Schematic plan of the of the ISIS facility

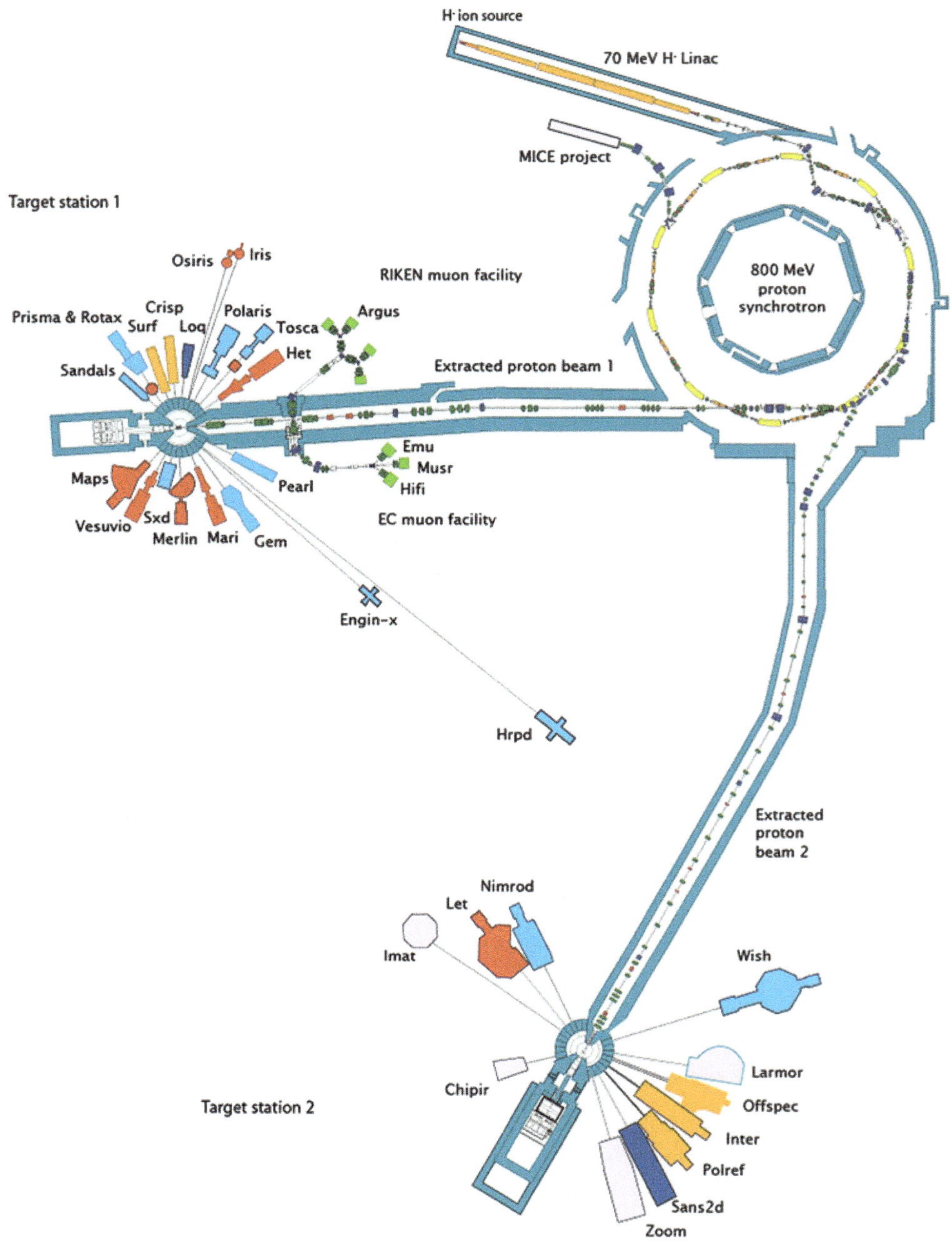

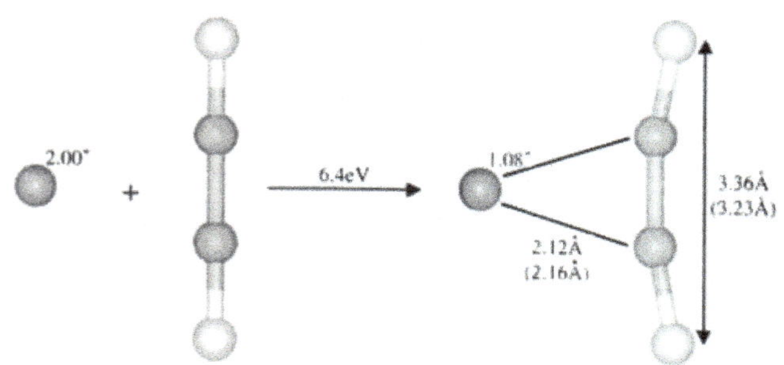

Fig. 3 Optimised geometries, Ni charge and binding energy for the interaction of one molecule of acetylene with a bare Ni^{II} ion, from Turner [36]

scattering techniques using the TOSCA and SANDALS instruments at ISIS, respectively, to investigate Ni–Ni and Ni–H distances of a RANEY nickel hydrogenation catalyst at 1 bar H_2 and ambient temperature. Further combined inelastic and elastic neutron diffraction studies were performed by Parker et al. [26] on the model oxidation catalyst, hydrous palladium oxide. A combination of X-ray and neutron diffraction, inelastic neutron scattering, electron microscopy, infrared spectroscopy and periodic-DFT calculation were used to comprehensively characterize hydrous palladium oxide and identify it as a nanocrystalline material of general formula $PdO{\cdot}H_2O$. Here, the

GEneral Materials diffractometer (GEM) at ISIS was used to determine the structure of PdO·H_2O. The GEM diffractometer can perform high intensity, high resolution experiments to study the structure of disordered materials and crystalline powders. It utilizes a highly stable detector array, which covers a very wide range in scattering angle (1.1°–169.3°) and a very large solid angle, leading to a wide dynamic range and a high count rate. The high real-space resolution of GEM approaches 0.1 Å due to the high maximum momentum transfer (55 $Å^{-1}$) [9]. The GEM diffractometer also offers the capability of diffraction within a high temperature gas flow cell up to 1,300 K and a variety of atmospheres and gas flow rates (Figs. 4, 5, [10]). This apparatus, in conjunction with the GEM diffractometer, was used to investigate the structural properties of anion deficient ceria,$CeO_{2\text{-}\delta}$, as a function of oxygen partial pressure. Here, analysis of the total scattering, together with reverse Monte Carlo (RMC) modeling highlighted preferential alignment of O^{2-} vacancies as pairs in the <111> cubic directions as the degree of non-stoichiometry increased [11].

Purely structural analysis of catalysis materials may also be undertaken at ISIS. In 2002, Fowkes et al. [8] undertook a study of copper (II) and copper (I) exchanged zeolite Y using the high resolution powder diffractometer (HRPD) at ISIS to reveal subtle changes in cation site occupancy, cation–framework interactions and framework relaxation. HRPD is the highest resolution neutron powder diffractometer of its type in the world with a main backscattering detector bank Δd/d resolution of ~4 × 10^{-4}, which allows resolution of peaks at d-spacings below 0.3 Å, and up to a maximum wavelength of 10–12 Å. Potential catalysis materials, such as metal organic framework (MOF) materials have also been studied at ISIS and in 2010, Yan et al. [37] used the GEM diffractometer at ISIS to study the preferential absorption sites for D_2 in the NOTT-115 and NOTT-116 MOF materials. High pressure diffraction studies may also be performed at ISIS using the PEARL diffractometer, which is a medium resolution high-flux diffractometer optimized for data collection from a Paris-Edinburgh pressure cell, permitting structural studies to be carried out at pressures up to 25 GPa and routinely up to ~10 GPa using ISIS tungsten carbide anvils. In 2005, Colligan et al. [7] utilized this instrument to investigate the zeolite natrolite in mixtures of perdeuterated methanol and water at pressures of up to 1.87 GPa.

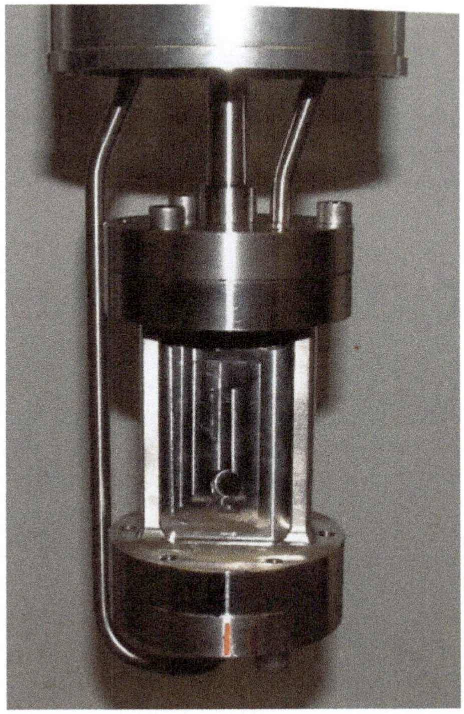

Fig. 5 The GEM diffractometer high temperature gas flow cell, capable of up to 1,300 K and a variety of atmospheres and gas flow rates

Fig. 4 The schematic diagram of the flow-through gas cell (with the *right hand portion* highlighting the central portion of the *left hand figure*), showing (*1*) oxygen and temperature sensor; (*2*) gas inlet; (*3*) gas outlet; (*4*) heater element; (*5*) heat shielding; (*6*) quartz sample holder; (*7*) sample; (*8*) quartz frit; (*9*) thermocouple, from [10]

Inelastic scattering

Inelastic neutron scattering (INS) covers a large range of both energy (equivalent to the microwave to the vacuum ultraviolet) and timescales (micro- to attoseconds). In all cases it is the energy difference, ΔE, between the incident and the scattered neutron that is required. With quasielastic scattering, $\Delta E \sim 0\ cm^{-1}$, the applications include measurements of diffusion constants in microporous materials [15], or the temperature of dissociation of hydrogen on a surface [24]. At higher energies, $100 \leq \Delta E \leq 4{,}000\ cm^{-1}$, neutron vibrational spectroscopy [22] has been used for the investigation of adsorbates [2] ranging from the molecular [21] to highly disordered hydrocarbonaceous overlayers [34].

ISIS is uniquely well-equipped for neutron vibrational spectroscopy. The four principle instruments for molecular vibrational INS at ISIS are TOSCA [6], MAPS [19], MARI [20] and MERLIN [5] covering an energy range of 0–16,000 cm^{-1}. For inelastic scattering, in addition to the total neutron flight time either the arrival time of the neutron at the sample (direct geometry spectrometer, MARI, MAPS and MERLIN) or the time for the neutron to travel from the sample to the detector must be known (indirect geometry spectrometer, TOSCA). This requires fixing either the incident energy or the final energy. TOSCA has a spectral range of −24–4,000 cm^{-1} with a resolution of ~1.25 % of the energy transfer for transitions below ca. 1,200 cm^{-1} and a precision of ~3 cm^{-1} below 2,000 cm^{-1}. The direct geometry instruments provide the opportunity to trade sensitivity and resolution and MARI produces a monochromatic neutron beam in the range of 72–8,060 cm^{-1} with a resolution of ~1 % of the incident energy, E_i, at the largest energy transfer. MAPS operates between 120 and 16,000 cm^{-1} with a resolution of up to 1.5 % E_i, while MERLIN is a very high count rate, medium energy resolution, spectrometer operating in the range of 50–16,000 cm^{-1}, with a resolution of 3–5 % E_i. Both MAPS and MERLIN have a very large array of position-sensitive detectors, on MERLIN this is nearly 180° in the horizontal plane and ±30° in the vertical plane, which provide high sensitivity for weakly scattering samples such as adsorbed species on catalysts. The two types of spectrometer are highly complementary [28], for modes at <1,200 cm^{-1}, TOSCA is the best resolution spectrometer in the world, while the direct geometry instruments are able to explore the 2,800–4,000 cm^{-1} region, where the characteristic C–H, N–H, and O–H stretch modes occur.

INS spectroscopy is not intrinsically a surface sensitive technique, high penetration of neutrons means that all the atoms in the sample contribute to the signal. Since the number of surface atoms is usually only a small fraction of the total number of atoms present, it is essential to maximize the signal from the surface species. In practice, this means that adsorbates must be hydrogenous, large samples are needed and high surface area is required. Sample sizes are typically in the 10–50 g range, a 1000-fold larger than is typically used for microreactor studies. For commercial catalysts, the sample size is generally not a concern and has the advantage that representative sampling is much less of a problem. For research materials, generating sufficient sample for an INS study can be a challenge. A flexible system that is capable of producing the samples for a range of heterogeneously catalyzed reactions has been jointly developed between ISIS and the University of Glasgow (Fig. 6) [35]. The technology is essentially the same as that used for microreactor studies but scaled-up to handle the larger quantities of sample and reactants required. The system can treat up to 30 g quantities of catalyst under representative reaction conditions of up to 823 K and pressures of up to 20 bar. Furthermore, the modularity of the design allows adaptation of the apparatus to different reaction conditions. The penetrating nature of neutrons means that there is a wide choice of materials for sample cells. Steel, quartz, Inconel™ (a Ni(70 %)/Cr/Fe alloy with minor amounts of other elements) and Zircaloy™ (95 % Zr, ~2 % Sn and small amounts of Nb, Fe, Cr, Ni) cells have all been used in a wide variety of designs, including flow-through and batch reactors (Fig. 7).

Zeolites are among the most important catalyst materials, but below ~1,800 cm^{-1} are opaque to infrared radiation; however, they are transparent to neutrons, thus the combination of INS to study the 'fingerprint region' with infrared spectroscopy to study the O–H stretch region is very powerful and enables access to the complete 0–4,000 cm^{-1} spectral range. This has been used to characterize hydroxyls and ammonium ions in Y zeolite [13, 14]

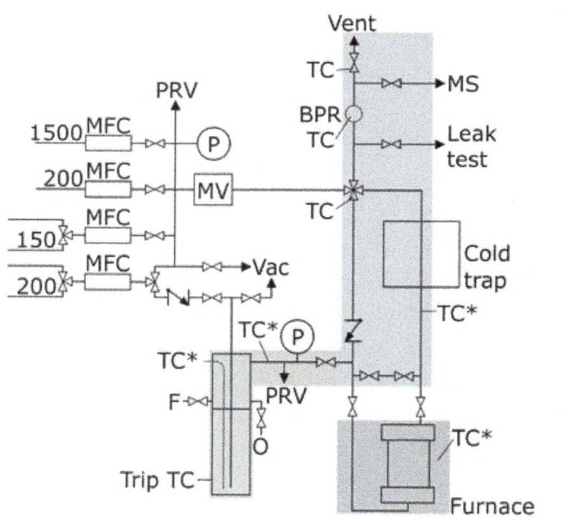

Fig. 6 Schematic of gas handling equipment showing arrangement of mass flow controllers [*numbers* indicate maximum flow rates for N_2 (sccm)], pressure relief valves (*PRV*), pressure sensors (*P*), mixing volume (*MV*), vacuum connections (*Vac*), thermocouple (*TC*), fill and overflow saturator lines (*F* and *O*), mass spectrometer (*MS*), and back pressure regulator (*BPR*). *Shaded areas* indicate heated gas lines. Reproduced from [35]

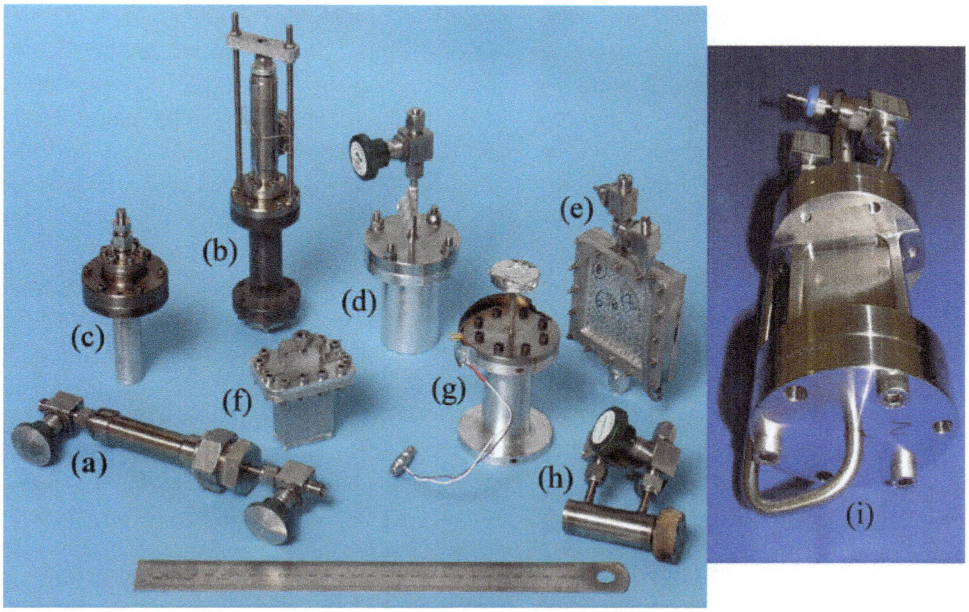

Fig. 7 *Left* vertically oriented catalyst cells as used on MAPS and TOSCA at ISIS: (*a*) high temperature (>600 K), flow-through, Swagelok™ sealed stainless steel cell, (*b*) high temperature gold wire sealed Zircaloy™ cell, (*c*) low volume and (*d*) high volume O-ring sealed aluminum cells for batch reactions, (*f*) aluminum cell for off-line prepared samples, (*g*) low temperature (323 K<) indium wire sealed aluminum cell, (*h*) flow-through, Conflat™ sealed stainless steel cell for use on MARI at ISIS. The scale in the foreground is 0.3 m. *Right* high pressure (20 bar), high temperature (873 K) Conflat™ sealed Inconel™ cell [36] for use on MAPS or TOSCA. The overall length including valves and pipe work is 0.28 m

and also to investigate the adsorption of short carbon chain (C_1–C_4) on cation-exchanged zeolite X [32, 33]. Using a combination of NMR, infrared and the TOSCA spectrometer at ISIS Schenkel et al. were able to identify the sorption structures for a number of C_1–C_4 *n*-alcohols, the strongest adsorption sites and a rationale for the difference in sorption behavior for the different cation-substituted materials.

Determination of surface-adsorbed species is important for all catalyst materials, albeit for varying reasons. For example, pyridine chemisorption is often used as a probe of surface acidity usually in conjunction with infrared spectroscopy. The use of neutrons greatly extends the spectral window available and in addition to the study of pyridine on alumina [18], where it is routinely used, it has also been possible to study pyridine adsorption on activated carbon, which is often used as a support material for precious metal catalysts [17]. Adsorbed species on used catalysts are also of considerable industrial interest. An examination of a 20 % Pd/C catalyst that had been used for Heck coupling reactions of bromobenzene and styrene to *trans*-stilbene produced the surprising result that the predominate surface species was *cis*-stilbene, even though a yield of about 90 % of *trans*-stilbene was measured in the product mixture [1].

In 2005, Lennon et al. with an industrial partner, carried out a combined infrared and INS study of the interaction of η-alumina and methanol using both the MARI and TOSCA spectrometers [21]. This reaction is important in the industrial scale manufacture of methyl chloride, where methanol and hydrogen chloride are combined over high surface area alumina catalysts. The combination of TOSCA and MARI INS instruments permits both high (MARI) and low (TOSCA) energy modes to be observed, and thus the complete spectroscopic range from 0–4,000 cm^{-1} to be investigated at high resolution. This combination of techniques permitted the investigators to determine the dominant surface species to be an adsorbed methoxy species. One of the problems associated with methyl chloride synthesis is that dimethyl ether, CH_3OCH_3, is formed as an unwanted by-product. As a result of this work, the industrial partner was able to modify the surface properties of the catalyst so as to drastically reduce the amount of dimethyl ether produced and the attendant cost of recycling it back to methanol. This is a clear demonstration of chemical understanding at the molecular level helping to rationalise improvements in a large scale industrial process with both financial and environmental benefits.

Recent work has examined the Lindlar catalyst (lead-modified palladium on calcium carbonate) and compared it to an unmodified catalyst (palladium on calcium carbonate) [4]. The Lindlar catalyst is remarkable, because it selectively reduces alkynes to *cis*-alkenes whereas the unmodified catalyst results in alkanes. The MAPS spectrometer is particularly suited to the study of supported catalysts as its enhanced sensitivity permits the study of the low concentrations of precious metal catalysts often found in commercial catalyst materials. Similar studies have been carried out by Ramirez-Cuesta et al. [30, 31] looking at zeolitic materials, Larese et al. [16] looking at binding of H_2 onto MgO and Albers et al. [3] looking at carbon-supported platinum–ruthenium fuel cell catalysts.

Reforming of methane to syngas (CO/H_2) is an important industrial process and the study of reforming catalysts, typically nickel/alumina, is of significant interest. However, deactivation by carbon build-up is a major problem. Utilizing the apparatus previously described (Fig. 6), a series of deactivated catalysts were prepared. Operating the MAPS spectrometer at an incident energy of 4,840 cm^{-1} to focus on the crucial C–H and O–H spectral region, Silverwood et al. [34] were able to identify the extent of overall carbon laydown on the catalyst and also to quantify the hydrogen associated with carbon and oxygen. This ability to quantify surface species is a unique capability of INS spectroscopy.

INS measurements are usually carried out at low temperature, <20 K. However, in favorable circumstances, by taking advantage of the capabilities of the MAPS spectrometer, it is possible to carry out in operando experiments. In an example of this, experiments were carried out in MAPS at room temperature under flowing 5 % CO/He (100 cm^3 min^{-1}) to investigate the low temperature oxidation of carbon monoxide over hydrous palladium oxide. This work identified that the active phase of PdO for CO oxidation is closely associated with hydroxyl groups [27].

To understand catalytic processes, it is often necessary to use model compounds and completely characterize the reactants and products. For these types of study TOSCA excels due to its spectral range and high resolution and has been used to study ethylidyne tricobalt nonacarbonyl [23], a model compound for the adsorbed ethylidyne that results from hydrogenation of ethene on (111) surfaces. Parker et al. [29] also recently completed a combined, simultaneous infrared and INS study of the C5 dienes and alkenes that result from hydrogenation of a C5 mixture over a Pd/Al_2O_3 catalyst. The mixture models a C5 raffinate cut from a refinery. In this study Parker et al. were able to compare INS, Raman and infrared spectra with calculated spectra for a number of species, and were able to show that the combination of spectroscopic techniques provided data that could be used for quantifiable, real-time gas-phase catalysis processes as well as molecular, isomer and conformer discrimination.

Conclusion

It can be seen that the use of neutron scattering techniques, either on their own, or in conjunction with other techniques can provide a range of information on catalysts and catalytic systems that is not available through the sole use of other, more routine, spectroscopic techniques. The large spectral range of the ISIS spectrometers, high resolution of ISIS diffractometers and high neutron cross section of hydrogen make ISIS a unique and world-leading facility for the study of catalysts and catalytic processes. The recent development of gas handling and specialist sample environments now allows scientists to test industrially relevant samples in environments that exactly mimic or closely approach those utilized in commercial processes. These in operando studies significantly increase the relevance of neutron scattering studies, permitting scientists to directly investigate commercially important processes and thus develop a greater understanding of catalytic mechanisms and processes, and thus produce catalysts and catalytic processes with enhanced properties.

References

1. Albers PW, Krauter JGE, Ross DK, Heidenreich RG, Kohler K, Parker SF (2004) Identification of surface states on finely divided supported palladium catalysts by means of inelastic incoherent neutron scattering. Langmuir 20:8254–8260.
2. Albers PW, Parker SF (2007) Inelastic incoherent neutron scattering in catalyst research. Adv Catal 51:99–132
3. Albers PW, Weber W, Kunzmann K, Lopez M, Parker SF (2008) Characterisation of carbon supported platinum–ruthenium fuel cell catalysts of different degree of alloying. Surf Sci 602: 3611–3616.
4. Albers PW, Möbus K, Frost CD, Parker SF (2011) Characterization of β-palladium hydride formation in the Lindlar catalyst and in carbon-supported palladium. J Phys Chem C 115:24485–24493.
5. Bewley RI, Eccleston RS, McEwen KA, Hayden SM, Dove MT, Bennington SM, Treadgold JR, Coleman RLS (2006) RLS MERLIN, a new high count rate spectrometer at ISIS. Physica B 385–386:1029–1031.
6. Colognesi D, Celli M, Cilloco F, Newport RJ, Parker SF, Rossi-Albertini V, Sacchetti F, Tomkinson J, Zoppi M (2002) TOSCA neutron spectrometer; the final configuration. Appl Phys A 74(Suppl.):S64–S66
7. Colligan M, Lee Y, Vogt T, Celestian AJ, Parise JB, Marshall WG, Hriljac JA (2005) High-pressure neutron diffraction study of superhydrated natrolite. J Phys Chem B 109:18223–18225.
8. Fowkes AJ, Ibberson RM, Rosseinsky MJ (2002) Structural characterization of the redox behavior in copper-exchanged sodium zeolite Y by high-resolution powder neutron diffraction. Chem Mater 14:590–602.
9. Hannon AC (2005) Results on disordered materials from the GEneral Materials diffractometer, GEM, at ISIS. Nucl Instrum Methods Phys Res A 551:88–107
10. Haynes R, Norberg ST, Eriksson SG, Chowdhury MAH, Goodway CM, Howells GD, Kirichek O, Hull S (2010) New high

temperature gas flow cell developed at ISIS. J Phys Conf Ser 251:012090.
11. Hull S, Norberg ST, Ahmeda I, Eriksson SG, Marrocchelli D, Madden PA (2009) Oxygen vacancy ordering within anion-deficient ceria. J Solid State Chem 182:2815–2821.
12. ISIS (2012) http://www.isis.stfc.ac.uk
13. Jacobs WPJH, van Santen RA, Jobic H (1994) Inelastic neutron scattering study of NH_4Y zeolites. J Chem Soc Faraday Trans 90:1191–1196
14. Jacobs WPJH, van Wolput JHMC, van Santen RA, Jobic H (1994) A vibrational study of the OH and OD bending modes of the Brönsted acid sites in zeolites. Zeolites 14:117–125
15. Jobic H, Theodorou DN (2007) Quasi-elastic neutron scattering and molecular dynamics simulation as complementary techniques for studying diffusion in zeolites. Microporous Mesoporous Mater 102:21–50.
16. Larese JZ, Arnold T, Frazier L, Hinde RJ, Ramirez-Cuesta AJ (2008) Direct observation of H_2 binding to a metal oxide surface. Phys Rev Lett 101:165302.
17. Lennon D, Lundie DT, Jackson SD, Kelly GJ, Parker SF (2002) Characterization of activated carbon using X-ray photoelectron spectroscopy and inelastic neutron scattering spectroscopy. Langmuir 18:4667–4673.
18. Lundie DT, McInroy AR, Marshall R, Winfield JM, Mitchell C, Dudman CC, Jones P, Parker SF, Lennon D (2005) An improved description of the surface acidity of eta-alumina. J Phys Chem B 109:11592–11601
19. MAPS (2012) http://www.isis.stfc.ac.uk/instruments/maps/
20. MARI (2012) http://www.isis.stfc.ac.uk/instruments/mari
21. McInroy AR, Lundie DT, Winfield JM, Dudman C, Jones P, Parker SF, Taylor JW, Lennon D (2005) An infrared and inelastic neutron scattering spectroscopic investigation on the interaction of η-alumina and methanol. Phys Chem Chem Phys 7:3093–3101
22. Mitchell PCH, Parker SF, Ramirez-Cuesta AJ, Tomkinson J (2005) Vibrational spectroscopy with neutrons: with applications in chemistry, biology, materials science, and catalysis. World Scientific, Singapore
23. Parker SF, Marsh NA, Camus LM, Whittlesey MK, Jayasooriya UA, Kearley GJ (2002) Ethylidyne tricobalt nonacarbonyl: infrared, FT-Raman, and inelastic neutron scattering spectra. J Phys Chem A 106:5797–5802.
24. Parker SF, Frost CD, Telling M, Albers P, Lopez M, Seitz K (2006) Characterisation of the adsorption sites of hydrogen on Pt/C fuel cell catalysts. Catal Today 114:418–421
25. Parker SF, Bowron DT, Imberti S, Soper AK, Refson K, Lox ES, Lopez M, Albers P (2010a) Structure determination of adsorbed hydrogen on a real catalyst. Chem Comm 46(17):2959–2961.
26. Parker SF, Refson K, Hannon AC, Barney ER, Robertson SJ, Albers P (2010b) Characterization of hydrous palladium oxide: implications for low-temperature carbon monoxide oxidation. J Phys Chem C 114:14164–14172.
27. Parker SF (2011) The role of hydroxyl groups in low temperature carbon monoxide oxidation. Chem Commun 47:1988–1990.
28. Parker SF, Lennon D, Albers PW (2011) Spectroscopy with neutrons: a review of new directions. Appl Spectrosc 65:1325–1341.
29. Parker SF, Siegel D, Hamilton NG, Kapitan J, Hecht L, Lennon D (2012) Characterization of C5 hydrocarbons relevant to catalysis. J Phys Chem A 116:333–346.
30. Ramirez-Cuesta AJ, Mitchell PCH, Ross DK, Georgiev PA, Anderson PA, Langmie HW, Book D (2007) Dihydrogen in zeolite CaX—an inelastic neutron scattering study. J Alloy Compd 446–447:393–396.
31. Ramirez-Cuesta AJ, Mitchell PCH, Ross DK, Georgiev PA, Anderson PA, Langmie HW, Book D (2007) Dihydrogen in cation-substituted zeolites X—an inelastic neutron scattering study. J Mater Chem 17:2533–2539.
32. Schenkel R, Jentys A, Parker SF, Lercher JA (2004ba) Investigation of the adsorption of methanol on alkali metal cation exchanged zeolite X by inelastic neutron scattering. J Phys Chem B 108:7902–7910
33. Schenkel R, Jentys A, Parker SF, Lercher JA (2004b) INS and IR and NMR spectroscopic study of C1–C4 alcohols adsorbed on alkali metal-exchanged zeolite X. J Phys Chem B 108:15013–15026.
34. Silverwood IP, Hamilton NG, Laycock CJ, Staniforth JZ, Ormerod RM, Frost CD, Parker SF, Lennon D (2010) Quantification of surface species present on a nickel/alumina methane reforming catalyst. Phys Chem Chem Phys 12:3102–3107.
35. Silverwood IP, Hamilton NG, McFarlane A, Ormerod RM, Guidi T, Bones J, Dudman MP, Goodway CM, Kibble M, Parker SF, Lennon D (2011) Experimental arrangements suitable for the acquisition of inelastic neutron scattering spectra of heterogeneous catalysts. Rev Sci Inst 82:034101.
36. Turner JFC, Done R, Dreyer J, David WIF, Catlow CRA (1999) On apparatus for studying catalysts and catalytic processes using neutron scattering. Rev Sci Instrum 70:2325.
37. Yan Y, Telepeni I, Yang SH, Lin X, Kockelmann W, Dailly A, Blake AJ, Lewis W, Walker GS, Allan DR, Barnett SA, Champness NR Schroder M (2010) Metal-organic polyhedral frameworks: high H-2 adsorption capacities and neutron powder diffraction studies. JACS 132:4092.

4

Gold catalysis: helping create a sustainable future

Nikolaos Dimitratos · Jennifer K. Edwards · Christopher J. Kiely · Graham J. Hutchings

Abstract In recent years there has been a general realisation that supported gold and gold bimetallic nanoparticles can be very effective for a broad range of redox reactions. In this paper we review the preparation of gold palladium nanoparticles using a sol-immobilisation methodology and show their effectiveness for the oxidation of benzyl alcohol and the direct synthesis of hydrogen peroxide.

Keywords Gold catalysis · Alcohol oxidation · Hydrogen peroxide synthesis

Introduction

Gold for many centuries has held a special place in the attention of the general public partly due its intrinsic value but also because of the beautiful jewelry and artwork that can be created from it. The reason for this is that gold is the most noble of metals and consequently is considered immutable. From a chemical perspective, gold has therefore relatively little inherent interest as it appears in its bulk form to have a very limited chemistry. However, a major discovery of the 1980s was that gold, when sub-divided down to the nanoscale, can be highly reactive and can activate small molecules. This has led to an explosion of interest in the chemistry of gold and, in particular, the synthesis and use of small gold-containing nanoparticles as redox catalysts.

On the basis of many studies, Au nanoparticles have been shown to be particularly effective for a broad range of oxidation and hydrogenation reactions as well as chemical synthesis [1–8]. These include the low temperature oxidation of CO [8], especially when used as part of the processes for purifying a fuel cell feedstock [9, 10], the synthesis of vinyl chloride by the hydrochlorination of ethyne [11], the selective oxidation of alkenes to epoxides [12, 13] and alcohols to aldehydes [14] as well as selective hydrogenation [15]. Recently, research in our groups has shown that the alloying of Pd with Au can enhance the activity of the nanoparticles, and such materials have been found to be particularly effective for the direct synthesis of hydrogen peroxide from its elemental constituents [16–20] and the oxidation of alcohols [21]. In this paper, we review an aspect of our work concerning the synthesis of supported Au–Pd alloy nanoparticles prepared by sol-immobilisation for two important classes of redox reaction; namely the direct synthesis of hydrogen peroxide and the oxidation of benzyl alcohol. These two reactions have very broad interest in the chemical community as they both display key features concerning selectivity control. In the case of hydrogen peroxide synthesis, it is important that the non-selective formation of water is minimised, and in the case of alcohol oxidation to the corresponding aldehyde it is essential that the over-oxidation to the carboxylic acid is minimised. By way of background information, hydrogen peroxide (H_2O_2) is a major commodity chemical [22] and has significant uses in cleaning and bleaching. At present, H_2O_2 is produced using an indirect process which involves the sequential hydrogenation and oxidation of an alkyl anthraquinone, thereby avoiding the

N. Dimitratos
Department of Chemistry, University College London, 20 Gordon Street, London WC1H 0AJ, UK

J. K. Edwards · G. J. Hutchings (✉)
School of Chemistry, Cardiff University, Main Building, Park Place, Cardiff CF10 3AT, UK
e-mail: hutch@cardiff.ac.uk

C. J. Kiely
Department of Materials Science and Engineering, Lehigh University, 5 East Packer Avenue, Bethlehem, PA 18015-3195, USA

potential for explosive contact between hydrogen and oxygen to occur [23]. The direct synthesis of hydrogen peroxide is not a reaction that is easy to control, since the formation of water is preferred under all conditions, and the quest to find a suitable catalyst for the direct reaction between H_2 and O_2 has remained a challenge for almost a century, and as such, it has attracted considerable attention. Supported Pd catalysts have been the focus of much of the research [24–32]. Our work on this topic [16–20] has shown that a combination of Pd with Au promotes both the activity and the selectivity for the direct H_2O_2 synthesis reaction. Furthermore, these catalysts are also very effective for the oxidation of alcohols, and once again the addition of Pd to Au increases both the selectivity and the activity [21]. The link between these two reactions is considered to be at the level of the possible surface intermediates present during the reaction, and it is thought that a surface hydroperoxy species may be responsible for the selective chemistry observed in both processes. In our initial studies [16–20], we used a wet co-impregnation method to prepare gold–palladium nanoparticles and concluded that either homogeneous random alloys or core–shell structures [19] could be formed. Based on this work, we hypothesised that small homogeneous Au–Pd nanoparticles were likely to be the active species in these catalysts and we set out to utilise a sol-immobilisation method [33] in an attempt to 'tailor-make' these nanoparticles. We reasoned that since the volume fraction of these ultra-small alloy particles formed using the co-impregnation method was quite small [19, 20], pre-forming colloidal nanoparticles should be more material efficient and hence overcome this problem. We found that this latter method can indeed produce well-defined particle size distributions [33]. However, the catalysts do not retain the selectivity displayed by the catalysts prepared by wet impregnation [16–20]. This paper reviews some of the essential structural features and reactivity traits of Au–Pd alloy catalysts prepared using a sol-immobilisation method.

A general preparation method for Au, Pd and Au–Pd nanoparticles using sol-immobilisation

In our studies that are highlighted in this brief review, we have used the following general method for catalyst generation by sol-immobilisation [33]. It presents a facile method of reproducibly forming either mono-metallic or bi-metallic nanoparticles. For the preparation of supported Au–Pd colloidal materials, an aqueous solution of $PdCl_2$ (Johnson Matthey) and $HAuCl_4{\cdot}3H_2O$ (Johnson Matthey) of the desired concentration was prepared. Poly-vinyl-alcohol (PVA) (1 wt% aqueous solution, Aldrich, MW = 10,000, 80 % hydrolyzed) and an aqueous solution of $NaBH_4$ (0.1 M) were also prepared. For example, a catalyst comprised Au–Pd nanoparticles with 1 wt% total metal loading on a carbon support (Aldrich G60) was prepared as follows: to an aqueous $PdCl_2$ and $HAuCl_4$ solution of the desired concentration, the required amount of a PVA solution (1 wt%) was added [PVA/(Au + Pd) (wt/wt) = 1.2]; a freshly prepared solution of $NaBH_4$ [0.1 M, $NaBH_4$/(Au + Pd) (mol/mol) = 5] was then added to form a dark-brown sol. After 30 min of sol generation, the colloid was immobilised by adding activated carbon (acidified at pH 1 by sulphuric acid) under vigorous stirring conditions. The amount of support material required was calculated so as to have a total final metal loading of 1 wt% and the molar ratio of Au:Pd was varied by adjusting the relative concentrations of the metals in solution. After 2 h the slurry was filtered, the catalyst washed thoroughly with distilled water (neutral mother liquors) and dried at 120 °C for 16 h. It should be noted that there are a range of protecting ligands other than PVA that can be used, as well as a range of alternative reducing agents. Variation in the reaction parameters can lead to different particle size distributions and represents an important way of controlling and varying the nature of the nanoparticles that can be obtained. For instance, by co-adding the Pd and Au precursors, this

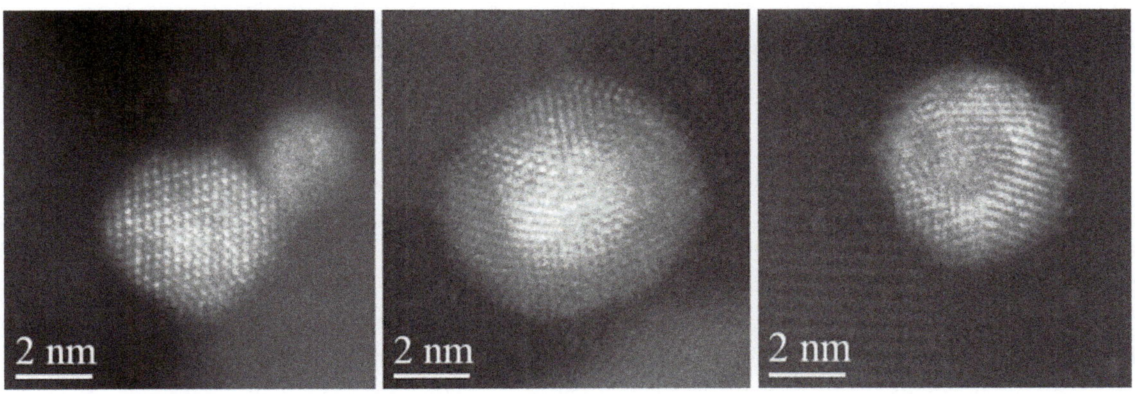

Fig. 1 STEM-HAADF images showing (*left*) a AuPd homogeneous alloy particles, (*middle*) a Au–Pd nanoparticle with a gold core and a palladium shell, and (*right*) a AuPd nanoparticle with a palladium core and a gold shell

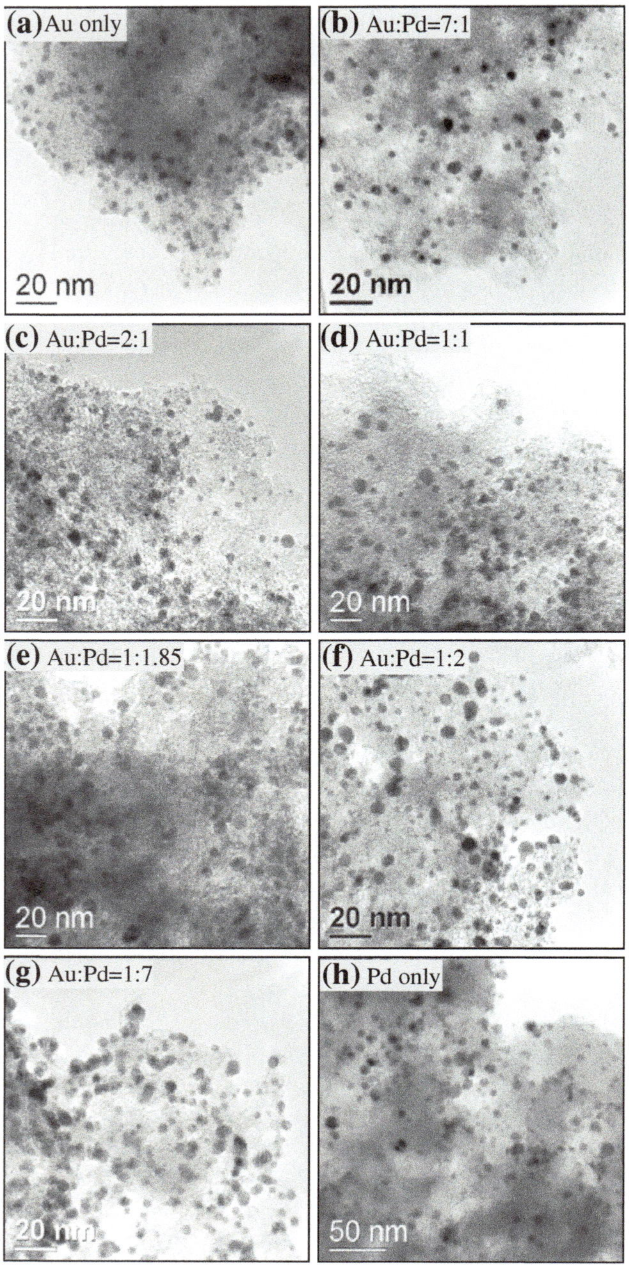

Fig. 2 Representative bright field TEM micrographs showing Au–Pd particles immobilised on an activated carbon support: **a** Au only, **b** Au:Pd = 7:1, **c** Au:Pd = 2:1, **d** Au:Pd = 1:1, **e** Au:Pd = 1:1.85, **f** Au:Pd = 1:2, **g** Au:Pd = 1:7, **h** Pd only. Reprinted with permission from reference [33]. Copyright (2010) American Chemical Society

method can be used to prepare sol-immobilised Au–Pd nanoparticles that comprise homogeneous alloys. However, by varying the order of addition of these reagents [34] it is possible to prepare nanoparticles that exhibit core–shell structures and examples of these are given in Fig. 1 together with an example of homogeneous Au–Pd alloy nanoparticles. The existence of core–shell morphologies is readily discerned due to the inherent Z-contrast of the Au and Pd in high angle annular dark field (HAADF) STEM images.

Characterisation of sol-immobilised Au, Pd and Au–Pd nanoparticles

One of the great advantages of the sol-immobilisation method is the flexibility that can be attained with respect to tuning the composition of the alloy nanoparticles. As an example, we prepared a range of Au, Pd and Au–Pd colloidal sols (Au:Pd = 1:0, 7:1, 2:1, 1:1, 1:1.85, 1:2, 1:7, 0:1) by varying the concentrations of the metal salts in the standard preparation method [33]. The reason for carrying out these experiments is that we considered that the active sites for the synthesis of hydrogen peroxide were associated with small nanoparticles that might be comprised mainly of Pd with only a very dilute amount of Au. Hence, by examining a wide range of compositions we could examine this hypothesis in some detail [33]. A set of representative bright field TEM micrographs is shown in Fig. 2 and the particle size distributions of the colloidal metal nanoparticles after immobilisation and drying on an activated carbon support are shown in Fig. 3. The median particles sizes for the monometallic Au and Pd colloids were found to be 3.5 and 5.4 nm, respectively. The median particle sizes for all the Au_xPd_y alloys were in the 2.9–4.6 nm range, and did not show any discernable systematic size trend with colloid composition. This shows that the method can be used to make a range of supported nanoparticles which have very similar particle size distribution, but have the potential for widely varying compositions. STEM-HAADF images of the alloy particles in the 2:1, 1:1 and 1:2 Au–Pd samples are shown in Fig. 4. In each case, the metal particles were found to be random homogeneous Au–Pd alloys with an f.c.c structure. Energy dispersive X-ray spectra acquired from areas containing ~100 metal particles showed both Au and Pd peaks suggesting intimate mixing of the two elements in the alloy samples. In addition, the relative Au to Pd peak areas from the ~100 particle average spectra qualitatively agreed with the nominal x and y values for each of the Au_xPd_y alloy compositions. However, it was noticeable that within specific Au_xPd_y samples, XEDS spectra obtained from individual particles did show considerable composition variations as a function of particle size. This is demonstrated by a set of HAADF images and corresponding XEDS point spectra from particles at the low, median and high ends of the particle size distribution for the Au:Pd (1:7) sample (Fig. 5). Whilst all the particles were homogeneous Au–Pd alloys, the smallest particles were Au-rich, whereas the largest ones tended to be Pd-rich. It is interesting to note that this systematic composition/size trend is diametrically opposite to the situation found for Au–Pd nanoparticles prepared by impregnation methods where the larger particles tended to Au-rich and the smaller Pd-rich.

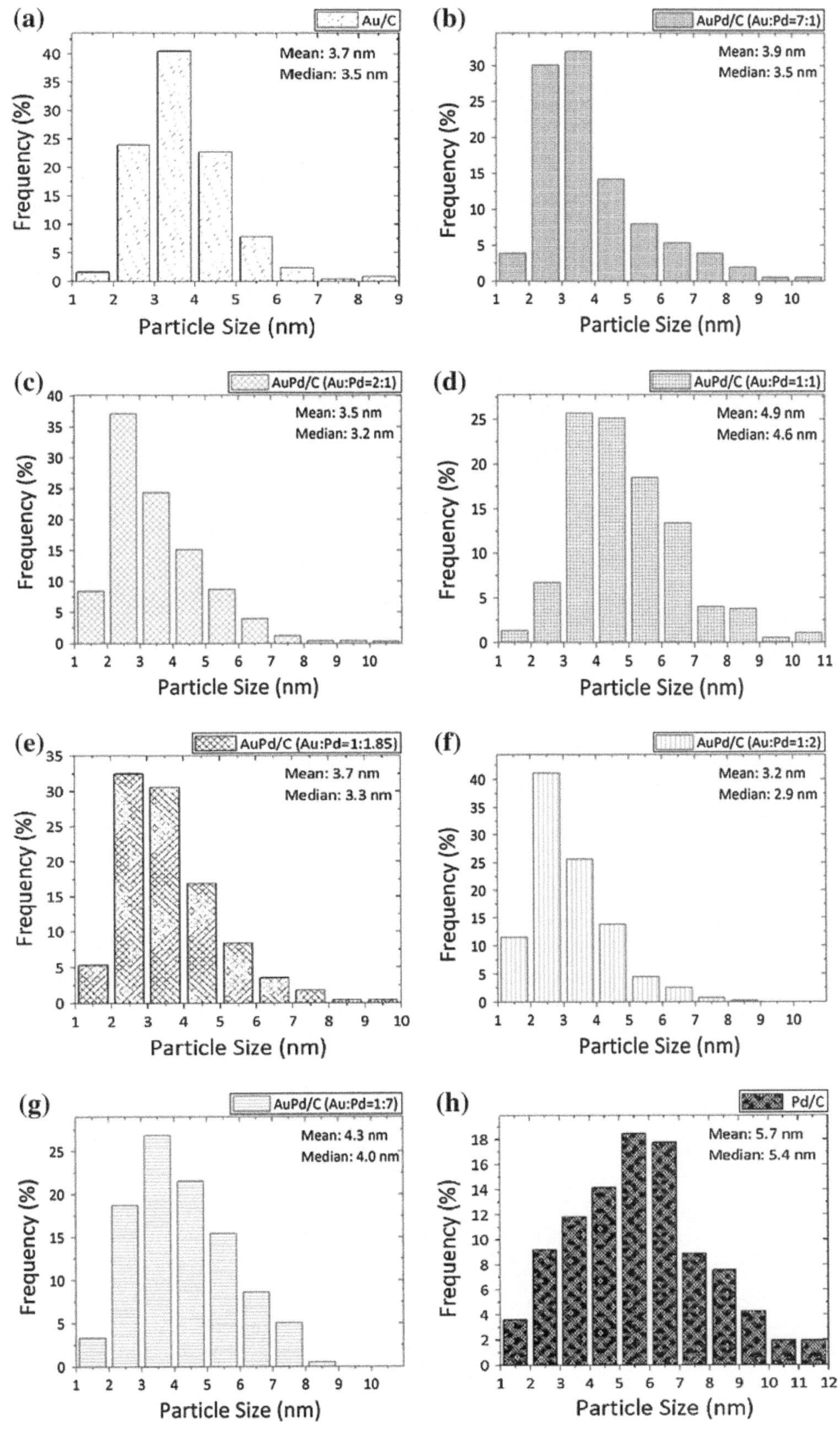

Fig. 3 Particle size distributions of the Au–Pd particles immobilised on an activated carbon support: **a** Au only, **b** Au:Pd = 7:1, **c** Au:Pd = 2:1, **d** Au:Pd = 1:1, **e** Au:Pd = 1:1.85, **f** Au:Pd = 1:2, **g** Au:Pd = 1:7, **h** Pd only. Reprinted with permission from reference [33]. Copyright (2010) American Chemical Society

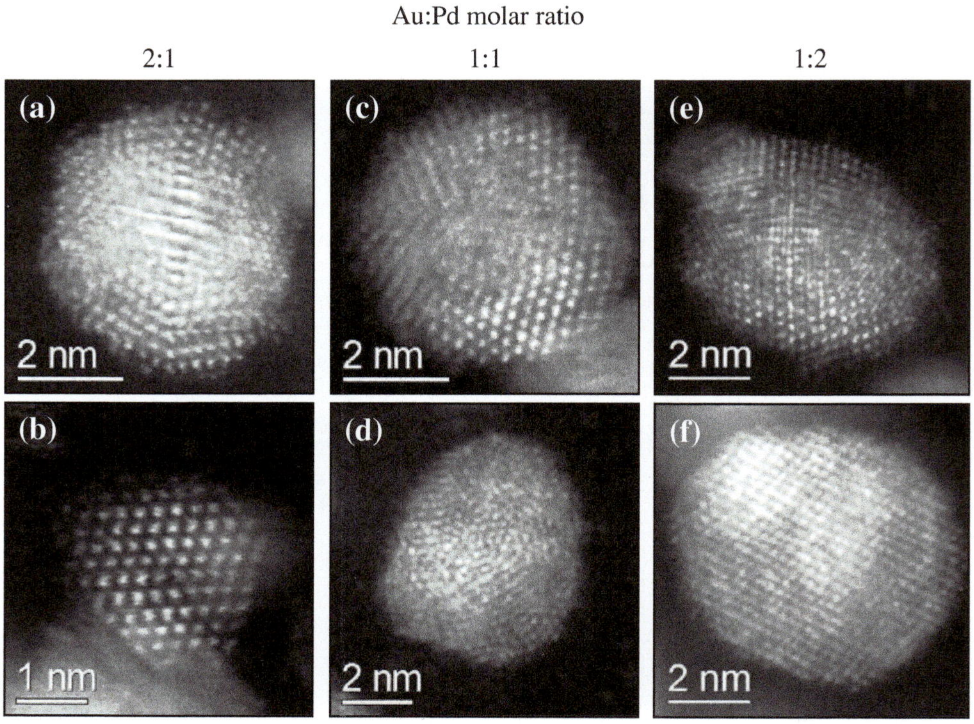

Fig. 4 Representative STEM-HAADF images showing Au–Pd homogeneous alloy particles. **a**, **b** Au:Pd = 2:1, **c**, **d** Au:Pd = 1:1, **e**, **f** Au:Pd = 1:2. Reprinted with permission from reference [33]. Copyright (2010) American Chemical Society

Hydrogen peroxide synthesis and hydrogenation

The supported Au–Pd catalysts were investigated for the direct synthesis of H_2O_2 and the results are shown in Fig. 6 [33]. The catalysts were found to be particularly effective for this reaction and the synthesis activity exhibited a maximum at Au:Pd = 1:2 molar ratio. In our previous studies using catalysts prepared using the impregnation method [16–20] we used an Au:Pd mol ratio of 1:1.85, which is very close to this optimum ratio. However, the catalysts prepared by sol-immobilisation are considerably more active than those prepared using the impregnation method [16–20]. This demonstrates that using the sol-immobilisation method we are able to utilise the precious metals much more effectively, as in the impregnation method a significant fraction very large particles are also generated and these have very limited catalytic activity.

In an earlier study [19] we observed that in the catalysts prepared using the impregnation method, the very smallest particles were mainly Pd and that 3–5 nm Au–Pd nanoparticles comprised typically 98 % Pd to 2 % Au. As the activity was considered to be associated with these small particles, this led us to conclude that for optimum activity we needed to find a method to fabricate such particles. However, the results of the sol-immobilised catalysts might suggest that this hypothesis is not necessarily the case, since the Pd-rich Au–Pd colloidal nanoparticles (i.e. Pd:Au $\geq$ 3) actually display a lower productivity than those with a Au:Pd = 1:2 molar ratio. It must also be taken into account that this effect could also be explained on the basis that the hydrogenation activity of the catalysts is expected to increase with Pd content and this somewhat cancels out the positive effect of adding Pd to the Au nanoparticles for the direct synthesis.

Benzyl alcohol oxidation

The catalytic oxidation of benzyl alcohol under free solvent conditions is an appropriate model reaction for exploring the catalytic performance of bimetallic Au–Pd catalysts and the results obtained for the conversion of benzyl alcohol to benzaldehyde are shown in Fig. 6. It is found that the catalytic performance in terms of reaction specificity is not dependent on the composition and all catalysts give high selectivities to benzaldehyde which is the primary product. Hence, the main outcome of alloying Au with Pd is to enhance activity. The activity of the mono-metallic 1 wt% Au/C sample was very poor, but on increasing the Pd content we observed a progressive increase in catalytic activity. This increase of activity reached a broad maximum with a Au:Pd molar ratio between 1:1 and 1:3. A further increase in the Pd content resulted in a progressive decrease in catalytic activity. It is important to note that even in the presence of a minor amount of gold or palladium (i.e. in the case of 1:7 and 7:1 Au:Pd molar ratios) a significant increase in the catalytic activity is observed as compared to the mono-metallic counterparts. It is apparent that the activity profile for benzyl alcohol oxidation has the same dependency on the

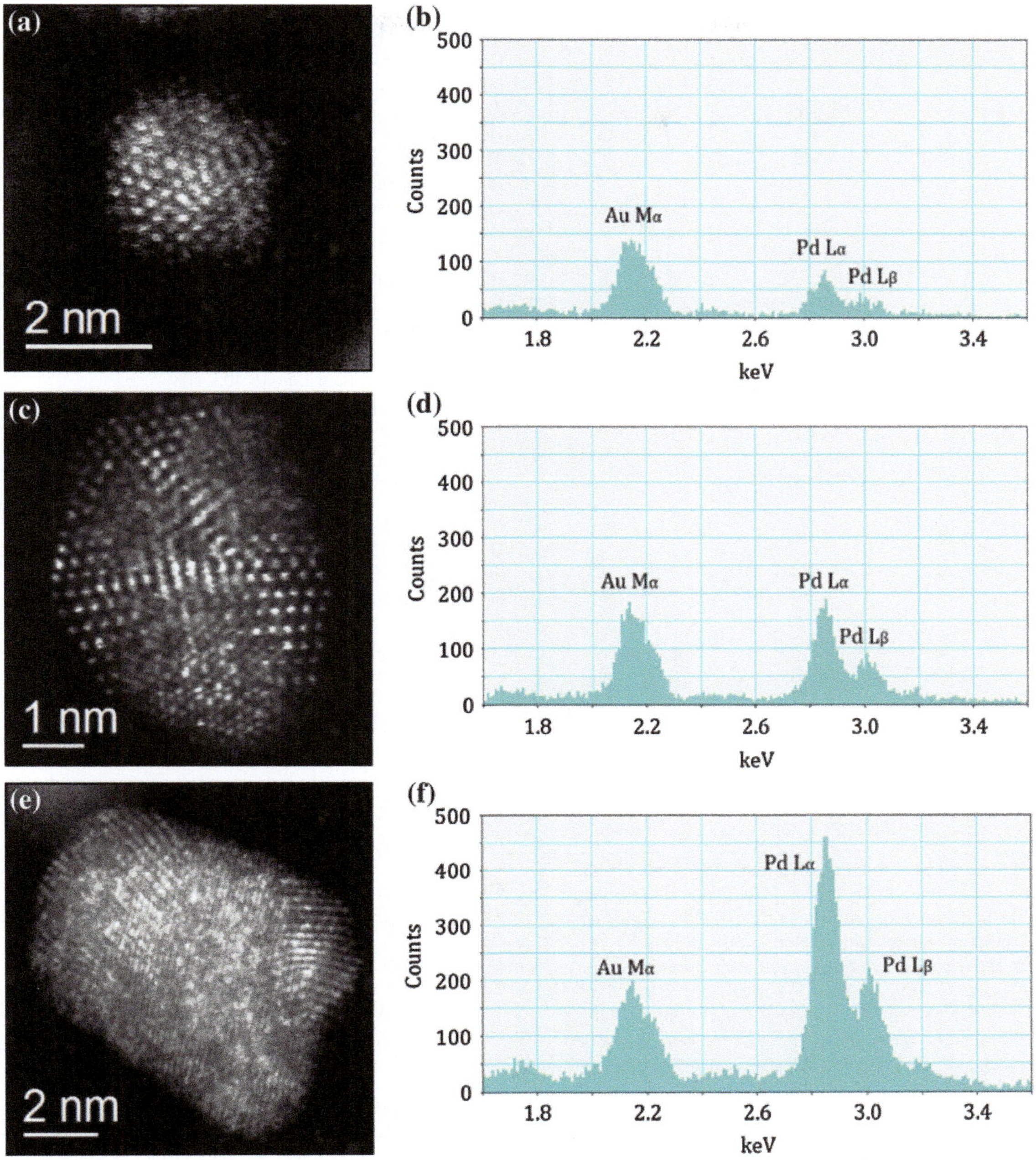

Fig. 5 A set of HAADF images and corresponding XEDS point spectra from particles at the low (**a**, **b**), median (**c**, **d**) and high (**e**, **f**) ends of the particle size distribution for the Au:Pd (1:7) sample. They demonstrate a systematic particle size/composition variation, where the smaller particles progressively become more Au-rich and Pd-deficient. Reprinted with permission from reference [33]. Copyright (2010) American Chemical Society

Au:Pd molar ratio as does the direct synthesis and hydrogenation of hydrogen peroxide.

Concluding comments

It is apparent that carbon-supported Au–Pd nanoparticles prepared by sol-immobilisation are very effective for the three reactions; namely, the direct synthesis of H_2O_2 and its subsequent hydrogenation and the selective oxidation of benzyl alcohol. For all these reactions we show that the optimum ratio of Au:Pd is ca. 1:1.85. As shown in Fig. 6 the catalytic activity per mol of metal of the Pd:Au series of catalysts for hydrogen peroxide synthesis, hydrogen peroxide hydrogenation and benzyl alcohol oxidation also shows a similar dependence on the Au:Pd molar ratio. This suggests that the active sites for these reactions could be common which is an interesting observation. However, it is also clear that there is considerable variation in the composition of the individual nanoparticles prepared by

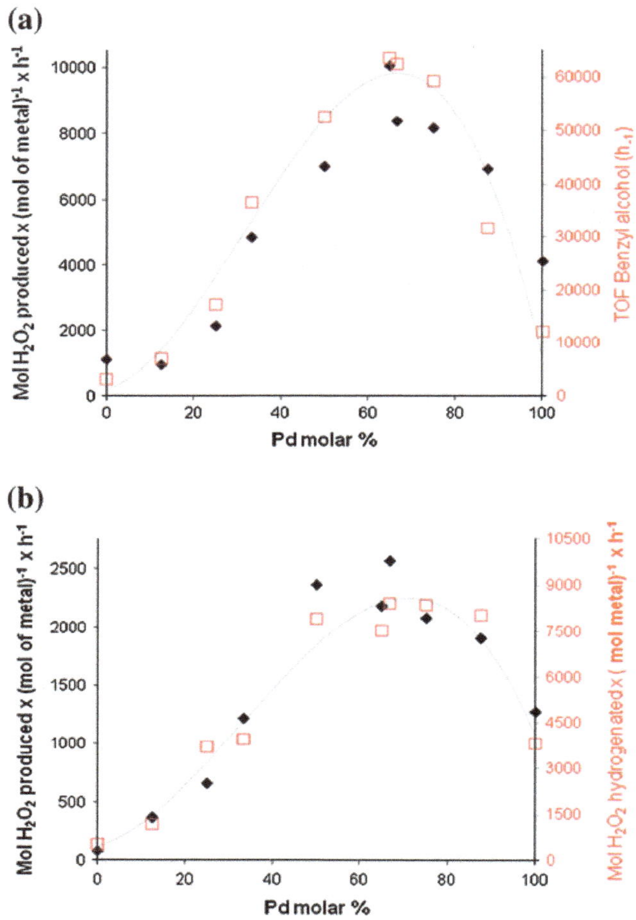

Fig. 6 Specific activity comparison, based on moles of metal present for benzyl alcohol oxidation, for hydrogen peroxide synthesis and hydrogen peroxide hydrogenation with a 1wt% Au–Pd/C catalysts prepared by sol immobilisation with Pd:Au molar ratio of 1.85. **a** *Black diamonds* H_2O_2 productivity calculated after 2 min reaction (*left axis*), *empty squares* TOF of benzyl alcohol calculated after 30 min reaction (*right axis*); **b** *black diamonds* H_2O_2 productivity calculated after 30 min (*left axis*), *empty squares* the rate of H_2O_2 hydrogenation calculated after 30 min. Reprinted with permission from reference [33]. Copyright (2010) American Chemical Society

sol-immobilisation and we are now exploring synthetic strategies that will help to overcome this problem since ideally we need to be able to synthesise bimetallic nanoparticles with both uniform particle size distributions and well-defined composition that does not vary from particle-to-particle.

References

1. Bond GC, Thompson DT (1999) Catalysis by Gold. Catal Rev Sci Eng 41:319–388
2. Bond GC, Thompson DT (2000) Gold-catalysed oxidation of carbon monoxide. Gold Bull 33:41–51
3. Haruta M (2004) Gold as a novel catalyst in the 21st century: preparation, working mechanism and applications. Gold Bull 37:27–36
4. Hashmi ASK (2004) Homogeneous catalysis by gold. Gold Bull 37:51–65
5. Meyer R, Lemaire C, Shaikutdinov ShK, Freund H-J (2004) Surface chemistry of catalysis by gold. Gold Bull 37:72–124
6. Hutchings GJ (2004) New directions in gold catalysis. Gold Bull 37:3–11
7. Hashmi ASK, Hutchings GJ (2006) Gold catalysis. Angew Chem Int Ed 45:7896–7936
8. Haruta M, Kobayashi T, Sano H, Yamada N (1987) Novel gold catalysts for the oxidation of carbon monoxide at a temperature far below 0 °C. Chem Lett 16:405–408
9. Landon P, Ferguson J, Solsona BE, Garcia T, Carley AF, Herzing AA, Kiely CJ, Golunski SE, Hutchings GJ (2005) Selective oxidation of CO in the presence of H_2, H_2O and CO_2 via gold for use in fuel cells. Chem Commun 3385–3387
10. Landon P, Ferguson J, Solsona BE, Garcia T, Al-Sayari S, Carley AF, Herzing A, Kiely CJ, Makkee M, Moulijn JA, Overweg A, Golunski SE, Hutchings GJ (2006) Selective oxidation of CO in the presence of H_2, H_2O and CO_2 using Au/Fe_2O_3 catalysts for use in fuel cells. J Mat Chem 16:199–208
11. Hutchings GJ (1985) Vapor phase hydrochlorination of acetylene: correlation of catalytic activity of supported metal chloride catalyst. J Catal 96:292–295
12. Sinha AK, Seelan S, Tsubota S, Haruta M (2004) A three-dimensional mesoporous titanosilicate support for gold nanoparticles: vapor-phase epoxidation of propene with high conversion. Angew Chem Int Ed 43:1546–1548
13. Hughes MD, Xu Y-J, Jenkins P, McMorn P, Landon P, Enache DI, Carley AF, Attard GA, Hutchings GJ, King F, Stitt EH, Johnston P, Griffin K, Kiely CJ (2005) Tunable gold catalysts for selective hydrocarbon oxidation under mild conditions. Nature 437:1132–1135
14. Abad A, Conception P, Corma A, Garcia H (2005) A collaborative effect between gold and a support induces the selective oxidation of alcohols. Angew Chem Int Ed 44:4066–4069
15. Corma A, Serna P (2006) Chemoselective hydrogenation of nitro compounds with supported gold catalysts. Science 313:332–334
16. Landon P, Collier PJ, Papworth AJ, Kiely CJ, Hutchings GJ (2002) Direct formation of hydrogen peroxide from H_2/O_2 using a gold catalysts. Chem Commun 2058–2059
17. Edwards JK, Solsona BE, Landon P, Carley AF, Herzing A, Kiely CJ, Hutchings GJ (2005) Direct synthesis of hydrogen peroxide from H_2 and O_2 using TiO_2-supported Au–Pd catalysts. J Catal 236:69–79
18. Edwards JK, Ntainjua E, Carley AF, Herzing AA, Kiely CJ, Hutchings GJ (2009) Direct synthesis of H_2O_2 from H_2 and O_2 over gold, palladium, and gold–palladium catalysts supported on acid-pretreated TiO_2. Angew Chem Int Ed 48:8512–8515
19. Edwards JK, Carley AF, Herzing AA, Kiely CJ, Hutchings GJ (2008) Direct synthesis of hydrogen peroxide from H_2 and O_2 using supported Au-Pd catalysts. Faraday Disc 138:225
20. Edwards JK, Solsona B, Ntainjua E, Carley AF, Herzing AA, Kiely CJ, Hutchings GJ (2009) Switching off hydrogen peroxide hydrogenation in the direct synthesis process. Science 323:1037–1041
21. Enache DI, Edwards JK, Landon P, Solsona-Espriu B, Carley AF, Herzing AA, Watanabe M, Kiely CJ, Knight DW, Hutchings GJ (2006) Solvent-free oxidation of primary alcohols to aldehydes using Au-Pd/TiO_2 catalysts. Science 311:362–365
22. Edwards JK, Hutchings GJ (2008) Palladium and gold-palladium catalysts for the direct synthesis of hydrogen peroxide. Angew Chem Int Ed 47:9192–9198

23. Hess HT (1995) Hydrogen peroxide. In: Kroschwitz I, Howe-Grant M (eds) Kirk-othmer encyclopaedia of chemical engineering, vol 13. Wiley, New York, p 961
24. Henkel H, Weber W (1914) Manufacture of hydrogen peroxide. US Patent 1108752
25. Van Weynbergh J, Schoebrechts JP, Colery JC (1995) Direct synthesis of hydrogen peroxide by heterogeneous catalysis, catalyst for the said synthesis and method of preparation for the said catalyst. US Patent 3265447706
26. Paparatto G, D'Aloisio R, De Alberti G, Furlan P, Arca V, Buzzoni R, Meda L (1999) New catalyst, process for the production of hydrogen peroxide and its use in oxidation processes. EP Patent 0978316A1
27. Zhou B, Lee L-K (2001) Catalyst and process for direct catalytic production of hydrogen peroxide. US Patent 6168775
28. Nystrom M, Wangard J, Herrmann W (2001) Process for producing hydrogen peroxide. US Patent 6210651
29. Lunsford JH (2003) The direct formation of H_2O_2 from H_2 and O_2 over palladium catalysts. J Catal 216:455–460
30. Dissanayake DP, Lunsford JH (2002) Evidence for the role of colloidal palladium in the catalytic formation of H_2O_2 from H_2 and O_2. J Catal 206:173–176
31. Dissanayake DP, Lunsford JH (2003) The direct formation of H_2O_2 from H_2 and O_2 over colloidal palladium. J Catal 214:113–120
32. Choudhary VR, Samanta C, Gaikwad AG (2004) Drastic increase of selectivity for H_2O_2 formation in direct oxidation of H_2 to H_2O_2 over supported Pd catalysts due to their bromination. Chem Commun 2054–2055
33. Pritchard J, Kesavan L, Piccinini M, He Q, Tiruvalam R, Dimitratos N, Lopez-Sanchez JA, Carley AF, Edwards JK, Kiely CJ, Hutchings G (2010) Direct synthesis of hydrogen peroxide and benzyl alcohol oxidation using Au-Pd catalysts prepared by sol immobilization. J Langmuir 26:16568–16577
34. Herzing AA, Watanabe M, Kiely CJ, Edwards J, Conte M, Tang ZR, Hutchings GJ (2008) Aberration corrected analytical electron microscopy studies of bimetallic nanoparticles. Faraday Discuss 138:337–351

Distinct new catalysts for utilization of ethylene gas to polymer and oligomer

Abdulhamid A. Alsaygh

Abstract Bi-nickel-center catalysts were prepared by Schiff-base condensation of 2,3-butanedione with 2,6-diisopropylaniline; 2-5-butylaniline, 2-isopropylaniline and 1,4-phenylene diamine(I). More Schiff-base ligands were prepared by condensation of 2,3-butandione with 2,6-diisopropylaniline; 2,4-6-trimethyl aniline and 4,4-diaminophenyl (benzidine)(II), and subsequent metathesis reaction with (DME)NiBr2. Bis(alpha diimine)nickel(II)dibromide complexes are suitable catalysts and are precursors for the polymerization of ethylene after activation with methylaluminoxane (MAO). In comparison with these complex analogs, mono-nickel-center catalysts, the new catalysts have much bigger molecules, illustrated by the distance between every two active centers (II and III). For bi-nickel-center catalysts, the catalyst's structural credibility had profound influence on the catalytic activity. When the substituent was diisopropyl; *t*-butyl or methyl, the catalysts demonstrated much higher catalytic activity than the corresponding mono-nickel-center catalysts. Catalyst (II) with phenyl bridge showed high activity compared with catalyst (III) with diphenyl bridge. The catalytic properties of these complexes and the character of the obtained polymers depend on the ligand structure of the used catalysts. Substituents on the arene moiety and/or the backbone of the ligand influence the polymerization reaction. Small aryl substituents result in the formation of low molecular weight oligomers, whereas bulky aryl substituents gave high molecular weight polyethylene. Catalysts are cheap and can be prepared easily with available starting material and stable in air. The effect of Al/Ni ratio, of reaction time, variation of ethylene pressure, and the effect of temperature on catalyst performance will be discussed.

A. A. Alsaygh (✉)
Petrochemical Institute at King Abdulaziz City for and Technology, PO Box 6086, Riyadh 11442, Saudi Arabia
e-mail: aalsaygh@kacst.edu.sa

Keywords Organometallic · Polymerizations · Late transition metals · Ethylene

Introduction

The area of olefin polymerization catalysis has endured a remarkable increase in research outcome over the past 20 years, with many academic and industrial research centers involving in the design of a new organometallic precatalysts for the controlled synthesis of polyolefin products. In the early 1970s researches on the Group 4 metallocenes gave more detailed understanding of the chemistry around of the activated sites [6]. In the early 1990s new classes of "post-metallocene" catalysts were developed which overcame the inherent problems in Group 4 cyclopentadienyl systems, by using late transition metal systems enabling the incorporation of polar comonomers into polyolefin materials, and gave a new morphology surface at low levels of comonomer incorporation and changed the bulk properties of the polyolefin at higher levels of incorporation [2, 5]. In the late 1990s, a new type of catalysts was developed, based on late transition complexes such as nickels' which are very active towards ethylene polymerizations coupled with, the last decade, breakthrough by utilizing such a system to do what is so called living polymerization of olefinic monomers. In this process, the absence of chain-transfer or chain-termination gives rise to a polyolefinic material with very narrow molecular weight distributions (typically <1.1), block copolymers, and polymers with novel topologies [3, 4, 7].

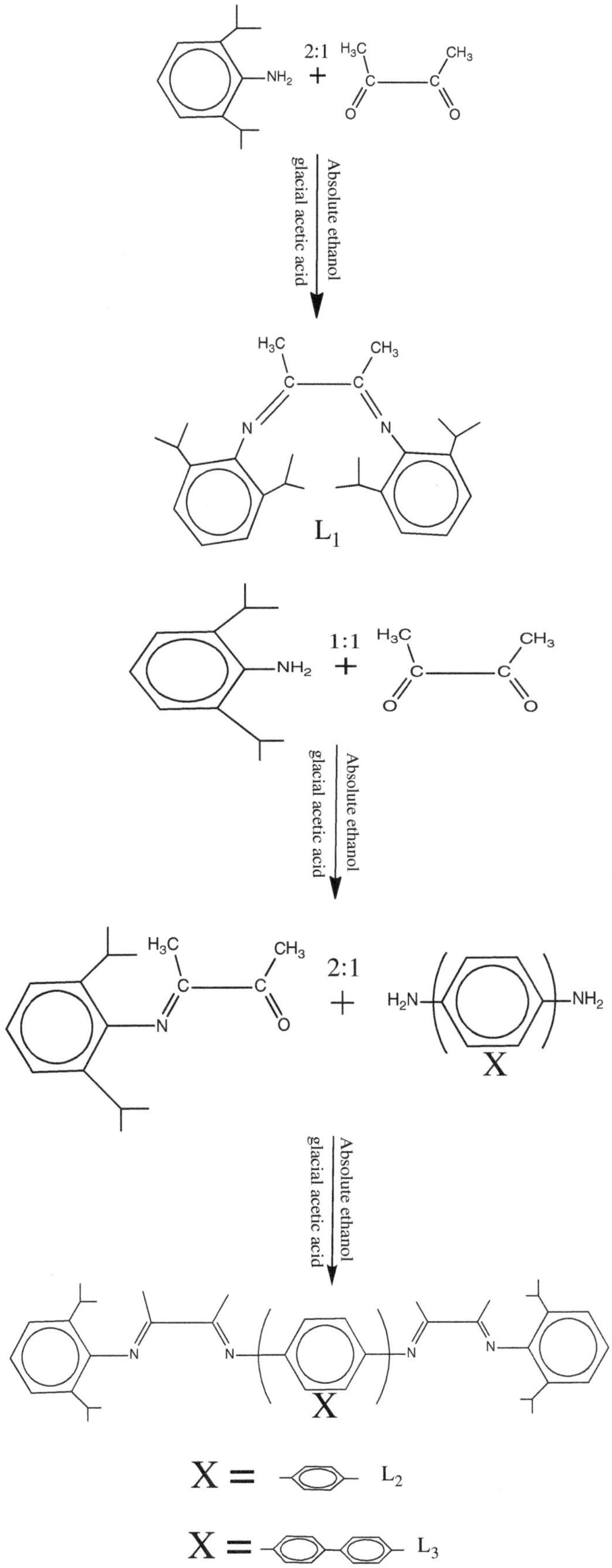

Scheme 1 Synthesis of ligands $\mathbf{L_1}$, $\mathbf{L_2}$ and $\mathbf{L_3}$

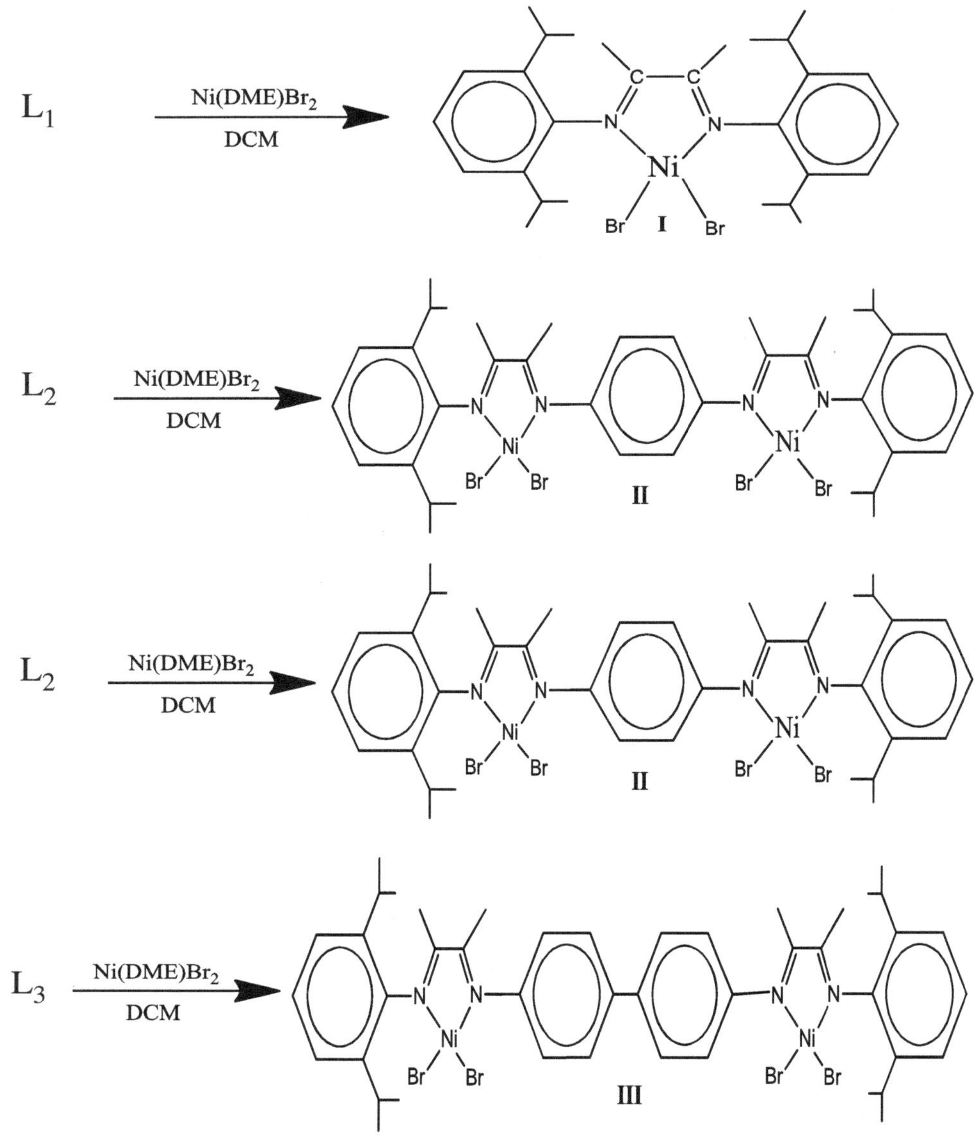

Scheme 2 Synthesis nickel complexes **1**, **2**, and **3**

Experimental

All manipulations of air and/or water-sensitive compounds were carried out under dry nitrogen atmosphere using a dry box or Schlenk line techniques. The molecular weight (MW) and polydispersity index (PDI) of PE were determined by GPC (PL-GPC220/FTIR) in 1,2,4-trichlorobenzene/THF using polystyrene columns as standard.

Synthesis of new ligands

The ligand $\mathbf{L_1}$ was synthesized by condensation of 2,3-butanedione with 2,6-di-isopropylaniline; 2-5-*t*-butylaniline, 2-isopropylaniline and 1,4-phenylene diamine. More Schiff-base ligands, $\mathbf{L_2}$ and $\mathbf{L_3}$, were prepared by condensation of 2,3-butandione with 2,6-diisopropylaniline; 2,4-6-trimethyl aniline and 4,4-diaminophenyl (benzidine)(II), (Scheme 1). For example, a 100-ml Schlenk flask was charged with 3.29 g (18.59 mmol) (2,6-di-isopropylaniline; 2-5-*t*-butylaniline, 2-isopropylanilin, or 2,4-6-trimethyl aniline), 1.60 g (18.59 mmol) 2,3-butanedione, 30 ml ethanol and 0.4 ml glacial acetic acid, the pale yellow solution was stirred for one day at room temperature (the intermediate products were not isolated). Then, 9.30 mmol of 4,4-diaminophenyl or benzidine, and more 30 ml of ethanol was added. The final solution was stirred for 1 week at room temperature which yielded an yellow precipitate. The yellow precipitate was isolated by filtration, washed with 2 × 10 ml ethanol, and dried in vacuum; 4.8 g yellow powder was obtained; yield, 63 %.

Synthesis of complexes **1**, **2** and **3**

The nickel complexes (Scheme 2) were prepared in good yield as shown below in Scheme 2.

General procedure for ethylene polymerization

A 1-l stainless steel reactor was backed out under N_2 flow for at least 1 h at 150 °C and subsequently cooled to room temperature for polymerization. After the reactor was

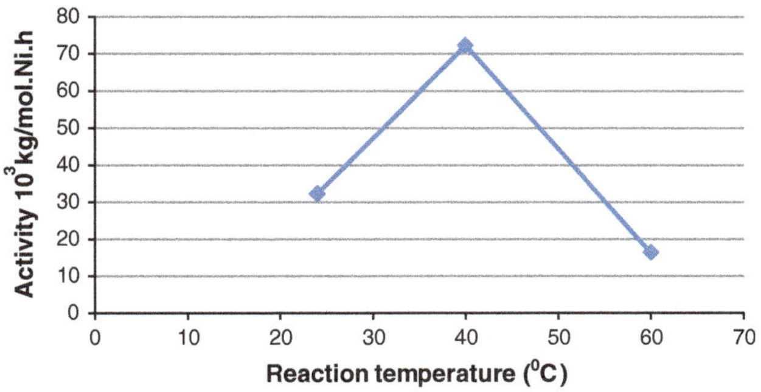

Fig. 1 A plot of reaction temperature versus the catalytic activity of catalyst **2** at ethylene pressure of 10 bars and polymerization time 1 h

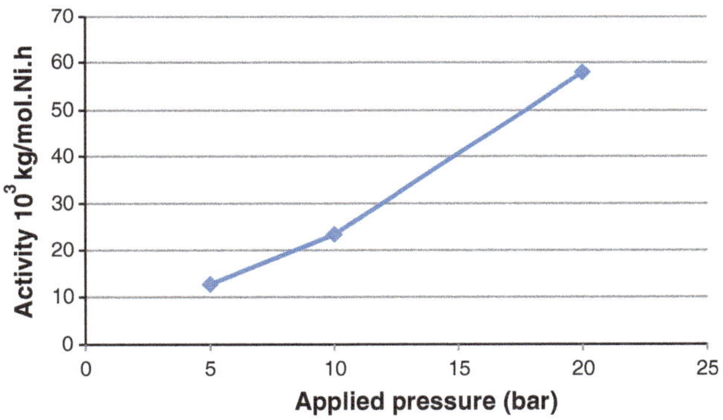

Fig. 2 A plot of applied pressure versus the catalytic activity of catalyst **2**. Polymerization time 1 h and temperature 24 °C

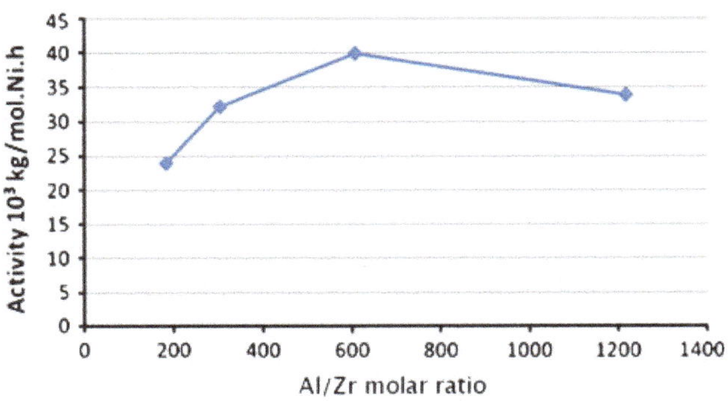

Fig. 3 A plot of Al/Ni ratio versus the catalytic activity of catalyst **2** at ethylene pressure of 10 bars, polymerization time 1 h and temperature 24 °C

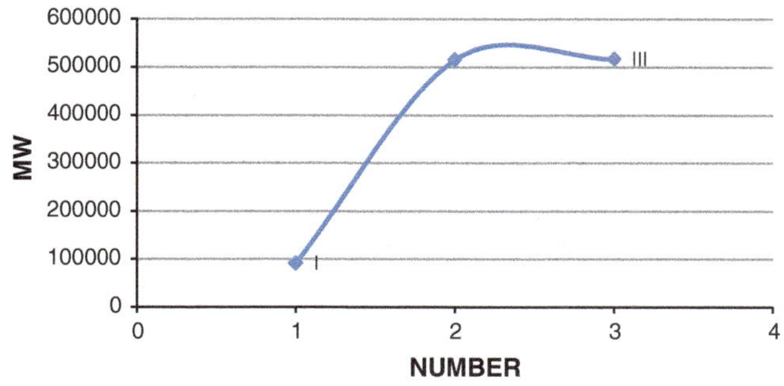

Fig. 4 A plot of molecular weight of polyethylene versus the substituent type

cooled, MAO dissolved in toluene (50 ml) was introduced into the reactor and stirred at reaction temperature. The ethylene was introduced into the reactor by back pressure of nitrogen for at least 1 h. The catalyst solution in toluene (or hexane) was then injected under N_2. The reactor pressure was maintained constant throughout the polymerization run by computer-controlled addition of ethylene. The polymerization time was between 20 and 60 min. Runs were terminated by venting off volatiles and the reactor contents were isolated, washed with aqueous HCl and methanol and the polymer produced was dried in a vacuum oven at 50 °C.

Results and discussion

Synthesis of new ligands and complexes

The goal of this collaborative research work is to develop and envisage a convenient method to prepare highly active new bi-nuclear alpha-diimine-nickel catalysts toward the catalytic performances for ethylene polymerization. The basic idea is to utilize substituents on the arene moiety and/or the backbone of the ligand to prepare new ligands and corresponding complexes. The structure around the metal may strongly influence the catalytic performances, such as catalytic activity. Under the guidance of this basic idea, three kinds of new catalysts including bi-nickel-center catalysts, were prepared and studied for catalytic ethylene polymerization in toluene using MAO as co-catalysts.

On treatment with MAO, all the three nickel complexes are active towards ethylene polymerization. They all showed strong temperature dependence of the catalytic activities. Furthermore, they later decrease monotonously with the increase of polymerization temperature (as shown in Figs. 1, 2). In addition, the catalyst's activity increases with the increase of the molar ratio, Al/Ni, as shown in Fig. 3 where the catalyst activity significantly increased; whereas, when the ratio value increased from 200 to 600 it has little or no effect upon any further increase. The most active was the new catalyst with diisopropyl substitute, the catalytic activity was high at 25 °C with Al(MAO)Ni ratio at 500. In general, the ratio of Al/Ni was 500:1 for the most of polymerization reported in this work.

Conclusion

The result of ethylene polymerization catalyzed by **1**, **2** and **3** complexes using MAO as a cocatalyst showed that the activity of complexes **2** and **3** is higher than that for **1** which could be explained by the fact that the ligand environment around the metal center has a profound effect on catalysts' productivity and product molecular weight. Changing the ligand environment results, generally, in a slight drop in productivity and in the molecular weight, as shown by Fig. 4. Confirming the conjugation and steric protection of the metal center are crucial for high activity [1].

Acknowledgments A.A. Alsaygh would like to give thanks and appreciations to King Abdulaziz City for Science and Technology for funding; to his counterparts at Oxford University represented by Peter Edwards and Vladimir Kuznetsov. Moreover, he would like to thank Prof Ibrahim M. Al-Najjar, and Mohammed H. Al-Musawi for their great contributions to this work.

References

1. Al-Saygh AA, Al-Najjar IM, Al-Musawi MH (2009) 2do SiLQ-COM, Maracaibo, Venezuela, from October 25 to 29
2. Britovsek GJP, Bruce M, Gibson VC, Kimberley BS, Maddox PJ, Mastroianni S, McTavish SJ, Redshaw C, Solan GA, Strömberg S, White AJP, Williams DJ (1999) Iron and cobalt ethylene polymerization catalysts bearing 2,6-Bis(imino)pyridyl ligands: synthesis, structures, and polymerization studies. J Am Chem Soc 121:(38)8728–8740
3. Helldorfer M, Backhaus J, Alt HG (2003) New bi-nuclear and multi-nuclear α-diimine/nickel catalysts for ethylene polymerization. Inorg Chim Acta 351(2003):34–42
4. Liu J-Y, Zheng Yi, Li Y-G, Pan Li, Li Y-S, Hu N-H (2005) Fe(II) and Co(II) pyridinebisimine complexes bearing different substituents on ortho- and para-position of imines: synthesis, characterization and behavior of ethylene polymerization. J Organomet Chem 690:1233–1239
5. Luo H-K, Schumann H (2005) Group 4 ansa-cyclopentadienyl-amido catalysts for olefin polymerization. J Mol Catal A Chem 227:153–161
6. McKnight AL, Waymouth RM (1998) The influence of the ligand structure on the properties of (α-diimine)nickel catalysts in the polymerization and oligomerization of ethylene. Chem Rev 98:2587–2598
7. Gibson VC, Spitzmesser SK (2003) Advances in non-metallocene olefin polymerization catalysis. Chem Rev 103:283–315

6

High alcohol synthesis (HAS) from syngas over supported molybdenum carbide catalysts

Hamid Al-Megren · Tiancun Xiao · Mohammad AlKinany · Saud Aldree · Yu Huang · Haoyi Chen · Peter P. Edwards · Vladimir Kuznetsov

Abstract In this work, different kinds of supported molybdenum carbide catalysts have been prepared using temperature programmed carburization method with 20 vol% CH_4/H_2. Cobalt and potassium were later loaded over the resultant carbides. The catalysts were evaluated on the new alcohols test rig under 70 bar pressure with syngas feed gas with a H_2/CO ratio of 2.0 at various temperatures. The carbide catalysts supported over boron-modified Al_2O_3 produced higher alcohols in significant yields. The best catalyst achieved 46.7 % selectivity to alcohol at 31.8 % conversion of CO. This corresponds to 199.7 g/alcohol/liter of catalyst/hour. In addition, 72.5 % selectivity to alcohol (93.1 % on CO_2 free basis) was achieved at lower temperature with a conversion of 14.1 %. Characterization results showed that the boron-modified supports result in smaller molybdenum carbide crystallite size which might be responsible for the high activity. Higher potassium content benefits the high alcohol selectivity, but affects the catalyst activity to some extent.

Keywords Higher alcohol synthesis · Supported molybdenum carbide · Syngas

H. Al-Megren · M. AlKinany · S. Aldree
Petrochemical Research Institute, King Abdulaziz City for Science and Technology, P. O. Box 6086, Riyadh 11442, Saudi Arabia

T. Xiao · Y. Huang · H. Chen
Guangzhou Boxenergytech Ltd, Guangzhou Boneng Nnengyuan Keji Youxian Gongsi, Guangzhou City, People's Republic of China

T. Xiao (✉) · P. P. Edwards · V. Kuznetsov
KOPRC, Inorganic Chemistry Laboratory, Oxford University, South Parks Road, Oxford, OX1 3QR, UK
e-mail: xiao.tiancun@chem.ox.ac.uk

Introduction

In the last 20 years, much attention has been paid to the synthesis of higher alcohols (HAS) from syngas because alcohols are being considered as a potential alternative synthetic fuels for automotive applications and as feedstock for the synthesis of a variety of industrial chemicals and polymers [1–4]. Syngas can be produced from all sorts of carbon containing feedstock such as biomass, natural gas, biogas and coal, which can be converted to alcohols by catalytic conversion. However, the catalytic conversion of syngas to alcohols remains challenging, and no commercial process exists as of today, although the research on this topic has been pursued for the past 90 years.

Transition metal carbides have been found to have the similar property to noble metal in many aspects, which have been extensively studied for high alcohol synthesis process. It has been shown that Mo–CO adsorption is weakened by C, leading to CO dissociation over Mo_2C. When promoted with alkali metals and/or Group VIII metals, the molybdenum carbide catalyst showed higher selectivity to alcohol under the typical operation conditions of 5–8 MPa, $H_2/CO = 2–1$, and 537–603 K.

So far, most studies on carbide catalysts systems are focused on bulk catalysts. Christensen reported "catalytic conversion of syngas into Higher Alcohols over WC and Mo_2C Carbide Catalysts", and found that K_2CO_3 addition increases the alcohol selectivity [5]. Wang Ning et al. [6] studied that Fe and K promoted Mo_2C catalyst for higher alcohol synthesis, and proposed that the Fe_3Mo_3Cx might be the active phase for the higher alcohol formation over the catalyst system. The effect of Fe/Mo ratio and the K_2CO_3 addition exerts a significant effect on the catalyst activity and selectivity [6, 7]. In contrast, Xiang's studies suggested that the modification of K/Mo_2C with Co and Ni gives higher selectivity to high alcohol selectivity with fair CO conversion [8–10].

An overview of the published papers on the molybdenum carbide catalysts for high alcohol synthesis showed that most research focuses on the bulk carbide catalyst, which showed higher selectivity to alcohol and relatively high activity, but from the practical view, the strong metallic property of the bulk molybdenum carbide does not have high strength and produce few pores in the catalyst particles when extruded into pellets for fixed bed applications. For industrial applications, it is desirable to have supported catalysts. However, so far, there have been very few studies published on the supported carbide catalyst for alcohol synthesis [11–14], and they almost totally focused on active carbon as the supports, which are also extensively studied on the molybdenum sulfide systems for alcohol synthesis [12, 15–18].

The alumina-supported molybdenum carbide-based catalysts are relatively new for alcohol synthesis; also, the supported catalysts are expected to have high strength and contain less metal. Various modifications of the catalyst are possible to improve the catalytic performance. Therefore, in this work, we have prepared series of alumina (varying the support property)-supported molybdenum carbide catalyst with different promoters. Their performances have been tested under various conditions and have been correlated with the preparation factors.

Experimental

Catalyst preparation

In this work, the HAS catalysts were prepared using CH_4/H_2 carburization method. Lots of studies have been reported on Mo-based carbide catalysts preparation by the carburization of bulk $CoMoC_x$ or NiMoCx [8–10], but few of them were on the supported Mo-based carbide catalysts. Here we prepare the supported carbides first and then modify with other additives and promoters. The preparation details are as follows.

The molybdenum precursor was prepared from ammonium molybdate, to which deionized water (DI) and hydrogen peroxide (H_2O_2) were added to increase the molybdate solubility. The molybdate solution was impregnated with supports (γ-Al_2O_3 or γ-Al_2O_3- B_2O_3, B_2O_3 content, 2.5 wt%) by pore volume impregnation method, e.g., incipient wetness at room temperature. The supported MoO_3/Al_2O_3 precursor was obtained after calcination at 500 °C, which is then converted into supported Mo_2C after the supported oxide was carburized using CH_4/H_2 to 700 °C with 20 vol% CH_4/H_2 stream, then cooled to room temperature naturally in H_2 stream, and passivated with static air for 24 h before it was unloaded from the reactor. The carburization process is monitored using TPR-TCD to study the effect of support on the supported molybdenum oxide reduction.

Finally, the desired modifier components or promoters' metal was added to the obtained Mo-based carbide support catalyst. It is noted that the cobalt salts were loaded onto the supported carbides after carburization to avoid sintering during the carburization process. The Mo carbide was formed due to the reaction between MoO_3 and CH_4. $Co(Ac)_2$ and K_2CO_3 were used as the promoters and modifiers.

The detailed preparation parameters of each catalyst are listed in Table 1.

Catalyst test and data analysis

The catalysts have been tested in a micro-reactor system. Each time, 2.73 ml HAS catalyst with 32.5 ml SiC was loaded in HAS reactor, plugged with quartz wool at both ends, and reduced with H_2 at 4.0 MPa at 404 ml/min at 350 °C (ramp 1 °C/min) for over 16 h.

The test conditions of the carbide catalysts are set as follows:

- Feed gas characteristic—syngas with H_2:CO ratio of 2 optionally with 5–10 % CO_2 with 5 vol% of N_2 as for internal calibration
- GHSV 2,000–4,000 H^{-1}
- Pressure 5.0–10.0 MPa
- Temperature 220–300 °C.

After the HAS catalyst was reduced with H_2 at 4.0 MPa and temperature cooled down to 200–220 °C from 350 °C,

Table 1 The physical properties and crystallite of the carbide catalysts

Sample no.	Composition	Surface area (m^2/g)	Mo_2C crystallite size (nm)
	1.6 % K 3.5 % Co 40 % Mo_2C	32.6	28.6
090615	1.6 % K 3.5 % Co 40 % Mo 2.5 % C/Al_2O_3	96.5	18.9
090714	1.6 % K 3.5 % Co 40 % Mo 2.5 % C/Al_2O_3–B_2O_3	78.8	14.2
090817	3.3 % K 3.5 % Co 40 % Mo/Al_2O_3	83.2	16.2
090824	3.3 % K 3.5 % Co 40 % Mo/Al_2O_3–B_2O_3	70.3	12.4
091019	4.6 % K 3.5 % Co 40 % Mo/Al_2O_3	59.8	15.4
090907	4.6 % K 3.5 % Co 40 % Mo/Al_2O_3–B_2O_3	62.5	9.3

syngas was introduced into reactor at 136 ml/min. It spent about 25–30 h after the reaction pressure was increased to 7.0 Mpa from 4.0 MPa, all H_2 was purged with syngas and reaction temperature reached 245 °C.

From the GC data of gas products, the compositions of H_2, CO, CO_2, CH_4, C_2–C_5 hydrocarbons can be analyzed, and the CO conversion, all gas products' selectivity and the selectivity to liquid are calculated. From another GC analysis of liquid products, we can obtain the compositions of alcohols, heavy hydrocarbons and H_2O. Then we get all products' selectivity and productivity.

Catalyst characterization

The catalyst crystalline structure was analyzed using powder XRD with a Philips X' PeRT Pro Alpha 1 diffractometer with Cu Kα radiation ($\lambda = 1.5406$ Å) operated at a tube current of 40 mA and a voltage of 45 kV. Data were collected over 2θ values from 10° to 80°, at a scan speed of 1° min^{-1}. The average crystal size was calculated using Scherer equation based on the strongest XRD peak of Mo_2C (101).

The Scherrer equation is expressed as follows:

$$\tau = \frac{K\lambda}{\beta \cos\theta}$$

where τ is the mean size of the ordered (crystalline) domains, which may be smaller or equal to the grain size; K is a dimensionless shape factor, which is about 0.9 in this work; λ is the 1.5406 Å; β is the line broadening at half the maximum intensive peak, e.g., plane (101), which has less overlap with the background diffraction peaks; and θ is the Bragg angle.

Results and discussion

Effect of support on the carburization of the molybdenum oxide

The carburization process of the bulk and supported molybdenum oxide was monitored using a TPR system and the changes of the products were detected using TCD. The results are shown in Fig. 1. It is seen that the carburization process starts at 200 °C, and most carburization process occurs at about 400 °C; this process contains two major peaks: the first at about 400 °C, during which reduction of MoO_3 by H_2 or CH_4 occurs, while the second peak at 650 °C, where complete carburization of MoO_2 or MoCx to Mo_2C occurs.

In case for the Al_2O_3-supported MoO_3 carburization, the first reduction of MoO_3 occurs at 500 °C, which showed that the supported MoO_3 is more difficult to reduce even when there are 40 wt% MoO_3 loaded over the support. The difficult reduction of the supported MoO_3 may result from the strong interaction between the MoO_3 and Al_2O_3. The second carburization peak appeared at 900 °C and the span range between the two peaks is much broader, suggesting that there may be different kinds of MoO_3 over the supports, the one close to the support surface would have a stronger interaction with support and more difficult to be carburized, leading to the high temperature carburization.

The crystalline structures of the prepared carbide catalysts were characterized using XRD and the results are shown in Fig. 2. It is noted that the aluminum plate was used as the sample holder, so the peaks at 48°, 44° and 64° are due to the diffraction of metallic Al. The diffraction peaks showed that in the bulk carbide, the main phase is Mo_2C, with hcp structure [6, 14, 19]. The peak at 40.5° is assigned to the plane [101] of Mo_2C, and the diffraction peak at 35.1° is ascribed to the diffraction of [002] of Mo_2C. Over the Al_2O_3-supported Mo_2C catalyst, the main phase is still Mo_2C, but the intensity of the diffraction is much weaker, suggesting a smaller crystallite size of Mo_2C over the alumina support. Over the B_2O_3-modified Al_2O_3 support, the resultant Mo_2C [101] diffraction peak is even broader, whereas the other peaks corresponding to [002] and [110] planes are weaker, suggesting that the modification of Al_2O_3 with B_2O_3 has some effect on the molybdenum carbide phase and dimensions. The difference of the supported carbide may be due to the pore volume difference. After B_2O_3 was added to the Al_2O_3 support, the pore volume changed from 0.753 to 0.65 cm^3/g, which can

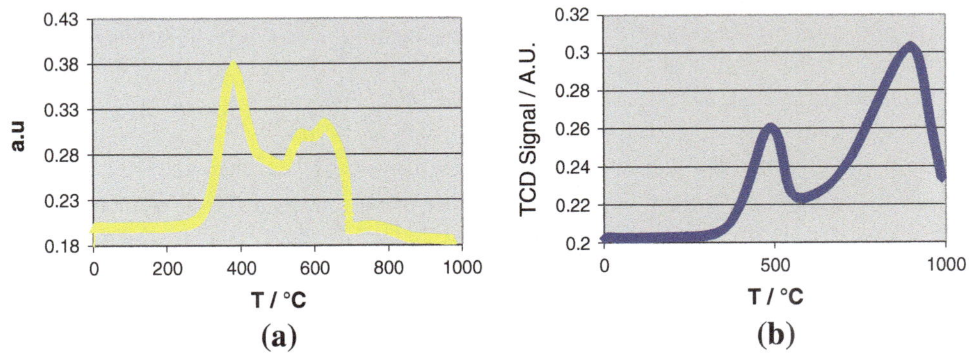

Fig. 1 The temperature-programmed carburization profiles of the **a** bulk MoO_3 and **b** 40 % MoO_3/Al_2O_3

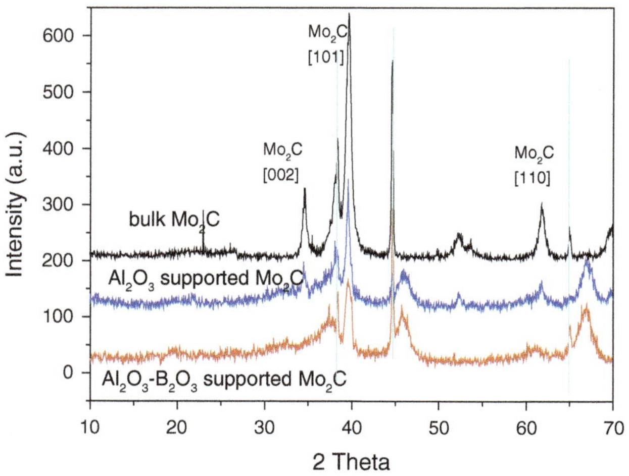

Fig. 2 XRD patterns of the bulk and different supported carbide catalysts. The measurement was carried out over an Al plate, which has strong diffraction peaks at 2θ of 38°, 44° and 65°

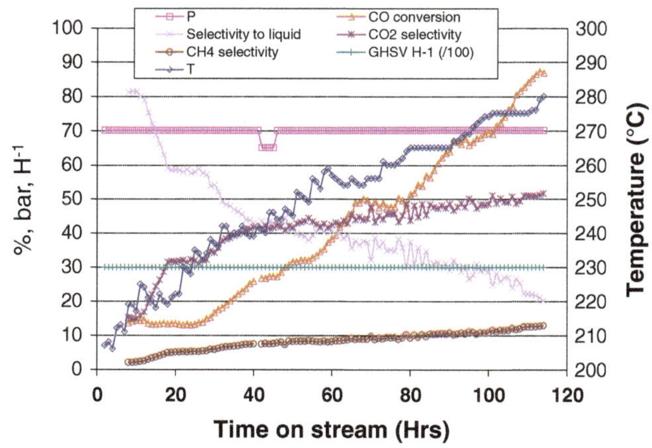

Fig. 3 The catalyst performance of 1.6 % K 3.5 % Co 40 % Mo_2C (bulk catalyst, powder) catalyst for syngas to alcohol. The catalyst test conditions are: 70 bar, H_2/CO ratio = 2, GHSV = 3,000 h^{-1}, internal calibration 5 vol% N_2 in syngas stream

restrict the cluster size of MoO_3, the precursor of the carbide.

The catalyst composition, its physical properties and crystallite size determined using XRD are shown in Table 1. The surface area of bulk Mo_2C catalyst is 32.6 m^2/g, which is due to it powder phase. The supported Mo_2C carbide catalysts modified with 1.6 and 3.5 wt% Co has a surface area of 78.8 m^2/g; the catalyst particle size was noted to be about 300 μm, due to the breakage of the extrudated Al_2O_3–B_2O_3 support. The pure Al_2O_3-supported Mo_2C modified with 1.6 wt% K and 3.5 % Co has slightly higher surface area, about 83.2 m^2/g. The crystallite size of Mo_2C is about 18.6 nm, while in the supported carbide sample, the crystallite size decreased to <20 nm. The Al_2O_3–B_2O_3-supported carbide has smaller Mo_2C crystallite, but overall the Mo_2C crystallite size decreases with the content of K modification, which is in agreement with the literature results [1, 5].

Catalyst test results

The bulk and supported molybdenum carbide catalysts have been tested for higher alcohol synthesis from syngas at various temperatures with constant pressure and space velocity. The bulk-modified molybdenum carbide catalysts performance for the HAS is shown in Fig. 3. At 200 °C, when the reaction started, CO conversion was above 10 %, methane selectivity is less that 4 %, but CO_2 selectivity is more than 14 %, while the selectivity to liquid is about 80 % at the initial reaction stage. The CO conversion was almost unchanged during the first 30-h reaction, while the temperature was raised to 230 °C. However, the liquid selectivity drops with the time on stream even the CO conversion does not increase. When the reactor temperature was increased to 240 °C, CO conversion increased to 25 %, CH_4 selectivity increased to 7.5 vol%, and CO_2 increased quickly to 40 vol%. In contrast, the liquid selectivity dropped to 40 wt%. At temperatures above 240 °C, the selectivity to liquid products dropped slowly, but selectivity to both CO_2 and CH_4, the undesirable products, increased gradually. When the reaction temperature is raised above 260 °C, CO conversion increased more rapidly, while the liquid selectivity decreased. In contrast, both CO_2 and CH_4 selectivities rose. When the CO conversion reached 75 % at 275 °C, liquid selectivity dropped below 30 %, but CH_4 selectivity rose above 10 % and CO_2 reached 50 %. However, the initial selectivity to liquid is higher than the bulk Mo_2C carbide system.

Over 1.6 % K 3.5 % Co 40 % Mo_2C/Al_2O_3 catalysts, the catalyst activity was lower than the bulk Mo_2C catalyst system; at 210 °C, the CO conversion was <5 %, but methane selectivity was more than 6 %, but almost no CO_2 detected in the products, and the selectivity to the liquid was more than 90 % at the initial reaction, but dropped to below 80 % when CO selectivity suddenly rose. Holding the reaction temperature at 220 °C for 10 h, the CO_2 selectivity increased to 20 %, but CH_4 also rose to 8 %. At 230 °C, the CO conversion was only 4.2 %, CO_2 selectivity was almost unchanged at 20 %, but liquid selectivity dropped to about 50 %. To reach a steady state, the reactor temperature was held at 245 °C for 50 h; during this period, CO conversion reached 10 %, which is obviously lower than the bulk catalyst, liquid selectivity and CO_2 selectivity all are about 33 %, while methane selectivity is about 9.2 %. Increasing the temperature to 255 °C increased CO conversion to 12 %, but the selectivity to liquid further dropped to below 30 % and CO_2 selectivity increased to 35 %. This suggests that the liquid selectivity drop may be due to disproportionation of CO into CO_2 and the increased yield of CH_4. The catalyst activity is much lower than the bulk catalyst, which may be due to the less active site over the catalyst.

When using Al_2O_3–B_2O_3 as the support, the resultant carbide catalyst performance for HAS from syngas is shown in Fig. 5. Interestingly, the catalyst showed relatively high activity for CO conversion at 220 °C, CO conversion goes up to 20 % for a while at 220 °C, with CH_4 selectivity less than 4 % and almost no CO_2 produced. The selectivity to liquid is more than 90 %, but dropped with the time on stream at 220 °C. This suggests that the catalyst active site or surface property might change quickly in the initial reaction stage, which does not lead to reach the steady state, and CO conversion dropped from nearly 20 to 6 % at 220 °C without changing the other conditions, and the methane selectivity rises gradually. When increasing the reaction temperature, the CO conversion increases slowly, but the CH_4 selectivity and CO_2 selectivity increase when the liquid selectivity decreases rapidly. At 245 °C, CO conversion increased to 10 %, but CO_2 selectivity increased to 22 %, and methane selectivity changed to 7 %; however, the liquid selectivity decreased to 62 %. When raising the reactor temperature to 255 °C, the liquid selectivity dropped to 53 %, CO_2 selectivity raised to 30 %, with CH_4 selectivity increase of 8.2 %. This suggests that increase of the temperature leads to more CO converting to CO_2, not favoring the liquid formation. This can be understood from the thermodynamic view. As the liquid alcohol synthesis is a strong exothermic reaction, while the conversion of CO to CO_2 is favored by high temperature. Increasing the reaction temperature to 265 °C leads to CO conversion to 20 %, but CO_2 selectivity to 38 %, and the liquid selectivity dropped to 42 %. The CH_4 selectivity slightly increases.

Comparison of Figs 3, 4 and 5 shows that the bulk molybdenum carbide catalyst has the highest activity for CO conversion, but its selectivity to liquid dropped rapidly with increasing temperature. The Al_2O_3-supported molybdenum carbide has the lowest activity, but the selectivity to liquid is slightly better than the bulk carbide. The modification of Al_2O_3 with B_2O_3 significantly improves the catalyst performance for liquid production from syngas. The catalyst activity is improved, while the selectivity to liquid is the best, and decrease slower than the Al_2O_3-supported carbide and bulk carbide catalysts. The comparison of the three catalysts' performance is shown in Table 2.

The liquid products from the HAS process were collected from the liquid trap regularly and analyzed using GC–MS. The results are shown in Table 2 after water is excluded. It is shown that the bulk molybdenum carbide catalyst contain 0.5 % liquid hydrocarbons, with methanol selectivity in the organic phase of about 64.1 wt%. There

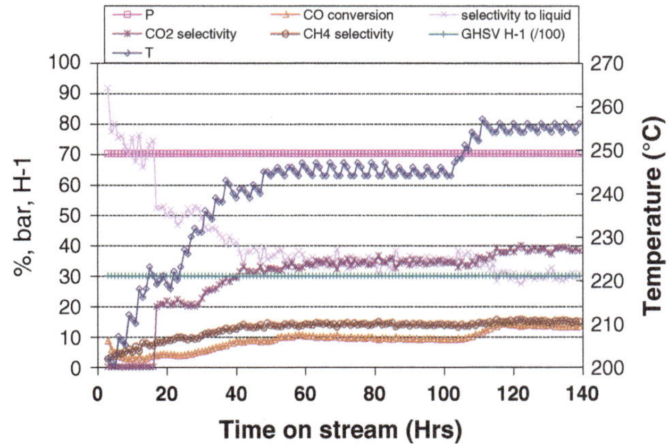

Fig. 4 The catalyst performance of 1.6 % K 3.5 % Co 40 % Mo_2C/Al_2O_3 catalyst for syngas to alcohol. The catalyst test conditions are: 70 bar, H_2/CO ratio = 2, GHSV = 3,000 h^{-1}

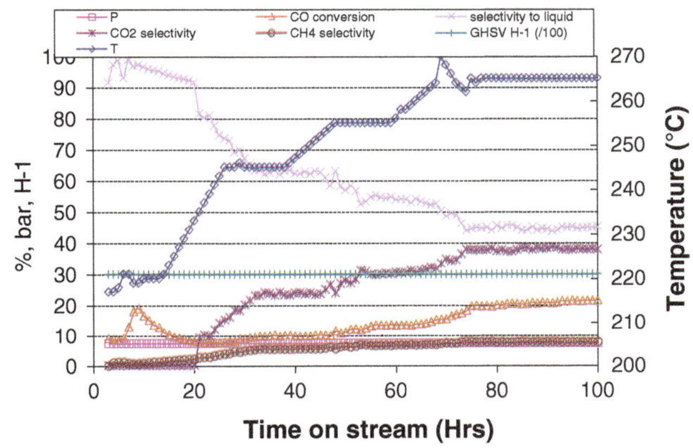

Fig. 5 The catalyst performance of 1.6 % K 3.5 % Co 40 % Mo_2C/Al_2O_3–B_2O_3 catalyst for syngas to alcohol. The catalyst test conditions are: H_2/CO ratio = 2, GHSV = 3,000 h^{-1}

the ethanol concentration in the organic products is 25.8 wt%, and propanol is 7.7 wt%. Butanol content is about 1.5 wt%, and pantanol about 0.1 wt%. The $C_2^+OH/MeOH$ is 0.53 over the bulk carbide catalyst, suggesting that nearly half of alcohol products in the organic phase is methanol.

Over the Al_2O_3-supported molybdenum carbide catalysts, HCx content is 0.1 wt%, but the methanol content is only 51.9 wt% less than that over the bulk carbide catalyst, the ethanol content increases to 31.5 wt%, and the other higher carbon alcohols from C3 to C5 all increase. The $C_2^+OH/MeOH$ is 0.94. Over the Al_2O_3–B_2O_3-supported molybdenum carbide catalyst, the methanol content further decreases to 50.5 % and almost no liquid hydrocarbon products were detected; the ethanol concentration increases to 32.3 wt% and propanol content decreases to some extent. The $C_2^+OH/MeOH$ ratio is 0.99, suggesting that nearly half of the organic products are higher alcohol products.

Table 2 The analysis of the liquid product from the HAS reaction over different carbide catalysts

	Liquid organic products composition in the HAS							$C_2^+OH/MeOH$
Catalyst	Heavy HCx	MeOH	EtOH	PrOH	BuOH	PeOH	Other	
1.6 % K 3.5 %Co 40 % Mo_2C	0.5	64.1	25.8	7.7	1.5	0.1	0.3	0.53
1.6 % K3.5 % Co 40 % Mo_2C/Al_2O_3	0.1	51.9	31.5	9.5	5.4	2.5	0.1	0.94
1.6 % K 3.5 % Co 40 % $Mo_2C/Al_2O_3–B_2O_3$	0	50.2	32.3	9.0	5.6	2.7	0.2	0.99

Effect of K content on the catalyst performance

To further increase the catalyst performance, Al_2O_3 and $Al_2O_3–B_2O_3$-supported Co and Mo_2C catalysts with 3.3 and 4.6 wt% potassium have been prepared and the catalyst properties are shown in Table 1. The higher K content seems to give rise to smaller molybdenum crystallites with slightly lower surface areas. The catalysts have been tested at different reaction temperatures and 70 bars with GHSV of 3,000 h^{-1}, $H_2/CO = 2$. The results of the effects of K level on two different supported catalysts performance are given in Fig. 6, Al_2O_3 and $Al_2O_3–B_2O_3$ are shown in Table 1. The results are shown graphically in Fig. 6. It is interesting to observe that the catalyst activity in terms of CO conversion and alcohol STY is higher with the increase of the potassium content in the catalyst, as higher K content gives rise to smaller molybdenum particle size, thus leading to more active sites. The higher activity of $Al_2O_3–B_2O_3$-supported carbide catalysts can also be explained with this assumption. Overall, the yield of higher alcohol increases with the rise of the reaction temperature, although it is shown that the alcohol selectivity decreased, but compensated with the increase of CO conversion with the reaction temperature.

Figure 7 shows the changes in the higher alcohol selectivity with the CO conversion over the different supported molybdenum carbide catalysts with various potassium contents. Overall, the alcohol selectivity drops with the conversion, although the $Al_2O_3–B_2O_3$-supported 3.8 wt% Co 40 wt% Mo_2C catalysts have slower decrease in high alcohol selectivity with increasing CO conversion. The catalyst has similar trends, even though the $Al_2O_3–B_2O_3$-supported catalyst has higher alcohol selectivity. When increasing the K content to 4.6 wt%, the Al_2O_3-supported carbide catalyst selectivity to alcohol dropped faster than the 3.3 % K 3.5 % Co 40 % Mo_2C/Al_2O_3. In contrast, over the 4.6 wt% K, 3.5 % Co, 40 % $Mo_2C/Al_2O_3–B_2O_3$ catalyst, the alcohol selectivity has lesser CO conversion than the ones over 3.3 % K 3.5 % Co 40 % Mo_2C/Al_2O_3.

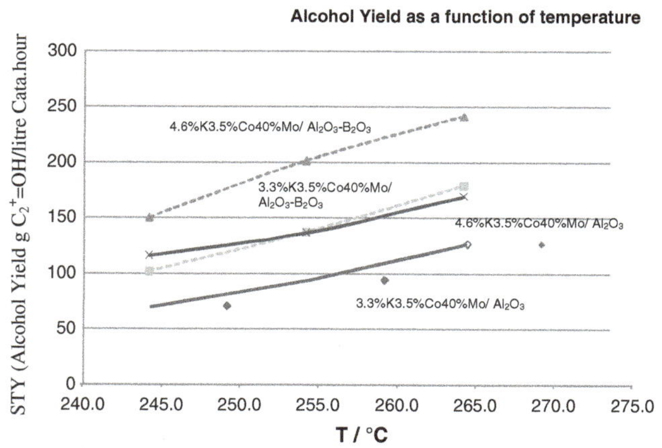

Fig. 6 The change in alcohol productivity with the temperature over higher K content of molybdenum carbide catalysts

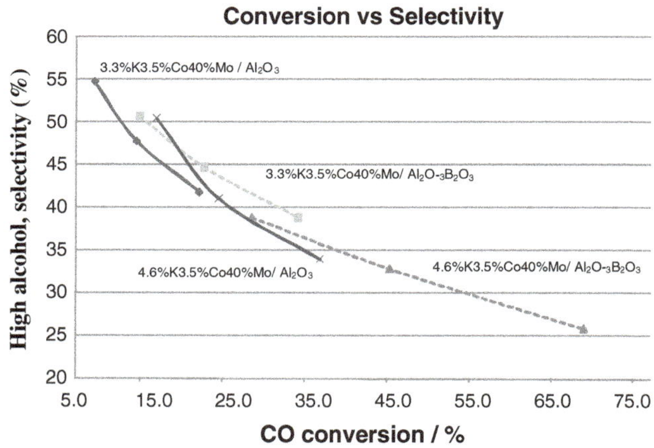

Fig. 7 The changes in the higher alcohol selectivity with the conversion over the different supported molybdenum carbides with different K concentrations

Catalyst stability and mass balances

To determine the long-term stability of the supported carbide, 3.3 % K 3.5 % Co 40 % Mo_2C/Al_2O_3 was tested continuously for 550 h (23 days). Initially, the catalyst was tested at three temperatures but the majority of the test continued at 245 °C (Fig. 8). The gap of the data between the time on stream of 180–220 h due to the holiday and the GC control was stuck; however, the reaction was continued.

It is clear that the stability of the catalyst is remarkably good. This is necessary for any commercial application. The material and mass balances were calculated for the 3.3 % K 3.5 % Co 40 % Mo_2C/Al_2O_3 catalyst at 245 °C. The results were 95.3 % for total mass balance, including 97.8 % hydrogen mass balance, 90.2 % carbon mass

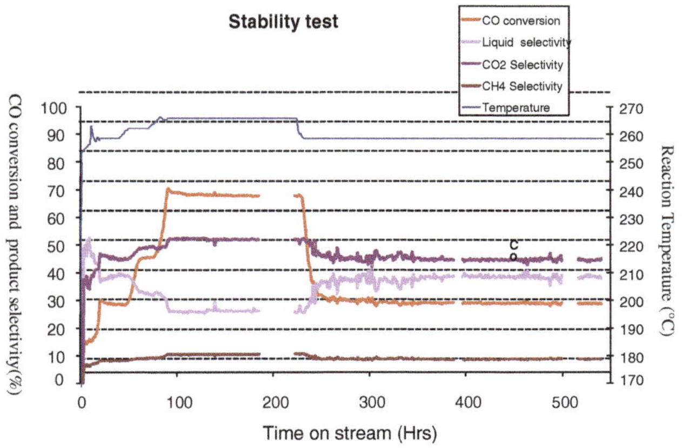

Fig. 8 The stability test of the 3.3 % K 3.5 % Co 40 % Mo 2.5 % C/Al_2O_3 catalysts

balance, 94.7 % oxygen mass balance and 95.1 % nitrogen mass balance. These numbers give confidence that the data produced are reliable. Further work is necessary to improve the C balance.

Conclusions

1. Different supported molybdenum carbide catalysts have been prepared using temperature-programmed carburization, and Co and K were added to the carbide catalyst using impregnation methods. The catalysts supported with Al_2O_3–B_2O_3 showed smaller Mo_2C crystallite size and lower surface area. However, the catalyst showed higher CO conversion and more selectivity to high alcohol synthesis from syngas. The bulk molybdenum carbide catalyst showed the higher activity, but also high selectivity to CO_2 and CH_4. Also, it is difficult to be shaped for fixed bed use.
2. The potassium content has a significant effect on the catalyst performance. Higher K content favors the yield of higher alcohol, and the promotion effect is more significant over the Al_2O_3–B_2O_3-supported carbide catalysts. This can be explained by the relatively smaller Mo_2C crystallites, which give higher Mo_2C dispersion over the catalysts.
3. The supported molybdenum carbide catalyst showed high stability for high alcohol synthesis with a fairly good mass balance. The high alcohol yield can be up to 199.7 g/alcohol/liter of catalyst/hour under the reaction conditions, nearly matching the supported rhodium catalyst under even harsher conditions [17, 20].

References

1. Zaman S, Smith KJ (2012) A review of molybdenum catalysts for synthesis gas conversion to alcohols: catalysts, mechanisms and kinetics. Catal Rev Sci Eng 54:41–132
2. Surisetty VR, Dalai AK, Kozinski J (2011) Alcohols as alternative fuels: an overview. Appl Catal A 404:1–11
3. Gupta M, Smith ML, Spivey JJ (2011) Heterogeneous catalytic conversion of dry syngas to ethanol and higher alcohols on cu-based catalysts. ACS Catal 1:641–656
4. Herman RG (2000) Advances in catalytic synthesis and utilization of higher alcohols. Catal Today 55:233–245
5. Christensen JM et al (2012) Catalytic conversion of syngas into higher alcohols over carbide catalysts. Ind Eng Chem Res 51: 4161–4172
6. Wang N et al (2010) Synthesis of higher alcohols from syngas over Fe/K/β-Mo2C catalyst. Catal Lett 136:9–13
7. Wang N et al (2010) Iron carbide promoted K/β-Mo2C for higher alcohols synthesis. Catal Today 158:241–245
8. Xiang M et al (2008) Synthesis of higher alcohols from syngas over Fischer-Tropsch elements modified K/β-Mo2C catalysts. Fuel 87:599–603
9. Xiang M et al (2007) Synthesis of higher alcohols from syngas over K/Co/β-Mo2C catalysts. Catal Commun 8:503–507
10. Xiang M et al (2007) Potassium and nickel doped β-Mo2C catalysts for mixed alcohols synthesis via syngas. Catal Commun 8:513–518
11. Shou H et al (2013) Use of infrared spectroscopy and density functional theory to study the influence of rubidium on alumina-supported molybdenum carbide catalyst for higher alcohol synthesis from syngas. J Catal 299:150–161
12. Surisetty VR, Eswaramoorthi I, Dalai AK (2012) Comparative study of higher alcohols synthesis over alumina and activated carbon-supported alkali-modified MoS2 catalysts promoted with group VIII metals. Fuel 96:77–84
13. Surisetty VR, Dalai AK, Kozinski JK (2012) Long term deactivation studies of novel catalysts for higher alcohols synthesis. Prepr Symp Am Chem Soc Div Fuel Chem 57:457
14. Wu Q et al (2011) Supported molybdenum carbides for higher alcohols synthesis from syngas. Prepr Symp Am Chem Soc Div Fuel Chem 56:153
15. Surisetty VR et al (2011) Structural characterization and catalytic performance of alkali (K) and metal (Co and Rh)-promoted MoS2 catalysts for higher alcohols synthesis. Appl Catal A 392:166–172
16. Surisetty VR, Dalai AK, Kozinski J (2011) Deactivation studies of alkali-promoted trimetallic Co-Rh-Mo sulfide catalysts for higher alcohols synthesis from synthesis gas. Energy Fuels 25: 580–590
17. Surisetty VR, Dalai AK, Kozinski J (2010) Effect of Rh promoter on MWCNT-supported alkali-modified MoS2 catalysts for higher alcohols synthesis from CO hydrogenation. Appl Catal A 381: 282–288
18. Li D et al (2005) Ni/ADM: a high activity and selectivity to C_2^+OH catalyst for catalytic conversion of synthesis gas to C1–C5 mixed alcohols. Top Catal 32:233–239
19. Shou H, Davis RJ (2011) Reactivity and in situ X-ray absorption spectroscopy of Rb-promoted Mo_2C/MgO catalysts for higher alcohol synthesis. J Catal 282:83–93
20. Surisetty VR, Dalai AK, Kozinski J (2010) Alkali-Promoted Trimetallic Co-Rh-Mo sulfide catalysts for higher alcohols synthesis from synthesis gas: comparison of MWCNT and activated carbon supports. Ind Eng Chem Res 49:6956–6963

7

Acidic and cesium salts of polyoxometalates with and without vanadium supported on MCM-41 as catalysts for oxidation of cyclohexane with H_2O_2

W. Trakarnpruk · J. Jatupisarnpong

Abstract Acidic and cesium salts of polyoxometalate (POM) with and without vanadium supported on MCM-41 were synthesized and their catalytic activities were investigated in the oxidation of cyclohexane with H_2O_2 under solvent-free condition. The products are cyclohexanol and cyclohexanone. The catalytic activity order is: CsV-POM/MCM > HVPOM/MCM > CsPOM/MCM ~ HPOM/MCM > CsVPOM > CsPOM catalysts. High conversion and selectivity to cyclohexanone over the supported catalysts are due to redox property and high surface area. These heterogeneous catalysts can be reused.

Keywords Polyoxometalate · Oxidation · Cyclohexane · Hydrogen peroxide

Introduction

Selective oxidation of cyclohexane yields cyclohexanol and cyclohexanone, important intermediates in the production of adipic acid and caprolactam, in the manufacture of nylon-6 and nylon-66 polymers [1]. Polyoxometalates (POMs) have been used as catalysts in various reactions, e.g. oxidation of organic compounds due to the variety of their compositions, sizes, shapes, redox potentials, acidity, and solubility [2–4]. Most of the homogeneous oxidation of cyclohexane catalyzed by polyoxometalates (POMs) was performed in acetonitrile solvent with H_2O_2, e.g. tetrabutylammonium salts of $[XW_{11}O_{39}]^{n-}$ and $[XW_{11}MO_{39}]^{(n-m)-}$ (X = P, Si, B and M^{III} = Fe, Mn), [5], $[M_4(H_2O)_2(PW_9O_{34})_2]^{10-}$, M^{II} = Co, Mn and M^{III} = Fe where $[Fe_4(H_2O)_2(PW_9O_{34})_2]^{6-}$ showed higher conversion of cyclohexane but the main product was cyclohexyl hydroperoxide [6]. Heteropolytungstic acid or HPW supported on MCM-41 [7–10] finds its place in catalysis over many organic reactions owing to the typical characteristics of the mesoporous materials of the MCM-41 type with highly ordered mesoporosity (pore size 2–10 nm) [11], large surface area, and high hydrothermal stability and acidity, making these materials as efficient catalysts in the synthesis and conversion of organic molecules. Transition metal-incorporated MCM-41 was used as catalyst for the oxidation of cyclohexane in the presence of solvent and H_2O_2 or *tert*-butyl hydroperoxide [12]. Ce-MCM-41 gave high selectivity to cyclohexanol in acetic acid as solvent and H_2O_2 as oxidant [13]. CeO_2/V-HMS using O_2 as oxidant exhibited 17.8 % conversion of cyclohexane but low selectivity to cyclohexanone (38 %) [14].

It has been reported that substitution of Cs^+ for H^+ in $Cs_xH_{3-x}PW_{12}O_{40}$ catalysts improved surface area and thermal stability compared to parent acids, the surface area is the largest when $x = 2.5$ [15]. A problem associated with the use of these Cs catalysts is that the salt particles disperse as a colloid in water and organic solvents, making it difficult to separate the salt from the reaction products by simple filtration [16]. Therefore they were supported on a carrier [17]. The POMs of the Keggin structure have molecular diameter of around 1.2 nm; it is therefore feasible to insert into the MCM-41 [18]. Introduction of vanadium into heteropolyacid improved the catalytic activity in oxidation due to its redox property [19, 20]. We have previously reported the catalytic activity of the tetrabutylammonium transition metal-substituted polyoxometalates supported on MCM-41 prepared by incipient wetness impregnation method and

W. Trakarnpruk (✉) · J. Jatupisarnpong
Department of Chemistry, Petrochemistry and Polymer Science, Faculty of Science, Chulalongkorn University, Bangkok 10330, Thailand
e-mail: wimonrat.t@chula.ac.th

found that the activity increased in the order V > Cu > Co > Fe [21].

The aim of this work was to study the oxidation of cyclohexane with H_2O_2 over acidic and cesium salts of polyoxometalates (with and without vanadium) supported on MCM-41. Hydrogen peroxide is the oxidant of choice as it is inexpensive, easily available, easy to handle and gives only water and oxygen as the products of decomposition.

Experimental

Materials

$H_3PW_{12}O_{40}$ (Wako Pure Chemical), Cs_2CO_3 (Fluka), $Na_2WO_4{\cdot}2H_2O$, Na_2HPO_4, $NaVO_3$ and oxalic acid (Sigma–Aldrich), H_2O_2 (30 wt% aqueous), cyclohexane (Merck), aqueous NH_3 (32 %, Merck), tetraethylorthosilicate (TEOS), cetyltrimethylammonium bromide (CTAB, 98 %, Fine Chem.) and triblock poly(ethylene oxide)–poly(propylene oxide)–poly(ethylene oxide) copolymer or P123 ($M = 5{,}800$) (Fluka) were used as received.

Catalysts preparation

MCM-41 [22]

CTAB and ammonia were dissolved in water at room temperature and then TEOS was added dropwise into the solution under stirring (300 rpm). The composition of SiO_2:CTAB:NH_4OH:H_2O was 1.0:0.12:8:114 (or 26 ml: 5 g: 57 ml: 235 g). After 2 h stirring, a gel was obtained. The gel was transferred into a Teflon bottle that was then heated at 110 °C for 96 h. The solid product was recovered by filtration, washed with deionized water until the filtrate was neutral, and finally dried. The removal of template was performed by refluxing the as-synthesized product in a mixture of methanol and HCl. The final product was separated, washed with methanol and dried.

$H_4PVW_{11}O_{40}$ [23]

10 g of $NaVO_3$ (dissolved in 100 ml of oxalic acid) and 6 g of Na_2HPO_4 (dissolved in 100 ml of distilled water) were mixed. 100 g of $Na_2WO_4{\cdot}2H_2O$ (dissolved in 150 ml of distilled water) was added, then the mixed solution was heated to 80 °C and then 60 ml of H_2SO_4 was slowly added to the solution. The solution was stirred for 8 h. After cooling the solution to room temperature, it was extracted with diethyl ether. The solvent was evaporated to obtain a solid product which was washed with distilled water and dried.

Cs salt of polyoxometalates

Cs salts of POM were prepared by adding the Cs_2CO_3 solution (0.47 M) dropwise to the $H_3PW_{12}O_{40}$ or $H_4PVW_{11}O_{40}$ solution (0.75 M) according to the literature [24]. The precipitate obtained was aged for 20 h at room temperature, then dried at 110 °C and calcined at 300 °C for 3 h. The products are denoted as CsPOM and CsV-POM, respectively. Their chemical compositions were characterized by ICP and XRF.

Cs salt of polyoxometalates supported on MCM-41

Cs salts of POM supported on MCM-41 were prepared by a two-step impregnation method [25] as the Cs salts of POM are not soluble in water and alcohol. First MCM-41 (2 g) was impregnated with Cs_2CO_3 (80 mg, 0.25 mmol) solution and the mixture was stirred for 12 h, then evaporated to dryness at 110 °C and calcined at 300 °C for 2 h. Then $H_3PW_{12}O_{40}$ or $H_4PVW_{11}O_{40}$ (15 and 20 wt%) was impregnated by the incipient wetness impregnation, using 1-butanol as a solvent to obtain CsPOM/MCM or CsV-POM/MCM, respectively. The resulting solids were dried and calcined at 350 °C for 3 h in air. For comparison, $H_3PW_{12}O_{40}$/MCM-41 (HPOM/MCM) and $H_4PVW_{11}O_{40}$/MCM-41 (HVPOM/MCM) with 20 % loading were prepared by incipient wetness impregnation method and calcined at 350 °C for 3 h in air. Percentages of Cs and W were determined by ICP.

Catalyst characterization

Specific surface area and pore volume were measured using BET method on a BELSORP-mini. The samples were degassed at 250 °C for 4 h.

XRD measurements were performed on Rigaku, DMAX 2002/Ultima Plus powder X-ray diffractometer with a monochromatic Cu Kα radiation.

The infrared spectra were obtained on a Nicolet FT-IR Impact 410 spectrophotometer with a pressed KBr pellet in the range of 3,000–400 cm^{-1}.

The NH_3-TPD of the samples was carried out in a CHEMBET-3000 (Quantachrome, USA) instrument. About 0.1 g powdered sample was taken inside a quartz U tube and degassed at 350 °C for 1 h with N_2 flow. The sample was then cooled to 30 °C and the gas flow was changed to NH_3 for 30 min. The NH_3-adsorbed sample was then purged with N_2 to remove any weakly adsorbed NH_3 on the catalyst surface. It was then heated under N_2 flow at a heating of 10 °C/min to record the spectra.

FT-IR spectra of adsorbed pyridine were also recorded to evaluate the acidic sites of the samples. The adsorption of pyridine was performed at 100 °C to avoid the influence of physisorption, and then the sample was evacuated at the same temperature for 30 min.

The amounts of Cs, V and W were analyzed by inductively coupled plasma emission (ICP, Perkin Elmer model PLASMA-1000). The P content was determined on ARL 8410 X-ray fluorescence spectrometer (XRF).

^{31}P MAS NMR experiments were performed with a Varian Mercury Plus NMR spectrometer operating at 121.4 MHz at ambient temperature using a 7 mm Varian CP/MAS probe and silicon nitride rotors and the samples were spun at a rate of 5 kHz. The spectra were referenced to H_3PO_4.

Oxidation of cyclohexane

Cyclohexane, catalyst and aqueous H_2O_2 (30 %) were added into a 60 ml Parr reactor. The reaction mixture was heated to 80 °C and stirred for 8 h. The catalyst was separated by filtration and the products were extracted with diethyl ether and dried, analyzed by GC [chromatograph Varian CP-3800 GC; CP-Sil (30 m × 0.25 mm) column].

Results and discussion

From the elemental analysis of the synthesized CsPOM and CsVPOM the analyzed percentages of elements can be matched with the chemical formula of $H_{0.5}Cs_{2.5}PW_{12}O_{40}$ and $H_2Cs_2PVW_{11}O_{40}$. CsPOM, anal (%): Cs, 10.62; W, 67.45; P, 0.92; calc (%): Cs, 10.76; W, 68.41; P, 0.96 and CsVPOM, anal (%): Cs, 11.10; W, 86.90; P, 1.17; V, 2.02; calc (%): Cs, 11.65; W, 84.82; P, 1.30; V, 2.14.

Surface area and pore volume of the catalysts are shown in Table 1. The introduction of large countercations such as Cs^+ into a heteropoly compound to replace the small H^+ increases specific surface areas of the CsPOM, resulted in the formation of micropores and mesopores. After loading with POM the reduction in the surface area and pore volume could be due to deposition of POM inside the mesochannels or dispersion on the surface of the MCM-41 support. Both surface area and pore volume decreased with increasing loading of POM. This is due to the inclusion of POM into the pore of one-dimensional mesopores of MCM-41 or by the agglomeration of POM on the external surface of the support. Table 1 also showed percentages of POM in the supported samples deduced from the chemical analysis of W content by ICP technique; they are close to the loaded values. The Cs:W mole ratios were calculated from percentages of analyzed Cs and W of both unsupported and supported catalysts. It was found that in the supported catalysts, Cs:W mole ratio decreased when increasing the percentage of loading of POM.

The N_2 adsorption–desorption isotherms of the samples are shown in Fig. 1. MCM-41 and the supported catalysts show a type IV isotherm according to IUPAC classification, given by mesoporous materials.

The total acidity of the catalysts was measured by temperature programmed desorption of ammonia (NH_3-TPD) and tabulated in Table 1. NH_3-TPD spectra of MCM, HPOM/MCM and CsPOM/MCM (not shown) reveal that the MCM-41 was weakly acidic due to surface Si–OH. It showed a broad peak about 100–300 °C. When these silanol groups were replaced with stronger acidic sites of HPOM the acidity increased. For the HVPOM/MCM when W atoms were partly substituted by V atoms, the acidity decreased. In case of the CsPOM/MCM, the acidity also decreased with substitution of Cs ion for protons [26]. However, the acidity of the CsPOM/MCM was higher than that of the unsupported CsPOM, as previously reported [27]. Compared between the HPOM/MCM and CsPOM/MCM, the acidity of the CsPOM/MCM was lower but not much, because it possessed high surface acidity.

Table 1 Characteristics of catalysts

Catalyst	% POM		Cs:W[b] mole ratio	S_{BET} (m^2/g)	V_{pore} (cm^3/g)	Acidity (mmol/g)
	Loaded	Analyzed[a]				
MCM	–	–	–	980	0.85	0.13
CsPOM	–	–	2.5:12	125	0.07	29.4
CsVPOM	–	–	2.0:12	120	0.07	24.8
CsPOM/MCM	15	14.8	1.7:12	745	0.68	32.4
	20	19.4	1.2:12	696	0.60	36.1
CsVPOM/MCM	15	14.7	1.9:12	765	0.67	28.6
	20	19.6	1.2:12	700	0.62	32.5
HPOM/MCM	20	19.5	–	677	0.71	40.3
HVPOM/MCM	20	19.7	–	668	0.68	37.9

[a] Deduced from the chemical analysis of W by ICP

[b] Deduced from the chemical analysis of Cs and W by ICP

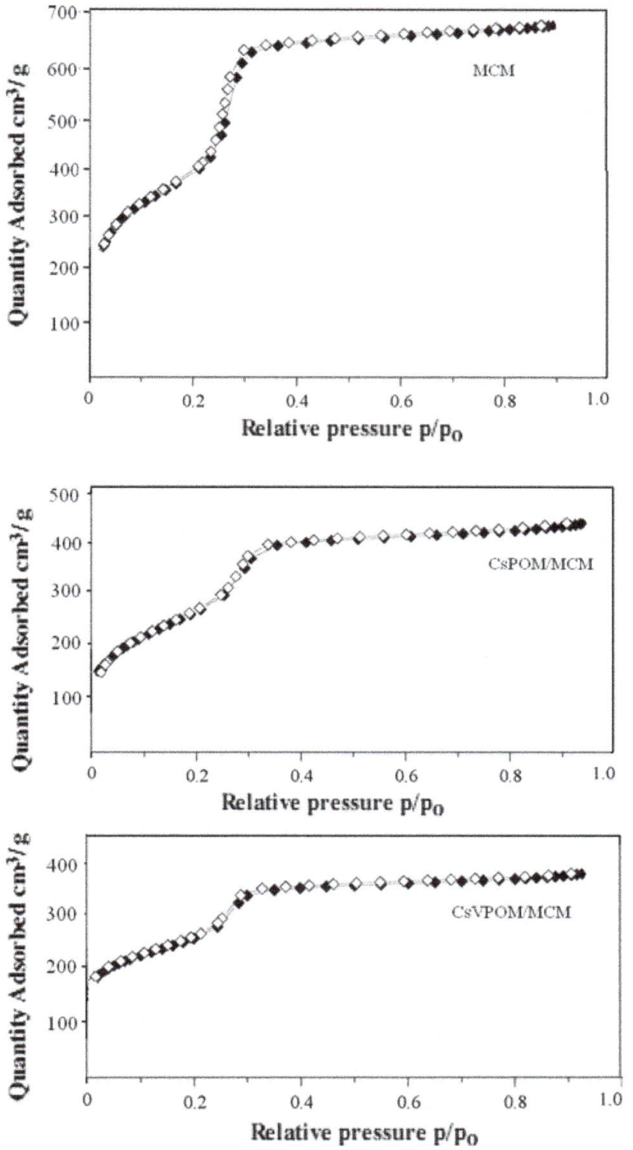

Fig. 1 N_2 adsorption–desorption isotherms of MCM-41, CsPOM/MCM and CsVPOM/MCM

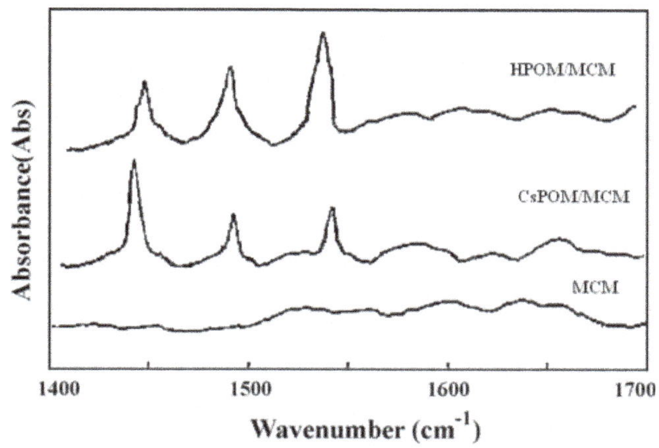

Fig. 2 IR spectra of adsorbed pyridine on the HPOM/MCM and CsPOM/MCM

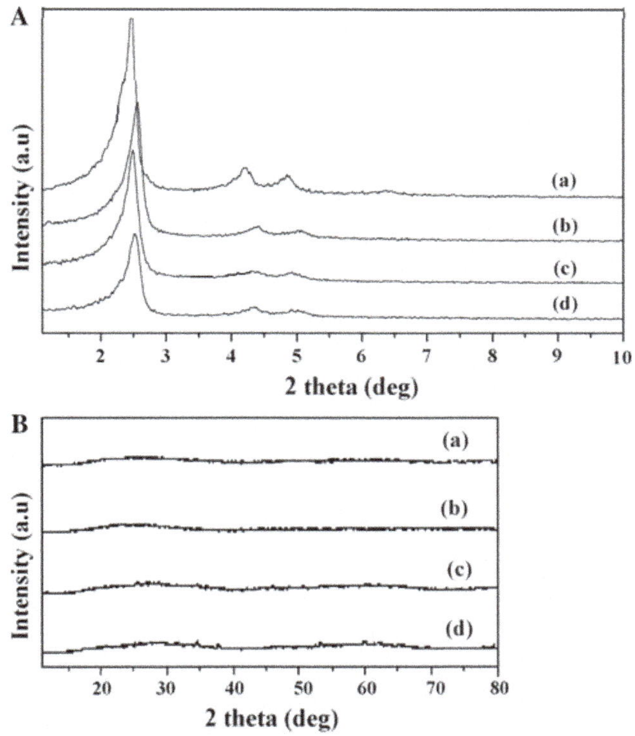

Fig. 3 Low and wide angle (**A** and **B**) XRD patterns of *a* MCM, *b* CsPOM/MCM, *c* CsVPOM/MCM and *d* HVPOM/MCM

To discriminate the type of acid sites of the supported catalysts, pyridine adsorbed FT-IR spectra of HPOM/MCM and CsPOM/MCM samples were taken. The spectra are presented in Fig. 2. A peak at 1,445 cm^{-1} is due to the pyridine adsorbed on Lewis acid sites (L), a peak at 1,545 cm^{-1} is due to adsorption on Bronsted acid sites (B) and a peak around 1,493 cm^{-1} is due to both Bronsted and Lewis acid sites [28]. The HPOM/MCM showed strong Bronsted acidity, higher Bronsted acid sites peak than Lewis acid sites peak. The CsPOM/MCM showed increased Lewis acidity.

In Fig. 3, the XRD pattern of the MCM-41 exhibits the ordered hexagonal structure, showing a strong peak at $2\theta = 2.2°$ due to (100) plane and small peaks due to (110) and (200) plane reflections within 5°. The XRD patterns of acidic and cesium POMs supported on MCM-41 still illustrate the characteristics of a typical mesoporous MCM-41 structure but with decreased intensity. This may indicate a structural distortion of the MCM-41 or more simply a reduction of its long range order. The absence of peaks belonging to crystalline POM at the high angles (compared to the XRD patterns of the HPOM and CsPOM without support shown in Fig. 4) indicated that POM is finely dispersed on the MCM-41 support or was due to low loadings (15–20 %). This result is not surprising since the

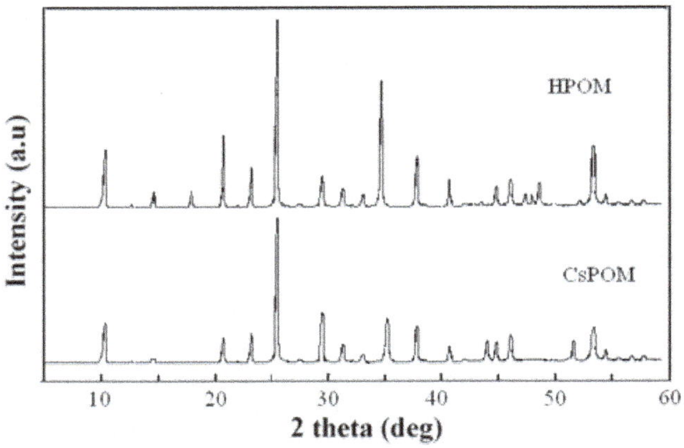

Fig. 4 XRD patterns of HPOM and CsPOM

MCM-41 has a pore size higher than POM crystals (1.2 nm) [29]. However, with higher loadings (above 60 %), it was reported that the reflections from the POM can be observed [10].

FT-IR spectra in the range of 400–2,200 cm^{-1} of HPOM and HPOM supported on MCM-41 are shown in Fig. 5. The characteristic FT-IR bands at 1,078, 985, 880 and 788 cm^{-1} are attributed to P–O, terminal bond (W–O), corner-sharing W–O–W and edge-sharing W–O–W, respectively. When the HPOM was supported on the MCM-41, these bands overlap with the framework bands of the MCM-41 in the same region. However, a tiny band can be seen at 882 cm^{-1}.

Figure 6 shows FT-IR spectra of CsPOM and CsPOM/MCM. The characteristic FT-IR bands of CsPOM at 1,080, 985, 890 and 800 cm^{-1} are attributed to P–O, terminal bond (W–O), corner-sharing W–O–W and edge-sharing W–O–W, respectively [24]. In the CsPOM/MCM, an

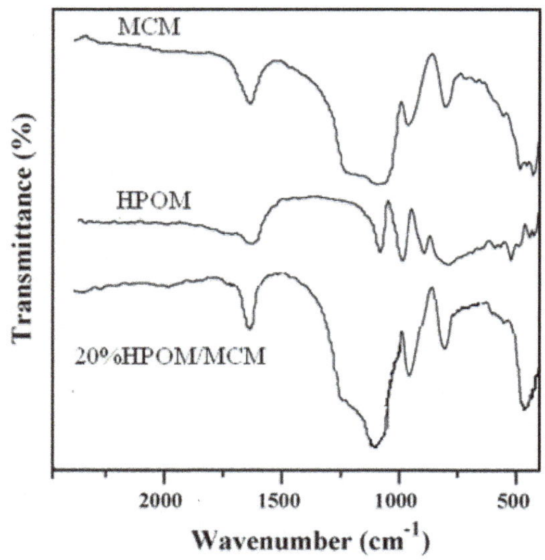

Fig. 5 FT-IR spectra of MCM, HPOM and 20 % HPOM/MCM

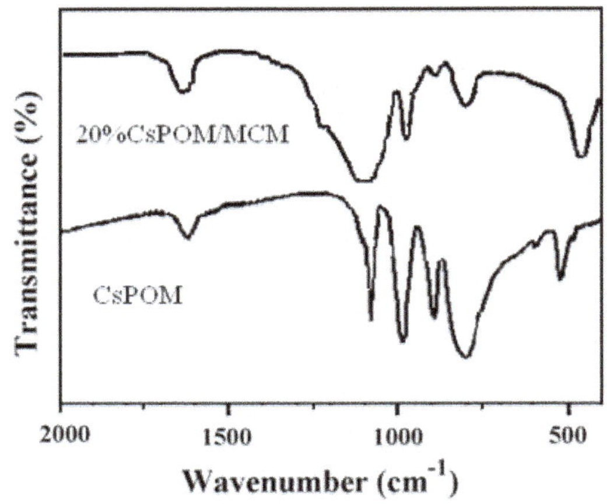

Fig. 6 FT-IR spectra of CsPOM and 20 % CsPOM/MCM

observed band at 889 cm^{-1} reveals that presence of CsPOM in the supported catalyst, as reported in the literature [27]. The FT-IR spectrum of the CsVPOM/MCM (not shown) was similar to that of the CsPOM/MCM.

Insight into the state of HPOM in the supported catalysts can be gained by $^{31}P\{^{1}H\}$ MAS NMR. As shown in Fig. 7, the HPOM exhibited a sharp resonance at −15.1 ppm [29], corresponding to the tetrahedral coordination of PO_4 in Keggin unit. The HPOM/MCM showed a strong peak at −15.4 ppm indicating that the Keggin structure of the HPW is retained after supporting in the MCM-41. The shift of peak is due to the interaction of HPOM with MCM-41,

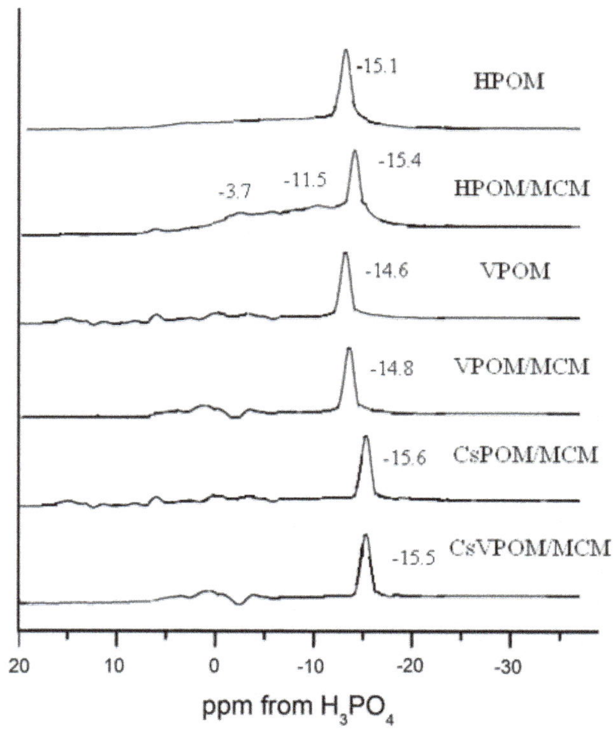

Fig. 7 $^{31}P\{^{1}H\}$ MAS NMR spectra of unsupported and supported catalysts

from $(SiOH_2)^+(H_2PW_{12}O_{40})^-$ species previously reported [9]. In addition two small peaks at −11.7 and 3.1 ppm are due to decomposition fragments (the HPOM/MCM sample was calcined at 350 °C). For the VPOM and VPOM/MCM the resonance appears at −14.6 and −14.8 ppm, respectively. Both the Cs-supported catalysts showed resonances at a little up-field shift (compared to the acid supported catalysts), −15.6 and −15.5 ppm for the CsPOM/MCM and CsVPOM/MCM, respectively, in agreement with the previous report [30]. No peak from decomposition fragment was detected, demonstrating their higher thermal stability than that of the HPOM/MCM.

The oxidation of cyclohexane was performed using H_2O_2 as oxidant in the presence and absence of catalyst. Blank reaction in the presence of MCM-41 did not show any products. The unsupported catalysts (CsPOM and CsVPOM) showed low conversion of cyclohexane. The catalytic activities of the supported catalysts were compared under the same reaction conditions and the results are presented in Table 2.

For the supported catalysts, HVPOM/MCM showed higher activity and selectivity to cyclohexanone than the HPOM/MCM under the same loading, 14 and 18 % conversion, and 66 and 70 % selectivity, respectively. The reduction potential of HVPOM was reported to be higher than that of the HPOM [31]. Correlation between reduction potential and oxidation activity of $H_nPV_xW_{12-x}O_{40}$ (x = 0–3) was also reported [32]. The remaining W content in the catalyst after the oxidation reaction was analyzed by XRF. The leaching of POM was calculated by the difference of W content in the catalyst before and after the reaction. The acidic POMs showed leaching into water (from aq. H_2O_2) used in this work.

For the insoluble Cs salt supported on MCM-41, the experimental results show that the CsPOM/MCM and HPOM/MCM with the same loading (20 %) have comparable activity. This might be due to difference in the nature of acidity of both supported catalysts. The HPOM/MCM has high Brønsted acidity while the CsPOM/MCM has high Lewis acidity. It was reported in the literature that the HPOM/Cs-MCM-41 is more active than the HPOM/MCM-41 for butane isomerization [30]. An advantage of using the Cs-supported catalysts is that there is no leaching of CsPOM from the support, demonstrating higher stability than the acidic catalysts. Higher selectivity to cyclohexanone was also obtained from the V-containing catalysts. This could be due to high redox property of V and the beneficial effect of redox metal V on the oxidation properties of heteropolyoxometallate compounds was already observed in other cases. It was reported that $[PVW_{11}O_{40}]^{4-}$ exhibited three reversible one-electron redox waves. The first wave is ascribed to the reduction of V^{5+} in the peripheral position, and the remaining two to the reduction of W^{6+} [33]. Another reason may be due to the surface of the MCM-41 support which can accumulate the produced cyclohexanol at the interfaces and, consequently, oxidized to cyclohexanone. The highest activity from the CsVPOM/MCM obtained in this work might be because of combined effect of surface area, role of V, Cs and acidity of the catalyst. The conversion obtained from this work is higher than that from our previous work over $(Bu_4N)_4PVW_{11}O_{40}$/MCM-41 [21] due to higher POM incorporated into the catalyst. The results of reusability show that there was a slight decrease in the activity of the CsPOM/MCM and CsVPOM/MCM at the 4th run (1 % drop).

When triphenylphosphine was added to the reaction products the GC analysis of this showed a sharp increase in the concentration of cyclohexanol, while the concentration of cyclohexanone substantially reduced. This proves that cyclohexane oxidation proceeds via the cyclohexyl hydroperoxide (intermediate) formation, which is found to be reduced by PPh_3. Addition of 2,6-di-*tert*-butyl-4-methylphenol, a radical scavenger into the reaction resulted in no product, showing that oxidation appears to be a radical process [5].

Table 2 Oxidation of cyclohexane (0.4 g catalyst, substrate 92 mmol, H_2O_2/cyclohexane molar ratio = 4, reaction time 8 h, temperature 80 °C)

Catalyst	% POM		Conversion[b] (%)	Selectivity[c] (%)	
	Loaded	After reaction[a]		–one	–ol
MCM	–	–	0	–	–
CsPOM[d]	–	–	5	52	48
CsVPOM[d]	–	–	9	58	42
HPOM/MCM	20	13.1	14	66	34
HVPOM/MCM	20	13.5	18	70	30
CsPOM/MCM	15	14.8	9	64	36
	20	19.2	13	68	32
	20[e]	19.0	12	66	34
CsVPOM/MCM	15	14.7	18	75	25
	20	19.1	22	78	22
	20[e]	18.9	21	77	23
VPOM/MCM[f]	16	15.0	19	76	24

[a] Deduced from the chemical analysis of W by ICP

[b] Based on the gas chromatographic peak areas

[c] Expressed as a percentage of the total products; –one = cyclohexanone, –ol = cyclohexanol

[d] With same mole of POM as in the 20 % supported catalyst

[e] At the 4th run

[f] $(Bu_4N)_4PVW_{11}O_{40}$ on MCM-41 under the same reaction condition [21]

Conclusions

In this study polyoxometalate supported MCM-41 was found to be an efficient catalyst for the oxidation of cyclohexane in the absence of added solvent using H_2O_2 as oxidant. Supported catalysts give higher cyclohexane conversion and cyclohexanone selectivity compared to bulk polyoxometalates. The advantage is easy separation and recycling.

References

1. Shylesh S, Prinson PS, Singh A (2007) Chromium-containing small pore mesoporous silicas: synthesis, characterization and catalytic behavior in the liquid phase oxidation of cyclohexane. Appl Catal A 318:128
2. Nezhadali A, Akbarpour M (2010) Selective oxidation of primary substituted aromatic amines to azoxy products using lacunary catalyses. Chinese Chem Lett 21:43
3. Kozhevnikov IV (1998) Catalysis by heteropoly acids and multicomponent polyoxometalates in liquid-phase reactions. Chem Rev 98:171
4. Mizuno N, Misono M (1998) Heterogeneous catalysis. Chem Rev 98:199
5. Simoes MMQ, Santos ICMS, Balula MSS et al (2004) Oxidation of cycloalkanes with hydrogen peroxide in the presence of Keggin-type polyoxotungstates. Catal Today 211:91–92
6. Santos ICMS, Gamelas JAF, Balula MSS et al (2007) Sandwich-type tungstophosphates in the catalytic oxidation of cycloalkanes with hydrogen peroxide. J Mol Catal A Chem 262:41
7. Kozhevnikov IV, Sinnema A, Janse RJJ, Pamin K, van Bekkum H (1994) New acid catalyst comprising heteropoly acid on a mesoporous molecular sieve MCM-41. Catal Lett 30:241
8. Ajaikumar S, Pandurangan A (2008) HPW and supported HPW catalyzed condensation of aromatic aldehydes with aniline: synthesis of DATPM derivatives. J Mol Catal 286:21
9. Blasco T, Corma A, Mart'ınez A, Martınez-Escolano P (1998) Supported heteropolyacid (HPW) catalysts for the continuous alkylationof isobutane with 2-butene: the benefit of using MCM-41 with larger pore diameters. J Catal 177:306
10. Khder ARS, Hassan HMA, El-Shall MS (2012) Acid catalyzed organic transformations by heteropoly tungstophosphoric acid supported on MCM-41. Appl Catal A Gen 77:411–412
11. Kozhevnikov IV, Kloetstra KR, Sinnema A et al (1996) Study of catalysts comprising heteropoly acid $H_3PW_{12}O_{40}$ supported on MCM-41 molecular sieve and amorphous silica. J Mol Catal A Gen 114:287
12. Sakthivel A, Selvam P (2002) Mesoporous (Cr)MCM-41: a mild and efficient heterogeneous catalyst for selective oxidation of cyclohexane. J Catal 211:134
13. Yao W, Chen Y, Min L et al (2006) Liquid oxidation of cyclohexane to cyclohexanol over cerium-doped MCM-41. J Mol Catal A Chem 246:162
14. Li J, Shi Y, Xu L et al (2010) Selective oxidation of cyclohexane over transition-metal-incorporated HMS in a solvent-free system. Ind Eng Chem Res 49:5392
15. Okuhara T, Watanabe H, Nishimura T et al (2000) Microstructure of cesium hydrogen salts of 12-tungstophosphoric acid Relevant to novel acid catalysis. Chem Mater 12:2230
16. Izumi Y (1997) Hydration/hydrolysis by solid acids. Catal Today 33:371
17. Dias AS, Lima S, Pillinger M (2006) Acidic cesium salts of 12-tungstophosphoric acid as catalysts for the dehydration of xylose into furfural. Carbohydr Res 341:2946
18. Tangestaninejad S, Mirkhani V, Moghadam M et al (2008) Hydrocarbon oxidation catalyzed by vanadium polyoxometalate supported on mesoporous MCM-41 under ultrasonic irradiation. Ultrason Sonochem 15(4):438
19. Serwicka EM, Broclawik E, Bruckman K, Haber J (1989) TPD/TPR and quantum chemical study of methanol interaction with $H_{3+n}PV_nMo_{12-n}O_{40}$ heteropolyacids. Catal Lett 2:351
20. Wang J, Hu C, Jian M et al (2006) Catalytic oxidation performance of the α-Keggin-type vanadium-substituted heteropolymolybdates: a density functional theory study on $[PV_nMo_{12-n}O_{40}]^{(3+n)-}$ (n = 0–3). J Catal 240:23
21. Jatupisarnpong J, Trakarnpruk W (2012) Transition metal-substituted polyoxometalates loaded MCM-41 as catalysts in the oxidation of cyclohexane and cyclooctane with H_2O_2. Mendeleev Commun 22:152
22. Zhang C, Zhou W, Liu S (2005) Synthesis and characterization of organofunctionalized MCM-41 by the original stepped templated sol–gel technology. J Phys Chem B 109(51):24319
23. Park DR, Song SH, Hong UG et al (2009) Redox properties and catalytic oxidation activities of polyatom-substituted $H_nPW_{11}M_1O_{40}$ (M = V, Nb, Ta, and W) Keggin heteropolyacid. Catalysts Catal Lett 132:363
24. Okuhara T, Nishimura T, Watanabe H et al (1992) Insoluble heteropoly compounds as highly active catalysts for liquid-phase reactions. J Mol Catal 74:247
25. Soled S, Miseo S, McVicker G et al (1997) Preparation of bulk and supported heteropolyacid salts. Catal Today 36:441
26. Rafiee E, Nejad FK, Joshaghani M (2011) $Cs_xH_{3-x}PW_{12}O_{40}$ heteropoly salts catalyzed quinoline synthesis via Friedländer reaction. Chinese Chem Lett 22:288
27. Parida KM, Rana S, Mallick S, Rath D (2010) Cesium salts of heteropoly acid immobilized mesoporous silica: an efficient catalyst for acylation of anisole. J Colloid Interf Sci 350:132
28. Karthikeyana G, Pandurangan A (2009) Heteropolyacid ($H_3PW_{12}O_{40}$) supported MCM-41: an efficient solid acid catalyst for the green synthesis of xanthenedione derivatives. J MolCatal A Chem 311:36
29. Alsalme AM, Wiper PV, Khimyak YZ, Kozhevnikova EF, Kozhevnikov IV (2010) Solid acid catalysts based on $H_3PW_{12}O_{40}$ heteropoly acid: acid and catalytic properties at a gas-solid interface. J Catal 276:181
30. Yang W, Billy J, Ben Taarit Y, Védrine JC, Essayem N (2002) $H_3PW_{12}O_{40}$ supported on Cs modified mesoporous silica: catalytic activity in *n*-butane isomerisation and in situ FTIR study. Comparison with microporous $Cs_xH_{3-x}PW_{12}O_{40}$. Cat Today 73:153
31. Song IK, Barteau MA (2004) Redox properties of Keggin-type heteropolyacid (HPA) catalysts: effect of counter-cation, heteroatom, and polyatom substitution. J Mol Catal A Chem 212:229
32. Park DR, Park S, Bang Y (2010) Reduction potential, UV–visible absorption edge energy, and oxidation catalysis of niobium-containing $H_{3+x}PW_{12-x}Nb_xO_{40}$ Keggin and $H_{6+x}P_2W_{18-x}Nb_xO_{62}$ Wells-Dawson heteropolyacid catalysts. Appl Catal A Gen 373:201
33. Himeno S, Takamoto M, Higuchi A et al (2003) Preparation and voltammetric characterization of Keggin-type tungstovanadate $[VW_{12}O^{40]3-}$ and $[V(VW_{11})O_{40}]^{4-}$ complexes. Inorg Chim Acta 348:57

8

Selective zeolite catalyst for alkylation of benzene with ethylene to produce ethylbenzene

Mohammed C. Al-Kinany · Hamid A. Al-Megren · Eyad A. Al-Ghilan · Peter P. Edwards · Tiancun Xiao · Ahmad. S. Al-Shammari · Saud A. Al-Drees

Abstract In this work, a selective catalyst of BXE ALKCAT zeolite has been developed with about 30 % of ZSM-5 balanced with kaolinite, and applied for gas-phase alkylation of benzene (BZ) with ethylene (E). The catalyst has been tested in a fixed-bed down-pass flow reactor under different conditions of temperatures ranging between 300 and 450 °C with BZ to E mole ratios ranging between 1:1, 3:1 and 6:1 under atmospheric pressure and space velocity ranges between 0.1 and 150 h^{-1}. The BXE ALKCAT zeolite catalyst has been characterized using: scanning electron microscope, X-ray diffraction, specific surface area, pore volumes, pore size distributions, X-ray photoelectron spectroscopy, and differential thermal analysis, and thermo-gravimetric analyses. Ethylbenzene was the main product of alkylation, and diethylbenzene isomers (*ortho-*, *meta-*, and *para-*) were the minor products. In the case of 1:1 mol ratio of BZ to E, the selectivity of EB about 85.5 % at highest conversion of BZ was obtained after 1 h of reaction on stream at 450 °C. A decrease in the temperature to 300 °C (with 1:1 mol ratio) caused the selectivity of EB to decrease to 73.0 %. EB and DEBs yields were found to increase with increasing the reaction temperature and decreasing the mole ratio of BZ to E. The conversion of BZ appeared to be depending strongly on mole ratio of BZ to E at a given temperature. The study has shown that the BXE ALKCAT zeolite is active as a catalyst for the alkylation reaction and selective to EB compared with other zeolite catalysts.

Keywords Alkylation · Ethylation · Ethylbenzene · Zeolite

M. C. Al-Kinany · H. A. Al-Megren (✉) ·
E. A. Al-Ghilan · S. A. Al-Drees
Petrochemicals Research Institute, King Abdulaziz City for Science and Technology, P.O.Box 6086, Riyadh 11442, Saudi Arabia
e-mail: almegren@kacst.edu.sa

P. P. Edwards · T. Xiao · Ahmad. S. Al-Shammari ·
S. A. Al-Drees
Inorganic Chemistry Laboratory, Wolfson Catalysis Centre, KACST-Oxford Petrochemicals Research Centre (KOPRC), University of Oxford, South Parks Road, Oxford OX1 3QR, UK

Introduction

Ethylbenzene is important in the petrochemical industry as an intermediate in the production of styrene, which in turn is used for making polystyrene, a common plastic material. In industry, EB is mainly manufactured by the alkylation of benzene with ethylene via two methods, i.e., the gas-phase method [1–5] and the liquid-phase method. The gas-phase method is the Mobil–Badger technology, which used mostly molecular sieve catalyst, e.g. ZSM families like ZSM-5 and ZSM-22, because of their unique advantages of highly selective, less toxic, environmentally friendly and readily reproducible in catalytic reactions [6, 7]. Another reason for ZSM-5 zeolite catalyst being used in alkylation of benzene with ethylene is that its proper pore size can increase the EB diffusion, while it prevents the polyethylbenzene (PEB) to diffuse through the catalyst [3, 4, 8, 9]. However, ZSM-5 zeolite has high acid strength and acid amount, which easily catalyses the carbon formation from ethylene [10, 11]. Therefore, in this process, the benzene to ethylene molar ratio is about 8–16 which increases the needed energy in the fraction unit for the separation of EB from benzene and transethylbenzene.

In addition, the gas-phase method normally is carried out under moderate pressure (1.0–20.8 MPa) and high temperature (573–773 K), which leads to higher energy consumption, more cooling systems and strict requirements for the apparatus. Many years' industrial operation results showed that the pure ZSM-5 based catalyst suffers from several disadvantages. For example, more byproducts are produced, especially toluene at about 1,000–2,000 ppm, which is much higher than the levels required by the downstream processes; the selectivity toward ethyl benzene is low, and the deactivation of the catalyst is so serious that it requires periodic regeneration. The by-products and the rapid deactivation of the ZSM-5 catalyst for the alkylation are believed to be due to its strong acidity.

Kaolinite has been widely used as binder and balanced material for zeolite catalyst, because it has global pore with 105 A mean pore size, which has little effect on the diffusion of the reactants products [12–14]. It also allows the easy process and extrudation of the acid catalyst. In this work, 30 wt% of ZSM-5 catalyst has been mixed with kaolinite and tested for benzene alkylation with ethylene at low benzene to ethylene ratio; the results show that the low content of ZSM-5 zeolite catalyst is suitable for low benzene to ethylene ratio and high stability. This may be promising to reduce the cycle and save energy.

Experimental

Catalyst preparation

The catalyst, BOX ALKCAT has been prepared using prilling process The main active component, e.g., ZSM-5 (Si/Al = 25) is mixed with kaolinite at 30:60 and 10 % of alumina as the binder. These are mixed with a small amount of water to form a slurry. The slurry is shaped into particles using a prilling machine (36MM, SZCX 160/45, made in China). The resultant catalyst particles contain 30 wt% of ZSM-5, which is then dried in static air at 450 °C for 5 h to remove the moisture and volatile impurities.

Chemicals and catalysts

All chemicals were analytical grade; benzene, ethylbenzene (Fluka Chemie 99.5 %), *o*-diethylbenzenes (Fluka Chemie 98.9 %), *m*-diethylbenzenes (Fluka Chemie 98.9 %), *p*-diethylbenzenes (Fluka Chemie 98.9 %), 1,3,5-triethylbenzene (Fluka Chemie 98 %) and 1,2,4,5-tetraethylbenzene (Fluka Chemie 98 %) were used directly without further purification. Ethylene gas (purity >99.95 %) was obtained from M/s. Abdullah Hashim for industrial gases. Hexane (Fluka Chemie) was of high grade and spectroscopically highly pure (purity >99.98 %).

Catalyst characterization

X-ray diffraction (XRD)

Measurements were conducted using Brucker diffractometer D8 which utilizes Ni-filtered CuKα radiation (λ = 1.54 Å). Diffraction patterns were obtained with X-Ray gun operated at 40 kV and 30 mA with a scan rate of 4°/min (2θ).

Scanning electron microscope (SEM)

The crystal size and morphology of BXE ALKCAT catalyst were determined with a FEI–NNL200, 5 kV and work distance 5.0 mm. The silicon and aluminum contents of the BXE ALKCAT zeolites were obtained using EDAX Ametek-Model 60040, 10 kV.

X-ray photoelectron spectroscopy (XPS)

XPS studies were recorded using JEOL JPS 9010MC photoelectron spectrometer, using MgKα (1,253.6 eV) radiation from an X-ray source operating at 10 kV and 20 mA, and the base pressure in the analysis was kept in the range from 5×10^{-10} to 1×10^{-9} mbar. The binding energies (BE) were referenced to the C1s level at 284.9 eV. A estimated error of ±0.1 eV can be assumed for all measurements.

Physisorption analysis

Textural properties were determined by nitrogen adsorption–desorption experiments. The isotherm was measured using a Micrometrics ASAP 2010 system. BET surface area, pore volume and pore size measurements studies were carried out using physisorption technique. The adsorption for nitrogen was measured at 77 K. Prior to the experiments, the samples were degassed under vacuum at 250 °C for 6 h. The surface area was calculated using the BET method based on adsorption data. The pore size distribution for mesopore was analyzed from desorption branch of the isotherm by the Parrett–Joyner–Halenda method and the pore size distribution for micropore was analyzed by HK method.

Differential thermal analysis (DTA) and thermo-gravimetric analyses (TGA)

DTA and TGA were recorded on Perkin Elmer (DTA-7) with thermal analysis controller TAC-7/DX. The catalyst samples were recovered after reaction and dried under vacuum before being analyzed by TGA.

Catalyst evaluation

The catalytic behavior of BXE ALKCAT zeolite catalyst for the alkylation of benzene with ethylene was studied in a conventional bench-top pilot plant, as shown in Fig. 1, fitted with a fixed-bed down-flow stainless steel reactor with an internal diameter of 5 mm and 35 cm long at atmospheric pressure and temperature ranging between 300 and 450 °C. The reactor was coupled to a mass flow meter to measure un-reacted ethylene. The reactor was heated in an electrical furnace, and the reactor's temperature was measured by a thermocouple located inside the furnace and was controlled by a temperature controller (Cole-Parmer Digi-sense).

Each time, 1.0 g of catalyst was loaded in the middle of the reactor. The feed stock of the alkylation reaction consisting of benzene and ethylene was introduced at the top of the reactor. Normally, the flow rate of ethylene was 2–22 ml/min with benzene to ethylene mole ratios ranging from 1:1 to 6:1, respectively, under atmospheric pressure. The flow rate of ethylene was adjusted through a separate thermal mass flow controller (Bronkhorst). The flow rate of benzene was controlled through a one channel syringe pump (Cole-Parmer). The alkylation reaction products were collected in a cooled condenser attached to the end of the reactor and was analyzed using a gas chromatograph.

The activity of the catalysts, yield and selectivity of alkylbenzenes (ethylbenzene, and *o*-, *p*- and *m*-diethylbenzene) and benzene conversion were studied at 300, 350, 400 and 450 °C by changing the benzene to ethylene mole ratios from 1:1 to 6:1 at each temperature. Blank reactor runs were conducted and no significant conversions were observed under the conditions of alkylation reaction.

Gas chromatographic analysis

Gas chromatographic analysis of the alkylation products was performed on Varian 3800 series instrument fitted with a flame ionization detector and a 50 m × 0.25 mm glass open tubular capillary PONA column. The column temperature was programmed as an initial temperature of 30 °C for 15 min, then 60 °C for 20 min (heating rate 1 °C/min) and finally 200 °C for 20 min (heating rate 2 °C/min).

The FID detector temperature was 250 °C and that of injector 250 °C (for manual injection). The maximum column temperature at which the stationary phase is stable is 200 °C. Flow rates of zero air (80 lb/in^2, 300 cm^3/min), hydrogen (40 lb/in^2, 30 cm^3/min) and He carrier gas (80 lb/in^2, 25 cm^3/min) were applied. Occasional checking showed that the flow rate was almost constant.

External standardization method is the analytical method which has been employed for the quantitative analysis of the alkylation products. The GC instrument was calibrated by analyzing the known composition of a prepared calibration mixture (standard mixture) of pure components of benzene, ethylbenzene, diethylbenzenes (*o*-, *p*-, and *m*-), triethylbenzene and tetraethylbenzene. GC chromatogram of standard mixture is shown in Fig. 2.

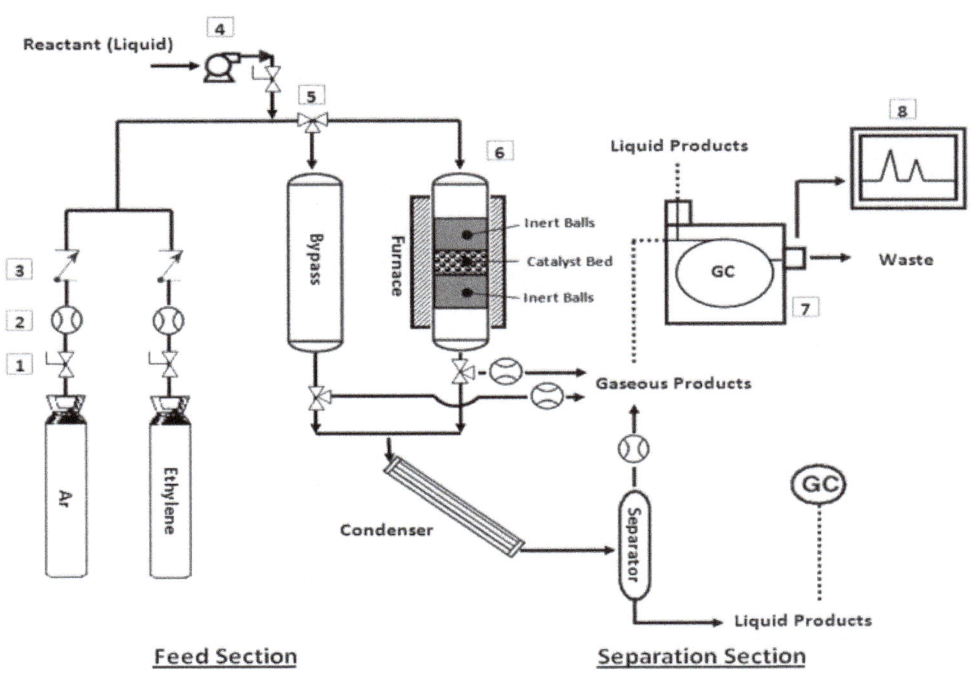

Fig. 1 Bench-top pilot plant fitted with fixed bed reactor. *1* two-way valve, *2* flow meter, *3* check valve, *4* pump, *5* three-way valve, *6* reactor *7* gas chromatography, *8* recorder

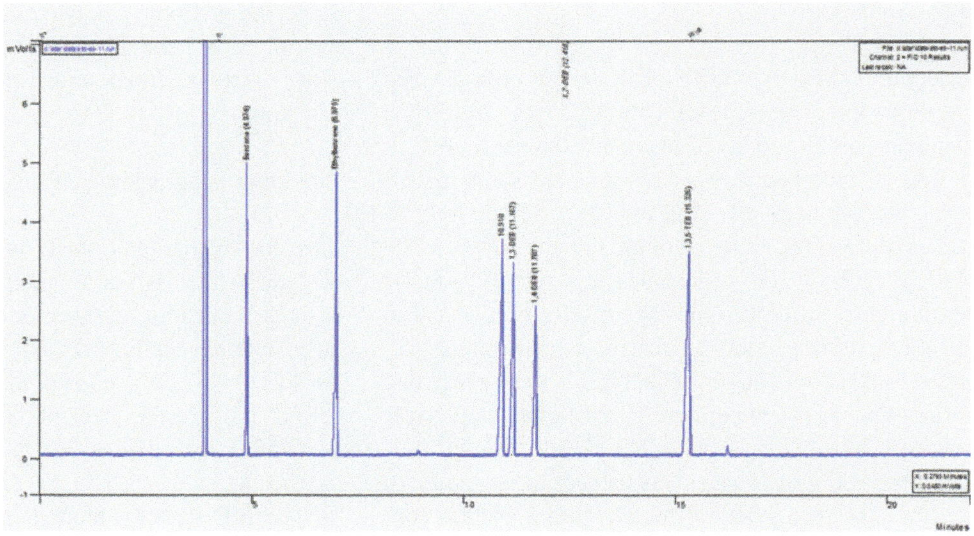

Fig. 2 Typical separation of the standard mixture of: solvent, benzene, ethylbenzene, diethylbenzenes (*o*-, *m*-, and *p*-), and triethylbenzene

Gas chromatography/mass spectroscopy analysis

Identification of the alkylation products was performed on Shimadzu GC/MS—QP2010. The GC fitted with PONA 50 m glass open tabular capillary column. The column temperature was programmed as an initial temperature of 30 °C for 15 min, then 60 °C for 20 min (heating rate 1 °C/min) and finally 200 °C for 20 min (heating rate 2 °C/min). The injector temperature was 250 °C. Flow rate of He (13 kPa, 155 ml/min) was applied: Ion source temperature 200 °C, interface temperature 250 °C and detector voltage 0.7 kV.

Results and discussion

Catalyst characterization

Physicochemical properties of BXE ALKCAT zeolite

Figure 3 shows SEM images of BXE ALKCAT zeolite. It can be seen from the figure that aggregates of spherical beads with irregular shape are prominent. The spherical do not have well-defined morphologies and most of the deformed spherical beads appear to be more than 1 μm in size.

BET surface area, pore volume and size of zeolite

N_2 adsorption/desorption analysis is a useful tool for examining textural characteristics of porous materials. The isotherms for BXE ALKCAT zeolite are displayed in Fig. 4.

The isotherm of the sample is a hysteresis loop at relatively high pressure (p/p_0), indicating that the crystal sizes are small and present a relatively high external surface area with little mesoporosity. The textural parameters of the corresponding sample have high BET surface area (228.5 m^2/g), pore volume (0.174 cm^3/g) and pore size (30.4 Å) which is attributed to porous kaolinite, where the pore of kaolinite can range very wide. The use of large pore kaolinite also helps the diffusion of the reactants, and dilutes the acidic sites, which is the goal to design the catalysts.

X-ray diffraction analysis

Figure 5 shows XRD pattern(s) of BXE ALKCAT zeolite. The high intensity of peaks in the XRD patterns indicated that the BXE ALKCAT zeolite samples are highly crystalline materials, in which the main phase is kaolinite and whose main diffraction peaks are seen at 2θ of 11° and 25°, which is in agreement with the catalyst composition. The rest main diffraction peaks all come from the calcined ZSM-5 sample. As the ZSM-5 only accounts for 30 wt% of the catalyst, its diffraction peaks are relatively weak.

X-ray photoelectron spectroscopy data of BXE ALKCAT zeolite

The XPS technique was applied to investigate the binding energies of the states of element between Si and O. The XPS data of curve fitting for the zeolite sample were recorded as shown in Figs. 6 and 7. All binding energy referred to C1s = 285.0 eV. Data for the elements detected were Si $2P_{1/2}$, and O 1s at binding energies of 102.1 and 532.1 eV.

It is shown that the surface compositions of the catalyst are mainly O, Si and Al, with Si as the dominant element, whose state is SiO_2. Aluminum existed as Al_2O_3 form. These results are in agreement with the composition of the

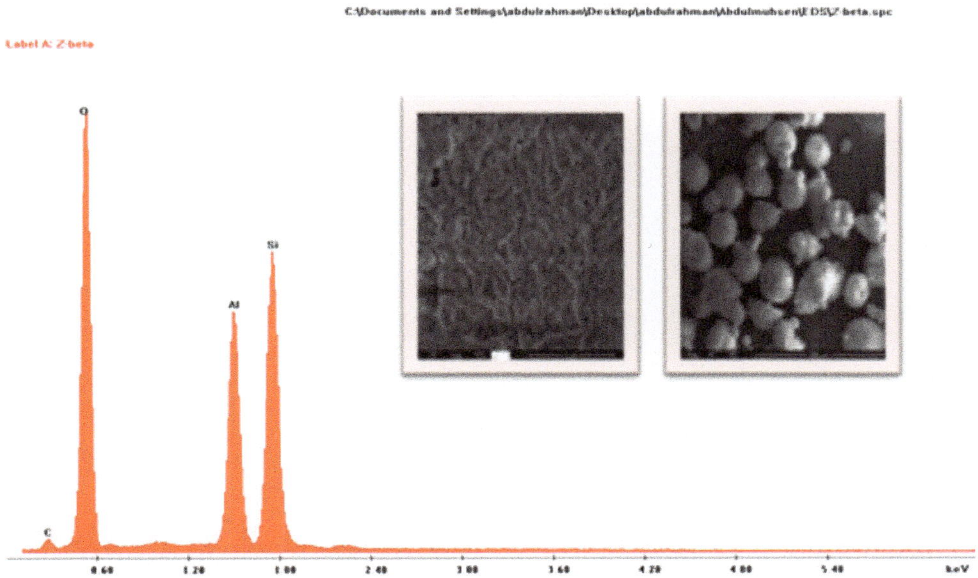

Fig. 3 SEM of BXE ALKCAT zeolites

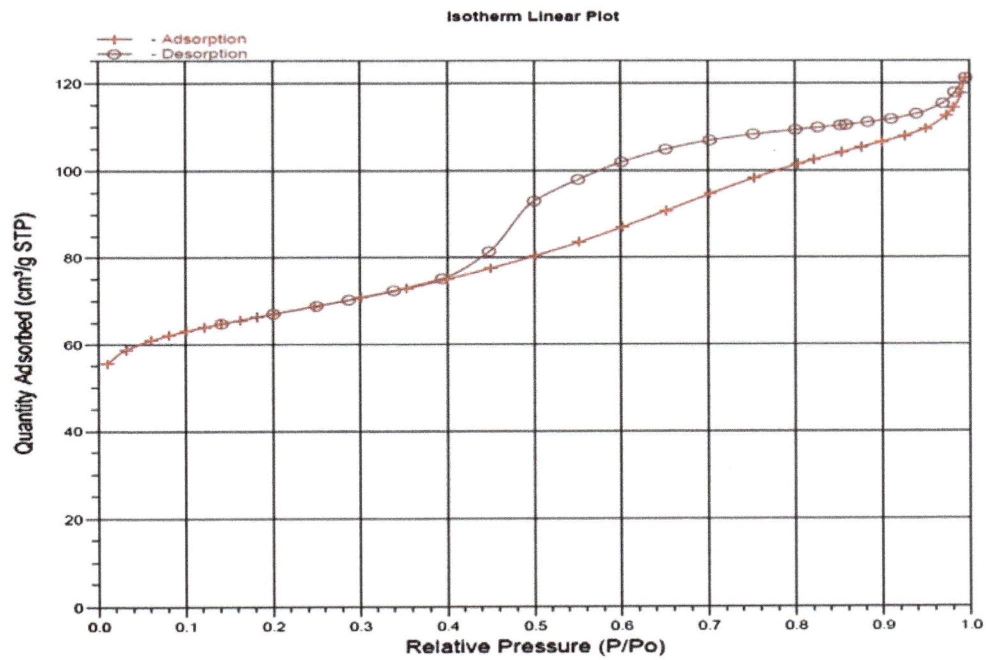

Fig. 4 BET surface area of zeolite

zeolite and kaolinite, and it is shown that the surface has less –OH groups, which is the main component of Bronsted acid site.

Differential thermal analysis of catalyst

In order to investigate the stability of zeolite catalyst at various temperatures, DTA was performed on zeolite by heating the sample up to 1,100 °C with a rate of heating 40 °C/min. The first sharp peak at 80 °C might result from the water or vapor desorption. The continuous endothermic curve is due to the loss of the hydroxyl group over the catalyst. Overall, it was found that the catalyst is thermally stable over a wide range of temperature ranging from 25 to 1,100 °C as shown in Fig. 8.

Catalyst evaluation

Effect of reaction temperature on benzene conversion

The alkylation of benzene with ethylene was carried out at 300, 350, 400 and 450 °C over BXE ALKCAT zeolite catalyst. The conversion of benzene and the product selectivity are presented in Figs. 9, 10, 11 and 12.

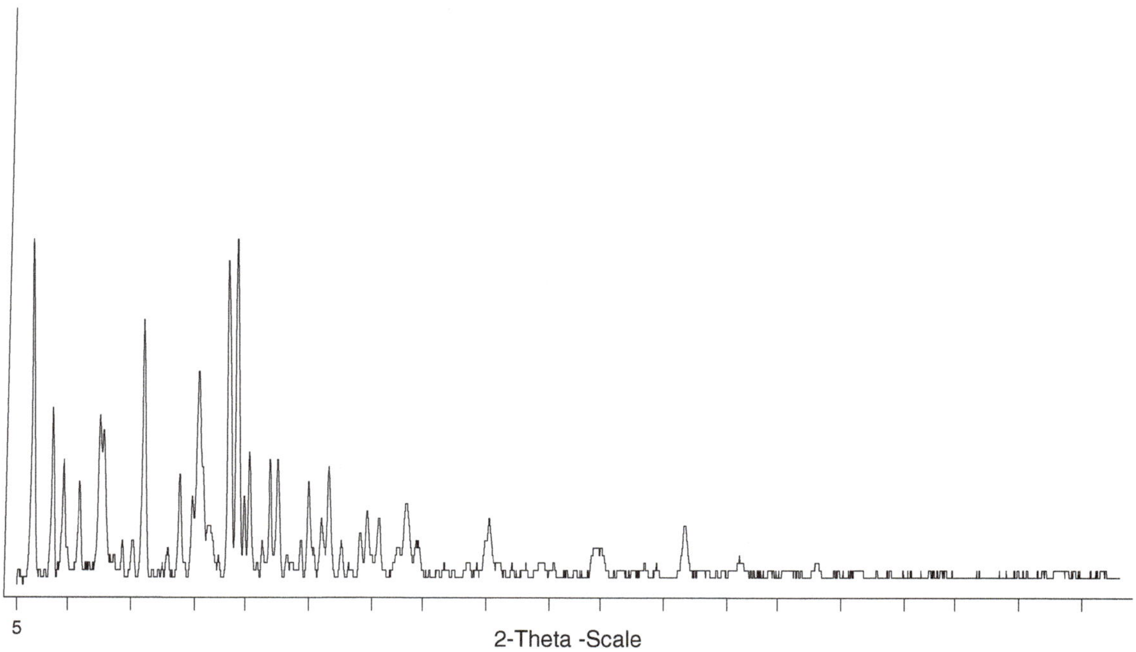

Fig. 5 XRD analysis of the ZSM-5/kaolinite catalyst for the alkylation reaction

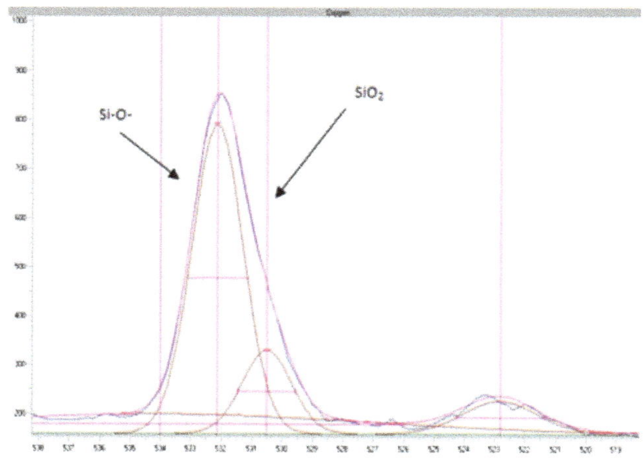

Fig. 6 Curve fitting XPS spectrum of O 1s of zeolite

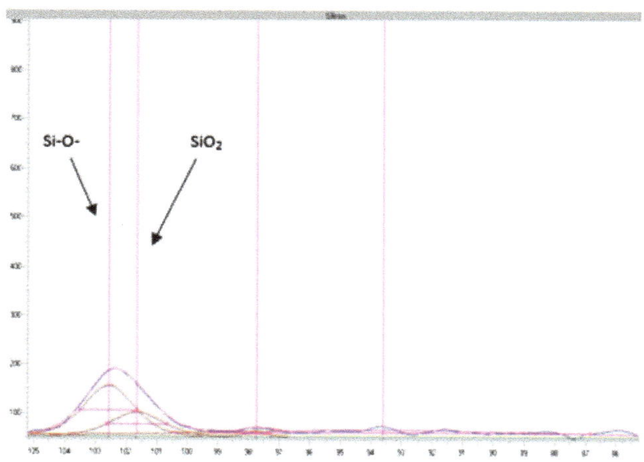

Fig. 7 Curve fitting XPS spectrum of Si $2P_{1/2}$ of zeolite

Conversion of benzene increases from 300 to 400 °C, and a slight decrease in the conversion of benzene was observed as the temperature was increased further up to 450 °C. Benzene conversions of approximately 6.5, 13, and 63 % were achieved at 300, 350 and 400 °C, respectively, while slightly decrease from 63 to 57 % as the temperature was increased from 400 to 450 °C, respectively, for a reaction time of 0.5 h, and mole ratio of BZ to E 1:1. This can be explained by the endothermicity of the alkylation reaction. High reaction temperature favors the conversion, and also the high temperature can increase the activity of the acidic site; thus, the reaction should be carried out at a relatively high temperature.

The effect of time-on-stream

The effect of time-on-stream on benzene conversion and EB selectivity was studied over BXE ALKCAT zeolite catalyst using a feed stock mole ratio ranging from 1:1 to 6:1 under variable temperatures. Reaction conditions and results are presented in Figs. 9, 10 and 11. It is shown that the BZ:E = 1:1, the BZ conversion drops more quickly than over the other catalysts, while practically no changes are observed in selectivity for EB and DEB isomers after about 2 h on stream. Increase of BZ:E decreases the benzene conversion, which is in fact due to the excessive feeding of benzene, to depress the carbon formation; but, the catalyst has a much stable conversion, which is due the depression of carbon formation by the benzene.

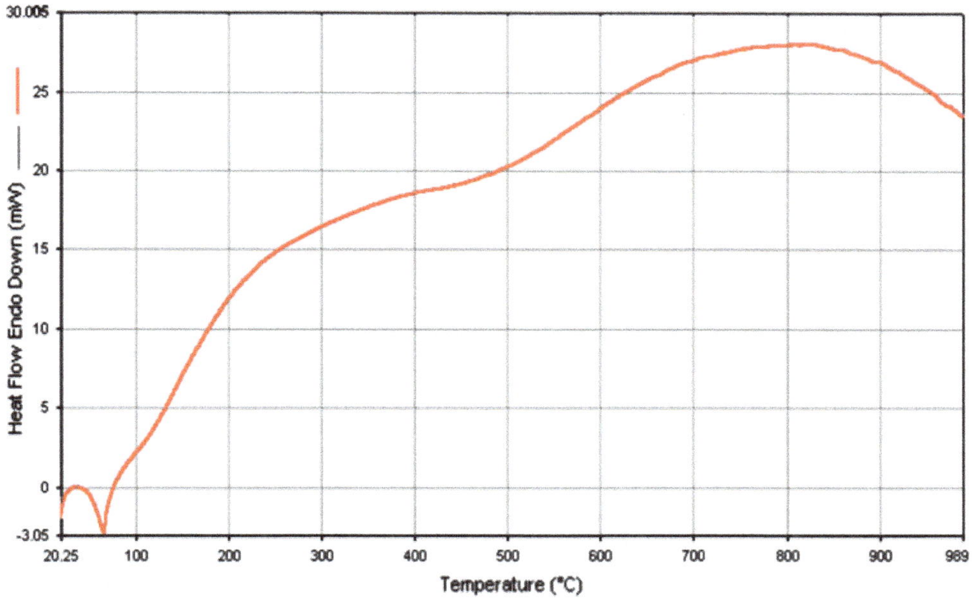

Fig. 8 Differential thermal analysis (DTA) thermogram of zeolite

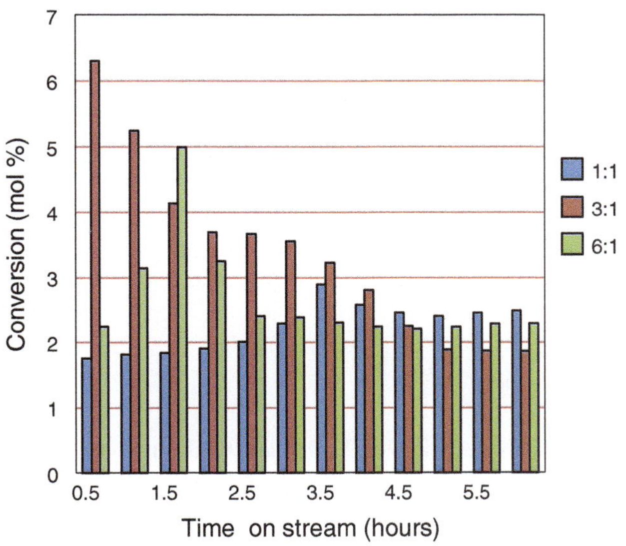

Fig. 9 Conversion of benzene with time for zeolite catalyst at 300 °C

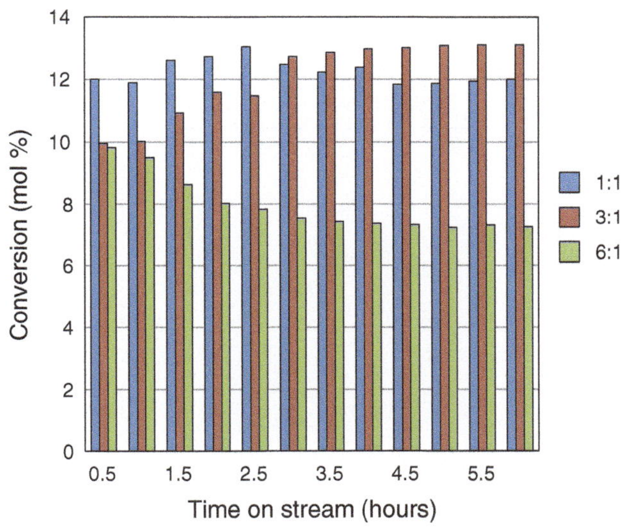

Fig. 10 Conversion of benzene with time for zeolite catalyst at 350 °C

Effect of reaction temperature on the yield and selectivity of ethylated products

The major product of the alkylation reaction over BXE ALKCAT zeolite catalyst is EB. The yield of EB increases as the temperature increases from 300 to 450 °C with different mole ratios of BZ to E: the reason has been given before. The maximum yield was obtained about 50 % with 1:1 molar ratio of BZ to E at 450 °C for the reaction time 0.5 h as shown in Fig. 13.

The maximum yield 17 % of DEB (*o*-, *m*-, and *p*-) isomers was obtained at 400 °C with mole ratio of BZ to E 1:1. There is variable trend of the formation of DEB isomers under the same mole ratio BZ to E and variable temperatures. The yield of DEB isomers increases as the temperature changes from 300 to 400 °C and then decreases as the temperature increased up to 450 °C as shown in Fig. 14. This is probably due to transalkylation of DEB isomers with benzene as shown in Scheme 1.

Based on these results, it is inferred that the increase of the reaction temperature makes more active sites available for the alkylation reaction, awhile more side reactions also occurs. However, overall, the EB yield increases with the temperature, due to high increase in the conversion rate.

Selectivity for EB was affected by temperature. The selectivity for EB decreased constantly as the temperature increased from 300 to 400 °C and then increases as the temperature increases up to 450 °C as shown in Fig. 15. The increase in selectivity is probably due to

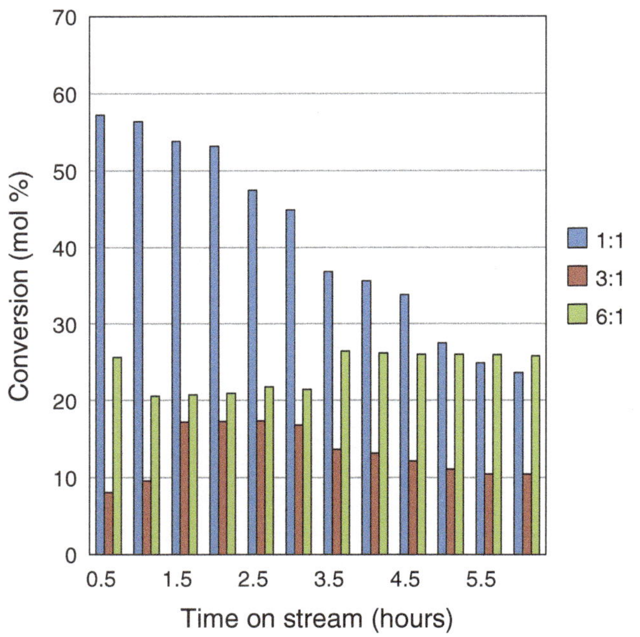

Fig. 11 Conversion of benzene with time for zeolite catalyst at 400 °C

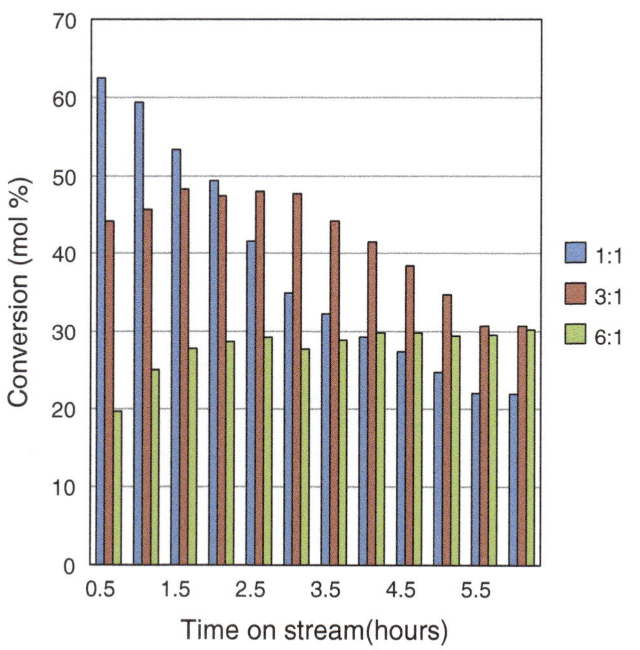

Fig. 12 Conversion of benzene with time for zeolite catalyst at 450 °C

transalkylation reactions between benzene and diethylbenzene (*o*-, *m*-, and *p*-) isomers as shown in Scheme 1.

The effect of feed stock mole ratio the yield and selectivity of ethylated products

The effect of feed stock mole ratio on ethylbenzene yield and product selectivity was studied by varying the feed mole ratios of BZ to E from 1:1 to 6:1 over BXE ALKCAT zeolite catalyst at variable temperatures, and the results are presented in Figs. 16, 17 and 18. The yield of ethylbenzene shows a significant decline as mole ratios of benzene to ethylene increased from 1:1 to 6:1 at temperatures of 300, 350, 400, and 450 °C. This is because of the increased dilution of ethylene by benzene. With lower ethylene content in the feed, the probability of ethylene interaction with benzene is reduced and a lower ethylbenzene yield is obtained. In addition, the excessive feeding of benzene would give a overall low conversion, even all the ethylene is consumed. The maximum of about 50 % yield was obtained with 1:1 molar ratio of BZ to E at 450 °C.

However, yield for the other alkylated products such as diethylbenzene (*o*-, *m*-, and *p*-) isomers decreased as the mole ratio of BZ to E increased. The increase in the yield of EB and diethylbenzene isomers results from a high conversion of BZ as the mole ratio of BZ to E decreases. At higher mole ratios of BZ to E, the accessibility of ethyl cations to yield diethylbenzene isomers by further alkylation of EB was reduced.

Selectivity for EB increased as the ratio of benzene to ethylene increased with lower conversion of benzene. The same trends were observed at temperatures of 300, 350, 400, and 450 °C. The maximum selectivity to EB (85 %) with higher conversion of benzene was obtained with a mole ratio of 1:1 at 450 °C, while 73 % with 1:1 at 300 °C.

Coke formation

In the alkylation of BZ with E, the deactivation of catalysts usually occurs as active sites are blocked by coke formation. The coking mostly results from the oligomerization of ethylene, because activated ethylene easily reacts with other ethylene molecules to form higher polymer. This is the reason the industry always use excessive benzene, which is to increase the ethylene benzene selectivity and depress the coke formation. The carbon balance was calculated as the following: (1) number of moles of carbon of benzene and ethylene in the feedstock, (2) number of moles of carbon of the unreacted ethylene, and reaction products (unreacted benzene, ethylbenzene, and *m*-, *p*-, *o*-diethylbenzenes), and (3) carbon deposited on the catalyst as obtained from TGA results (Table 1). It was found that the errors are ranging from 5 to ±10 %.

Conclusion

A kaolinite-supported ZSM-5 zeolite catalyst has been prepared using prilling process. The catalyst has stable

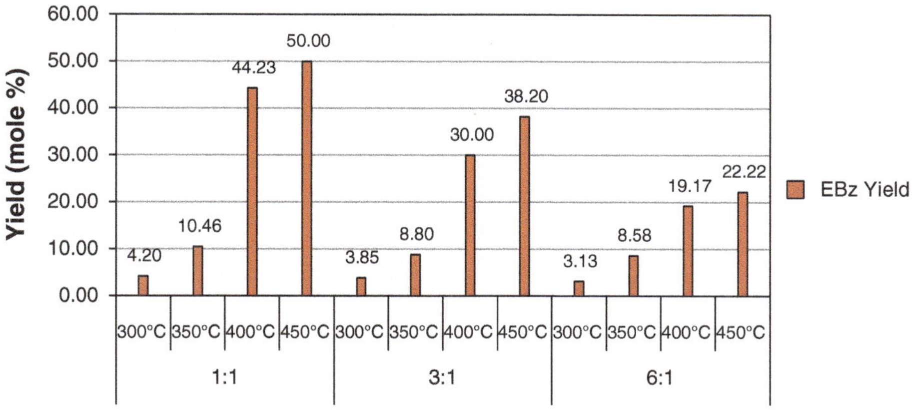

Fig. 13 Yield of EB at variable temperatures using different mole ratio of BZ:E under atmospheric pressure

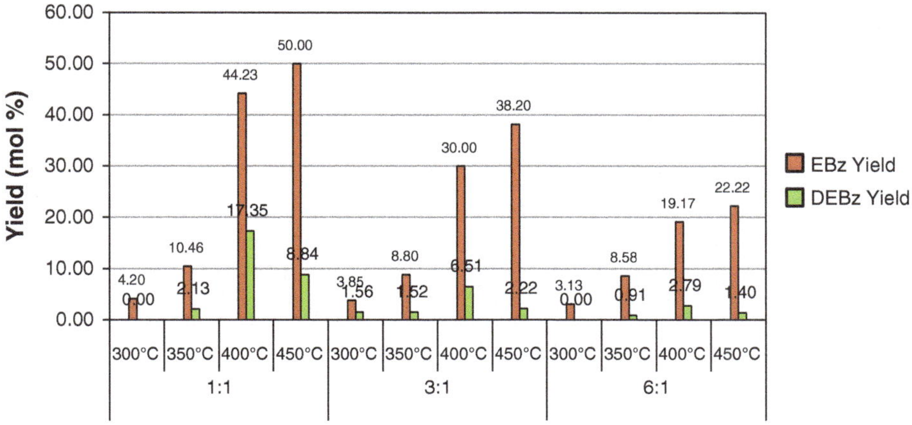

Fig. 14 Yield of EB and DEB at variable temperatures using different mole ratio of BZ:E under atmospheric pressure

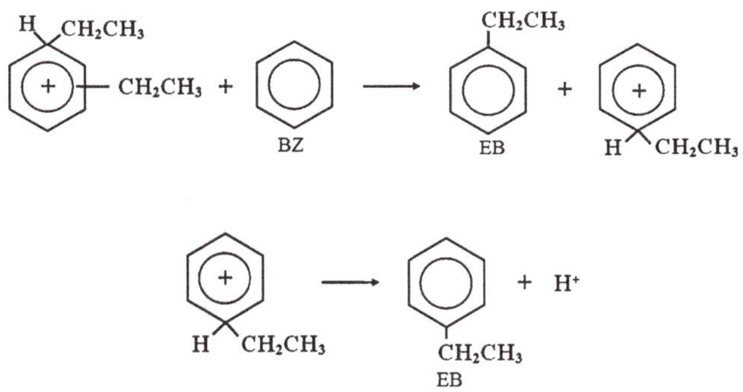

Scheme1 Transalkylation reaction of diethylbenzene isomers with benzene

structure, with surface mainly composed of SiO_2. The catalyst has been tested in the alkylation of the benzene with ethylene. It is found that the catalyst showed high selectivity to ethylbenzene at low temperature and high benzene to ethylene ratio.

The prepared catalyst has high stability at high benzene to ethylene ratio, and high yield of ethylbenzene. The increase of reaction temperature gives higher benzene conversion, but a low with more side products. Higher benzene to ethylene ratio helps to depress the side reaction, but lead to lower conversion of benzene due to the excess feeding.

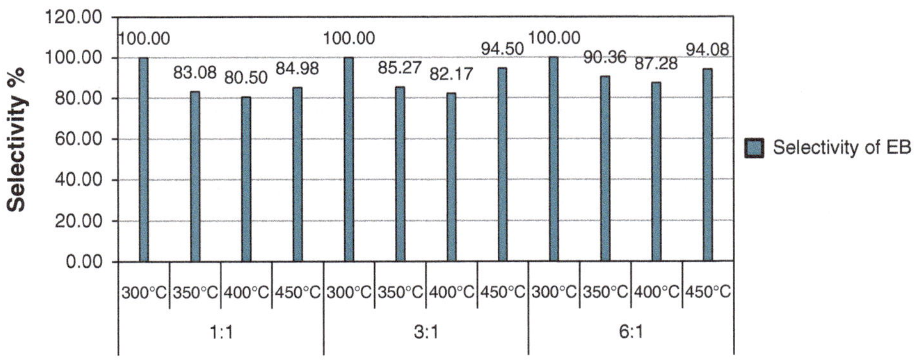

Fig. 15 Selectivity of EB at variable temperatures using different mole ratio of BZ:E under atmospheric pressure

Fig. 16 Yield of EB using different mole ratio of BZ:E at variable temperatures and under atmospheric pressure

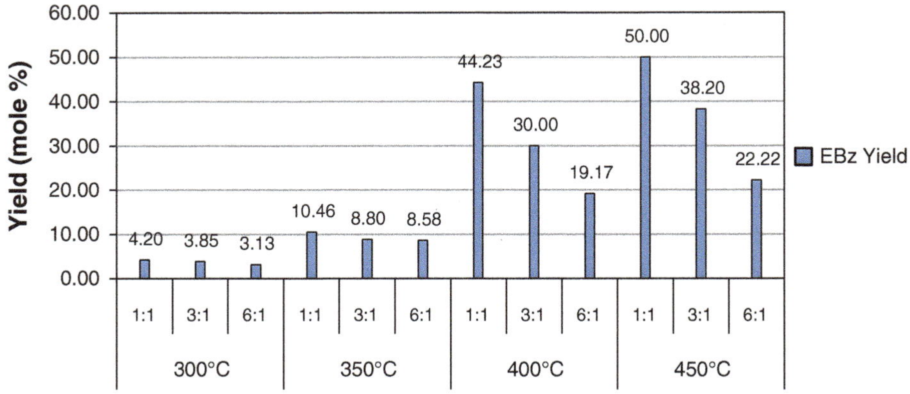

Fig. 17 Yield of EB and DEB using different mole ratio of BZ:E at variable temperatures and under atmospheric pressure

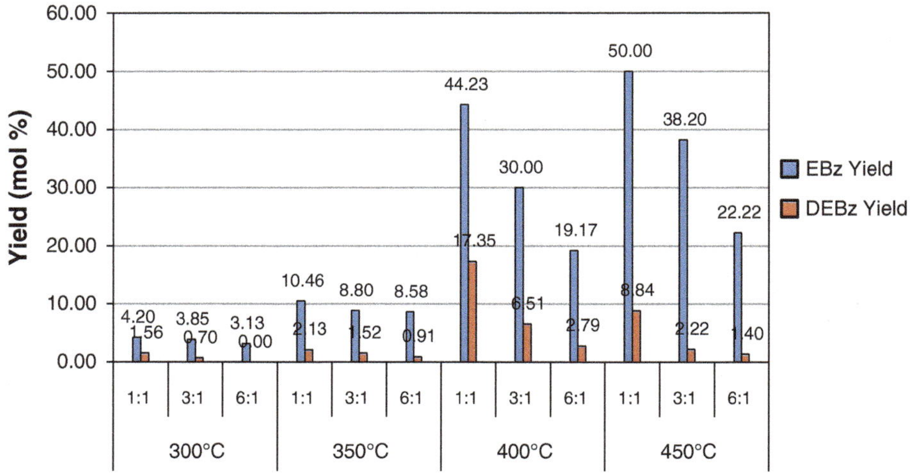

Fig. 18 Selectivity of EB using different mole ratio of BZ:E at variable temperatures and under atmospheric pressure

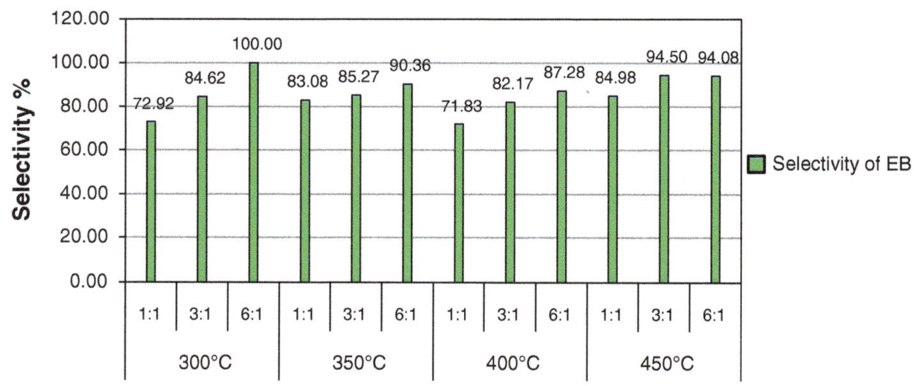

Table 1 Thermal gravimetric analysis (TGA) results for carbon deposition balancing with 6:1 mol ratio at variable temperature

	Starting weight (g)	Un-subtracted weight (%)	Final weight (g)	Balance (g)
Zeolite[a]	11.238	85.85130	9.648	1.59
Zeolite[b]	10.282	85.60697	8.802	1.48
300 °C; Zeolite[a] 6:1	10.658	83.47701	8.907	1.751
300 °C; Zeolite[b] 6:1	10.275	82.67406	8.497	1.778
350 °C; Zeolite[a] 6:1	10.366	84.58893	8.769	1.597
350 °C; Zeolite[b] 6:1	9.429	84.25536	7.955	1.474
400 °C; Zeolite[a] 6:1	11.191	88.01292	9.861	1.33
400 °C; Zeolite[b] 6:1	12.830	87.37062	11.217	1.613

[a] Heating rate: 10 ml/min
[b] Heating rate: 40 ml/min

There is carbon deposition occurred over the BXE ALKC at catalysts, which accounts for about 10 % of carbon, even the catalyst is still active, which suggest that the carbon may not be the poison for the catalyst active site.

The optimal operation conditions for the BXE ALKCAT are relatively high temperature 350–400 °C and high benzene to ethylene ratio to give high yield and high selectivity of ethylbenzene over the catalyst.

Acknowledgments We would like to thank Dr. Turki bin Saud bin Mohammad Al Saud, the Vice President for Research Institutes for his valuable support and funding the joint research. Also, I would like to thank the alkylation research team: Eng. Khalid S. Al-Ghamdi, Abdullah J. Al-Ghamdi, Sami D. Al-Dress, Waleed A. Al-Suwaylih and Sami D. Al-Zahrani.

References

1. Al-Kinany MC, Al-Khowaiter SH (1998) Ethylation and isopropylation of benzene using superacid catalyst. In: Proceedings of 15th World Petroleum Congress. vol 2, pp 830–834
2. Cejka J, Wichterlova B, Bednarova S (1991) Alkylation of toluene with ethene over H-ZSM-5 zeolites. Appl Catal A 79(2): 215–226
3. Hansen N et al (2008) Theoretical investigation of benzene alkylation with ethene over H-ZSM-5. J Phys Chem C 112(39): 15402–15411
4. Hansen N et al (2010) Quantum chemical modeling of benzene ethylation over H-ZSM-5 approaching chemical accuracy: a hybrid MP2:DFT study. J Am Chem Soc 132(33):11525–11538
5. Kartal OE, Onal I (2008) Synthesis of ZSM-5 from modified clinoptilolite and its catalytic activity in alkylation of benzene to ethylbenzene. Chem Eng Commun 195(8):1043–1057
6. Bassler EJ et al (1999) Upgrade by catalytic process technology. Hydrocarbon Eng 4(7):36–40
7. Dwyer FG (1981) Mobil/Badger ethylbenzene process—chemistry and catalytic implications. Chem Ind (Dekker) 5:39–50
8. Ebrahimi AN et al (2011) Modification and optimization of benzene alkylation process for production of ethylbenzene. Chem Eng Process 50(1):31–36
9. Hansen N et al (2009) Analysis of diffusion limitation in the alkylation of Benzene over H-ZSM-5 by combining quantum chemical calculations, molecular simulations, and a continuum approach. J Phys Chem C 113(1):235–246
10. Liu J et al (1999) The carbon depositing behavior and its kinetic research in benzene alkylation process over high silicate ZSM-5 zeolite. J Therm Anal Calorim 58(2):375–381
11. Song Y et al (2006) Coke burning behavior of a catalyst of ZSM-5/ZSM-11 co-crystallized zeolite in the alkylation of benzene with FCC off-gas to ethylbenzene. Fuel Process Technol 87(4): 297–302
12. Hassan M, El-Shall H (2009) Texture and microstructure of thermally treated acid-leached kaolinitic clays. Adsorpt Sci Technol 27(7):671–684
13. Suraj G et al (1997) The effect of micronization on kaolinites and their sorption behavior. Appl Clay Sci 12(1–2):111–130
14. Westlake DJ, Atkins MP, Gregory R (1985) The use of layered clays for the production of petrochemicals. Acta Phys Chem 31(1–2):301–308

9

The time-on-stream stability of some selected bifunctional nanoporous-based catalysts in *n*-heptane hydroisomerisation

Faisal M. Alotaibi · Raed H. Abudawood · Hamid A. Al-Megren · Mohammed C. Al-Kinany · Arthur A. Garforth

Abstract In this work, some commercial nanoporous-based catalysts, such as USY, beta and mordenite zeolites loaded with Platinum metal acting as bifunctional catalysts, were used for hydroisomerisation experiments in a fixed-bed reactor at the atmospheric pressure and at feed space time 5.12 h^{-1} to hydroisomerise *n*-heptane over a temperature range of 210–270 °C. The study aimed to evaluate the changes with time-on-stream in the catalytic activity, product selectivity and manner of deactivation of metal-loaded zeolite catalysts, at constant contact time of 5.13 h^{-1} and a hydrogen-to-*n*-heptane molar ratio fixed at 9. Various analytical techniques were used to characterise fresh and aged catalysts. Results show that pore architecture is the most important factor affecting coke formation and deactivation in zeolite catalysts, and those catalysts with three-dimensional pore structures lacking cavities or cages were best able to resist deactivation. In addition, it was found that those catalysts with high Si/Al ratios and those which had been acid-leached or steamed showed better activity, higher selectivity towards isomeric products and better time stability. Moreover, the balance between the number of metal sites and the number of acid sites played an important role in determining the activity, selectivity and stability of the bifunctional catalysts.

Keywords The time-on-stream stability · Alkanes hydroisomerisation · Catalyst deactivation · Zeolites · Bifunctional catalysts · Nanoporous catalysts

Introduction

Catalysts that are being tested for industrial processes must adhere to strict criteria for the reaction which is being targeted. For example, they need to have high levels of both activity and selectivity, together with a long period of stability. One of the most common problems which occur is deactivation at some point during the process, which can reduce the levels of activity, of selectivity, or of both. One of the possible contributory causes is carbonaceous deposits over the catalyst, which is quite common in hydrocarbon conversion processes. The type of catalyst, in conjunction with the reaction conditions, will affect the success of the catalyst. Its lifetime is a crucial feature when working with zeolites, and extending the catalyst lifetime from minutes to days will improve success rates dramatically [1, 2]. The cost of catalyst deactivation is very high in industrial processes; therefore, catalysts with superior time stability are very valuable. Large pore zeolites, especially high-silica zeolites, are attractive catalysts for reforming-type reactions, and the time stability of these catalysts would be one of the important factors in determining their use in such reactions [2].

Despite extensive studies over the last 20 years into the formation and nature of coke on zeolites, there is still much discussion and controversy on many questions. Rollmann and Walsh were among the first to show that coke

F. M. Alotaibi (✉) · H. A. Al-Megren · M. C. Al-Kinany
Petrochemicals Research Institute, King Abdulazeez City for Science and Technology (KACST), PO Box: 6086, Riyadh 11442, Saudi Arabia
e-mail: alotaibi1975@gmail.com; fmsalotaibi@kacst.edu.sa

R. H. Abudawood
Research and Development Center Saudi Aramco, Dhahran 31311, Saudi Arabia

A. A. Garforth
School of Chemical Engineering and Analytical Science, The University of Manchester, Manchester M13 9PL, UK

formation is a shape-selective reaction. They noted that coke yield in molecular shape-selective zeolites (small and intermediate pore) is at least an order of magnitude lower than in the large pore materials [3, 4]. However, the effect of zeolite pore structure is concerned with more than just steric restrictions on the formation of coke precursors. The contact time of the organic molecule with the active sites is determined by the rate of diffusion of these molecules and therefore by the characteristics of the diffusion path inside the zeolite crystallites: the length related to the crystallite size, the size of the pore apertures, the size of the channel intersections and the acid site density. The choice of pore structure of the zeolite must therefore ensure that the spaces in the vicinity of the acid sites are big enough to permit the intermediates of the desired reaction to form yet small enough to limit the formation of coking intermediates by strict constraints. In addition, the diffusion of the desirable molecules must be quick enough for the reactant transformation to be restricted to the formation of the desired product [5, 6]. The study by Gopal and Smirniotis demonstrated that three-dimensional zeolites without supercages, such as zeolite beta, when they have the correct level of acid site density and metal loading, are ideal catalysts for reforming-type reactions, due to their time stability. Catalyst performance was greatly affected by acid site density and the researchers were able to identify an optimum range for obtaining good stability [2, 7]. Zhang and Smirniotis found that the zeolite pore structure played a very important role in the reforming of industrial naphtha (rich in naphthenes) and that an appropriate combination of the zeolite pore structure and its acidity, controlled via dealumination, gave superior time-on-stream stability for the reactions investigated. Thus, zeolites with channel intersections (cavities) of comparable size with the zeolite apertures did not favour coke formation. Therefore, they found that the dealuminated ZSM-12 demonstrated unique TOS stability for reforming types of reactions and concluded that this behaviour resulted from (a) its pore structure, which does not favour coke formation, and (b) the balance of its acidity. In terms of the zeolite structure and its effect on coke formation, there are two extreme cases, in the first case, coke precursors can be trapped and enlarged in the supercages, resulting in a blocking of the pores, since the zeolite has relatively large cavities or supercages in comparison with the opening of the channels. Whereas for the other type of zeolites, the channel intersection is comparable in size with the opening of the zeolite aperture, making it difficult for coke precursors to accumulate and form further large molecules in the channel, because of the space constraint [8]. The effects of acidity and of the pore structure are difficult to tell apart, but as a rule, zeolites with higher acid site densities deactivate faster as a result of higher conversion severity, which ultimately gives higher coke yields. In reforming reactions, when carried over bifunctional catalysts, the coke precursors are hydrogenated over the metal function and become desorbed. This is advantageous because the coke production rate is much lower than that over an acidic catalyst [9].

At present, as new gasoline standards require the content of aromatics and olefins to be reduced, the hydroisomerisation of paraffins with chain lengths above C6 is extremely advantageous. For this reason, much research has been done to find effective catalysts which can encourage the selectivity of branched paraffins in the hydroisomerisation of these paraffins. *n*-C7 is one of the most important fractions of naphtha; due to its low octane number, it is separated and diverted to a continuous catalyst regeneration (CCR) unit, where hydrocracking occurs under severe conditions, resulting in the production of some light gaseous hydrocarbons. Thus, the hydroisomerisation of *n*-C7 to mono- and multibranched isomers is an important area which merits research. The reactions of hydrocarbons over acidic zeolite catalysts are accompanied by the formation of coke, which causes catalysts to deactivate and leads to changes in product distribution [10–13].

In this work, the time-on-stream (TOS) behaviour and deactivation rates of some commercial USY, beta and mordenite zeolites that have been loaded with Platinum metal, with a wide range of Si/Al ratios, pore sizes and acidity characteristics, are evaluated in the hydroisomerisation of *n*-C7 when reaction conditions such as temperature, pressure and contact time are varied. *n*-C7 was used as a representative model compound for naphtha. A range of techniques was used to characterise fresh and aged catalysts under the conditions of each reaction.

Experimental

Catalyst preparation

Some commercial USY, beta, and mordenite samples were provided by different suppliers [14–18], as shown in Table 1. These zeolite supports were designed and modified by various post-synthesis techniques so that they could be used for multiple purposes; for instance, high alkane hydroconversions such as cracking, isomerisation and alkylation. They were supplied in powder form without binder materials.

The catalysts listed in Table 1 were loaded with 1 wt% platinum via the ion exchange technique in their ammonium form using an amino complex of platinum. This technique was used by Weisz and Frillete [19] to exchange platinum into zeolite A. In the loading procedure, 1 g of each catalyst in the ammonium form was mixed with 0.0175 g of

Table 1 List of catalyst supports investigated in the TOS study

	Zeolite trade name	Zeolite type	Post-synthesis modification applied	Supplier
1	CBV712	USY	Steamed and acid-leached	Zeolyst International
2	CBV720	USY	Severely steamed and acid-leached	Zeolyst International
3	USY-A	USY	Mildly steamed	Crosfield
4	CP814E	BEA	As-synthesised	Zeolyst International
8	CBV21A	MOR	As-synthesised	Zeolyst International
9	640HOA	MOR	Mildly steamed and acid-leached	Tosoh Corporation
10	660HOA	MOR	Moderately steamed and acid-leached	Tosoh Corporation
11	690HOA	MOR	Severely steamed and acid-leached	Tosoh Corporation

tetramine platinum (II) chloride [$Pt(NH_3)_4Cl_2 \cdot H_2O$] (Aldrich, 99.5 % purity), and dissolved in 50 ml of deionised water, then the pH of the mixture was adjusted to more than 9 by adding ammonia solution (32 %, Aldrich) such that it resulted in 1 wt% platinum loading on the zeolite sample, assuming a complete uptake of platinum in the catalyst. The mixture was left stirring at room temperature for 24 h, after which it was centrifuged and washed with 10 times its volume of deionised water to ensure that the catalyst would be free of any residual chloride ions. The washed catalyst was then dried overnight at 110 °C. Elemental analysis was used to confirm loadings at 1 wt% and that the samples were free of chloride ions.

Catalyst characterization

Characterization techniques used were: ICP optical emission spectroscopy (ICP-OES), X-ray diffraction (XRD), ^{27}Al and ^{29}Si MAS-solid-state NMR, pyridine adsorption Fourier transform infrared spectroscopy, ammonia temperature programmed desorption (NH_3-TPD), scanning electron microscopy and energy dispersive X-ray analysis (SEM), transmission electron microscopy (TEM), X-ray photoelectron spectroscopy (XPS), hydrogen chemisorption of platinum, BET surface area analysis by nitrogen adsorption, and thermogravimetric analysis (TGA).

Catalyst loading and activation

Catalysts were pelletized and sieved to obtain particles in the size range of 250–420 μm prior to loading into the reactor. A sample (1 g) of catalyst was sandwiched in the reactor between two layers of inert glass beads held in place by glass wool to form the catalytic bed. The catalyst was activated in situ by first removing ammonium via calcination with flowing air (50 ml min^{-1}) at 500 °C for 2 h at a heating rate of 3 °C min^{-1}. Following the removal of the ammonium, the catalyst was then reduced in a hydrogen (100 ml min^{-1}) atmosphere at 450 °C for 4 h, which resulted in the conversion of platinum ions into their metallic form. As fast heating rates can cause agglomeration of platinum ions, which leads to a poor catalyst performance, a slow temperature ramping rate of 3 °C/min was used during calcination.

Catalyst testing

n-Heptane (*n*-C7) with >99 % purity (Sigma-Aldrich) was fed to the catalyst bed using an HPLC pump, and hydrogen of high purity was introduced to give an H_2/*n*-C7 molar ratio of 9. The deactivation behaviour of zeolite catalysts was investigated during the catalytic conversion of *n*-C7 at reaction temperatures of 210, 230, 250 and 270 °C, with fresh catalyst was used for each run at each of the reaction temperatures and at atmospheric pressure using the glass fixed-bed flow reactor system loaded with 1 g of 1 % Pt loaded catalyst, and weight hourly space velocity (WHSV) was 5.13 h^{-1}.

Most of the deactivation runs for each catalyst lasted for at least 5 days. All gas and liquid samples were collected once a day, except in some cases where the sample could be collected more frequently. Gas samples were collected in a 1-L gas bag and the gas outlet flow rate was measured regularly using a bubble meter in order to perform a mass balance on the unit. The liquid samples collected were analysed by injecting 0.1 μL into a Varian 450 GC. After each reaction, the reactor was reduced to ambient temperature and purged with nitrogen overnight to remove any residual *n*-C7. The used or spent catalysts were collected after the reaction and stored in sealed containers for further analysis. Products were analyzed using a Varian 450 GC equipped with a 50 m × 0.32 mm i.d. PLOT Al_2O_3/KCl capillary column fitted to an FID detector for gas products and a CP-Sil PONA CB optimised gasoline column fitted to an FID detector for liquid products.

Results and discussion

Deactivation behaviour of CBV712 USY zeolite

CBV712 USY is produced via mild acid leaching of CBV600 USY which is prepared by steaming of CBV300 (Na-Y) as the starting material and which has been ammonium-exchanged for over 80 % [20]. Figures 1 and 2 show, respectively, the XRD pattern and SEM microgram of CBV712 USY.

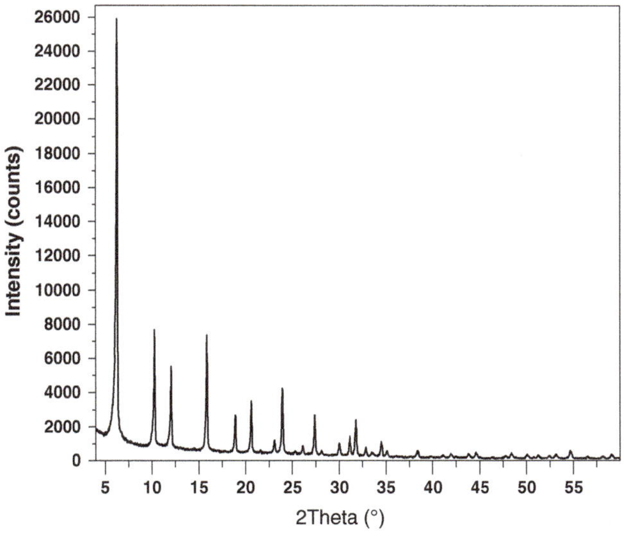

Fig. 1 XRD pattern of CBV712 USY

Fig. 2 SEM micrograph of CBV712 USY

The XRD pattern is consistent with that reported in the literature for ultrastable zeolite Y phase [21], thus confirming the USY phase. The SEM image shows sharp-edged cubic crystallites of zeolite Y phase. These varied in size from 1 to 3 μm and were combined with an amorphous phase which normally surrounds the zeolite Y crystals after the dealumination process. This may be a consequence of the acid leaching treatment which followed dealumination and resulted in USY zeolite with less extra-framework aluminium (EFAL) species (the amorphous phase).

Figures 3 and 4 show solid-state NMR spectra for ^{27}Al and ^{29}Si, respectively. In ^{27}Al MAS NMR analysis of the fresh CBV712 USY sample, as delaminated and partially acid-leached zeolites, Fig. 3 shows unique resonance

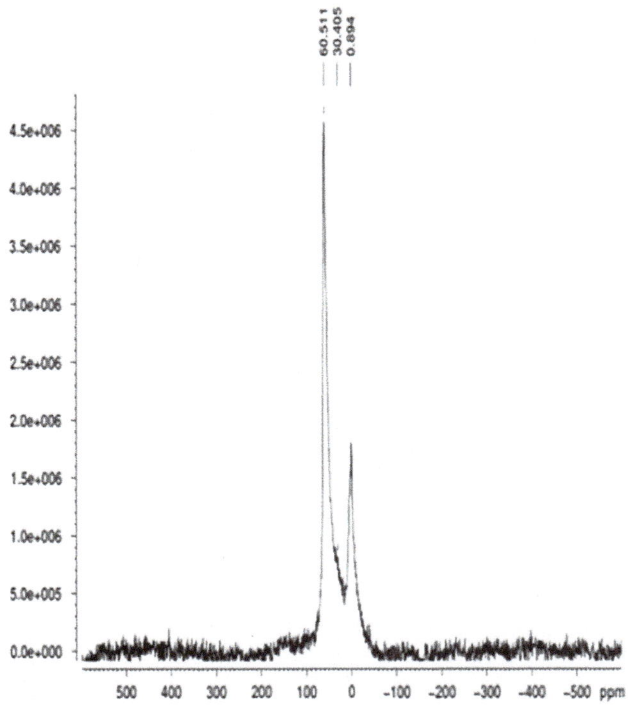

Fig. 3 ^{27}Al MAS NMR spectra of CBV712

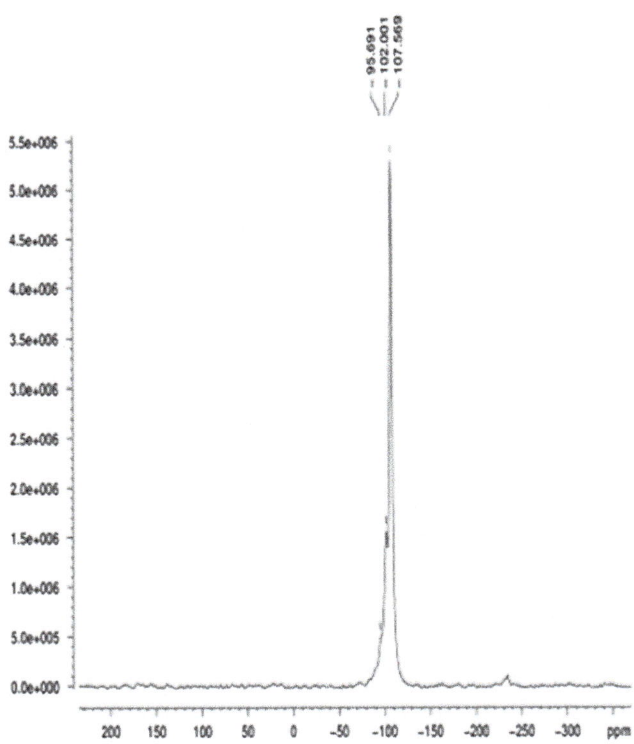

Fig. 4 ^{29}Si MAS NMR spectra of CBV712

spectra for three different chemical environments of aluminium. The first environment is assigned to in-framework aluminium in tetrahedral coordination $Al(OSi)_4$ with a chemical shift of approximately 60 ppm. The second

Table 2 In-house characterisation of CBV712 USY

	Before ion exchange				After ion exchange with Pt precursor as tetramine platinum (II) chloride ($Pt(NH_3)_4Cl_2)\cdot H_2O$			
Elemental analysis (bulk)	Si/Al mole ratio		5.93		Si/Al mole ratio	5.93	Pt wt%	0.98
XPS[a] atomic % conc	Si	Al	Si/Al	Pt	Pt was found in its metallic form on the surface			
	26	4.27	6.09	0.04				
XRD	Crystallinity				Related to in-house as-synthesised Y			
	87.50							
NMR	Si/Al mole ratio for the framework				Three chemical environments for Si-NMR: Si(0Al), Si(1Al), and Si(2Al)			
	8.34				Lewis acid sites are existed: $Al(OSi)_4$, $(Al(H_2O)_5)^{3+}$, and $(Al(H_2O)_6)^{3+}$			
Ammonia TPD[b]	NH_3 desorption through TPD/mmol g^{-1}				Ammonia adsorption calorimetry followed by temperature programmed desorption			
	0.70 (±0.01)							
Hydrogen chemisorption[b]	Metal dispersion %	Metallic surface area $m^2\ g^{-1}$ metal			Metallic surface area $m^2\ g^{-1}$ sample		Crystallite size nm	
	38.44	94.95			0.93		2.95	
BET[b]	Surface area $m^2\ g^{-1}$	Pore volume $cm^3\ g^{-1}$			Pore size Å		Shows both micro and mesoporosity	
	816.26	0.45			42.75			
TGA for coke content in spent catalyst (mass loss wt% in air) after various reaction temperatures	210 °C	Top		0.30	250 °C		Top	1.47
		Mid		0.12			Mid	1.26
		Bot		0.12			Bot	0.88
	230 °C	Top		0.23	270 °C		Top	4.89
		Mid		0.17			Mid	4.78
		Bot		0.12			Bot	4.87

[a] LPD Laboratory Services Ltd.

[b] MCA Services

environment is assigned to the five-coordinated EFAL with a chemical shift of about 30 ppm. The third environment is assigned to non-framework aluminium in octahedral coordination with a chemical shift of around 0 ppm. Information about the Si atoms coordinated (via oxygen) with neighbouring T atoms (Si, Al) is provided by ^{29}Si MAS NMR. Figure 4 shows the ^{29}Si MAS NMR spectra of CBV712 USY. The ^{29}Si MAS NMR spectrum exhibits peaks at around −107, −102 and −95 ppm which were assigned to different configurations of SiO_4 (4Si,0Al), SiO_4 (3Si,1Al), SiO_4 (2Si,2Al), respectively. Table 2 shows the in-house characterisation of CBV712 USY using various techniques.

Figure 5a–d illustrates the effects of TOS on the C7 conversion, hydroisomerisation, hydrocracking and cyclic product yields at reaction temperatures ranging from 210 to 270 °C.

Initially, 30 min after introducing the feed to the reactor at temperatures of 210, 230, 250 and 270 °C, the overall conversion was 20, 47, 75 and 88 wt%, respectively. At the lowest temperature of 210 °C, the catalyst initially lost about 3 wt% of its activity, then, after 60 h on stream, a constant value was achieved. On the other hand, the deactivation rate at 230 °C for the fresh catalyst was slightly more rapid, as shown in Figs. 4, 5 and 6a. The overall conversion 30 min after introducing the feed to the reactor was 47 wt%, which decreased after 240 min on stream to 45 wt%, then decreased slowly, with the deactivation rate ranging between 2 and 4 wt% every 24 h until the reaction terminated after about 3 days (4,440 min). At the higher temperatures of 250 and 270 °C, it showed a rapid fall in catalyst activity during the initial few hours, followed by a constant deactivation rate being reached, which ranges from 3 and 5 wt% every 24 h until the reaction terminated.

Selectivity towards the hydroisomerisation products ranged from 99 to 100 % during the deactivation run at the lower temperatures of 210 and 230 °C, from the introduction of the feed to the reactor until both reactions terminated. On other hand, selectivity towards cracked and cyclic products was very low and fluctuated in a very narrow range of 0.5 %, as a function of TOS. The initial selectivity towards the hydroisomerisation products was 95 and 77 % at 30 min after introducing the feed to the reactor, which increased throughout the TOS to 98 and 97 wt% after 720 min at temperatures of 250 and 270 °C,

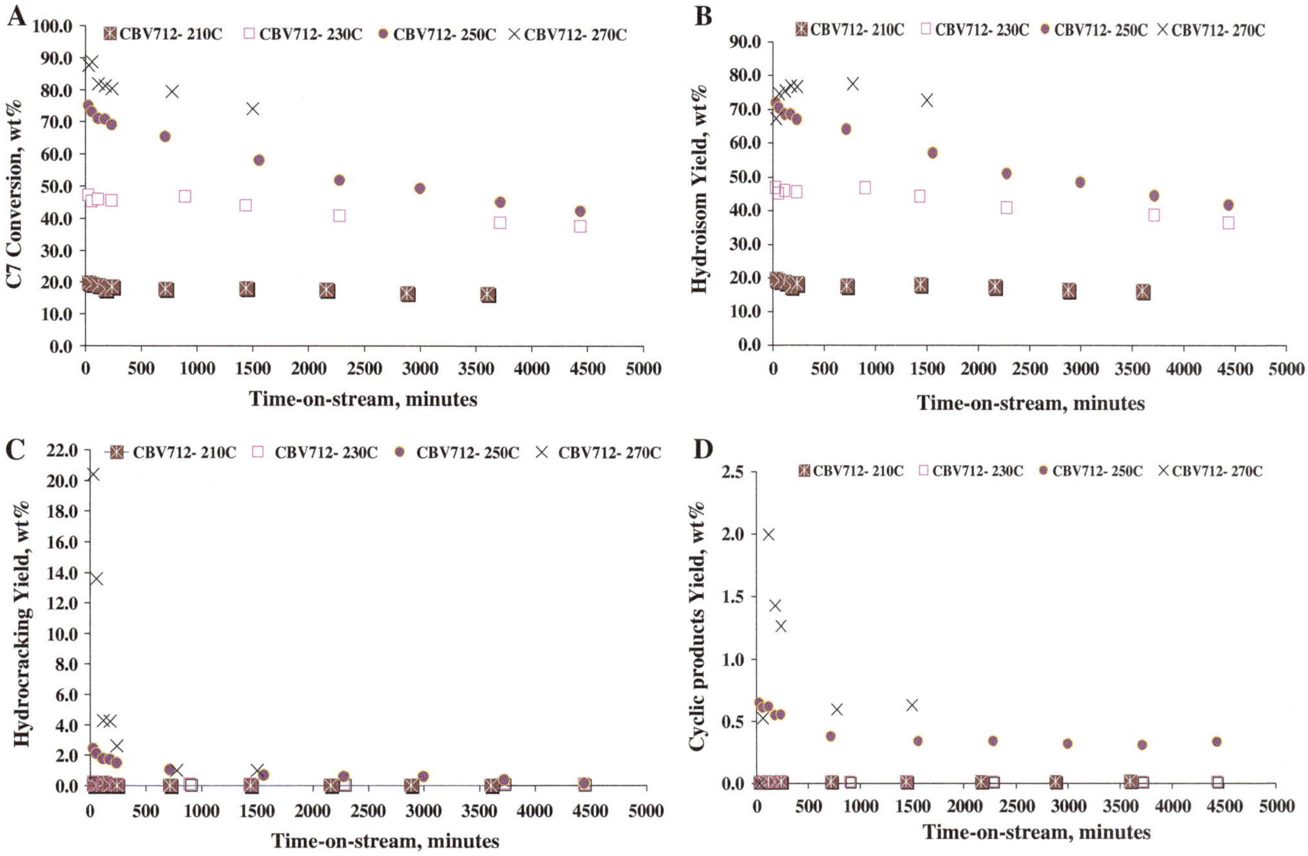

Fig. 5 Overall conversion (**a**), total yield of C7 isomers (mono + di + tribranched isomers) (**b**), total hydrocracking yield (**c**), and total yield of the cyclic products (**d**) as a function of TOS for CBV712 USY at various reaction temperatures. WHSV = 5.13 h^{-1}; H_2/C7 (mol/mol) = 9; total pressure = 1 atm; temperature = 210, 230, 250 and 270 °C; 1 wt% Pt loading

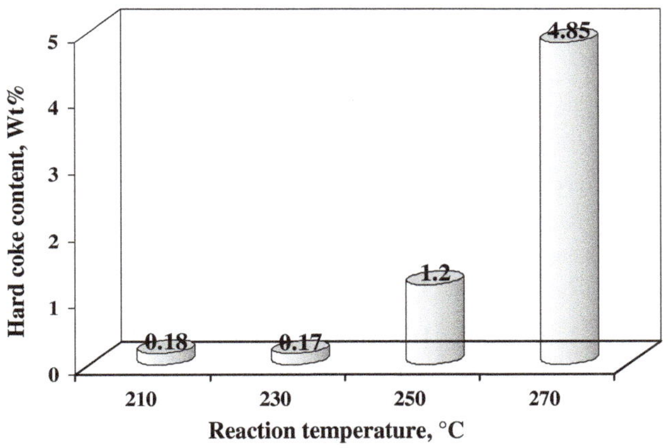

Fig. 6 Average hard coke content over the aged CBV712 USY bed

respectively. Meanwhile, selectivity towards cracked products was quite high at the higher temperatures during the initial few hours of the deactivation reaction. The highest cracked product selectivity was 23 % at 270 °C, 30 min after the introduction of the feed to the reactor, after which it decreased rapidly throughout the TOS to 3 wt% after 240 min. The yield of cyclic products at 250 and 270 °C was higher than at lower temperatures.

It can be observed that the total conversion increased dramatically when increasing reaction temperature in the range of 210–270 °C, with the maximum isomer yield obtained with the CBV712-based catalyst was around 77 wt% at 270 °C, with the predominant conversion being hydroisomerisation. However, the selectivity for C7 isomers at lower temperatures was very high throughout the deactivation reactions of 210 and 230 °C, yielding a greater proportion of branched C7 as a result of the thermodynamic equilibrium. The selectivity of cracked products initially increased at higher temperatures (250–270 °C) during the first few hours of the deactivation reaction, then decreased rapidly throughout the TOS, whereas the selectivity of cyclic products was slightly higher at 250 and 270 °C than at the lower temperatures of 210 and 230 °C.

The CBV712 was steamed and acid-leached to ensure the microporosity and mesoporosity of the networks within its structure; resulting in a high surface area and pore volume, as shown in Table 2. This resulted in a rapid loss in C7 hydroisomerisation conversion activity with TOS, particularly at higher temperatures, that can be explained by considering the Brønsted and Lewis acid sites of the CBV712. The Brønsted acidity is predominant, with acid sites within its structure being stronger, resulting in the majority of the aluminium ions incorporated in a tetrahedral coordination, as seen in the ^{27}Al MAS NMR and TPD analysis. In addition, it can be concluded that the catalyst CBV712 has a high acid site density and a mild strength of acidity due to the existence of both types of acid sites, resulting in the catalyst exhibiting high activity, particularly at higher temperatures, resulting in increased activity for the cracked products as coke precursors in the initial few hours of the deactivation run, ultimately leading to coke formation, which preferentially deactivated the strongest acid sites which contribute to the cracking reaction.

Figure 6 shows the TGA results of the average of the hard coke content deposited over the top, middle and bottom zones of the aged CBV712 USY catalyst bed after unloading the spent catalyst from the atmospheric glass reactor. It can be clearly seen that the top zones had the highest hard coke content in the catalyst bed at each reaction temperature, whereas there was slightly less coke in the middle and bottom zones. This suggests that the top zone of the catalyst bed contributed the most activity and was where the reaction took place. In addition, the high coke content in the top zone is to some extent indicative of the high acidity of this catalyst, affecting its stability and thus increasing the rates of deactivation and coking. It is evident that there is a clear correlation between the temperature and the yield of cracked products obtained, which relates to the existence of stronger acid sites and to the hard coke content deposited in the catalyst cages at different reaction temperatures. The maximum hard coke value of 4.85 wt%, which was found at the highest reaction temperature of 270 °C, was 27 times higher than that at the lower temperatures of 210 and 230 °C.

Research has shown that the ratio between the number of Pt atoms (n_{Pt}) and the number of acid sites (n_A) plays an important role in determining the activity, selectivity and stability of the bifunctional catalyst, which must be well-balanced to behave as an 'ideal' hydroconversion catalyst. It has been found that the optimum limits for the n_{Pt}/n_A ratio range between 0.03 and 0.17, and the greater the value of n_{Pt}/n_A, the smaller the number of chemical steps by which the cracked products can be generated and then oligomerated to form the coke molecules, hence, the slower the coke formation and the higher the catalyst stability [22–24]. The n_{Pt}/n_A ratio for CBV712 was calculated to be 0.072, which is within the optimum limits specified above, although its time stability was not quite as good at 250 and 270 °C. A possible explanation for this is that the above suggestion is not empirical in all reaction conditions and that it can be proven that CBV712 did not behave as an ideal hydroconversion catalyst at higher temperatures,

Table 3 In-house characterisation of CBV720 USY

	Before ion exchange				After ion exchange with Pt precursor as tetramine platinum (II) chloride ($(Pt(NH_3)_4Cl_2).H_2O$)			
Elemental analysis (bulk)	Si/Al mole ratio		14.99		Si/Al mole ratio	14.99	Pt wt%	1.08
XPS[a] atomic % conc	Si	Al	Si/Al	Pt	Pt was found in its metallic form on the surface			
	28.55	1.76	16.22	0.02				
XRD	Crystallinity				Related to in-house as-synthesised Y			
	92.12							
NMR	Si/Al mole ratio for the framework				Two chemical environments for Si-NMR: Si(0Al) and Si(1Al)			
	31.2				Two Lewis acid sites existed: $Al(OSi)_4$ and $(Al(H_2O)_6)^{3+}$			
Ammonia TPD	NH_3 desorption through TPD/ mmol g^{-1}				Ammonia adsorption calorimetry followed by temperature programmed desorption			
	0.48 ± 0.01							
BET[b]	Surface area (m^2 g^{-1})	Pore volume (cm^3 g^{-1})			Pore size Å		Shows both micro- and mesoporosity	
	880.17	0.50			25.33			
TGA for coke content in spent catalyst (mass loss wt% in air) after different reaction temperatures	210 °C	Top		–	250 °C		Top	0.53
		Mid		–			Mid	0.48
		Bot		–			Bot	0.16
	230 °C	Top		0.39	270 °C		Top	0.46
		Mid		0.11			Mid	0.17
		Bot		0.20			Bot	0.11

[a] LPD Laboratory Services Ltd.

[b] MCA Services

although its stability was very good in comparison with its deactivation behaviour at lower temperatures.

The platinum dispersion of CBV712 determined by hydrogen chemisorption is given in Table 2. The low value (38.44 %) could have resulted from the formation of a considerably larger amount of coke, and it has been reported that the acid site density also affects the metal/acidity balance of the catalyst and metal dispersion, such that zeolites with Si/Al ratios between 15 and 40 had a good balance between the two functions and showed good stabilities [2].

Deactivation behaviour of CBV720 USY zeolite

CBV720 USY is obtained by steaming CBV600 for a second time at higher temperature and leaching with a mineral acid to produce the rest of the CBV-700 series, such as CBV740, CBV760 and CBV780. The preparation procedures performed by Zeolyst International are described by Remy et al. [20, 25]. Table 3 shows the in-house characterisation of CBV720 USY using various techniques.

Figure 7 shows TEM images of 1 wt% platinum loaded by an ion exchange method on CBV720 USY, and an identical activation method used during catalyst calcination by air and platinum reduction by pure hydrogen was used prior to the TEM analysis. A few very large (~25 nm) Pt particles were seen within the bulk of the zeolite support and at higher magnifications, Pt particles of 2 or 3 nm in diameter were seen in varying quantities throughout the USY support, with slightly smaller particles occurring at or near the surface.

The effect of TOS on C7 conversion, hydroisomerisation, hydrocracking and cyclic product yields at reaction temperatures ranging from 230 to 270 °C is shown in Fig. 8a–d, respectively.

Thirty minutes after the introduction of feed to the reactor, overall conversions were 11, 35, and 62 wt% at 230, 250 and 270 °C, respectively. At the lowest temperature of 230 °C, the catalyst lost about 1 wt% of its activity, and then reached a constant value after 60 h on stream. On the other hand, the deactivation rate of the fresh catalyst at 250 °C was slightly more rapid, as shown in Fig. 8a, with the overall conversion being 35 wt% 30 min after the introduction of feed to the reactor, decreasing with TOS to 33 wt% after 240 min, and then decreasing slowly, with the deactivation rate ranging from 1 to 2 wt% every 24 h until the reaction terminated, indicating reasonably good time stability. A rapid fall in catalyst activity occurred during the first few hours at the highest temperature of 270 °C, and then reached a constant deactivating value, ranging between 3 and 5 wt% every 24 h until the reaction terminated.

At the lower temperatures of 230 and 250 °C, selectivity towards hydroisomerisation products ranged from 97 to 98 % throughout the deactivation run, from the introduction of the feed to the reactor until both reactions terminated, while selectivity towards cracked products ranged from 0.05 to 0.4 % and from 0.04 to 0.3 %, respectively. Cyclic products reached high maximum values of 2.5 and 1 % at 230 and 250 °C, respectively, with the lowest value of 0.04 % observed at 270 °C. The maximum isomer yield obtained with the CBV720-based catalyst was around 61.8 wt% at 270 °C, where hydroisomerisation conversion was predominant. Cracked product selectivity increased initially at the higher temperatures of 250 and 270 °C during the initial few hours of the deactivation reaction, and then decreased rapidly throughout the TOS, such that selectivity of cyclic products was slightly higher at 230 and 250 °C than at the higher temperature of 270 °C.

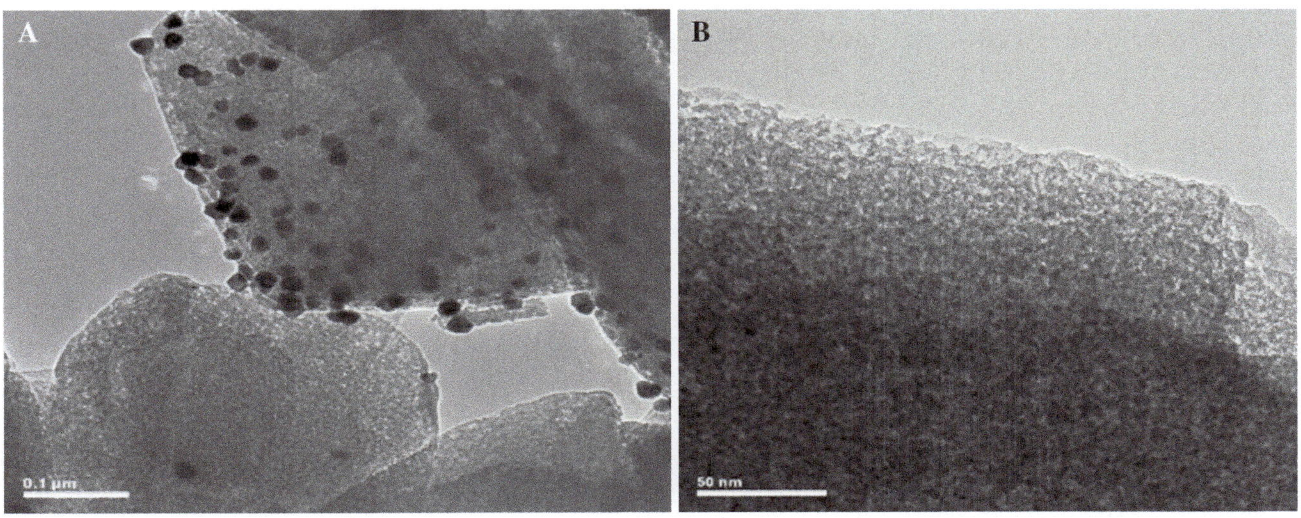

Fig. 7 TEM images of 1 wt% Pt/CBV720 USY

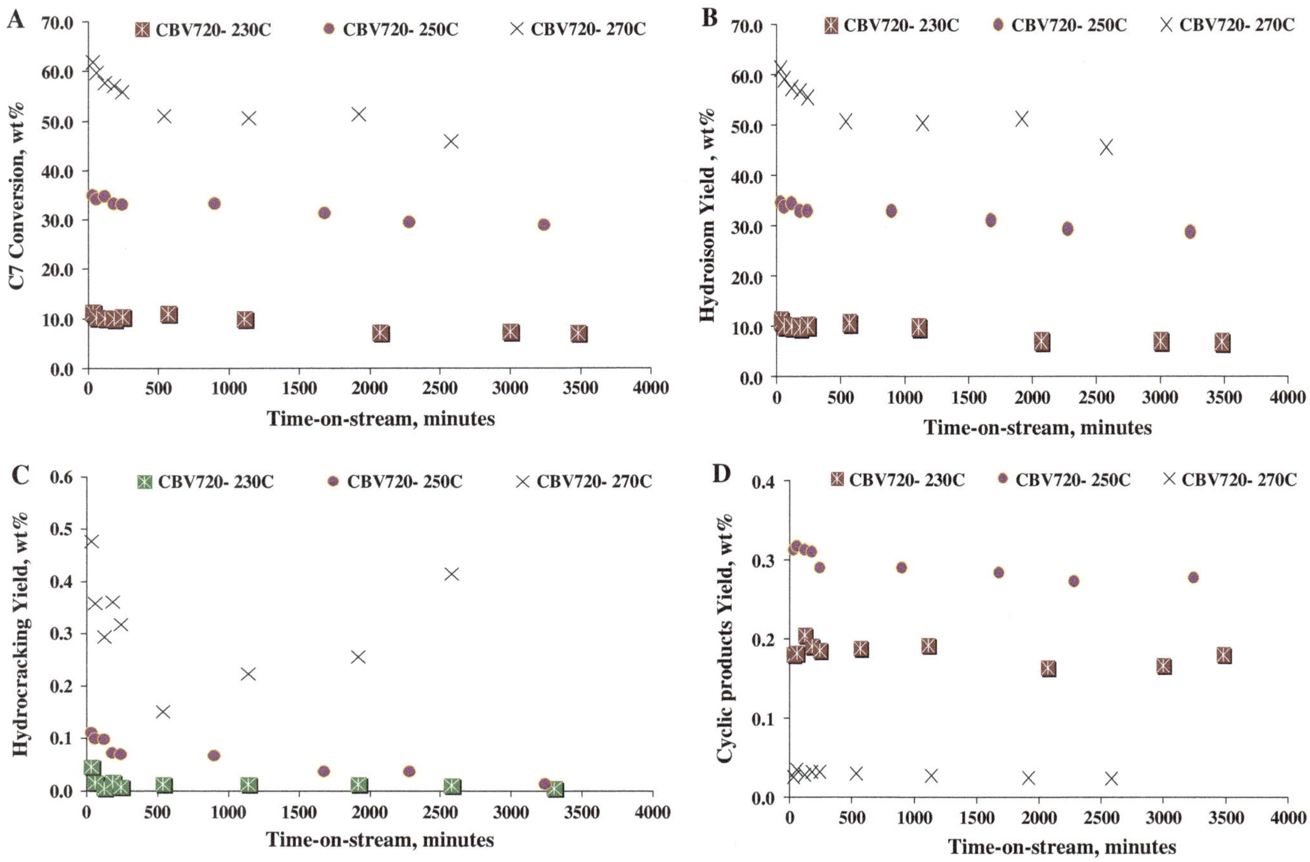

Fig. 8 Overall conversion (**a**), total yield of C7 isomers (mono + di + tribranched isomers) (**b**), total hydrocracking yield (**c**), and total yield of the cyclic products (**d**) as a function of TOS for CBV720 USY at various reaction temperatures

As CBV720 was produced via a severe dealumination process and leached by a mineral acid, the resultant catalyst has a small number of acid sites. This explains the lower activity of CBV720 accompanied by increased time stability compared with CBV712. Furthermore, removing the EFAL species by acid leaching, which can lead to enhanced acidity through the formation of Lewis acid sites, reduced the catalyst activity. Brønsted activity was predominant due to the stronger acid sites within the structure, and it has been hypothesised that these EFAL species increase the activity of the zeolite by stabilising the negative charge that is present in the zeolite structure after protons are removed, resulting in a synergy with nearby Brønsted acid sites and improved activity [26–28].

Due to the low acidity, there was less cracking activity, resulting in the catalyst having a higher stability against deactivation by coking, which is the case for zeolites with high Si/Al ratios [10, 12]. As with CBV712, it is obvious that the top zone had the highest hard coke content in the catalyst bed at each reaction temperature, whereas there was slightly less coke in the middle and bottom zones, as shown in Table 3. This suggests that the top zone of the catalyst bed contributed most activity and is where the reaction took place. In addition, the coke content in the top zone of this catalyst is much lower than in the case of CBV712, indicating the lower acidity of CBV720, which affects the stability of the catalyst, resulting in much slower deactivation and coke formation. In addition, the high surface area and large pore volume shown in Table 3, due to the more severe dealumination process and acid leaching, reduced the hard coke content and thus the deactivation rate.

The n_{Pt}/n_A ratio for CBV720 was calculated to be 0.115, indicating a balanced catalyst that performed as an 'ideal' hydroconversion catalyst. The high value of n_{Pt}/n_A meant that the number of chemical steps by which the cracked products could be generated and then oligomerated to form the coke-maker molecules was small, which in turn led to slower coke formation and higher catalyst stability. Despite large (~25 nm) Pt particles being seen within the surface of the catalyst support in the TEM images shown in Fig. 7, most of the Pt clusters, seen at higher magnifications, ranged between 2 and 3 nm across in varying quantities throughout the USY support, which indicates how well platinum is dispersed through the structure due to the cleaner pores and fewer EFAL species.

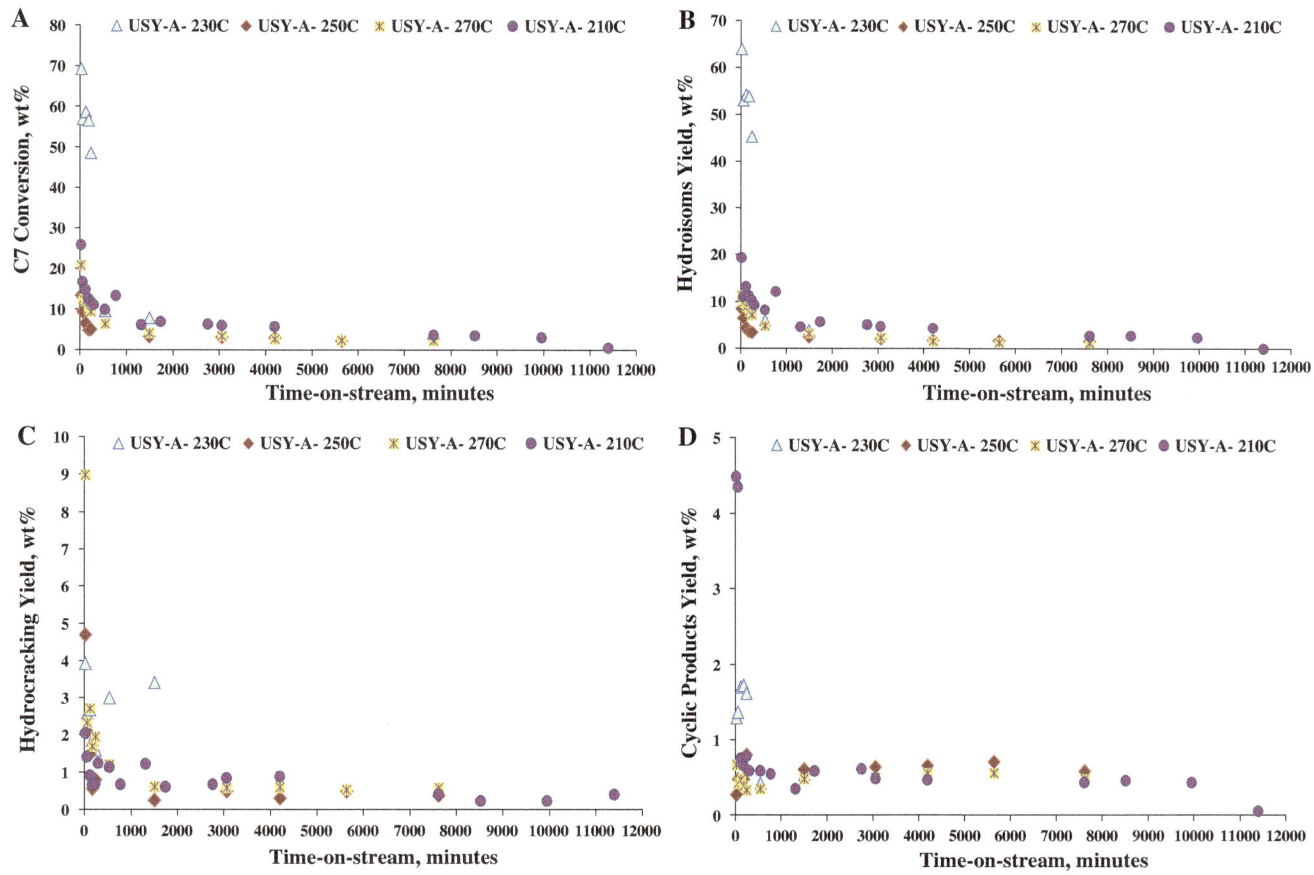

Fig. 9 Overall conversion (**a**), total yield of C7 isomers (mono + di + tribranched isomers) (**b**), total hydrocracking yield (**c**), total yield of the cyclic products (**d**) as a function of TOS for USY-A at various reaction temperatures

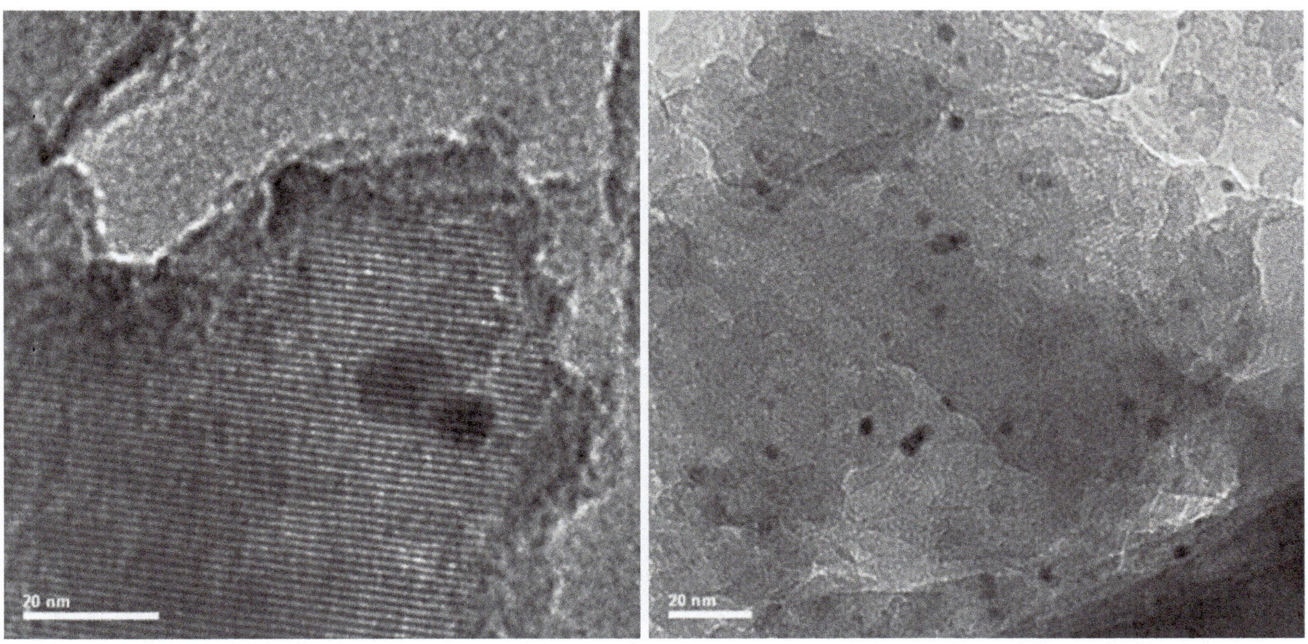

Fig. 10 TEM images for coked 1 wt% Pt/USY-A

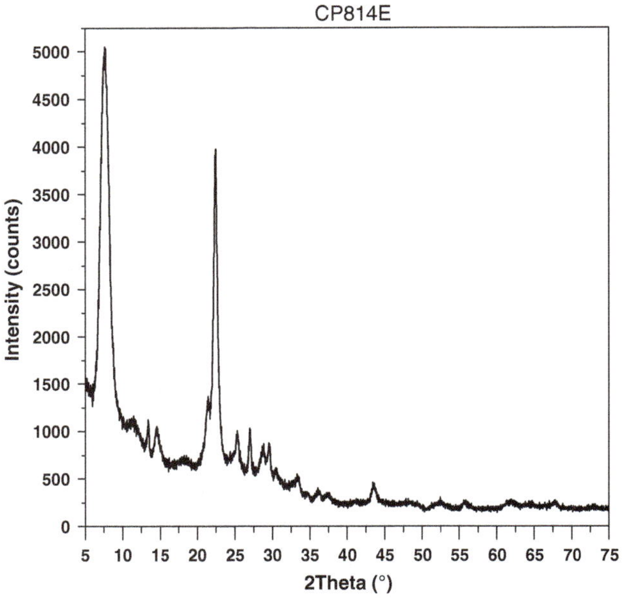

Fig. 11 XRD pattern of CP814E Beta

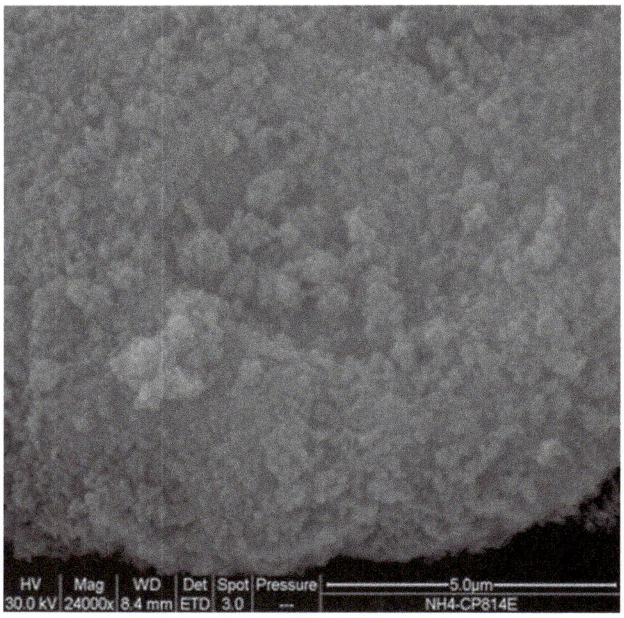

Fig. 12 SEM micrograph of CP814E Beta

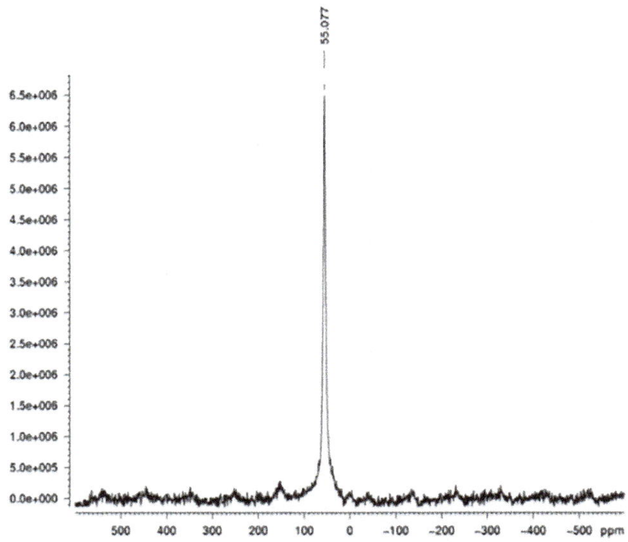

Fig. 13 ^{27}Al MAS NMR spectra of CP814E Beta

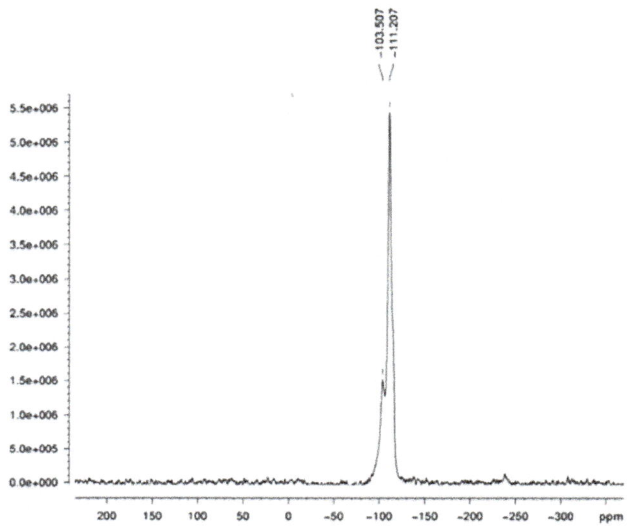

Fig. 14 ^{29}Si MAS NMR spectra of CP814E Beta

Deactivation behaviour of USY-A zeolite

USY-A was synthesised using a pilot steaming rig supplied by the Crosfield company. Table 4 shows the in-house characterisation of USY-A using a variety of techniques. As this was a low Si/Al molar ratio catalyst, there were three chemical environments for Al confirming that USY-A was produced by a less severe dealumination process in order to achieve a lower Si/Al molar ratio and had not been acid-leached to remove EFAL species, giving it the same Si/Al ratio as via synthesis. Furthermore, four configurations of Si atoms were found within its structure, explaining the high total acidity for ammonia TPD.

The effect of TOS on the C7 conversion, hydroisomerisation, hydrocracking and cyclic product yields, respectively, at reaction temperatures of 210 to 270 °C, is shown in Fig. 9a–d, where the same amount of catalyst was used at each temperature.

Initially, the overall conversion at 210 °C was 25.8 wt% after 30 min, decreasing sharply with TOS to 9.87 wt% after 540 min. Conversion decreased steeply with a deactivation rate of around 2 to 0.46 wt% after 190 h, at which the 1 wt% Pt/USY-A was completely deactivated, as can be seen in Fig. 9a. The yields of isomers decreased more sharply from 19.26 wt% to approximately 8.58 wt% after 540 min on stream, then decreased slowly, whereas the

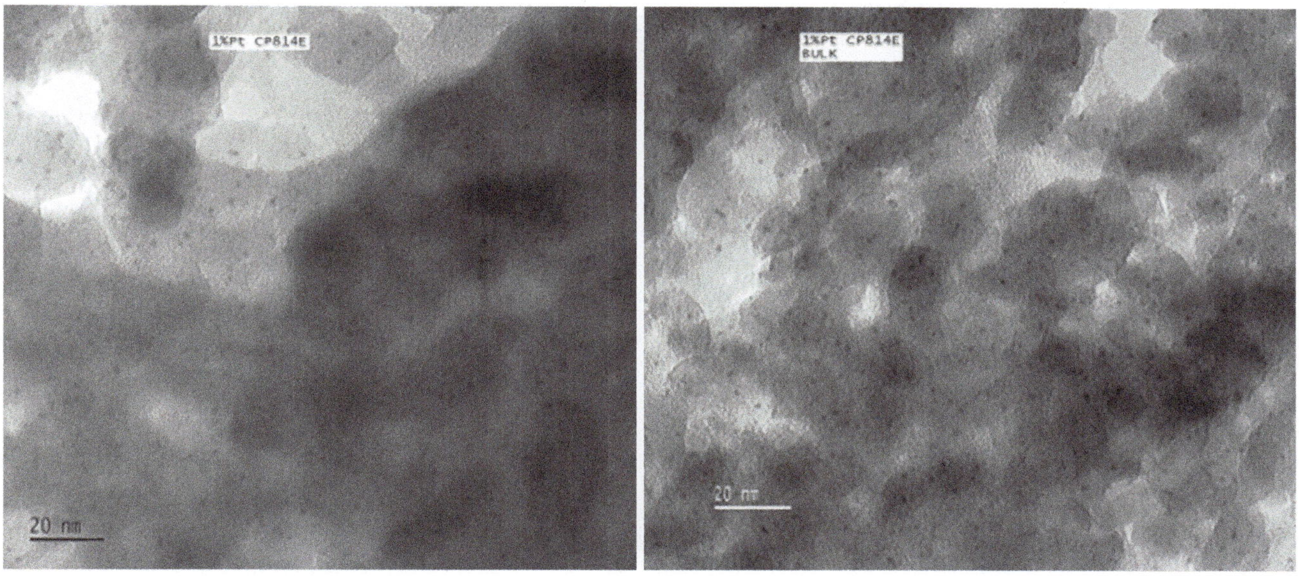

Fig. 15 TEM images of 1 wt% Pt/CP814E Beta

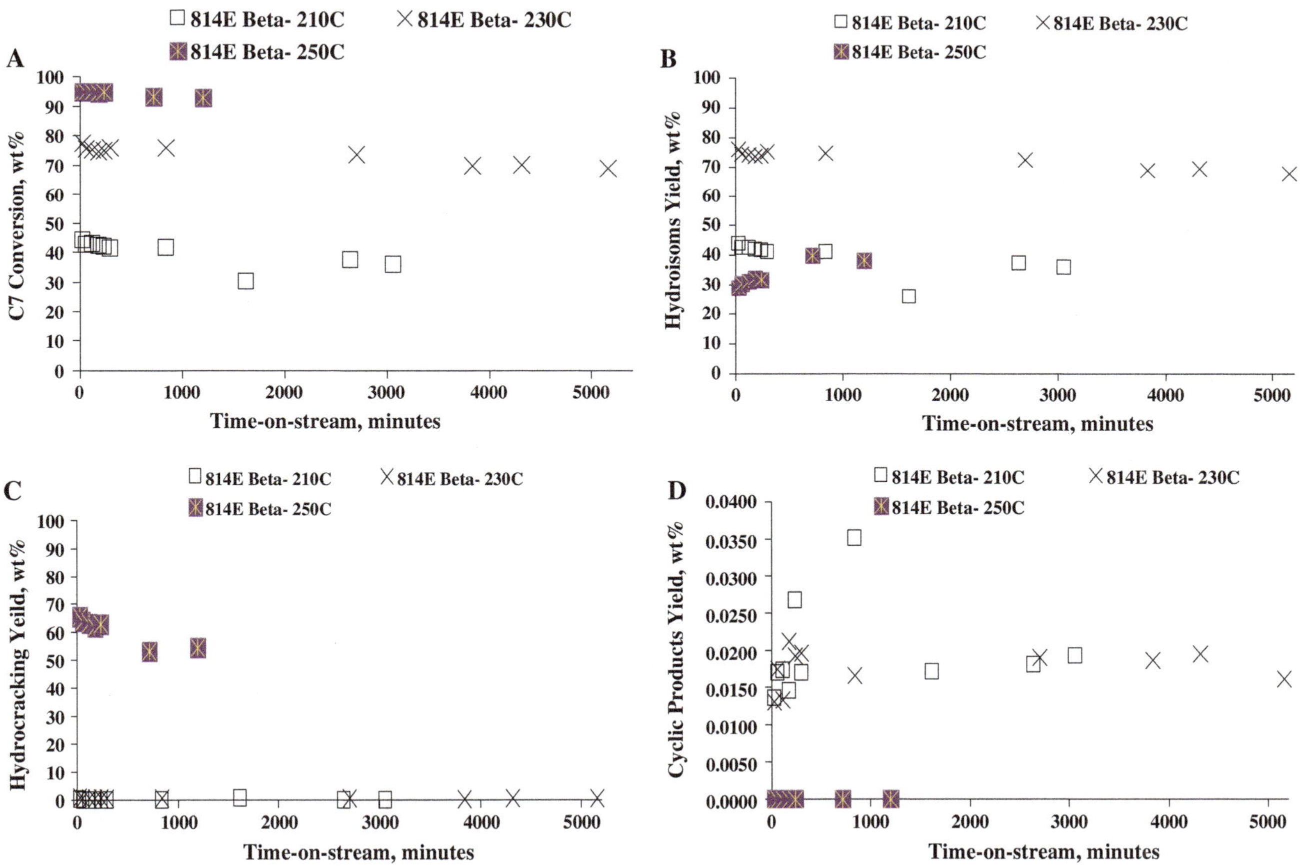

Fig. 16 Overall conversion (**a**), total yield of C7 isomers (mono + di + tribranched isomers) (**b**), total hydrocracking yield (**c**), and total yield of the cyclic products (**d**) as a function of TOS for CP814E at various reaction temperatures

yields of cracked products decreased more steeply with increasing TOS. The same TOS experiment was repeated after regenerating the catalyst and performed at the same reaction temperature of 210 °C for 166 h (approximately 7 days) in order to evaluate the stability of the catalyst against coking and to check its reversibility using the same reaction conditions; almost identical deactivation behaviour was recorded for this experiment. Significantly rapid

Table 4 In-house characterisation for USY-A

	Before ion exchange	After ion exchange with Pt precursor as tetramine platinum (II) chloride ($(Pt(NH_3)_4Cl_2).H_2O$
Elemental analysis (bulk)	Si/Al mole ratio 2.87	Si/Al mole ratio 2.87; Pt wt% 1.01
XPS[a] atomic % conc	Si 24.07; Al 10.86; Si/Al 2.22; Pt 0.03	Traces of Pt was found on the surface
XRD	Crystallinity 99.16	Related to in-house as-synthesised Y
NMR	Si/Al mole ratio for the framework 7.11	Four chemical environments for Si-NMR: Si(0Al), Si(1Al), Si(2Al) and Si(2Al) Lewis acid sites existed: $Al(OSi)_4$, $(Al(H_2O)_5)^{3+}$ and $(Al(H_2O)_6)^{3+}$
Pyridine-TPD/FTIR	Brønsted acid site concentration (B) (mmol g^{-1} $_{catalyst}$) 0.97	Lewis acid sites (L) concentration (mmol g^{-1} $_{catalyst}$) 0.22; L/B ratio 0.22
Ammonia TPD	NH_3 desorption through TPD/mmol g^{-1} 0.86 ± 0.01	Ammonia adsorption calorimetry followed by temperature programmed desorption
BET[b]	Surface area (m^2 g^{-1}) 755.67; Pore volume (cm^3 g^{-1}) 0.42	Pore size Å 27.99; Shows both micro- and mesoporosity
TGA for coke content in spent catalyst (mass loss wt% in air)	Top 2.37	Mid 2.06; Bot 1.11
	TGA analysis done after deactivation reactions at 210, 230, 250 and 270 °C	

[a] LPD Laboratory Services Ltd.

[b] MCA Services

deactivation behaviour occurred at the higher temperatures of 230 to 270 °C, with greater cracking activity during the initial few minutes, resulting in a very high rate of coke formation, which also accounted for the poor performance.

The higher concentration of Brønsted than Lewis acid sites resulted in a high total acidity and the predominance of Brønsted acidity, as shown in Table 4, meaning that cracking activity was high and coke formation very rapid during the initial few hours. It has been shown that the EFAL species generated during dealumination by steaming can be cationic, neutral, monomeric, oligomeric or polymeric and hydroxylated or not. These species can, therefore, either positively increase the acid strength by the interaction between bridging hydroxyl groups and neighbouring small EFAL species with Lewis acidity, or negatively by exchange of protonic sites (EFAL cationic species) and by pore blockage (polymeric species) [16].

Table 4 shows the TGA results for the hard coke content deposited over the aged USY-A top, middle, and bottom catalyst bed zones after unloading the spent catalyst from the atmospheric glass reactor and after the deactivation reactions at 210, 230, 250 and 270 °C. Coke was found predominantly in the top zone of the bed, suggesting pore blockage by excessive cracking (which probably occurred at higher temperatures) and reduced diffusion limitation compared with other zones.

As can be observed in Fig. 10, showing TEM images of spent USY-A, it is likely that the catalytic coke components were essentially formed and trapped within the channels of the catalyst micropores, so the first direct effect on the active sites was either by poisoning or by blocking of access to reactant molecules. Coke can also have an indirect effect, one that needs to be removed during catalyst regeneration and requires an oxidative treatment under severe conditions. This frequently has detrimental effects on the dealumination and degradation of the zeolite framework and the sintering of supported metals.

Some agglomerates of platinum particles can be observed within or upon the micropore channels of the coked USY-A, in Fig. 10, with the most likely explanation for this being catalyst regeneration prior to each deactivation run at each reaction temperature may have resulted in the sintering of supported platinum. Furthermore, as shown in the XPS analysis, some traces of platinum atoms were detected on the outer surface of the catalyst, indicating that some kind of agglomeration would have occurred towards the surface.

Table 5 In-house characterisation of CP814E Beta

	Before ion exchange				After ion exchange with Pt precursor as tetramine platinum (II) chloride ($Pt(NH_3)_4Cl_2).H_2O$			
Elemental analysis (bulk)	Si/Al mole ratio		13.21		Si/Al mole ratio	13.21	Pt wt%	1.07
XPS[a] atomic % conc	Si	Al	Si/Al	Pt	Pt was not found on the surface			
	25.77	2.39	10.78	0				
XRD	Crystallinity %				As was reported in the literature [8]			
	100							
NMR	Si/Al mole ratio for the framework				Two chemical environments for Si-NMR: Si(0Al) and Si(1Al)			
	13.45				Lewis acid sites are not present			
Ammonia TPD[b]	NH_3 desorption through TPD/mmol g^{-1}				Ammonia adsorption calorimetry followed by temperature programmed desorption			
	0.63 (± 0.01)							
Hydrogen chemisorption[b]	Metal dispersion %	Metallic surface area m^2 g^{-1} metal			Metallic surface area m^2 g^{-1} sample		Crystallite size, nm	
	30.67	75.74			0.81		3.69	
BET[b]	Surface area m^2 g^{-1}	Pore volume cm^3 g^{-1}			Pore size Å		Shows both micro- and mesoporosity	
	577.48	0.85			72.70			
TGA for coke content in spent catalyst (mass loss wt% in air) after various reaction temperatures	210 °C	Top		1.05	250 °C		Top	–
		Mid		0.20			Mid	–
		Bot		0.37			Bot	–
	230 °C	Top		0.85	270 °C		Top	–
		Mid		1.04			Mid	–
		Bot		0.94			Bot	–

[a] LPD Laboratory Services Ltd.

[b] MCA Services

Deactivation behaviour of CP814E beta zeolite

The XRD pattern for the fresh sample of CP814E (Fig. 11) is consistent with that reported in the literature for beta zeolite phase [21], and the SEM image of the sample (Fig. 12) shows nano-crystallites with no clear boundaries, and crystal particles clustered into agglomerates to form larger particles of different sizes. Similar observations on the CP814E morphology have been reported in the literature with the nano-sized crystals having different sizes depending on the synthesis or modification history [29, 30].

Figures 13 and 14 show the solid-state NMR spectra for ^{27}Al and ^{29}Si, respectively, with the NMR spectrum showing an ^{27}Al peak centred at 56.08 ppm in the region corresponding to the tetrahedrally coordinated aluminium, and no additional peaks were found which could be attributed to aluminium atoms in either octahedral or pentahedral coordinated EFALs. The ^{29}Si MAS NMR spectra (Fig. 14) shows two peaks at 111.21 and 103.51 ppm, which can be assigned to the Si(0Al) and Si(1Al) configurations, respectively, and the same spectral interpretation for both ^{27}Al and ^{29}Si MAS NMR has been reported in the literature [31, 32]. Moreover, as beta zeolite can be synthesised with Si/Al ratios between 10 and 100 [32], it can be concluded that CP814E is an as-synthesised beta zeolite and no post-synthesis modifications have been used [33]. Table 5 shows the in-house characterisation of CP814E using a range of different techniques. As shown in Table 5, a value of 13.45 was obtained for the framework Si/Al ratio using ^{29}Si MAS NMR, which concurred with the measurement given by chemical analysis of 13.21, which confirms that CP814E is an as-synthesised beta zeolite and no post-synthesis modification was used [33]. However, the Si/Al ratio of 10.78 on the surface ($\sim$5 nm) using XPS was significantly lower than the expected value of 13.45 as given in ^{29}Si MAS NMR or chemical analysis.

Since the metal cluster size in the bifunctional zeolite catalysts has a significant effect on the type of reaction catalysed by the metal function [34, 35], Fig. 15 shows TEM images of CP814E with 1 wt% platinum loaded by ion exchange. Prior to TEM analysis, activation and platinum reduction using air and hydrogen, respectively, were used to ensure comparable results with previous analyses. Fine platinum particles ($\sim$2 nm) could be seen within the grain of the beta zeolite support, even though hydrogen chemisorption indicated a low platinum dispersion as shown in Table 5. Furthermore, no platinum in the metallic form could be seen on the catalyst subsurface ($\sim$5 nm), a fact that was confirmed by XPS.

The effect of TOS experiments on the C7 conversion, hydroisomerisation, hydrocracking, and cyclic product yields at different reaction temperatures from 210 to 250 °C is shown in Fig. 16a–d, respectively, with fresh catalyst used for each experimental run at the different reaction temperatures.

Initially, the overall conversion was 44, 77, and 95 wt% after 30 min from the point when the feed was introduced to the reactor at temperatures of 210, 230, and 250 °C, respectively. At 210 °C, the newly activated catalyst lost about 1–2 wt% of its activity and then reached a pseudo-stable state for the initial 60 min on stream. The initial conversion at 230 °C for the newly activated catalyst remained unchanged for about 2 days on stream, and then slightly decreased over time, with an overall conversion of 77 wt% after 30 min decreasing with the TOS to 74 wt% after 2,700 min, and then decreased slowly within the deactivation rate range of 1 to 2 wt% every 24 h until the reaction was finally terminated after about 86 h. On the other hand, the selectivity towards hydroisomerisation products was almost 99 % during the deactivation run at the 210 and 230 °C from the point of the feed being introduced into the reactor until both reactions terminated. In addition, selectivity towards cracked and cyclic products was very low, fluctuating in a very narrow range as a function of TOS in the range of 0.03 to 1 %.

It can be seen that CP814E (Fig. 16) demonstrated a significantly higher cracking activity at the reaction temperature of 250 °C than would be expected from its Si/Al ratio, with the initial conversion at 250 °C for the new activated catalyst being 95 wt% after 30 min, decreasing with the TOS to 93 wt% until the reaction was terminated after 2 days. The initial selectivity towards cracked products was high (70 %) compared to the hydroisomerisation products (30 %) for the C7 isomers. However, the cracking activity decreased with time and the C7 isomer production was seen to increase at the level-off stage, indicating that the reaction could not be conducted at temperatures higher than 270 °C as cracking activity would become the dominant reaction.

It has been reported that beta zeolite with a three-dimensional pore structure combined with no cavities or cages provided the best structure to resist deactivation due to coking with acid site density, metal/acidity balance and metal dispersion of the catalyst affecting the amount of coke deposited [2]. The methane-to-propane (C1/C3) ratio was employed as to provide an approximate measure of the activity of the metal function vs. acidic function, and is used in many industrial processes to follow the balance between activities of both functions. Methane is produced due to hydrogenolysis of the alkanes over the metal

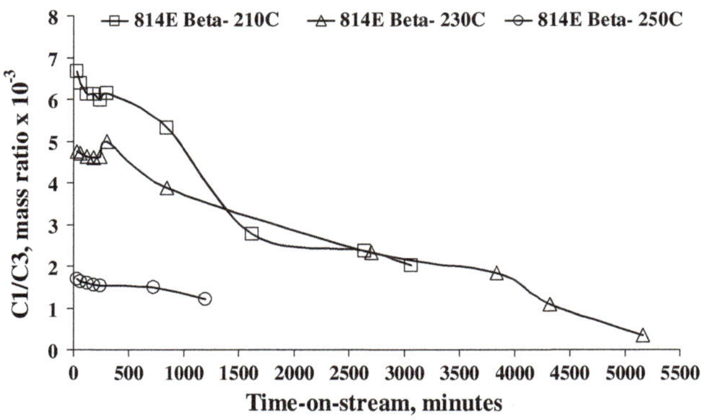

Fig. 17 Variation of the C1/C3 mass ratio as a function of TOS of CP814E

function whilst cracking on the acid sites produces propane almost exclusively [2]. The variation in C1/C3 mass ratio with TOS at various reaction temperatures is shown in Fig. 17, where the acid and metal functions were initially active at the temperatures of 210 and 230 °C, meaning that constant C1/C3 ratios were obtained. However, deactivation of the acid and metal functions caused the C1/C3 ratio to decrease rapidly, indicating a change in the balance between the two functions with TOS that may be due to the deposition of coke on both functions. The metal function deactivated faster than the acid function, though it was not significantly severe to cause a drop in *n*-C7 conversion.

The hard coke content shown in Table 5 at 210 and 230 °C was in the range of 0.2 and 1.05 wt%, indicating that the lower dispersions in CP814E result in the rate of deposition being strongly related to the initial activity of the acid function, the reaction temperature and the balance between the activities of metal and acid functions. CP814E loaded with 1 wt% Pt by ion exchange showed a low C1/C3 ratio indicating a good balance between the metal and acidic functions and thus it demonstrated satisfactory stability time.

Deactivation behaviour of some selected commercial mordenite zeolites

Figure 18a shows the SEM micrograms of CBV21A MOR. From the SEM images, it can be seen that CBV21A MOR contains larger crystallites up to 500 nm in length, accompanied by some smaller debris and agglomerates. The literature indicates that no synthesis route for nanocrystalline mordenite is so far available and that synthesis parameters such as silicon precursors, water content and $Na_2O/SiO_2/Al_2O_3$ ratios strongly affect mordenite morphology in terms of the size and aspect ratios of the crystals [36–38]. Figure 18b–d shows the SEM micrograms of 640HOA MOR, 660HOA MOR and 690HOA MOR,

Fig. 18 SEM micrographs of CBV21A MOR (**a**), 640HOA MOR (**b**), 660HOA MOR (**c**) and 690HOA MOR (**d**)

respectively. From the SEM images, it can be seen that the catalysts contained sub-micron sized crystallites that had agglomerated into larger particles. Table 6 shows the in-house characterization of these catalysts using various different techniques. ^{27}Al and ^{29}Si solid-state NMR spectra for 640HOA, 660HOA and 690HOA catalysts have one ^{27}Al peak at about 55 ppm, which is assigned to tetrahedrally coordinated aluminium, with two more peaks that could be attributed to aluminium atoms as EFAL for 640HOA and 660HOA, whereas 690HOA showed no evidence of EFAL. This means that these three catalysts underwent dealumination of different severity, either by acid leaching, by steaming or both. In the ^{29}Si MAS NMR spectra, two peaks at about 112 and 106 ppm were assigned to Si(0Al) and Si(1Al) configurations, respectively. It is also evident that the intensity of the chemical environment of Si(0Al) increases as the Si/Al mole ratio increases and hence also as dealumination severity increases.

As dealuminated catalysts, they showed greater total pore volume, which means greater mesoporosity alongside the microporosity, hence larger surface area and smaller pore size compared with as-synthesised catalyst CBV21A MOR, as shown in Table 6. Furthermore, a good match was obtained for the Si/Al mole ratio of 690HOA by NMR and XPS, which supports the observation that there was no EFAL on the surface of this catalyst.

Well-dispersed platinum particles in the size range of 1–3 nm were obtained for the CBV21A MOR catalyst loaded with 1 wt% platinum by the ion exchange method. An activation method identical to that used during catalyst calcination by air and platinum reduction by pure hydrogen was used prior to the TEM analysis, as shown in the TEM images in Fig. 19.

TOS behaviour during C7 hydroisomerisation conversion was examined for the following commercial and in-house mordenite catalysts: CBV21A, 640HOA, 660HOA, and 690HOA. All of the above catalysts were loaded with 1 wt% Pt by the ion exchange method, as described above. Two very acidic commercial catalysts—1 wt% Pt/640HOA and 1 wt% Pt/660HOA deactivated very rapidly during the initial few seconds, due to a rapid coking that may have poisoned the acid sites of these catalysts and changed the effective acid site density needed for an ideal bifunctional conversion. As a consequence, no TOS results were reported for these catalysts. Figure 20 shows the effects of

Table 6 Characterization of zeolite MOR samples determined

Bulk (Si/Al) mol/mol	Framework (Si/Al) mol/mol	Crystallinity %	Bulk Pt wt%	(Si/Al) atomic conc % (XPS)	Pt Atomic Conc % (XPS)	Acidity mmol g^{-1}	Surface area $m^2\ g^{-1}$	Pore volume $cm^3\ g^{-1}$	Pore size Å
CBV21A MOR									
10.49	10.15	100	1.04	–	–	–	458.90	0.23	28.026
640HOA MOR									
–	12.0	100	–	–	–	–	–	–	–
660HOA MOR									
–	10.93	97	–	–	–	1.08	540.56	0.28	29.38
690HOA MOR									
114.38	17.88	108	0.97	16.50	0.04	0.10	559.94	0.28	24.48

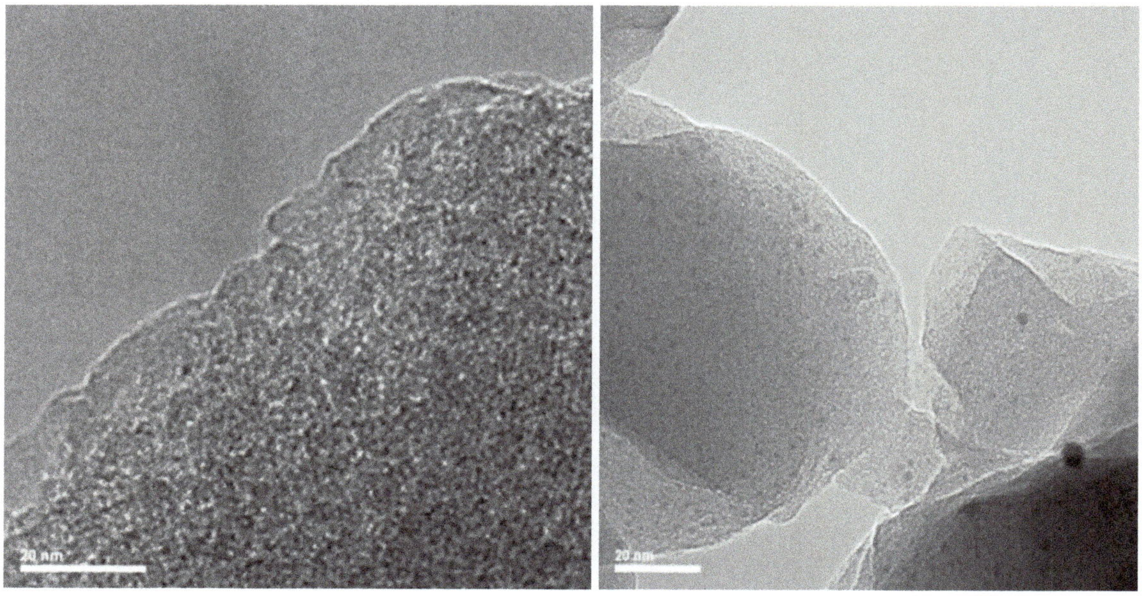

Fig. 19 TEM images of 1 wt% Pt/CBV21A MOR

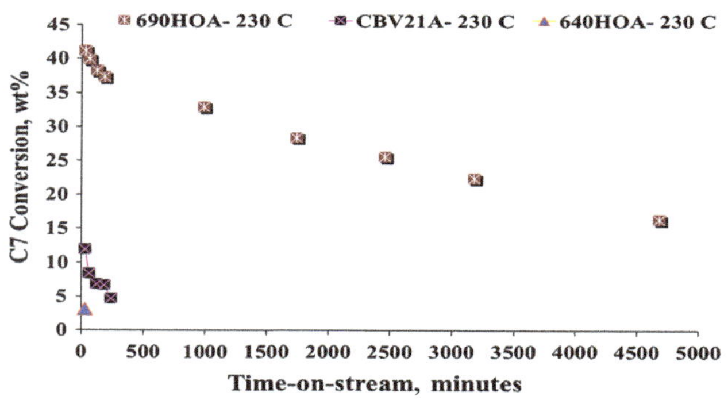

Fig. 20 Overall conversion as a function of TOS for commercial mordenite catalysts. WHSV = 5.13 h^{-1}; H_2/C7 (mol/mol) = 9; total pressure = 1 atm; temperature = 230 °C; 1 wt% Pt loading

TOS at a reaction temperature of 230 °C on C7 conversion, for the following catalysts: 1 wt% Pt/CBV21A and 1 wt% Pt/690HOA.

Initially, the overall conversions were 12, and 41 wt% after 60 min from the point when the feed was introduced to the reactor at 230 °C, for the 1 wt% Pt/CBV21A and 1 wt% Pt/690HOA mordenite catalysts, respectively. A rapid deactivation occurred for the 1 wt% Pt/CBV21A catalyst with high cracking activity, resulting in selectivity towards the cracked products at 24 %, compared with 71 % for the isomer products, and thereafter the catalyst was almost deactivated after 240 min on stream. On the other hand, greater TOS stability was shown by the 1 wt% Pt/690HOA mordenite catalyst. However, the 1 wt% Pt/690HOA catalyst deactivated continuously from the point when the feed was introduced to the reactor at the slower rate of 3–5 wt% every day, before achieving a pseudo-stable state at the end of the reaction. In addition, selectivity towards C7 isomers and cyclic products for both catalysts was in the range of 97–98 % and 2–3 %, respectively.

Faster deactivation and the mono-dimensional pore system of 1 wt% Pt/CBV21, coupled with a higher density of strong acid sites and as-synthesised zeolite, may have caused the lower activity, which may also be attributable to platinum crystallites blocking the pore openings, as previously reported [39, 40]. There are side pockets in the pore topology of mordenite that match the 12MR and 8MR crossing. The higher number of Brønsted acid sites in the cavities also increases the sensitivity of the 8 MR pores to coking and they can easily fill up and become blocked by carbonaceous deposits [41]. The average lifetime of the carbocations on the surface can be lengthened by strong acidity and narrow pore diameter, while the diffusion of the branched products can be slowed down. Each of these factors will help to crack the tertiary carbocations that have formed during isomerisation prior to desorption and will assist in the re-adsorption and cracking of the branched paraffins before they leave the pores and enter the gas stream [42].

There are various possible explanations for the slower deactivation of 1 wt% Pt/690HOA mordenite. First, it may be due to the medium acid strength of the catalysts and their reduced acid site density. Alternatively, it may be due to the balance between acid and metal active sites. It could also be ascribed to the existence of mesopores and to the slight increase in the pore volume, which may be beneficial to the diffusion of bulky moieties. Finally, it may be due to post-synthesised mordenite having smaller crystallites. The last two characteristics can be seen as reducing the resistance of the product molecules to formation and diffusion, resulting in less cracking and deactivation of the catalyst caused either by blockages at the site or by plugging of channels. The method of dealumination applied to the final lattice Si/Al ratio influences catalyst activity. For example, steam dealumination leaves the lattice with aluminium removed as EFAL, whereas samples that are practically free of EFAL can result from acid leaching, so EFAL may have a negative effect by blocking pores and preventing the Pt from being well dispersed during the ion exchange process. This may explain the deactivation of the 1 wt% Pt/690HOA steamed mordenite catalysts.

Conclusion

It can be concluded from this investigation of the deactivation of a range of commercial bifunctional nanoporous-based catalysts during the hydroisomerisation of normal heptane (*n*-C7) under identical reaction conditions. The main cause of deactivation of acid zeolite catalysts is generally coking, as coke compounds become trapped in the zeolite pores and can poison the acid sites or block their access. Two factors have been found to affect strongly the rate of coke formation: the structure and openings of the pores, and the density of strongly acid sites. There is a clear correlation between the yield of cracked products obtained, the existence of stronger acid sites and the formation of hard coke. Thus, lower acidity results in the catalyst being more stable against deactivation by coking, which is the case for zeolites with high Si/Al ratios and those which have been acid-leached. The number of strong acid sites can be reduced by dealumination and acid leaching, thus increasing mesoporosity inside the zeolite crystallites and so reducing the limitations to diffusion, resulting in shorter residence times, which facilitates desorption of the products. Furthermore, the balance between the number of metal sites and the number of acid sites plays an important role in determining the activity, selectivity and stability of bifunctional catalysts.

It was found that zeolites with channel intersection cavities of comparable size to their apertures, like zeolite beta, were minimally affected by coke formation; thus the size of the channel intersection should be only slightly larger than that of the zeolite aperture, such that the coke precursors are unable to remain trapped inside the cavities and form bulky carbonaceous compounds. On the other hand, coke precursors can be trapped and become enlarged in the supercages, thus blocking the pores of a catalyst such as zeolite Y, which has relatively large cavities or supercages in comparison with the openings of the channels. Moreover, zeolites with nano-crystalline structures are more stable with respect to coking and have better TOS behaviour, due to the shorter time it takes reactants and products to diffuse into and out of the smaller crystals in the structure.

References

1. Guisnet M, Costa L, Ribeiro FR (2009) Prevention of zeolite deactivation by coking. J Mol Catal A Chem 305:69–83
2. Gopal S (2003) Synthesis, Modification, Characterization and Catalytic Studies of Zeolite based Bifunctional Catalysts for Hydroisomerization Reactions. Dissertation, University of Cincinnati, Cincinnati, Ohio
3. Rollmann LD (1977) Systematics of shape selectivity in common zeolites. J Catal 47:113–121
4. Rollmann LD, Walsh DE (1979) Shape selectivity and carbon formation in zeolites. J Catal 56:139–140
5. Guisnet M, Magnoux P (1989) Coking and deactivation of zeolites—influence of the pore structure. Appl Catal 54:1–27
6. Guisnet M, Magnoux P (1992) Composition of the carbonaceous compounds responsible for zeolite deactivation—modes of formation. Zeolite Microporous Solids Synth Struct React 352: 437–456

7. Gopal S, Smirniotis PG (2002) Deactivation behavior of bifunctional Pt/H-zeolite catalysts during cyclopentane hydroconversion. J Catal 205:231–243
8. Zhang W, Smirniotis PG (1998) Dealuminated zeolite-based composite catalysts for reforming of an industrial naphthene-rich feedstock. Appl Catal A 168:113–130
9. Walsh DE, Rollmann LD (1977) Radiotracer experiments on carbon formation in zeolites. Effect of aromatic content of hydrocarbon mixtures. J Catal 49:367–375
10. Stöcker M (2005) Gas phase catalysis by zeolites. Microporous Mesoporous Mater 82:257–292
11. Sie ST (2008) Isomerization. In: Ertl G, Knözinger H, Schueth F, Weitkamp J (eds) Handbook of Heterogeneous Catalysis, 2nd edn. Wiley-VCH Verlag GmbH, Weinheim, pp 2809–2829
12. Akhmedov VM, Al-Khowaiter SH (2007) Recent advances and future aspects in the selective isomerization of high *n*-alkanes. Catal Rev 49:33–139
13. Guisnet M, Ribeiro FR (2011) Deactivation and regeneration of zeolite catalysts. Imperial College Press, London
14. Zeolyst International. http://www.zeolyst.com/our-products/standard-zeolite-powders/zeolite-y.aspx. Accessed 29 May 2011
15. Tosoh Corporation. http://www.tosoh.com/Products/tcdzeo.htm#hzc. Accessed 26 June 2011
16. Rawlence DJ, Earl GJ (1993) Materials having novel structure factors for catalysis of hydrocarbon transformation and development of selective processes. BRITE-EURAM 4633, 5th Meeting Report, Elf Solaize
17. Abudawood R, Alotaibi F, Garforth A (2011) Hydroisomerization of *n*-heptane over Pt-loaded USY zeolites. Effect of steaming, dealumination and the resulting structure on catalytic properties. Ind Eng Chem Res 50:9918–9924
18. Abudawood R (2010) Hydroisomerization of Alkanes over Metal-Loaded Zeolite Catalysts. Dissertation, University of Manchester
19. Weisz PB, Frilette VJ (2002) Intracrystalline and molecular-shape-selective catalysis by zeolite salts. J Phys Chem 64:382
20. Remy MJ, Stanica D, Poncelet G, Feijen EJP, Grobet PJ, Martens JA, Jacobs PA (1996) Dealuminated H–Y zeolites: relation between physicochemical properties and catalytic activity in heptane and decane isomerization. J Phys Chem 100: 12440–12447
21. Treacy MMJ, Higgins JB (2001) Collection of simulated XRD powder patterns for zeolites. Elsevier, Amsterdam
22. Guisnet M, Ribeiro FR (2011) Deactivation and regeneration of zeolite catalysts. Imperial College Press, London, pp 204–205
23. Guisnet M, Alvarez F, Giannetto G, Perot G (1987) Hydroisomerization and hydrocracking of *n*-heptane on PtHzeolites. Effect of the porosity and of the distribution of metallic and acid sites. Catal Today 1:415–433
24. Alvarez F, Rebeiro FR, Perot G, Thomazeau C, Guisnet M (1996) Hydroisomerization and hydrocracking of alkanes—7. Influence of the balance between acid and hydrogenating functions on the transformation of *n*-decane on PtHY catalysts. J Catal 162: 179–189
25. Morin S, Ayrault P, Gnep NS, Guisnet M (1998) Influence of the framework composition of commercial HFAU zeolites on their activity and selectivity in *m*-xylene transformation. Appl Catal A 166:281–292
26. Alvarez F, Giannetto G, Guisnet M, Perot G (1987) Hydroisomerization and hydrocracking of *n*-alkanes: 2. *n*-Heptane transformation on a Pt-dealuminated Y zeolite—comparison with a Pt-Y zeolite. Appl Catal 34:353–365
27. Gauw F (2002) Kinetic Studies of Alkane Hydroisomerization over Solid Acid Catalysts. Dissertation, Eindhoven University of Technology
28. Guisnet M, Ribeiro FR (2011) Deactivation and regeneration of zeolite catalysts. Imperial College Press, London, p 27
29. Petushkov A, Giorvanni M, Larsen SC (2011) From nanoparticles to hierarchical structures: controlling the morphology of zeolite beta. Microporous Mesoporous Mater 143:97–910
30. Trombetta M, Busca G, Storaro L, Lenarda M, Casagrande M, Zambon A (2000) Surface acidity modifications induced by thermal treatments and acid leaching on microcrystalline H-BEA zeolite. A FTIR, XRD and MAS-NMR study. Phys Chem Chem Phys 2:3529–3537
31. Maier SM, Jentys A, Lercher JA (2011) Steaming of zeolite BEA and its effect on acidity: a comparative NMR and IR spectroscopic study. J Phys Chem 15:8005–8013
32. Park JN, Wang J, Hong SI, Lee CW (2005) Effect of dealumination of zeolite catalysts on methylation of 2-methylnaphthalene in a high-pressure fixed-bed flow reactor. Appl Catal A 292:68–75
33. Jia C, Massiani P, Barthomeuf DJ (1993) Characterization by infrared and nuclear magnetic resonance spectroscopies of calcined beta zeolite. Chem Soc Faraday Trans 89:3659–3665
34. Blomsma E, Martens JA, Jacobs PA (1997) Isomerization and hydrocracking of heptane over bimetallic bifunctional PtPd/H-beta and PtPd/USY zeolite catalysts. J Catal 165:241–248
35. Bowen TC, Vane LM (2006) Ethanol, acetic acid, and water adsorption from binary and ternary liquid mixtures on high-silica zeolites. Langmuir 22:3721–3727
36. Van Laak ANC, Sagala SL, Zečević J, Friedrich H, de Jongh PE, de Jong KP (2010) Mesoporous mordenites obtained by sequential acid and alkaline treatments—catalysts for cumene production with enhanced accessibility. J Catal 276:170–180
37. Zhang L, van Laak ANC, de Jongh PE, de Jongh KP (2009) Synthesis of large mordenite crystals with different aspect ratios. Microporous Mesoporous Mater 126:115–124
38. Van Laak ANC (2011) Post-Synthesis Modifications on Zeolites for Improved Accessibility and Catalytic Performance. Dissertation, Utrecht University
39. Wang ZB, Kamo A, Yoneda T, Komatsu T, Yashima T (1997) Isomerization of *n*-heptane over Pt-loaded zeolite/beta catalysts. Appl Catal 159:119–132
40. Lee JK, Rhee HK (1997) Characteristics of Pt/H-beta and Pt/H-mordenite catalysts for the isomerization of *n*-hexane. Catal Today 38:235–242
41. Van Donk S (2002). Adsorption, Diffusion and Reaction Studies of Hydrocarbons on Zeolite Catalysts. Dissertation, Utrecht University
42. Chica A, Corma A, Miguel PJ (2001) Isomerization of C5–C7 *n*-alkanes on unidirectional large pore zeolites: activity, selectivity and adsorption features. Catal Today 65:101–110

10

Noble metal ion-substituted ceria as efficient oxygen storage catalysts for clean combustion of fuels

Rajagopalan Vijayaraghavan

Abstract Three-way catalysts, a component of automobile converters, can control auto exhaust emissions by its capacity of converting CO, un-burnt hydrocarbon and oxides of nitrogen (NO_x) into less harmful CO_2, H_2O and N_2 simultaneously. This process is efficient only when the *A*/*F* ratio is at 14.7. To widen the ratio for better chemical control, oxygen storage catalysts (OSC) based on CeO_2, capable of three-way action, are employed in converters along with conventional catalysts. In this article, the enhanced activity of noble metal ion incorporated/dispersed CeO_2 catalysts towards three-way action over conventional metal particles dispersed (in a matrix) catalysts is illustrated. The better performance of the *metal ion dispersed catalysts* is attributed to better dispersion of active metal ion sites over the reducible matrix like CeO_2, leading to many-fold increase in the number of active sites.

Keywords Oxygen storage capacity · Pd^{2+}/CeO_2 catalysts · CO oxidation · XPS · Temperature programmed reduction

Introduction

To abate pollution due to emissions from automobiles, auto exhaust catalysts containing noble metals such as Pt, Pd and Rh dispersed in γ-Al_2O_3 as the support have been employed in converters. γ-Al_2O_3 is mixed with 30 % by weight of $Ce_{0.7}Zr_{0.3}O_2$ as an oxygen storage material [1]. Although ceria-based catalysts as oxygen storage catalysts (OSC) are known since 1980s for the three way action, various approaches are adopted to widen/control the *A*/*F* ratio at 14.7 where the concentrations of oxidizing and reducing species are equal. Systems monitor the dynamic *A*/*F* ratio but monitoring through chemical control is a more versatile method. In this method, the operating window is widened by the OSC through the storage of oxygen in the oxygen-rich condition and releasing oxygen in the lean condition in situ. In this context, CeO_2 has been an important oxygen storage material in TWC [2, 3]. The reactions are

$$CeO_2 \rightarrow CeO_{2-\delta} + \delta/2O_2 \text{ (oxygen lean region)}$$
$$CeO_{2-\delta} + \delta/2O_2 \rightarrow CeO_2 \text{ (oxygen-rich region)}$$

The amount of oxygen released in the first reaction and the oxygen consumed in the second reaction is referred to as the oxygen storage capacity of the ceria material [4, 5]. The reversible intercalation of oxygen is derived from its structural feature.

Ceria crystallizes in fluorite structure (CaF_2) in which the Ce^{4+} ions form the face centred cubic (FCC) structure with the tetrahedral holes occupied by oxide ions and the octahedral holes remaining vacant. The unit cell is shown in Fig. 1.

Oxygen deficient phase of $CeO_{2\text{-}\delta}$ has been studied using X-ray diffraction (XRD), temperature programmed reduction (TPR) and magnetic measurements [6, 7]. Noble metals such as Pt, Pd and Rh, bimetallic Pt–Rh are dispersed as nanocrystalline metal particles or metallic ions into the matrix of OSC materials with gamma-alumina support ($Ce_{0.7}Zr_{0.3}O_2$ with alumina)

R. Vijayaraghavan (✉)
School of Advanced Sciences, Centre for Nano Materials, VIT University, Vellore 632 014, India
e-mail: rvijayaraghavan@vit.ac.in

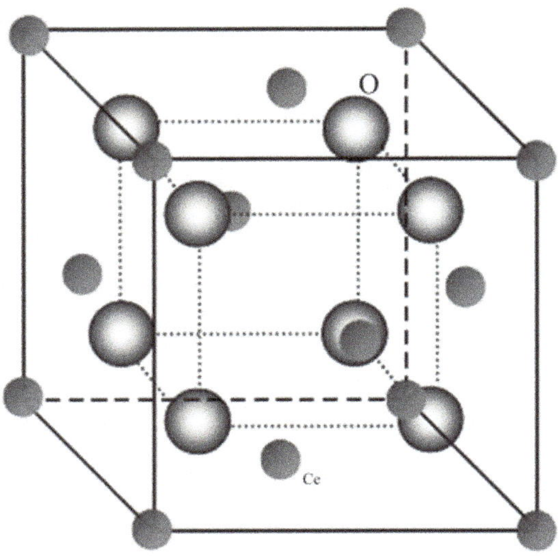

Fig. 1 Unit cell of CeO_2

Experimental

Synthesis of noble metal ion substituted ceria is carried out by solution combustion method [8]. A typical procedure is as follows. Stoichiometric quantities of ceric ammonium nitrate, H_2PdCl_6 and a fuel, urea or oxaldihydrazide are dissolved in minimum amount of water. This solution mixture is introduced into a preheated furnace kept at 500 °C. The mixture undergoes controlled combustion reaction resulting in the crystalline product within 5 min of the reaction. The products are characterized by powder XRD recorded with a Bruker D8-Advanced machine with a slow scan. X-ray photoelectron spectroscopy (XPS) was carried out by ESCA-3 Mark II VG Scientific Spectrometer using Al Kα radiation (1,486.6 eV). Binding energies reported are with respect to C (1S) at 285 eV and were measured with a precision of $\neq 0.2$ eV. The catalytic reactions were done in a temperature programmed reaction system equipped with a quadrupole mass spectrometer SX 200 (VG Scientific Ltd, England) and a Gas Chromatograph (Mayura Analytical, India) equipped with chromosorb 101 column using a thermal conductivity detector. Catalysts were placed between glass wool plugs in the centre of a quartz tube which is inserted into the furnace appropriately. The furnace is controlled by a Eurotherm temperature controller connected by a chromel–alumel thermocouple. The gaseous products were sampled through a fine leak valve via a differential pumped sampling chamber to an ultra high vacuum (UHV) system employing a quadrupole mass spectrometer. Gases were analysed by GC. Mass balance confirmed the amount of evolved gases.

Results and discussion

Figure 2 shows the XRD pattern of nanocrystalline CeO_2 by combustion method. It reveals the formation of nanocrystalline CeO_2 with crystallite size of about 15–20 nm, calculated using Scherrer formula. Pd loading into ceria is carried out by the same combustion method starting from corresponding Ce and Pd salts.

Pd incorporated ceria exhibits sharper peaks indicating bigger crystallite size (100–110 nm). The cubic lattice parameters of CeO_2 and $Ce_{0.98}Pd_{0.02}O_{2-\delta}$ are 5.413 (2) Å and 5.4107(3) Å respectively, indicating the incorporation of smaller Pd^{2+} into the site of bigger Ce^{4+}. The diffraction patterns clearly show the absence of Pd metal or PdO within the detectable limits of XRD. The XRD patterns are the same before and after CO oxidation reaction.

The XPS spectrum of Pd^{2+} substituted CeO_2 along with that of Pd metal, PdO and $PdCl_2$ is shown in Fig. 3. The binding energies of Pd 4d peak in 2 % doped Pd/CeO_2 are close to that of Pd 4d states in $PdCl_2$. This indicates that Pd in ceria is present as Pd^{2+}, in contrast to metal particles dispersed in a matrix like Al_2O_3. The redox matrix of CeO_2 facilitates the existence of ionic form of Pd in the support and such ionic species can enhance the number of active sites for the catalytic reactions.

The XPS spectrum of Ce (Fig. 4) in 2 % Pd-doped CeO_2 shows satellite peaks (marked) corresponding to CeO_2 with Ce in the +4 oxidation state [9] in the as-prepared catalyst indicating indirectly that oxygen vacancies are created on substitution of Pd^{2+} into Ce^{4+} site into the lattice of CeO_2.

To understand the oxygen storage capacity of Pd/CeO_2 catalyst, H_2-TPR is carried out where the uptake of

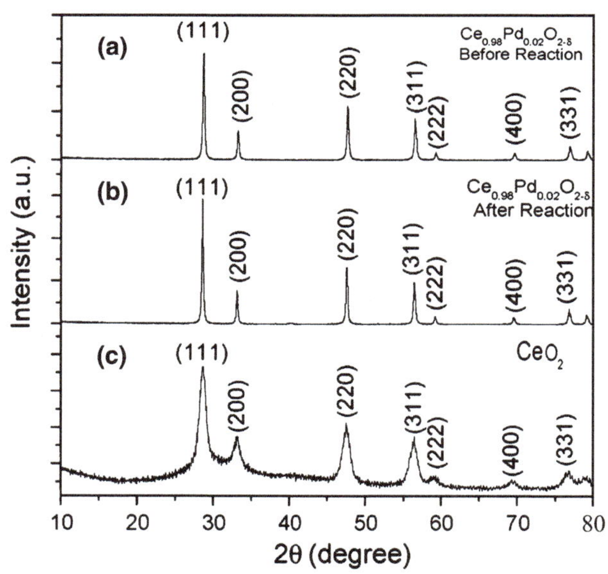

Fig. 2 XRD pattern of combustion synthesized nanocrystalline CeO_2 and Pd ion-doped CeO_2

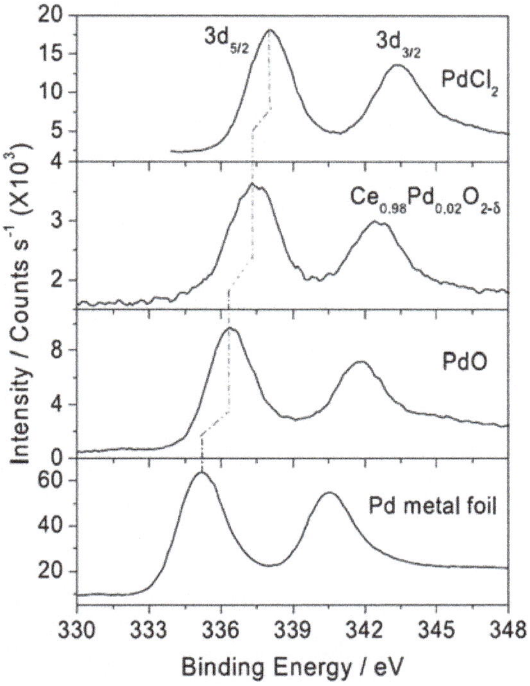

Fig. 3 XPS spectrum of Pd ion-doped ceria (from Ref. [10], with the permission of the corresponding author)

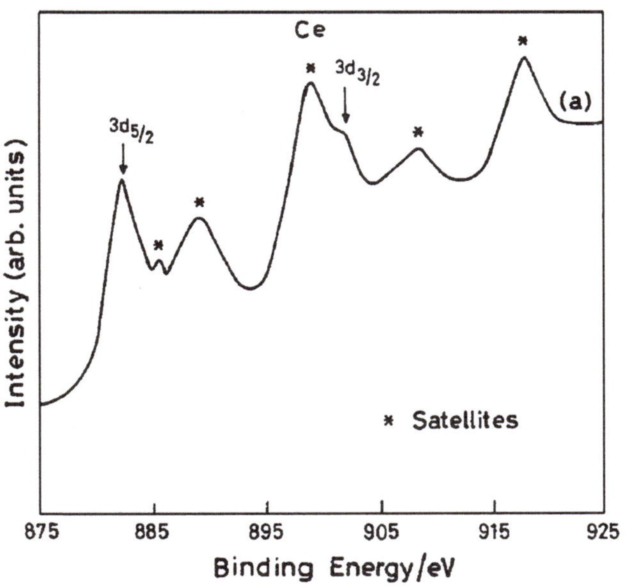

Fig. 4 XPS spectrum of Ce in Pd ion-doped ceria. (from Ref. [10], with the permission of the author)

hydrogen is measured as a function of temperature. It is depicted in Fig. 5. Pure CeO_2 shows the uptake from about 350 °C until a peak at 500 °C, whereas the Pd ion-doped CeO_2 uptakes hydrogen at a much lower temperature of 65 °C, a situation favourable for OSC.

These ionic catalysts are tested for various catalytic reactions. CO oxidation reaction is shown in Fig. 6. It is

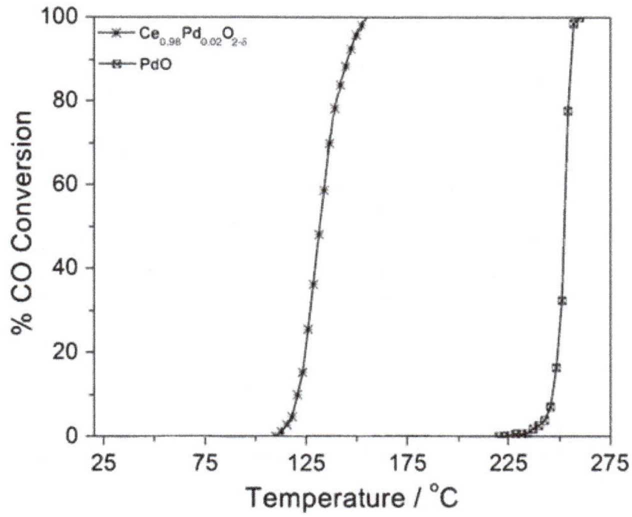

Fig. 5 Hydrogen TPR of OSC (from Ref [10], with the permission of the corresponding author)

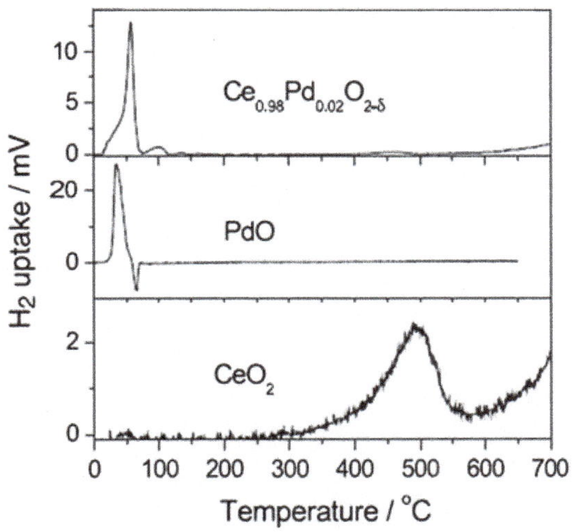

Fig. 6 CO oxidation reaction of Pd ion dispersed ceria along with PdO (from Ref. [10], with the permission of the corresponding author)

very evident from the figure that Pd ion-substituted ceria exhibits the highest CO conversion with the lowest activation energies compared to PdO revealing that Pd ion-doped CeO_2 is an efficient OSC. The rates of conversion are 20–30 times higher than the corresponding metal particles impregnated catalysts [10] and their activation energies could be obtained from Arrhenius plots of the temperature dependence of reaction rates. These catalysts have been tested for NO_x reduction also [11]. Using first principle density functional theory (DFT), it is shown that CO adsorption in these catalysts occurs at the ionic sites of Pd^{2+}, with a decrease in the net energy of the system [12].

Conclusion

Pd^{2+}-dispersed CeO_2 oxygen storage catalysts have been synthesized by solution combustion method resulting in noble metal ions dispersed OSC. We have shown that these ionically dispersed catalysts are catalytically more active towards exhaust emission reactions like CO oxidation.

Acknowledgments The author thanks KOPRC for the invitation to 2nd KACST-Oxford forum meeting at Riyadh in 2012.

References

1. Gandhi HS, Graham GW, McCabe RW (2003) Automotive exhaust catalysts. J Catal 216:433–442
2. Yao HC, Yu Yao YF (1984) Ceria in automotive exhaust catalysts I. Oxygen storage. J Catal 86:254–258
3. Taylor KC (1993) Nitric oxide catalysis on automotive exhaust system. Catal. Rev. Sci. Eng. 35:457–481
4. Kaspar J. Graziani, Fornasia P (2000) Ceria containing three way catalysts. In: Gschneidner KA Jr, Eyring L (eds) Hand book on the physics and chemistry of rare earths, vol 29. Elsevier, Amsterdam, pp 159–267
5. Kaspar J, Fornasiero P, Hickey N (2003) Automotive catalytic converters: present status and some perspective. Catal Today 77:419–449
6. Ranga Rao G (1999) Influence of metal particles on the reduction properties of ceria based materials studies by TPR. Bull Mater Sci 22:89–94
7. Ranga Rao G, Kaspar J, Merioni S, Monte RD, Graziani M (1994) NO decomposition over partially reduced metallized CeO_2 –ZrO_2 solid solution. Catal Lett 24:107–112
8. Vijayaraghavan R et al (unpublished results on ceria based perovskite catalysts)
9. Sarma DD, Hegde MS, Rao CNR (1981) Study of surface oxidation of rare-earth metals by photoelectron spectroscopy. J Chem Soc Faraday Trans 277:1509–1520
10. Bera Parthssarathy, Hegde MS (2010) Recent advances in auto exhaust catalysts. J. Indian Inst Sci 90–2:299–325
11. Hegde MS, Madras Giridhar, Patil KC (2009) Noble metal ionic catalysts. Acc Chem Res 42:704–712
12. Dutta G, Gupta A, Waghmure UV, Hegde MS (2011) CO adsorption in ionic Pt, Pd and Cu sites in Ce1-xMxO2-δ (M = Pt2+, Pd2+, Cu2+). J Chem Sci 123:509–516

11

Direct oxidation of cyclohexane to adipic acid using nano-gold catalysts

A. Alshammari · A. Köckritz · V. N. Kalevaru · A. Bagabas · A. Martin

Abstract Adipic acid (AA) is one of the most important commercially available aliphatic dicarboxylic acids, which has a large industrial application in the manufacture of Nylon-6 and Nylon-66. The present work demonstrated a green chemistry route for the direct oxidation of cyclohexane (CH) to AA using supported nano-gold catalyst in one-step reaction. The catalysts were prepared in two steps and were then characterized by different methods such as ICP, BET surface area, XRD, SEM, TEM, etc. Catalytic tests were carried out using Parr autoclave in the temperature range of 100–170 °C, and the products were analyzed by gas chromatograph. Among all investigated catalysts, TiO_2 supported nano-gold exhibited the superior activity compared to all other tested catalysts. The conversion of CH and the selectivity of AA obtained over TiO_2 supported catalyst were 16.4 % and ca. 21.6 %, respectively, without a noticeable change in the catalyst stability. In summary, it is possible to produce AA with ca. 8 % yield using nano-gold supported on TiO_2.

Keywords Adipic acid · Cyclohexane · Nano-gold · Titanium dioxide

A. Alshammari (✉) · A. Bagabas
National Nanotechnology Research Centre, King Abdulaziz City for Science and Technology, King Abdullah Road, P.O. Box 6086, Riyadh 11442, Saudi Arabia
e-mail: aalshammari@kacst.edu.sa

A. Köckritz · V. N. Kalevaru · A. Bagabas · A. Martin
Leibniz-Institut für Katalyse e.V. an der Universität Rostock, Albert-Einstein-Str. 29a, 18059 Rostock, Germany

Introduction

The selective oxidation of cyclohexane (CH) to adipic acid (AA) is an industrially important reaction for manufacturing various valuable materials such as polyamides (e.g., nylon 6,6), polyurethanes, polyesters, plasticizers, intermediates for pharmaceuticals and insecticides, etc. [1, 2]. AA is also used in medicine and food industry for different applications [3, 4]. The current commercial production process of AA on a commercial scale is made up of two steps. The first step results in a mixture of cyclohexanone and cyclohexanol (i.e., KA—K for ketone and A for alcohol) at around 150 °C and at 10–20 bar of air using a cobalt or a manganese catalyst [2, 5]. KA can also be obtained by phenol hydrogenation, as depicted in Scheme 1. The second step, is an oxidation of KA into AA using nitric acid [6]. This method is environmentally harmful, and it is costly and energy demanding. The generation and liberation of NO_X gases, resulted from the reduction of nitric acid, harm the environment while recycling >90 % of un-reacted CH increases the production cost and the energy demand. Therefore, it is a must to develop a novel, environmentally benign for the production of AA.

Besides the commercial process, there are other routes for producing AA. For examples, AA can be obtained by direct oxidation of CH using hydrogen peroxide, by carbonylation of butadiene, by dimerization of methyl acrylate, or by fermentation of glucose (Scheme 1). In spite of these various routes, the direct oxidation of CH to AA in one step using O_2, as an oxidant, is indeed an economic and environmentally friendly approach, which is the core of this work.

The use of heterogeneous, solid catalyst in the direct oxidation of CH to AA is also known from the prior art. For example, iron phthalocyanine encapsulated in Y-zeolite was applied as a catalyst for the direct oxidation of CH to AA [7]. However, this process strongly suffers from much longer induction periods, i.e., the catalyst requires about 300 h to reach CH conversion of ca. 35 % and needs 600 h to get higher amounts of adipic acid in the product stream, which makes the process commercially unattractive. Furthermore, efforts were also made by various researchers to use gold-based catalysts for the direct oxidation of CH to AA, but to the best of our knowledge, all such attempts went unsuccessful until now. For instance, various gold catalysts such as Au/graphite, Au/MCM-41 [8], Au/SBA-15, and SiO_2 [9] supports were applied for the said reaction, which gave only cyclohexanol and cyclohexanone as major products without any adipic acid in the product stream. Using such catalyst systems, the conversion of CH was varied in the range from 6 to 20 %, but almost no formation of adipic acid was reported. However, the selectivity of both cyclohexanol and cyclohexanone products together were found to be in the range of 17–90 % depending on the catalyst system used and the reaction conditions applied.

On the whole, the catalyst systems used in the prior art are completely different from ours. The goal of the present paper is therefore to provide a direct method for producing AA in a single step with acceptable selectivity from the oxidation of CH using effective and potential catalyst compositions. The work also aims to supply an easy method for preparing the catalyst and its use in the said oxidation reaction.

Experimental

Materials

Tetrachloroauric acid (99 %, Fluka), tri-sodium citrate dehydrate (99.5 %, Fluka), and tannic acid (98 %, Aldrich) were commercially available and were used as received. Commercially available metal oxides were also used. Deionized water (18.2 MΩ cm) was obtained from a Milli-Q water purification system (Millipore).

Catalyst preparation

The preparation of supported gold nanoparticles (AuNPs) on different metal oxides was performed in two steps. The first step involves the preparation of colloidal AuNPs by the reduction of $HAuCl_4$ (0.15 mmol) in aqueous solution using 1 % tannic acid, 1 % sodium citrate, and K_2CO_3. The second step involves the impregnation of the colloidal AuNPs with a suitable oxidic support (e.g., CaO, MgO, ZrO_2, Al_2O_3, and TiO_2). After impregnation (step 2), the slurry was vigorously stirred for 2 h at room temperature and then the excess solvent was removed on rotary evaporator. The obtained solid was washed three times with water, and then oven dried 120 °C for 16 h. The dried sample was finally calcined at 350 °C for 5 h in air.

Catalyst characterization

Inductively coupled plasma (ICP)

ICP results of catalysts were obtained by PERKIN ELMER instrument (model: optima 3000XL), using a microwave pressure digestion (MDS 200; CEM) with hydrofluoric and aquaregia.

Thermal analysis

Thermogravimetric and differential thermal analysis (TG–DTA) were performed with a TG apparatus (Shimadzu TGA-50). Catalysts of about 10 mg were heated from 20 to 900 °C at a rate of 5 °C/min and a flow rate of 25 mL/min in air atmosphere.

BET surface area

BET surface areas (BET-SA) and pore size distribution of the catalysts were obtained on Micrometrics Gemini III-2375 instrument by N_2 physisorption at 77 K. Prior to the measurements, the known amount of the catalyst was evacuated for 2 h at 150 °C to remove physically adsorbed water.

X-ray diffraction (XRD)

X-ray diffraction results of the catalysts were recorded on a STADI P (STOE) setup with transmission geometry and equipped with a Ge primary monochromator with CuK_α radiation in the 2θ ranges from 5° to 60° and a position-sensitive detector. The diffraction peaks of the crystalline phase were compared with those standard compounds reported in the JCPDS Date file.

X-ray photoelectron spectroscopy (XPS)

X-ray photoelectron spectroscopy analysis was carried out in an ESCALAB 220iXL spectrometer after exposure the sample to air at room temperature. Measurements were carried out at constant energy of 150 eV (survey), 25 eV (for quantification) and without charge compensation. Monochromatic Al-Kα radiation (1,486.6 eV) was applied as the X-ray source. The binding energy scale was calibrated with pure and clean Cu, Ag, and Au samples.

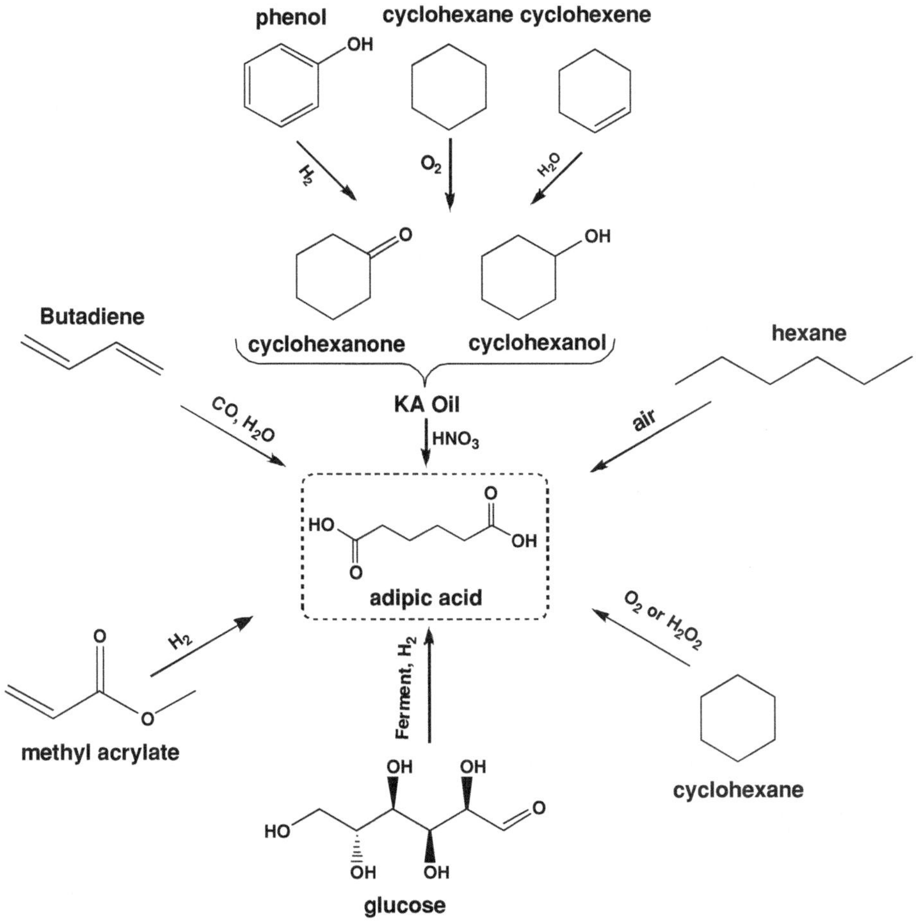

Scheme 1 Summary of the different pathways for AA production

Transmission electron microscopy (TEM)

TEM analysis was carried out using JEM-2100F high-resolution electron microscopy at a voltage of 200 kV. Sample preparation of supported AuNPs samples were dispersed in water/methanol and treated with ultrasound for 5 min, and then deposited on a carbon coated grid.

Catalytic measurement

Activity tests were carried out under pressure using a Parr autoclave (100 mL) according the procedure described below. In a typical experiment, the reaction mixture consists of 0.3–0.4 g of supported gold catalyst, 5 mL of cyclohexane (CH), 25 mL of acetonitrile as solvent, and 0.1 g of *tert*-butyl hydroperoxide (TBHP), unless otherwise stated. These components were taken in an autoclave and flushed three times with O_2 before setting the initial reaction pressure of O_2 to 10 bar. Concerning the start-up procedure, it was performed with the O_2 line opened, and as the O_2 was consumed, it was replaced from the cylinder, which maintains the overall pressure constant. The stirring speed of reaction mixture was set to 1,500 rpm in general and the reaction was performed at 150 °C for 4 h unless otherwise stated. At the end of the reaction, the solid catalyst was separated by centrifugation.

Results and discussion

Catalysts characterization

Thermal analysis

Thermogravimetric (TGA) analysis of uncalcined Au catalysts supported on different metal oxides (e.g., MgO, CaO, ZrO_2, TiO_2, Al_2O_3, etc.) are presented in Fig. 1a. It is clear from the figure that the stability of Au catalyst is depended on the nature of the supports used. It should be noted that the precursors of MgO and CaO are $Mg(OH)_2$ and $Ca(OH)_2$. All studied catalysts have shown that the weight loss mainly occurs in two stages. First stage occurred in the temperature range from r.t. to 150 °C, which is due to the loss of water molecules weakly bound to the material. Second stage observed between 150 and 350 °C, seems to be due to dehydroxylation and also to the decomposition of reducing agents used. The weight loss observed on the whole temperature range from 5 to

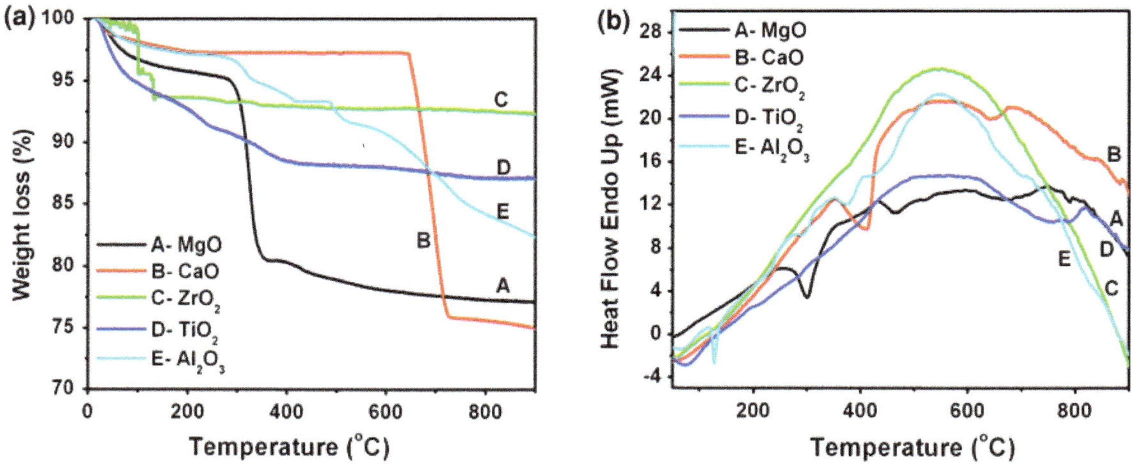

Fig. 1 TGA (**a**) and DTA (**b**) profiles of Au catalysts supported on different metal oxide carriers

~20 %, which however depends on the type of support used. Among all catalysts, alumina supported one is losing weight continuously with rise in temperature, while all others except CaO supported one showed only a marginal weight loss in the temperature range from 350 to 900 °C. The sudden weight loss observed in case of MgO supported catalyst at around 330 °C can be attributed to the conversion of $Mg(OH)_2$ into MgO [10]. Interestingly, CaO supported solid displayed quite stable behavior up to ca. 630 °C and then showed abruptly ca. 20 % wt. loss at this temperature. This is undeniably due to transformation of $Ca(OH)_2$ into CaO. Holgado et al. (1992), also observed different stages of weight loss in case of $Ca(OH)_2$ decomposition to CaO, the first one at 323 °C (3 % wt. loss) and the second one at 623 °C (wt. loss of 17 %) and the third one at 880 °C (wt. loss of 7 %). Our results are in good agreement with those reported by Hologado et al. [11]. In addition, AuNPs supported on reducible carriers (e.g., ZrO_2, TiO_2) are found to have the lowest weight loss in the whole temperature range among all investigated samples, which is about 6 and 11 %, respectively.

In addition, differential thermal analysis (DTA) patterns of the gold catalysts were also carried out, and the results are given in Fig. 1b. The initial weight loss observed by TGA for the all catalysts from 20 to 300 °C are also evidenced by a small endothermic transition in the DTA curve. Such endothermic effect is observed from 300 to 450 °C on the whole in almost all samples, which can be ascribed to the loss of water by dehydroxylation, decomposition of metal precursors to form oxides and also due to decomposition of organic reductants. Almost all catalysts, exhibited a broad exothermic peak in the range of 470 to ca. 700 °C, which seems to be due phase transformation of supports (e.g., TiO_2 (anatase) to TiO_2 (rutile), monoclinic ZrO_2 to tetragonal ZrO_2, etc. It is also more likely that such exothermic peaks are due to the formation of metal oxide phase of the used support. XRD analysis of these also gave some supporting information on such assumption, which will be discussed later on.

BET surface areas

The specific surface areas and pore volumes of different pure supports and gold catalysts supported on different metal oxides are presented in Table 1. It is clear from the table that the surface areas and pore volumes are observed to change significantly by changing the nature of support materials. It is obvious that the surface area of most supports were drastically decreased after impregnation into the AuNPs. Nevertheless, for both TiO_2 and Al_2O_3, an addition of AuNPs did not show a noticeable effect on the values of BET surface areas. These results indicate that the AuNPs are highly dispersed on the support without any sintering effect. Such good dispersion is also confirmed by XRD and TEM, which will be shown later.

X-ray diffraction analysis

The XRD patterns of the fresh Au catalysts are displayed in Fig. 2. Different oxide supports have been used such as

Table 1 BET-SA and pore volumes of gold catalysts on various oxide supports

Entry	Catalyst	BET-SA (m^2/g)		Pore vol. of cat. (cm^3/g)
		Pure support	Catalyst	
1	Au/MgO	59	34	0.09
2	Au/CaO	37	27	0.07
3	Au/ZrO_2	53	39	0.1
4	Au/TiO_2	47	43	0.12
5	Au/Al_2O_3	265	261	0.80

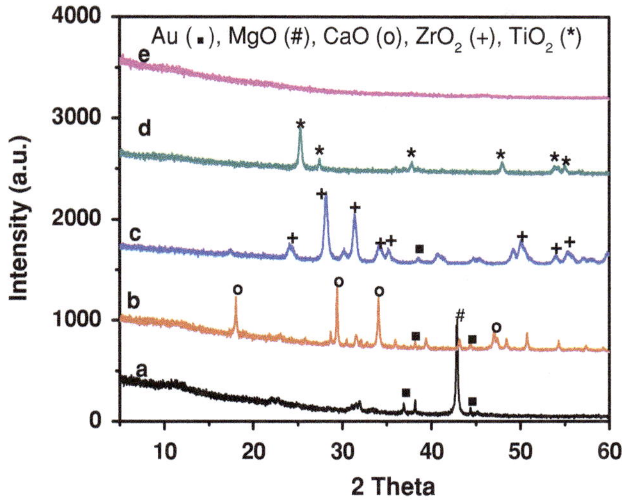

Fig. 2 XRD patterns of fresh 1 % Au catalysts with different supports A—(*a* MgO, *b* CaO, *c* ZrO_2, *d* TiO_2, and *e* Al_2O_3)

Table 2 Binding energy and the surface atomic ratio of supported gold catalysts

Catalyst	Binding energy (eV)				Au/M atomic ratio
	Au ($4f_{7/2}$, $4f_{5/2}$)	M	O	C	Au
Au/MgO	83.3	458.8	531.5	286.9	0.009
	87.4	472.0	538.9	288.2	
Au/CaO	83.5	350.5	534.4	285.2	0.008
	87.6		538.7	287.4	
Au/ZrO_2	83.6	182.2	529.8	285.3	0.011
	87.6	184.9	531.2	289.1	
Au/TiO_2	83.4	458.9	529.9	285.7	0.016
	87.6	472.0	532.5	289.2	
Au/Al_2O_3	83.7	74.5	523.3	285.4	0.012
	87.6		531.1	288.3	

MgO, CaO, ZrO_2, TiO_2, and Al_2O_3. In general, the XRD pattern of Au shows the typical diffraction pattern of metallic Au by the presence of two diffraction lines one at $2\theta = 38.2°$ ($d = 2.35$ Å) and the other at $2\theta = 43.4°$ ($d = 2.03$ Å). Among all studied catalysts, Al_2O_3 powder supported Au catalysts is found to be X-ray amorphous, which is in agreement with BET surface area data (261 m^2/g). The XRD of the fresh Au/TiO_2 sample also showed no patterns due to Au, indicating that the gold particle size is less than 3 nm or Au has been doped into TiO_2 lattice. This observation lent good support to the observations occurred by TEM that confirmed the size of Au found to be less than 3 nm. However, the sample reveals reflections that corresponding to TiO_2. In contrast, the XRD results of the ZrO_2 supported Au catalysts showed that a weak reflections correspond Au phase besides the typical zirconia reflections (e.g., monoclinic phase). Moreover, gold catalysts supported on MgO and CaO exhibited the gold metal phases in addition to intense reflections that corresponding to MgO and CaO. Hints on the formation of bigger AuNPs are also provided by TEM and SEM (not shown here).

X-ray photoelectron spectroscopy

The summary of XPS results such as binding energy (B.E.), atomic ratios obtained from supported gold catalysts are given in Table 2. It can be seen that the B.E. values of Au $4f_{7/2}$ and $4f_{5/2}$ spectral lines of Au are varied over a narrow range from 83.3 to 83.7 eV, and 87.4 to 87.6 eV, respectively. These values clearly indicate that the gold in the catalysts is present mainly in metallic form, i.e., Au(0). This is good agreement with reported results in the literature [12, 13]. The most striking feature here is the clear enrichment of Au in the near-surface-region particularly in the Au/TiO_2 solid compared to others as evidenced by the Au/metal ratios. The values of spectral lines correspond to other components of the catalysts, e.g., supports, O as well as C are also summarized for better comparison in Table 2.

TEM study

TEM studies were used to explore such properties since the catalytic activity of gold catalyst is strongly depended on its particle size. It reported that the nature of the support can affect the size, morphology, and dispersion of Au particles

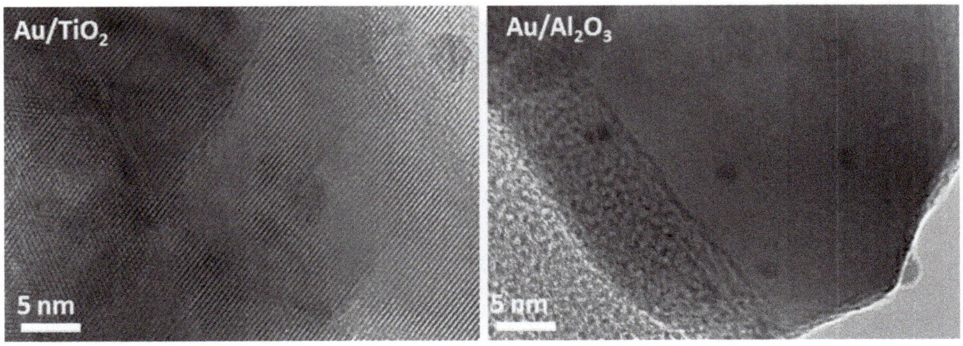

Fig. 3 TEM images of AuNPs supported on TiO_2 and Al_2O_3

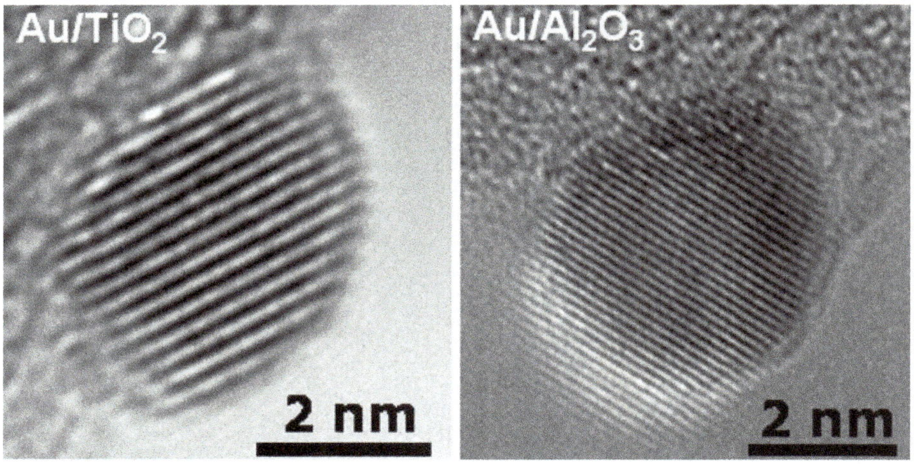

Fig. 4 HRTEM images of the AuNPs catalysts supported on TiO_2 and Al_2O_3 carriers

via metal-support interactions [14]. Representative of selected TEM images of Au catalysts supported on different metal oxides carriers (e.g., TiO_2, Al_2O_3) are presented in Fig. 3. It is obvious that the resulting Au particles show almost spherical shape dispersed on the supports and the nature of support has considerable influence on the particle size. Among all catalysts, TiO_2 and Al_2O_3 carriers found to give particles size with a narrow size distribution in the range from 1 to 5 nm. However, in case of non-reducible supports (e.g., MgO, CaO, not shown here), there is a considerable increase in the Au particles, which varies in the range from 1 to 10 nm, while the morphology of the particles remained unaltered. The sequence of increasing mean Au particle size for the gold catalysts supported on different carriers are $Au/TiO_2 < Au/Al_2O_3 < Au/ZrO_2 < Au/CaO < Au/MgO$, suggests a dependency on the reducibility of the carrier. Nevertheless, in terms of activity, Au/TiO_2 and Au/Al_2O_3 are found to be the best catalysts. Therefore, these supports are much considered in our discussion.

Table 3 The inter planar spacing and diffraction planes of AuNPs supported on TiO_2 carrier

d-spacing calculated from HRTEM (Å)	*d*-spacing in bulk Au (Å)	Miller indices (*hkl*) assignment
2.35	2.35	111
2.06	2.04	200
1.43	1.44	220
1.18	1.17	222

Table 4 The interplanar spacing and diffraction planes of AuNPs supported on Al_2O_3 carrier

d-spacing calculated from HRTEM (Å)	*d*-spacing in bulk Au (Å)	Miller indices (*hkl*) assignment
2.38	2.35	111
1.41	1.44	220
1.16	1.17	222

High-resolution TEM

In order to control the nanocrystal shape of Au and further to understand the growth mechanism, high-resolution TEM (HRTEM) was performed. This technique has provided us information regarding the nature of the crystal faces. HRTEM images of AuNPs supported on TiO_2 and Al_2O_3 are selected as examples and shown in Fig. 4. These images obviously revealed that the metal particles are almost spherical and most of them are highly diffused into the matrix of the both supports. However, some of them are also located on the surface of the support. From HRTEM image of AuNPs supported on Au/TiO_2 and Au/Al_2O_3 solids, one can clearly observe the crystal planes of Au. The lattice plane fringes of the AuNPs are used to calculate the *d*-spacing values, and were compared with those of bulk Au (the values in Tables 3 and 4 correspond to images of Au/TiO_2 and Au/Al_2O_3 of Fig. 4), indicating the formation of Au nanocrystals with spherical lattice. These results confirmed that precursor of $HAuCl_4$ was reduced to Au^0.

Catalysts evolution

Prior to perform the catalytic tests, some blank tests were carried out initially under similar conditions as that of real tests to determine whether the oxidation of CH can occur in the absence of catalyst and/or TBHP, in particular according to a radical mechanism. Such blank tests (in the absence of catalyst but in the presence of TBHP, and also in the absence of both TBHP and catalyst) showed almost negligible CH conversion of only ~2 % after 4 h of reaction and no AA was found in the product stream. This result clearly implies that (1) no significant reaction takes place under the conditions applied and (2) the nature of catalyst plays a key role on the performance. The influence

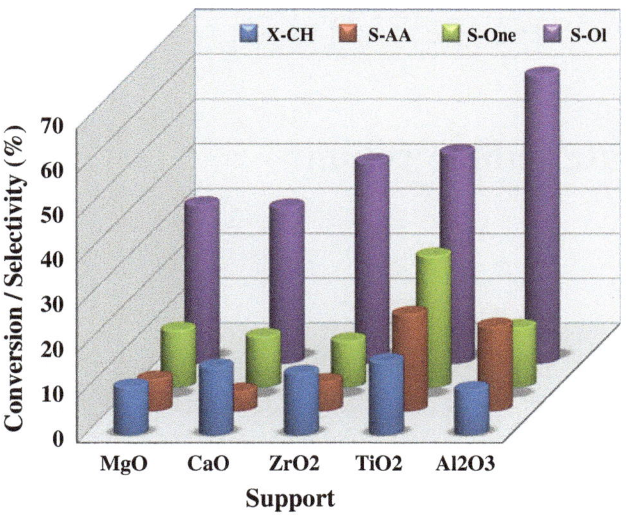

Fig. 5 Effect of support on the oxidation of cyclohexane (CH) over AuNPs/SU catalysts (SU = MgO, CaO, ZrO_2, TiO_2, Al_2O_3), X-CH, conversion of cyclohexane; S-AA, selectivity to adipic acid; S-One, selectivity to cyclohexanone; S-Ol, selectivity to cyclohexanol

of oxide supports on the oxidation activity of CH to AA is presented in Fig 5. It is obvious that the nature of support has a significant influence on the conversion of CH as well as the selectivity of products. TiO_2 supported AuNPs exhibited the superior activity compared to all other catalysts. In addition, these results provide further evidence that the smaller size of Au particle is necessary and is also a crucial parameter for improving the performance of the catalysts. Again in this case, the MgO & CaO supported catalysts gave the poorest performance. Over these two catalysts, the highest amount of undesired side-products such as CO and CO_2 (S = up to 35 %) is observed. Therefore, it can be stated that MgO and CaO are not really suitable supports for this reaction. The conversion of CH and the selectivity of AA obtained over TiO_2 supported catalyst is 16.4 % and ca. 21.6 %, respectively [15]. Several efforts were also made in the literature by different research groups to used gold-based catalysts for the direct oxidation of cyclohexane to adipic acid, but not even 1 % yield of adipic acid could be achieved from those investigations.

Conclusions

In this study, AuNPs supported on TiO_2 was determined to be an optimum catalyst for the direct oxidation of cyclohexane to adipic acid. We have also proved that the size of Au plays a key role on the catalytic performance. In particular, it is proved that the production of adipic acid with 8 % yield through the one-pot oxidation of cyclohexane is indeed a remarkable achievement of this study, with reasonable catalyst stability.

Acknowledgments The authors gratefully thank King Abdulaziz City for Science and Technology (KACST) for financing this work. The authors would also like to thank M. Ababtain (KACST) for TEM analysis and Schneider and Radnik (LIKAT) for solid analyses.

References

1. Castellan A (1991) Industrial production and use of adipic acid. Catal Today 9:237–254
2. Schuchardt U, Cardoso D, Sercheli R, Pereira R, Cruz RS, Guerreiro MC, Mandelli D, Spinace EV, Pires EL (2001) Cyclohexane oxidation continues to be a challenge. Appl Catal A Gen 211:1–17
3. Horn HJ, Holland EG, Hazleton LW (1957) Safety of adipic acid as compared with citric and tartaric acid. Agric Food Chem 5:759–762
4. Roew R (2009) Adipic acid: handbook of pharmaceutical excipients, Pharmaceutical Press, London, pp 11–12
5. Sheldon RA, Kochi JK (1981) Metal-catalyzed oxidations of organic compounds. Academic Press, New York
6. Saji PV, Ratnasamy C, Gopinathan S (2002) Process for the oxidation of cyclohexane to adipic acid. U.S. Patent 6,392,093, B1
7. Starzyk FT (1994) Oxidation of cyclohexanone and cyclohexane to adipic acid by iron-phthalocyanine on zeolite Y. Stud Surf Sci Catal 84:1419–1424
8. Lu G, Zhao R, Qian G, Qi Y, Wang X, Suo J (2004) A highly efficient catalyst Au/MCM-41 for selective oxidation cyclohexane using oxygen. Catal Lett 97:115–118
9. Xie J, Wang Y, Li Y, Wei Y (2011) Self-assembly preparation of Au/SiO_2 catalyst and its catalysis for cyclohexane oxidation with air. Reac Kinet Mech Catal 102:143–154
10. Lee MH, Park DG (2003) Preparation of MgO with high surface area, and modification of its pore characteristics. Bull Korean Chem Soc 24:1437–1452
11. Hologado MJ, Rives V, Román SS (1992) Thermal decomposition of $Ca(OH)_2$ from acetylene manufacturing: a route to supports for methane oxidative coupling catalysts. J Mater Sci Lett 11:1708–1710
12. Hutchings GJ, Hall MS, Carley AF, Landon P, Solsona BE, Kiely CJ, Herzing A, Makkee M, Moulijn JA, Overweg A, Fierro-Gonzalez JC, Guzman J, Gates BC (2006) Role of gold cations in the oxidation of carbon monoxide catalyzed by iron oxide-supported gold. J Catal 242(1):71–81
13. Schimpf S, Lucas M, Mohr C, Rodemerck U, Brückner A, Radnic J, Hofmeister H, Claus P (2002) Supported gold nanoparticles: in-depth catalyst characterization and application in hydrogenation and oxidation reactions. Catal Today 72:63–78
14. Bond GC (1997) Characterization of solid catalysts. In: Ertl G, Knoezinger H, Weitkamp J (eds) Handbook of heterogeneous catalysis, vol 2. VCH, Weinheim, p 752
15. Alshammari A, Köckritz A, Kalevaru VN, Martin A (2012) Method for catalyst preparation and its use in the oxidation of saturated hydrocarbons to carboxylic acids. US Patent 13/269,988, A1

12

Modeling an industrial sodium bicarbonate bubble column reactor

Ataallah Soltani Goharrizi · Bahador Abolpour

Abstract This paper deals with the study of the gas–liquid mass transfer, coupled with chemical reactions, the gas–liquid-solid mass transfer, and crystallization. Sodium bicarbonate is produced in a bubble column reactor which contains a solution of carbonate and bicarbonate and carbon dioxide gas injected into this column. In this mathematical modeling, a mole balance has been instituted on flows and components through the bubble column. A population balance is utilized to obtain the nucleation and growth formula for the solid phase. Danckwerts theory is utilized for mass transfer between gas and liquid phases. This model can predict the effects of several parameters on the production and size distribution of sodium bicarbonate crystals and also conversion of carbon dioxide. The mathematic simulator model results are compared with the experimental results to valid this model. Effects of different parameters on the production and size distribution of sodium bicarbonate crystals and also absorbent of carbon dioxide are investigated.

Keywords Sodium bicarbonate · Bubble column · Triple phase mass transfer · Crystallization · Modeling

List of symbols

a	Growth rate coefficient
b	Growth order (g)
B_0	Nucleation (#/s kg solution)
C_{CO_2e}	Carbon dioxide concentration in liquid bulk (mol/m^3)
c	Nucleation rate coefficient
C_{CO_2i}	Carbon dioxide concentration in gas–liquid interface (mol/m^3)
d	Magma density order
d_B	Average of bubbles diameter (m)
D_{CO_2}	Carbon dioxide molecular diffusion coefficient (m^2/s)
D_g	Gas phase radial dispersion coefficient (m^2/s)
D_l	Liquid phase radial dispersion coefficient (m^2/s)
d_R	Column diameter (m)
E	Enhancement factor
e	Nucleation order
g	Gravity (m/s^2)
G	Molar velocity of gas (mol/m^2 s)
G_0	Growth rate (μm/s)
G_{FR}	Gas flow rate (m^3/s)
h	Column height (m)
H	Henry constant (Kmol/atm m^3)
I	Ionic content of solution (kg ion/m^3)
I_0	Height of column that sodium bicarbonate reachs to saturation concentration
k	Constant rate of first order reaction (1/s)
K_l	Liquid phase mass transfer coefficient (m/s)
L	Molar velocity of liquid (mol/m^2 s)
L_{FR}	Liquid mass flow rate (kg/s)
L_{nu}	Size of crystals (μm)
M_T	Magma density (g crystal/kg solution)
N	Flux of mass transfer (mol/m^2 s)
n	Population density (no./μm kg solution)
n_{Dis}	Flux of mass transfer by dispersion (mol/m^2 s)
no	Number of nucleons that born at dz
P_0	Gas pressure at bottom of column (atm)
P_{CO_2}	Carbon dioxide partial pressure at gas phase (atm)

A. S. Goharrizi · B. Abolpour (✉)
Department of Chemical Engineering, Faculty of Engineering, Shahid Bahonar University of Kerman, Jomhoori Blvd, 76175 Kerman, Iran
e-mail: bahadorabolpor1364@yahoo.com

P_{CO_2e} Carbon dioxide partial pressure at liquid bulk (atm)
P_{CO_2i} Carbon dioxide partial pressure at gas–liquid interface (atm)
Q Liquid flow rate (m^3/s)
r Rise of nucleons from dz to bottom of the column (m^3)
R Gases constant (J/g mol K)
S Molar velocity of solid (mol/m^2 s)
T Liquid temperature (K)
t Time
U_g Gas phase velocity (m/s)
U_l Liquid phase velocity (m/s)
V Volume of element (m^3)
w Weight fraction of component in liquid phase
Δw Supersaturation (g $NaHCO_3$/kg solution)
x Mole fraction of components in liquid phase
$x^*_{NaHCO_3}$ Mole fraction of sodium bicarbonate at supersaturation
y Mole fraction of components in gas phase
z Height of column (m)

Greek symbols

α_g Gas–liquid interface (m^2/m^3)
α_s Solid–liquid interface (m^2/m^3)
δ Liquid surface tension (N/m)
ε_g Gas holdup
ε_l Liquid holdup
μ_j jth moment of population density
μ_l Liquid dynamic viscosity (N s/m^2)
ρ_{H_2O} Water density (kg/m^3)
ρ_l Liquid density (kg/m^3)
υ_l Liquid kinematic viscosity (m^2/s)

Subscripts

i At inlet point position of element
I_0 At I_0 point position
in At inlet point position of column
o At outlet point position of element

Introduction

Sodium bicarbonate crystals have a low solubility in water. These crystals decompose to carbonate and carbon dioxide at a temperature of more than 60 °C. Bubble column reactors are used to produce sodium bicarbonate. These columns have high rate of mass transfer, small volumes, and low costs of operation. Industrial crystallization processes involve a large number of complex physical and chemical phenomena. To control the quality of the product, it is essential to understand the role of each parameter in the nucleation and growth processes.

Absorbance of carbon dioxide in a solution of sodium bicarbonate and sodium carbonate, in a lab scale has been simulated by Ref. [1]. Those experimentations were done in different temperatures and catalyst concentrations and the rate of gas–liquid interface changes, liquid mass transfer coefficient, and reaction constant rate were computed using Danckwerts model [2]. Comparing experimental and model results, they concluded that gas film resistance has no effect on mass transfer and can be neglected. The sodium bicarbonate settlement process in industrial scale bubble columns is investigated by Ref. [3]. They used computational fluids dynamic's method and compared results with operating data.

Estimation of sodium bicarbonate crystal size distributions of such bubble column reactors is presented in the previous study [4]. Using a constant nucleation and growth rate of sodium bicarbonate crystals ($B_0 = 26.685 \times M_T^{0.42} \times \Delta w^{1.3}$, $G_0 = 1.381 \times 10^{-4} \Delta w^{1.53}$ that experimentally were calculated by Saberi et al. [5], at an unsteady-state condition) in different operating condition decreased the accuracy of model results. The main purpose of the present study is modeling and simulating sodium bicarbonate production process in bubble columns reactors. Flow in these reactors is three phase (solid, liquid and gas) and conjugated with a set of chemical reactions. Therefore, the respective model is based on the three-phase mass balance, Danckwerts theory for mass transfer from gas phase to liquid, absorbance of CO_2 with mass transfer and chemical reaction. This mathematical model can predict the effects of several parameters (i.e. liquid temperature, gas pressure, and mole fraction of carbon dioxide in gas phase) on the production and size distribution of sodium bicarbonate crystals, and also conversion of carbon dioxide.

In this research, the production and size distribution of sodium bicarbonate crystals, and conversion of carbon dioxide in the bubble column reactor of Shiraz Petrochemical Complex are studied. This study contains a triple-phase modeling to simulate the operation of this bubble column. In this model, mole and population balances are utilized to obtain the nucleation and growth formula for the solid phase. This is a novel method to obtain the nucleation and growth rates of produced sodium bicarbonate crystals in different operating conditions. It must be noted that, the production of sodium bicarbonate crystals in this bubble column is a complex operation and simulation of this phenomenon can be used to optimize the production and size distribution of these crystals.

Process description

The process of sodium bicarbonate producing is shown in Fig. 1. This bubble column has the inner diameter of 1.2 m

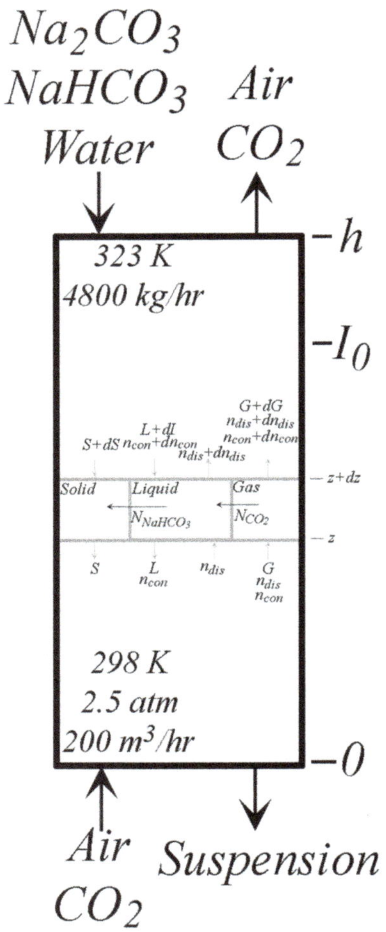

Fig. 1 Production of sodium bicarbonate in bubble column

with the height of 22 m. The temperature of operation is assumed to be constant in the length of column due to low heat of reaction and using of a water jacket around it. The solution contains water, sodium carbonate and sodium bicarbonate. This solution enters the column from its upper side and a gaseous mixture of carbon dioxide and air is continuously injected to the column from its bottom. The following equilibrium reactions can be established in the column:

$$CO_2(g) \leftrightarrow CO_2(l) \quad (1)$$

$$CO_2(l) + OH \leftrightarrow HCO_3 \quad (2)$$

$$HCO_3^- + OH^- \leftrightarrow CO_3^{2-} + H_2O \quad (3)$$

$$Na^+ + HCO_3^- \rightarrow NaHCO_3. \quad (4)$$

When the concentration of produced sodium bicarbonate reaches its solubility limit in the solution (solution becomes saturated from sodium bicarbonate), reaction (4) takes place and the crystals begin to form in the column [6]. The column operating conditions is shown in Fig. 1. The solubility of $NaHCO_3$ in the presence of Na_2CO_3 is given in the Ref. [7] as below:

$$\text{Log}(x^*) = 6.71535 - \frac{843.0681}{T} - 2.24336 \times \log(T). \quad (5)$$

Model formulation

The governing and mass transfer equations are obtained using a one-dimensional differential control volume in the length of reactor, as shown in Fig. 1. Since there exists a rotational flow in this column, a radial dispersion model is applied to model the gas and liquid phases. Reference [8] used a circulation cell to study a mixing process.

Flux of carbon dioxide from gas to liquid phase and flux of sodium bicarbonate from liquid to solid phase affect the sodium bicarbonate concentration in the liquid phase. Therefore, the quantity of supersaturation is a function of these fluxes in the bubble column.

Gas phase mole balance

A mole balance for gas phase around the element yields the following differential governing equation:

$$\frac{dG}{dz} = -N_{CO_2} \times \alpha_g \quad (6)$$

that:

$$\alpha_g = \frac{6 \times \varepsilon_g}{d_B} \quad (7)$$

and mole balance for carbon dioxide results in the following differential equation:

$$\frac{d}{dz}(G \times y_{CO_2}) - \frac{\varepsilon_g \cdot D_g}{U_g} \times \frac{d^2}{dz^2}(G \times y_{CO_2}) = -N_{CO_2} \times \alpha_g \quad (8)$$

that gas velocity is given by:

$$U_g = \frac{4 \times G_{FR}}{\pi \cdot d_R^2 \cdot \varepsilon_g} \quad (9)$$

It is assumed that gas velocity is constant in the column. The first and second terms of this equation are derived from fluid bulk motion and radial dispersion in gas phase (that is raised from bubbles motions with different velocities), respectively.

Liquid phase mole balance

The column is divided by two zones. The first is from top to a point that sodium bicarbonate achieves to saturation concentration (I_0 height in Fig. 1), and the second is from this point to bottom. Inlet solution has a saturation concentration of sodium bicarbonate, at operating conditions illustrated in Fig. 1. Therefore, crystallization occurs in the whole length of the column, and the produced sodium bicarbonate goes to solid phase. If inlet solution has a saturation or supersaturation concentration of sodium bicarbonate, the first zone will be removed and the second zone covers the entire column ($I_0 = 0$). The reaction of sodium bicarbonate production is a fast reaction in high temperatures. Therefore, it is assumed that the rate of carbon dioxide consumption reaction be equal with its absorbency rate. Therefore, the molar velocity of liquid phase and components is calculated using carbon dioxide absorbance and reaction stoichiometry.

Mole balance in the first zone (which has no solid) can be written in the following form:

$$\frac{dL}{dz} = 0 \tag{10}$$

Mole balance for sodium carbonate, sodium bicarbonate, and water in this zone can be written in the following forms:

$$\frac{\varepsilon_l \cdot D_l}{U_l} \times L \times \frac{d^2}{dz^2}(x_{Na_2CO_3}) + L \times \frac{d}{dz}(x_{Na_2CO_3}) = N_{CO_2} \times \alpha_g \tag{11}$$

$$\frac{\varepsilon_l \cdot D_l}{U_l} \times L \times \frac{d^2}{dz^2}(x_{NaHCO_3}) + L \times \frac{d}{dz}(x_{NaHCO_3}) = -2 \times N_{CO_2} \times \alpha_g \tag{12}$$

$$\frac{\varepsilon_l \cdot D_l}{U_l} \times L \times \frac{d^2}{dz^2}(x_{H_2O}) + L \times \frac{d}{dz}(x_{H_2O}) = N_{CO_2} \times \alpha_g \tag{13}$$

Mole balance in the second zone (that produced sodium bicarbonate is inserted to solid phase) can be written as the following differential equation:

$$\frac{dL}{dz} = N_{NaHCO_3} \times \alpha_s \tag{14}$$

and for each component we have:

$$\frac{\varepsilon_l \cdot D_l}{U_l} \times \frac{d^2}{dz^2}(L \times x_{Na_2CO_3}) + \frac{d}{dz}(L \times x_{Na_2CO_3}) = N_{CO_2} \times \alpha_g \tag{15}$$

$$\frac{\varepsilon_l \cdot D_l}{U_l} \times \frac{d^2}{dz^2}(L \times x_{NaHCO_3}) + \frac{d}{dz}(L \times x_{NaHCO_3}) = N_{NaHCO_3} \times \alpha_s - 2 \times N_{CO_2} \times \alpha_g \tag{16}$$

$$\frac{\varepsilon_l \cdot D_l}{U_l} \times \frac{d^2}{dz^2}(L \times x_{H_2O}) + \frac{d}{dz}(L \times x_{H_2O}) = N_{CO_2} \times \alpha_g. \tag{17}$$

The first and second terms of recent equations are derived from the radial dispersion in liquid phase (that is raised from the motion of water by bubbles motion) and fluid bulk motion, respectively. The velocity of liquid is given with the following equation:

$$U_l = \frac{4 \times L_{FR}}{\pi \cdot d_R^2 \cdot \varepsilon_l \cdot \rho_l}. \tag{18}$$

Solid phase mole balance

The first zone of the column produces no solid, therefore:

$$S = 0. \tag{19}$$

The produced sodium bicarbonate in the second zone is inserted to the solid phase. Therefore we have:

$$\frac{dS}{dz} = -N_{NaHCO_3} \times \alpha_s. \tag{20}$$

It is assumed that crystals have a same velocity as the liquid flow. Then, the number of nucleons that born at an element, rising of nucleons from an element to the bottom of column and molar velocity of solid are given by the following equations, respectively:

$$\text{no} = \frac{B_0 \cdot \rho_l \cdot \pi \cdot d_R^2 \cdot (dz)}{4} \tag{21}$$

$$r = \frac{\pi}{6}\left[\frac{G \times 10^{-6}}{U_l} \times (h - z)\right]^3 \tag{22}$$

$$S = \frac{4 \times \rho_s}{\pi \cdot d_R^2 \cdot M_s} \times \left[\sum(\text{no} \times r)\right]. \tag{23}$$

The boundary conditions of the bubble column can be written as below:

$$\frac{d}{dz}(x_{Na_2CO_3}) = 0,\quad \frac{d}{dz}(x_{NaHCO_3}) = 0,\quad \frac{d}{dz}(x_{H_2O}) = 0,\ G = G_{in},\quad y_{CO_2} = y_{CO_2in} \quad \text{at} \quad z = 0 \tag{24}$$

$$x_{Na_2CO_3} = x_{Na_2CO_3 I_0},\ x_{H_2O} = x_{H_2O I_0},\ x_{NaHCO_3} = x{*}_{NaHCO_3},\ L = L_{I_0},\ S = 0 \quad \text{at}\, z = I_0 \tag{25}$$

$$x_{Na_2CO_3} = x_{Na_2CO_3in},\ x_{NaHCO_3} = x_{NaHCO_3in},\ x_{H_2O} = x_{H_2Oin},\ \frac{d}{dz}(G \times y_{CO_2}) = 0,\ L = L_{in} \quad \text{at} \quad z = h. \tag{26}$$

It is necessary to obtain pressure distribution to describe the volume fractions of components in the gas phase. The pressure distribution in bubble column is calculated by the following simple linear equation:

$$P(z) = P_0 - \rho_l \times (1 - \varepsilon_g) \times z \tag{27}$$

Danckwerts obtained carbon dioxide absorbance rate using investigation of mass transfer with chemical reaction as the following equation:

$$N_{CO_2} = E \times K_l \times (C_{CO_2i} - C_{CO_2e}) \quad (28)$$

where E calculates mass transfer coefficient with chemical reaction and can be obtained from the following relationship:

$$E = \sqrt{1 + \frac{D_{CO_2} \cdot k}{K_l^2}} \quad (29)$$

By attention to the carbon dioxide physical absorbance in the solution, its absorbance rate is defined as follow:

$$N_{CO_2} = E \times H \times K_l \times (P_{CO_2i} - P_{CO_2e}) \quad (30)$$

The gas phase resistance can be ignored by attention to the circulation in bubbles and high gas mixing in this column. Therefore:

$$N_{CO_2} = E \times H \times K_l \times (P_{CO_2} - P_{CO_2e}). \quad (31)$$

The carbon dioxide partial pressure also can be ignored in liquid bulk, because carbon dioxide reacts with the sodium carbonate in liquid phase, rapidly. Therefore, carbon dioxide concentration in liquid phase is very low [9].

Reference [5] used an industrial scale bubble column scrubber to absorb simulated carbon dioxide gas and obtained precipitation kinetics. Precipitates of sodium bicarbonate were occurred using sodium bicarbonate and sodium carbonate solution as an absorbent to react with the carbon dioxide gas in the scrubber under an aqueous solution condition. Measured crystal size distributions at certain times were used to calculate the nucleation and growth rate of sodium bicarbonate crystals, which is proportional to an experimental nucleation and growth rate equation. But, these equations were gotten at unsteady-state operating conditions of bubble column. Those cannot be utilized at steady-state operating conditions. A population balance was used to determine the crystallization kinetic and obtain the value of growth and nucleation rate parameters. An empirical power law form is assumed for the growth rate:

$$G_0 = a \times \Delta w^b. \quad (32)$$

The nucleation rate is considered as:

$$B_0 = c \times M_T^d \times \Delta w^e. \quad (33)$$

Population balance equation for the crystallizer is defined as the following equation [10]:

$$\frac{\partial n(L_{nu}, t)}{\partial t} + \frac{\partial (G_0(L_{nu}, t) \cdot n(L, t))}{\partial L_{nu}} = -\sum_k \frac{n_k(L_{nu}, t) \cdot Q_k}{V}. \quad (34)$$

In this equation, it is assumed that, agglomeration is negligible (which is also confirmed by Saberi et al. experimental data [5]). It is assumed that the growth rate of crystals is independent of their size. This assumption simplifies upper equation for each steady-state element as follow:

$$G_0 \times \frac{\partial n}{\partial L_{nu}} = \sum \frac{n_o \cdot Q_o}{V} - \sum \frac{n_i \cdot Q_i}{V}. \quad (35)$$

Reference [11] describes determination of growth and nucleation rate by moment transformation of the population balance as below. The jth moment of the population density with respect to crystal size is defined as:

$$\mu_j = \int_0^\infty L_{nu}^j \times n \times dL_{nu}. \quad (36)$$

Applying moment transformation on (33), and use $B_0 = n(0) \times G_0$ results:

Table 1 The hydrodynamics and mass transfer parameters in bubble column

Parameter	Equation	Reference
ε_g	$\frac{\varepsilon_g}{(1-\varepsilon_g)^4} = 0.2 \times \left(\frac{9.8 \times d_R \cdot \rho_l}{\delta}\right)^{\frac{1}{8}} \times \left(\frac{g \cdot d_R^2 \cdot \rho_l^2}{\mu_l^2}\right)^{\frac{1}{12}} \times \left(\frac{U_g}{\sqrt{d_R \cdot g}}\right)$ (40)	[12]
D_g	$D_g = 5 \times d_R \times \frac{U_g}{\varepsilon_g}$ (41)	[13]
D_l	$D_l = 2.7 \times d_R^{1.4} \times \left(0.219 \times U_g^{0.77} + 0.122 \times U_g^{0.1}\right)$ (42)	[14]
K_l	$K_l \times \alpha_g = 0.6 \times \frac{D_{CO_2}}{d_R} \times \left(\frac{\mu_l}{D_l \cdot \rho_l}\right)^{0.5} \times \left(\frac{g \cdot \rho_l}{\delta}\right)^{0.62} \times \left(\frac{g}{(\mu_l/\rho_l)^2}\right)^{0.31} \times \varepsilon_g^{1.1}$ (43)	[15]
H	$Log\left(H/H_w\right) = -K_s \times I;\ K_s = 0.06, I = 6.2$ (44)	[16]
H_w	$\mathrm{Log}(H_w) = \frac{1{,}140}{T} - 5.3$ (45)	[16]
d_B	$d_B = 26 \times d_R \times \left(\frac{g \cdot d_R^2 \cdot \rho_l}{\delta}\right)^{-0.5} \times \left(\frac{g . D_l^3}{\upsilon_l^2}\right)^{-0.12} \times \left(\frac{U_g}{\sqrt{g \cdot d_R}}\right)^{-0.12}$ (46)	[15]
D_{CO_2}	$D_{CO_2} = 5.35 \times 10^{-0.12} \times \frac{T}{(\mu_l/10^{-3})^{1.035}}$ (47)	[17]
k	$k = \left[2.2 \times 10^7 \times \exp\left(-\frac{71{,}500}{R \cdot T}\right)\right] - \left[3.2 \times 10^{-6} \times \rho_{H_2O} \times \exp\left(-\frac{13{,}800}{R \cdot T}\right)\right]$ (48)	[18]

$$j \times G_0 \times \mu_{j-1} + (0)^j \times B_0 = \sum \frac{\mu_{jo} \cdot Q_o}{V} - \sum \frac{\mu_{ji} \cdot Q_i}{V} \tag{37}$$

Thus, the upper equation for ($j = 0, 1$) becomes:

$$B_0 = \sum \frac{\mu_{0o} \cdot Q_o}{V} - \sum \frac{\mu_{0i} \cdot Q_i}{V} \tag{38}$$

$$G_0 = \left(\sum \frac{\mu_{1o} \cdot Q_o}{V} - \sum \frac{\mu_{1i} \cdot Q_i}{V} \right) \Big/ \mu_0 \tag{39}$$

Using these two equations obtains the values of growth and nucleation rate parameters at steady-state operation conditions of bubble column. The hydrodynamics and mass transfer parameters in bubble column are given by the listed equations in Table 1.

Solution procedure

After mathematic modeling, the obtained equations must be solved to simulate the column. In this simulation, liquid flow rate, mole fractions, and liquid temperature at the top of column, and gas flow rate, mole fractions, and gas pressure at the bottom of column are obtained. Also, the values of growth and nucleation rate parameters at steady-state operation conditions of bubble column are obtained. Therefore, liquid physical properties can be obtained. Then the Eqs. (9), (18), (40) can be solved numerically to obtain U_g, U_l and ε_g, respectively. Now, all of the Eqs. (5), (7), (27), (29), (31) and (41)–(48) can be solved to obtain x^*, α_g, $P(z)$, E, N_{CO_2} and all of the hydrodynamics and mass transfer parameters except ε_g, respectively. Solving the Eqs. (21)–(23) obtains no, r and S, respectively. Now, the Eqs. (6) and (8) can be solved by the boundary conditions (24d, e) and (26d), respectively to find N_{CO_2}. Finally, Eqs. (21)–(23), (38) and (39) can be solved numerically by the respective boundary conditions to find all of the unknown parameters.

Results and discussion

The results of model are compared with the experimental results that presented by Saberi et al. [5] for the bubble column reactor of Shiraz Petrochemical Complex to valid our simulator model. Figure 2 shows the sodium bicarbonate supersaturation in the length of reactor at steady-state operating condition and Fig. 3 shows the variation of the final supersaturation of sodium bicarbonate in the bottom of the column with time. Figure 4 shows the sodium bicarbonate magma density in the length of reactor at steady-state operating condition and Fig. 5 shows the variation of the final magma density of sodium bicarbonate

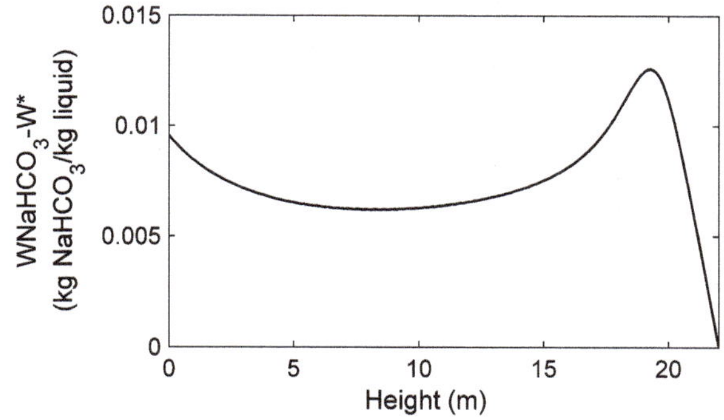

Fig. 2 Supersaturation in the length of column by simulator model

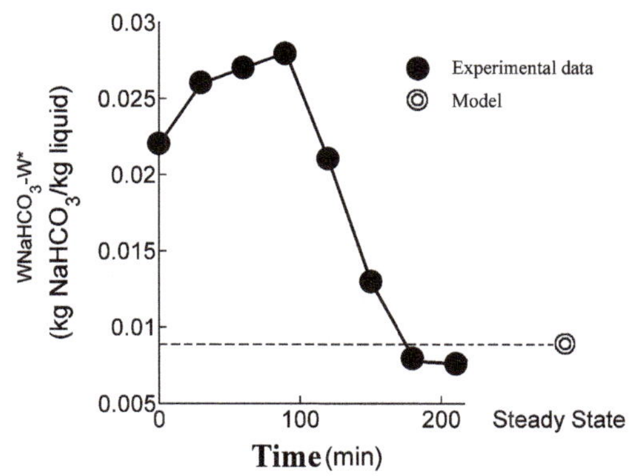

Fig. 3 Supersaturation at the bottom of column vs. time

in the bottom of the column with time. As shown in these figures, the results of model are in good agreement with the experimental data at steady-state operating conditions. Figure 6 shows the conversion of carbon dioxide in the length of column. As shown in this figure, the conversion of carbon dioxide in the top of the column is 50 % which is accorded with the measured value by Saberi et al. [5]. Figure 7 shows a comparison between the predicted values of cumulative mass undersize by model in the steady-state operating condition illustrated in Fig. 1 and unsteady-state experimental data that obtained by Saberi et al. [5]. As shown in this figure there is a suitable agreement between the results of model and experimental data.

The sodium bicarbonate supersaturation in liquid phase is a function of difference of carbon dioxide (from gas phase to liquid phase) and sodium bicarbonate (from liquid phase to solid phase) fluxes. Flux of carbon dioxide is also a function of concentration of carbon dioxide in gas phase that is decreased in the length of column. Flux of sodium bicarbonate is a function of nucleation and growth of crystals that depends on supersaturation and magma density. Increasing of sodium bicarbonate supersaturation increases the flux of sodium bicarbonate. But, increasing of flux of sodium bicarbonate decreases the sodium bicarbonate supersaturation. Hence, supersaturation of sodium bicarbonate has a behavior as shown in Fig. 2.

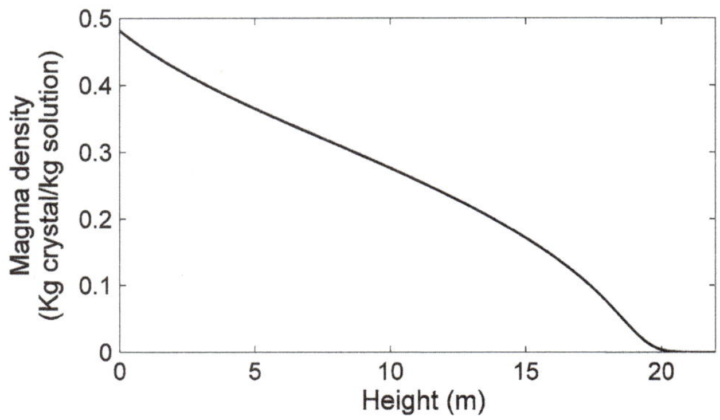

Fig. 4 Magma density in the length of column by simulator model

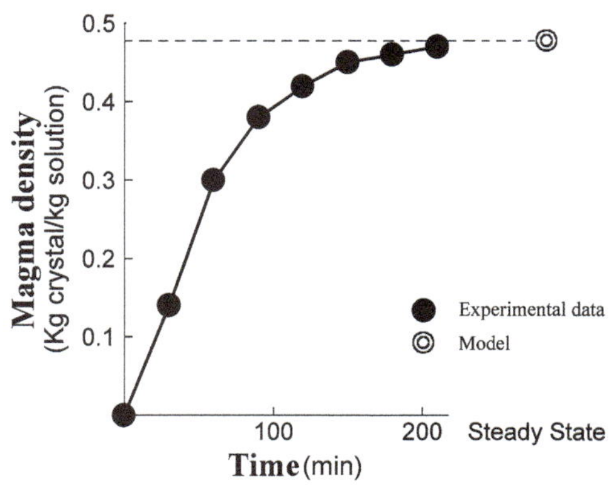

Fig. 5 Magma density at the bottom of column vs. time

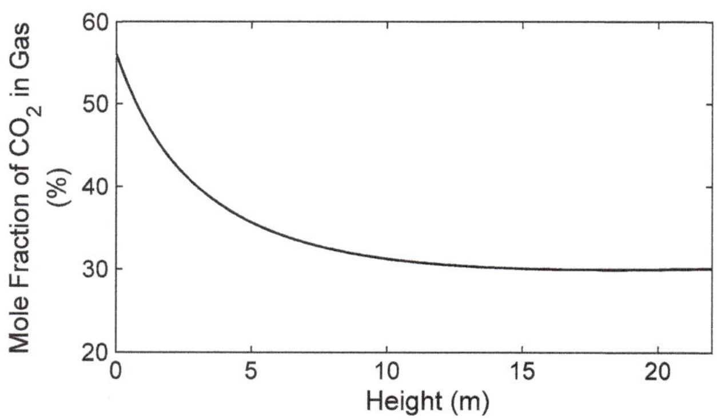

Fig. 6 Molar fraction of carbon dioxide in gas in the length of column by simulator model

Using Saberi et al. nucleation and growth formulas [5] as the first guess and mathematical methods, nucleation and growth rate for the operating conditions illustrated in Fig. 1 are obtained as following equations:

$$B_0 = 26.398 \times M_T^{0.42} \times \Delta w^{1.31} \quad (49)$$

$$G_0 = 1.333 \times 10^{-4} \times \Delta w^{1.53}. \quad (50)$$

These two equations are too similar to the obtained experimental formulas by Saberi et al. ($B_0 = 26.685 \times M_T^{0.42} \times \Delta w^{1.31}$, $G_0 = 1.381 \times 10^{-4} \times \Delta w^{1.53}$) [5]. It is a validation for our model. It is mentioned that the supersaturation of sodium bicarbonate has a complex self dependency. Increasing of supersaturation increases the rate of nucleation and growth rate. Increasing of crystal production decays the liquid supersaturation. A change in the operating conditions can change the liquid supersaturation and also changes this complex dependency of nucleation and growth rate to the liquid supersaturation. Therefore, the nucleation and growth rate are different functions of liquid supersaturation for different operating conditions. For other different operating conditions the fitted constants of Eqs. (32) and (33) are listed in Table 2.

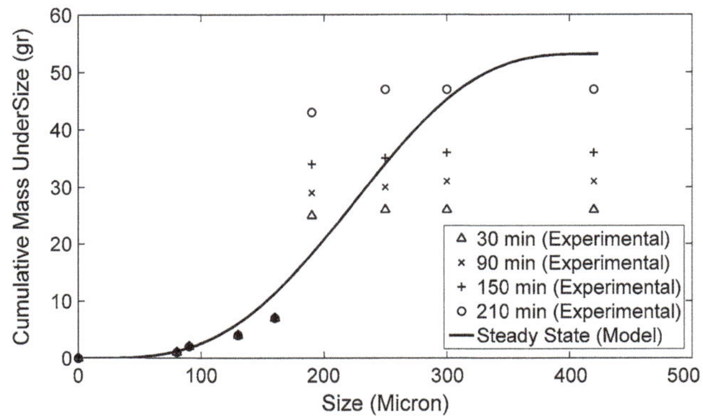

Fig. 7 Cumulative mass undersize at the bottom of column

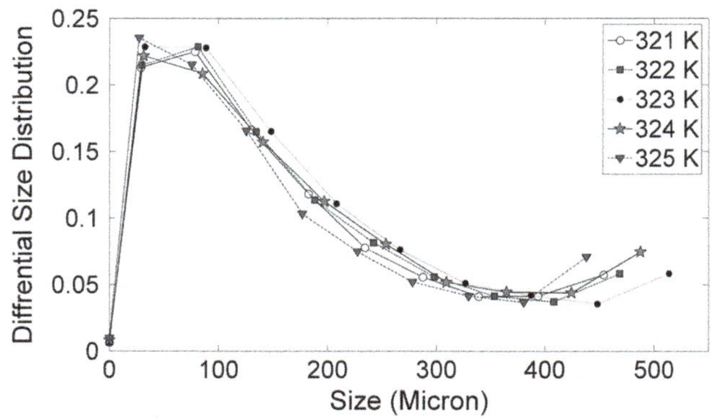

Fig. 8 Predicted produced sodium bicarbonate crystals size distribution in different liquid temperatures

Table 2 The optimized values of the kinetic parameters of sodium bicarbonate precipitation

		$a \times 10^4$	b	c	d	e
Liquid temperature (K)	322	1.295	1.53	22.437	0.42	1.31
	324	1.384	1.53	33.09	0.42	1.31
Inlet gas pressure (atm)	2	1.346	1.53	27.557	0.42	1.31
	3	1.073	1.53	19.946	0.42	1.31
Inlet carbon dioxide mole fraction (%)	50	1.344	1.53	28.026	0.42	1.31
	60	1.266	1.53	22.585	0.42	1.31

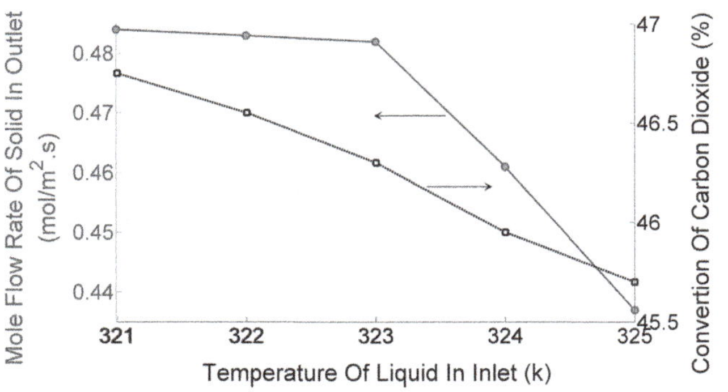

Fig. 9 Predicted percent of carbon dioxide conversion and magma density of sodium bicarbonate crystals in different liquid temperatures

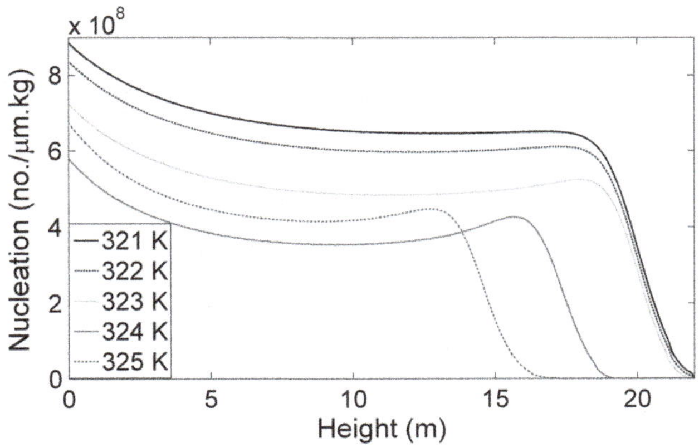

Fig. 10 Nucleation of sodium bicarbonate crystals in the length of column at different temperature

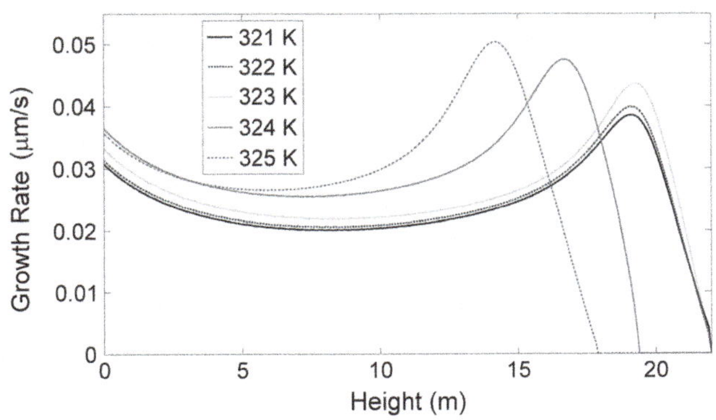

Fig. 11 Growth rate of sodium bicarbonate crystals in the length of column at different temperature

Effect of liquid temperature

Figure 8 shows the predicted produced sodium bicarbonate crystals size distribution in different liquid temperatures. As shown in this figure, increasing of liquid temperature from 321 to 323 K increases the maximum size of crystals. In the temperature range from 323 to 325, this effect is reversed. But the maximum size of crystals is increased in

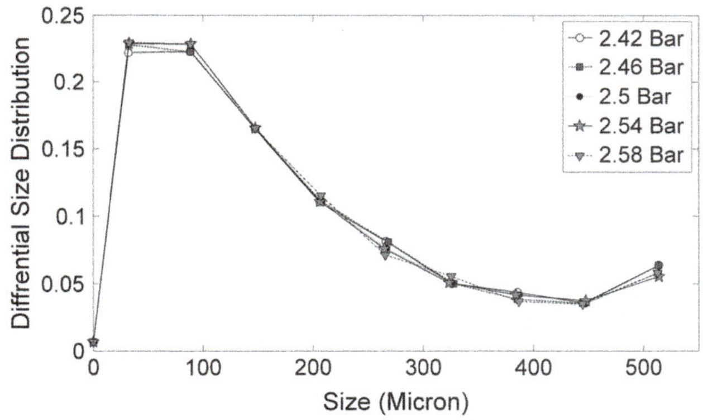

Fig. 12 Predicted produced sodium bicarbonate crystals size distribution in different gas pressures

this range of temperature. Cause of this effect of temperature is complex. Increasing of liquid temperature increases the production rate of $NaHCO_3$ by increasing the rate of reactions (1–4), and also increases the solubility of $NaHCO_3$ in water by attention to Eq. (5).

Figure 9 shows the predicted percent of carbon dioxide conversion and magma density of sodium bicarbonate crystals in different liquid temperatures. It is seen that the carbon dioxide conversion and also the magma density of sodium bicarbonate crystals are decreased with increasing liquid temperature. The reason is increasing of the reaction rate and therefore more carbon dioxide consumption with increasing temperature. It must be noted that increasing of the rate of sodium bicarbonate disintegration which increases the disposing of carbon dioxide could not amend this carbon dioxide decrement when liquid temperature is increased.

Figures 10 and 11 show the nucleation and growth rate of sodium bicarbonate crystals in the length of column at different temperatures, respectively. As shown in these figures, temperature has an important effect on the nucleation and growth rate of crystals. More liquid temperature causes more liquid supersaturation by increasing the reaction rate and also $NaHCO_3$ disintegration. Increasing of liquid supersaturation increases the nucleation and growth rate of crystallization, as mentioned in the Eqs. (32) and (33). Since, production of solid crystals consumes the sodium bicarbonate of the liquid phase, increasing of the nucleation and growth rate decay amount of supersaturation of this phase. Therefore, amount of supersaturation affects self, indirectly. This is the cause of different trends in the liquid temperature 324 and 325 K, as seen in Figs. 10 and 11.

Effect of gas pressure

Figure 12 shows the crystals size distribution of predicted produced sodium bicarbonate in different gas pressures.

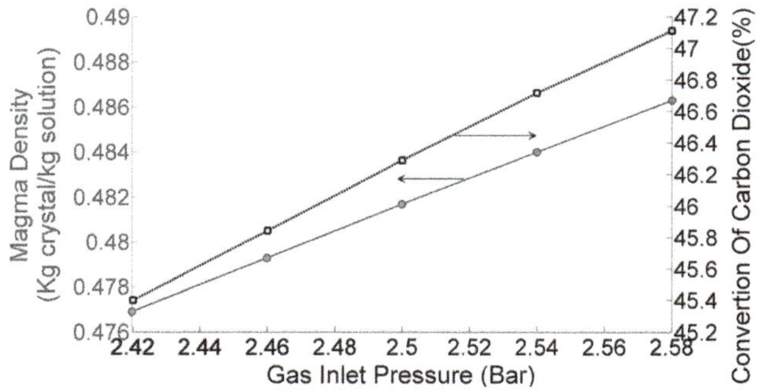

Fig. 13 Predicted percent of carbon dioxide conversion and magma density of sodium bicarbonate crystals in different gas pressures

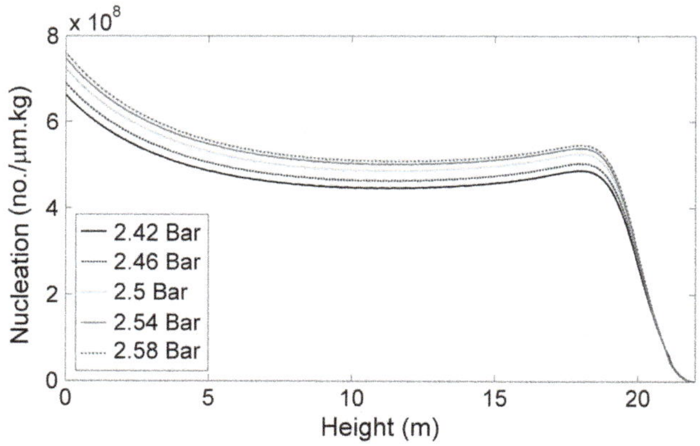

Fig. 14 Nucleation of sodium bicarbonate crystals in the length of column at different gas inlet pressure

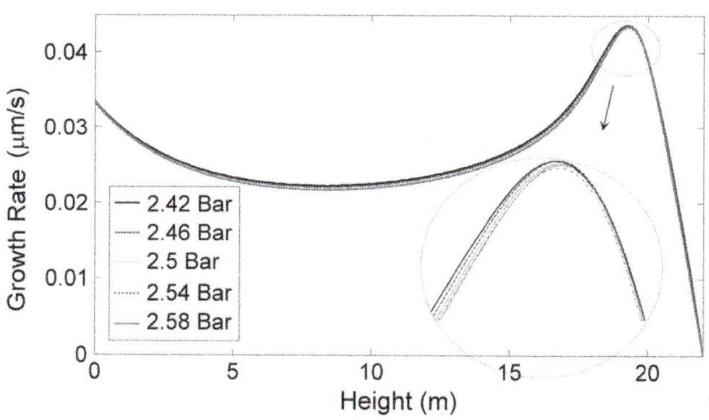

Fig. 15 Growth rate of sodium bicarbonate crystals in the length of column at different gas inlet pressure

Increasing of gas pressure decreases the size of bubbles and increases the number of them. More bubbles have a more gas–liquid interface, but cause a more mixing in the liquid phase. These two factors counteract each other in the crystal production. Therefore, increasing of the gas pressure has no significant effect on the crystals size distribution, as shown in this figure.

Figure 13 shows the predicted percent of carbon dioxide conversion and also magma density of sodium bicarbonate

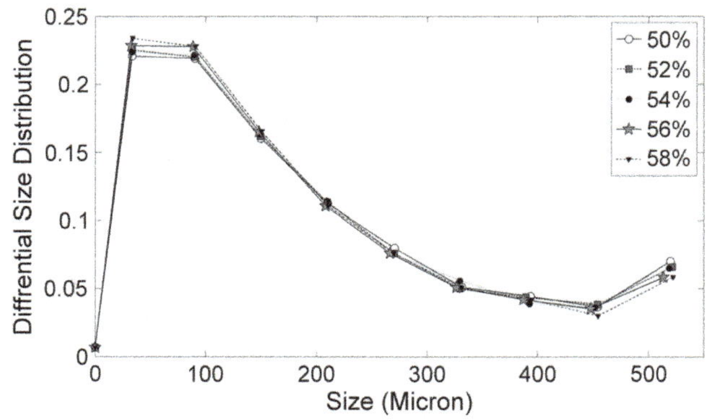

Fig. 16 Predicted produced sodium bicarbonate crystals size distribution in different carbon dioxide mole fractions

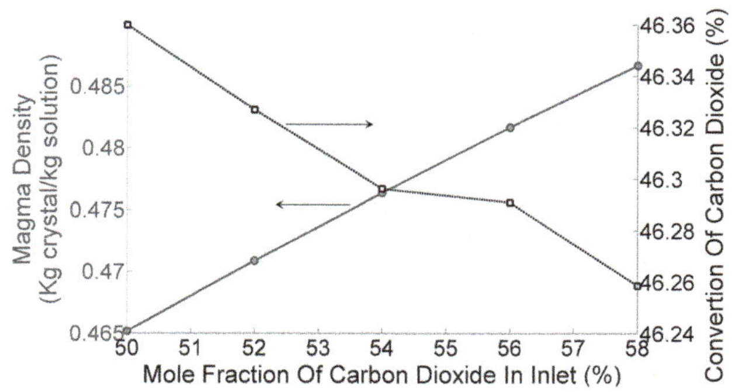

Fig. 17 Predicted percent of carbon dioxide conversion and magma density of sodium bicarbonate crystals in different carbon dioxide mole fractions

crystals in different gas pressures. This figure shows that the percent of carbon dioxide conversion and also mole flow rate of sodium bicarbonate crystals are increased with increasing gas pressure. Gas density is increased and bubbles size is decreased by increasing gas pressure. These are the cause of increasing the carbon dioxide conversion and also production rate of sodium bicarbonate crystals.

Figures 14 and 15 show the nucleation and growth rate of sodium bicarbonate crystals in the length of column at different gas inlet pressure, respectively. As shown in these figures, increasing of the inlet gas pressure increases nucleation, but has no effect on the growth rate of crystals. More gas–liquid interface and more mixing in the liquid phase are the causes of these effects on the nucleation and growth rate of crystals, respectively.

Effect of carbon dioxide mole fraction

Figure 16 shows the predicted produced sodium bicarbonate crystals size distribution in different mole fraction of carbon dioxide in the gas phase. Increasing of carbon dioxide mole fraction in the gas phase has no significant effect on the crystals size distribution, as shown in this figure. Increasing of carbon dioxide flow rate increases the

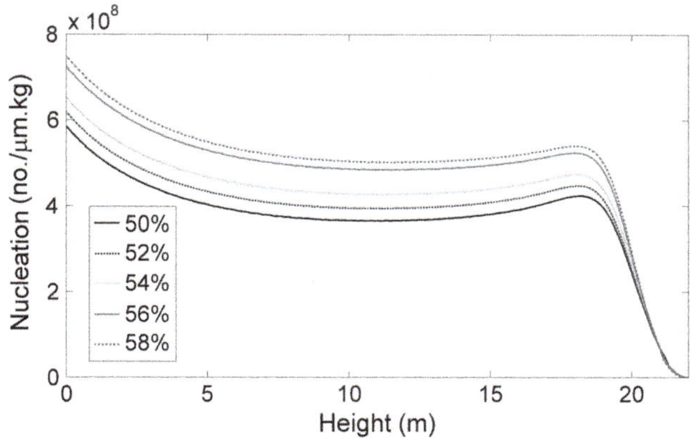

Fig. 18 Nucleation of sodium bicarbonate crystals in the length of column at different carbon dioxide mole fractions

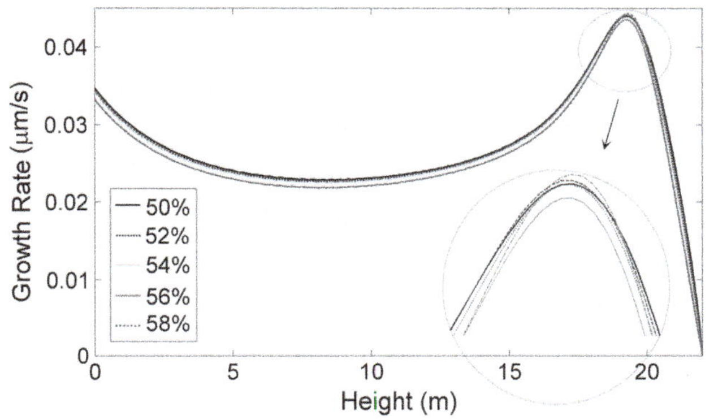

Fig. 19 Growth rate of sodium bicarbonate crystals in the length of column at different carbon dioxide mole fractions

bubbles size and decreases the number of bubbles which decreases the gas–liquid interface. On the other hand, this increment decays mixing of liquid phase. These two factors counteract each other in the crystal production.

Figure 17 shows the predicted percent of carbon dioxide conversion and also magma density of sodium bicarbonate crystals in different carbon dioxide mole fractions. As shown in this figure, the percent of carbon dioxide conversion is decreased but mole flow rate of sodium bicarbonate crystals is increased with increasing mole fraction of carbon dioxide in the gas phase. Increasing of the amount of carbon dioxide in the gas phase increases the production of sodium bicarbonate crystals. Decreasing of the gas–liquid interface decreases the rate of carbon dioxide transfer from gas phase to liquid, therefore, decreases the conversion of this component in the gas phase.

Figures 18 and 19 show nucleation and growth rate of sodium bicarbonate crystals in the length of column at different carbon dioxide mole fractions, respectively. As shown in these figures, inlet carbon dioxide mole fraction increases nucleation, but has no significant effect on the growth rate of crystals. The complex behavior of supersaturation is the cause of these effects on the nucleation and growth rate of crystals.

Conclusion

Results of mathematical simulation of bubble column show that, there are optimized operating conditions for a suitable crystal size distribution, or more carbon dioxide conversion and produce sodium bicarbonate crystals. The percent of carbon dioxide conversion and also mole flow rate of sodium bicarbonate crystals are decreased with increasing of liquid temperature and decreasing of gas pressure. The percent of carbon dioxide conversion is decreased but mole flow rate of sodium bicarbonate crystals is increased with increasing of mole fraction of carbon dioxide in gas phase. The nucleation and growth rate of crystals are changed at different operating condition, and there are different formulas at different temperatures, especially. Finally, the model results are compared with experimental results. This comparison confirms the accuracy of our model for simulating the sodium bicarbonate bubble columns.

References

1. Benadda B, Prost M, Ismaily S, Bressat R, Otterbein M (1994) Validation of the gas-lift capillary bubble column as a simulation device for a reactor by the study of CO_2 absorption in Na_2CO_3/$NaHCO_3$ solutions. Chem Eng Proc 33:55–59
2. Danckwerts PV (1970) Gas–liquid reactions. McGraw-Hill, New York
3. Haut B, Halloin V, Cartage T, Cockx A (2004) Production of sodium bicarbonate in industrial bubble columns. Chem Eng Sci 59:5687–5694
4. Soltani Goharrizi A, Abolpour B (2012) Estimation of sodium bicarbonate crystal size distributions in a steady-state bubble column reactor. Res Chem Intermed 38(7):1389–1401
5. Saberi A, Soltani Goharrizi A, Ghader S (2009) Precipitation kinetics of sodium bicarbonate in an industrial bubble column crystallizer. Cryst Res Technol 44(2):159–166
6. Hou TP (1942) Manufacture of soda, 2nd edn. Reinhold, New York
7. Broul M, Nyvlt J, Sohnel O (1981) Solubility in inorganic two component system. Elsevier, New York
8. Joshi JB, Sharma MM (1979) A circulation cell model for bubble columns. Trans Inst Chem Eng 57:244–251
9. Wylock CE, Colinet P, Cartage T, Haut B (2008) Coupling between mass transfer and chemical reactions during the absorption of CO_2 in a $NaHCO_3$–Na_2CO_3 brine: experimental and theoretical study. Int J Chem React Eng 6:A4
10. Randolph AD, Larson MA (1988) Theory of particulate processes. Academic Press, New York

11. Tavare NS (1995) Industrial crystallization: process simulation, analysis and design. Plenum press, New York
12. Akita K, Yoshida F (1973) Gas hold-up and volumetric mass transfer coefficient in bubble columns. Ind Eng Chem Proc Des Dev 12:76–80
13. Pavlica RT, Olson JH (1970) Unified design method for continuous contact mass transfer operations. Ind Eng Chem 62:45–58
14. Wendt R, Steiff A, Weinspach PM (1983) Fiussigphasenruck Vermischung In Blasen Saulen-Reaktoren. Chemie-ingenieurtechnik 55:796–797
15. Akita K, Yoshida F (1974) Bubble size, interfacial area, and liquid mass transfer coefficient in bubble columns. Ind Eng Chem Proc Des Dev 12:84–91
16. Danckwerts PV, Sharma MM (1966) The absorption of carbon dioxide into solutions of alkalis and amines. Chem Eng 44:244–280
17. Tamimi A, Rinker EB, Sandall OC (1994) Diffusion coefficient for hydrogen sulfide, carbon dioxide, and nitrous oxide in water over the temperature range 293–368 K. J Chem Eng Data 39(2):396–398
18. Gallagher PK, Brown ME, Kemp RB (1998) Handbook of thermal analysis and calorimetry. Elsevier, New York

13

The catalytic stability of some selected bifunctional nanoporous-based catalysts in the hydroisomerisation of n-C7 and the effect of post-synthesis modification techniques

Faisal M. Alotaibi · Raed H. Abudawood · Hamid A. Al-Megren · Mohammed C. Al-Kinany · Essam H. Jamea · Arthur A. Garforth

Abstract In this work, some commercial and in-house nanoporous-based catalysts, such as USY, beta and mordenite zeolites, and mesoporous aluminosilicate molecular sieves such as MCM-48 and SBA-15, loaded with metals and acting as mono- and bimetallic bifunctional catalysts, were used for hydroisomerisation experiments in a fixed-bed reactor at pressures between 1 and 15 bar and at feed space time ranging from 2.57 to 10.26 h^{-1} (35.14–140.6 kg s mol^{-1}) to hydroisomerise *n*-heptane over a temperature range of 210–270 °C. The effect of post-synthesis treatments of micro- and mesoporous catalysts was examined, regarding their activity, selectivity and stability, such as acid and steam dealumination techniques, acid leaching via a chelating agent, bimetal loading techniques, different platinum loading methods, and composite or hybrid catalyst generation. Results show that pore architecture is the most important factor affecting coke formation and deactivation in zeolite catalysts. It was found that those catalysts with high Si/Al ratios and those which had been acid-leached or steamed showed better activity, higher selectivity towards isomeric products and better time stability. Moreover, the balance between the number of metal sites and the number of acid sites played an important role in determining the activity, selectivity and stability of the bifunctional catalysts. Higher metal loading improves catalytic stability, due to a better balance and closeness of the catalytic functions. Moreover, the bimetallic catalyst improves the formation of smaller metal particles and better dispersion, which may affect selectivity and stability.

Keywords The time-on-stream stability · Alkanes' hydroisomerisation · Catalyst deactivation · Zeolites · Bifunctional catalysts · Nanoporous catalysts · Bimetallic catalysts

F. M. Alotaibi (✉) · H. A. Al-Megren · M. C. Al-Kinany
Petrochemicals Research Institute, King Abdulaziz City for Science and Technology (KACST), P.O.Box 6086, Riyadh 11442, Saudi Arabia
e-mail: fmsalotaibi@kacst.edu.sa

R. H. Abudawood
Research and Development Center, Saudi Aramco, Dhahran 31311, Saudi Arabia

E. H. Jamea
SABIC Technology Center, Saudi Basic Industries Corporation, Riyadh 11551, Saudi Arabia

A. A. Garforth
School of Chemical Engineering and Analytical Science, The University of Manchester, Manchester M13 9PL, UK

Introduction

Tailoring catalysts to optimise their chemical, physical and mechanical features is the most widely used method of limiting a reduction in catalytic activity. With relatively slow coke formation, the fine tuning of the strength and density of acid sites in zeolites is essential to achieve a sufficiently high activity for the desired reaction. Furthermore, limitations to the diffusion of reactant, product and coke precursor molecules can be reduced by the creation of mesoporosity within zeolite crystals to produce nano-sized crystals, which will slow down coke formation. Appropriate selection of operating conditions is a crucial factor in promoting desorption of the desired products and thus limiting deactivation of the catalyst [1].

Despite the inevitable loss of catalytic activity (catalyst deactivation) in most industrial processes, there are many ways to minimise this process and thus extend the lifetime of a catalyst. As the old saying goes, "an ounce of prevention is worth a pound of cure" [2]. The zeolite pore architecture has been found to play an important role for both the deactivation rate as a result of coking and also the composition of the carbonaceous deposits, such that zeolites with relatively large supercage pore systems, such as Y zeolites, are inherently favourable to coking reactions, which in turn leads to faster deactivation. In contrast, beta zeolite exhibits much better coking resistance, even though both zeolites have three-dimensional systems of channels [3].

The Pt/H-MOR catalyst is the most widely recognised example of a zeolitic isomerisation catalyst. It was first developed to isomerise smaller alkanes such as *n*-pentane and *n*-hexane for the Shell Hysomer process and can bring about a high isomer yield [4]. However, due to the strong tendency of middle paraffin isomerisation (C7–C9 carbon chain length) to undergo cracking, no hydroisomerisation process has yet been used industrially [5]. Strongly acidic zeolites such as H-mordenite have weaknesses as isomerisation catalysts due to the rapidity of their deactivation, particularly by olefinic carbonaceous species. There is therefore a great incentive for academic and industrial researchers to discover and develop new catalysts that are active, selective and able to resist deactivation [6, 7]. There have been detailed descriptions in the literature of how post-synthesis treatments by steam and acid of several zeolites can reduce the concentration of lattice aluminium and improve thermal/hydrothermal stability and catalytic performance whilst retaining the zeolite crystal structure. It must also be noted that mesopores form as a result of the dealumination of H-mordenite, significantly diminishing the coke's deactivating effect [8, 9].

The vast majority of microporous zeolites have pore diameters below 15 Å, leading to very high mass transfer limitations; hence, the intracrystalline diffusion of species inside the pores of zeolites is severely limited, which affects their catalytic activity and significantly reduces their product selectivity, because the increased residence time of the products in the zeolite pores favours undesired reactions such as over-cracking. Because of this limitation, there has been much work to find creative ways to eliminate the mass transfer problem. Moreover, the formation of mesopores can favour the formation of more branched or bulkier products. Evidently, generating a network of mesopores in zeolites crystals would offer numerous advantages for a great number of reactions [10]. Mesoporous aluminosilicate molecular sieves, such as SBA-15 and MCM-48, which have high surface area, large pore diameters and narrow pore size distribution and which possess uniform hexagonal and cubic one- and three-dimensional pore systems, respectively, are thus supposedly more resistant to pore blocking, which allows faster diffusion of reactants. However, they have been shown to be inferior to microporous zeolites in their acid strength and despite their larger pore diameters, which make them more able to deal with bulky molecules, they do not show higher activity than zeolites, e.g. in heavy oil cracking and the catalytic conversion of large molecules in vacuum gas oil and distillation residues. Thus, many significant improvements in both the hydrothermal stability and acidity of mesostructured aluminosilicates have been recently reported [11–13].

In this work, the time-on-stream (TOS) behaviour and deactivation rates of some commercial USY, beta and mordenite zeolites and mesoporous aluminosilicate molecular sieves such as Al-MCM-48 and Al-SBA-15 that have been loaded with platinum metal, with a wide range of Si/Al ratios, pore sizes and acidity characteristics, are evaluated in the hydroisomerisation of n-C7 when reaction conditions such as temperature, pressure and contact time are varied. n-C7 was used as a representative model compound for naphtha. A number of post-synthetic methods have been adopted to improve the catalytic properties of zeolites, such as acidity, porosity and thermal and hydrothermal stability, as these modifications are hard to achieve directly during zeolite synthesis. The main methods are ion exchange, dealumination by acid leaching or acid chelating, and stabilising through hydrothermal treatment (steaming), mono- and bimetallic effects.

Experimental

Catalyst preparation

Some commercial USY, beta, mordenite, and mesoporous aluminosilicates samples were provided by different suppliers [14–20], as shown in Table 1. These zeolite supports were designed and modified by various post-synthesis techniques so that they could be used for multiple purposes; for instance, high alkane hydroconversions like cracking, isomerisation and alkylation. They were supplied in powder form without binder materials. The in-house mordenite zeolite sample was synthesised with different properties from those of the commercial mordenite samples, then post-synthesis modified by acid and steam dealumination techniques under different treatment conditions.

USY-A, USY-C, 320HOA and 350HOA samples were subjected to post-synthetic modification by treating with AHFS to remove any EFAL. Three series of bifunctional catalysts supported on 350HOA USY were also prepared: monometallic types containing only platinum or nickel were prepared by the ion-exchange method, while the method used for bimetallic ones containing both platinum and nickel was simultaneous competitive ion exchange.

Table 1 List of catalyst supports investigated in the TOS study

	Zeolite trade name	Zeolite type	Post-synthesis modification	Supplier
1	USY-A	USY	Mildly steamed	Crosfield
2	USY-C	USY	Severely steamed	Crosfield
3	320HOA	USY	Moderately steamed	Tosoh corporation
4	350HOA	USY	Moderately steamed and acid leached	Tosoh corporation
5	CP814E	BEA	As synthesised	Zeolyst international
6	Al-MCM-48	Mesoporous aluminosilicates	Aluminated by grafting	Claytec
7	Al-SBA-15	Mesoporous aluminosilicates	Aluminated by grafting	Claytec

Finally, two series of dealuminated mordenite catalysts were prepared by acid leaching under different conditions and by applying the steaming dealumination technique at different temperatures to the in-house synthesised mordenite in its ammonium form.

Hydrothermal synthesis of mordenite zeolite

The recipe for mordenite zeolite was obtained from a verified synthesis handbook [21]. Gel calculations to determine the required weights of each reactant were performed using the synthesis molar gel composition of 6 Na_2O:Al_2O_3:30 SiO_2:780 H_2O. Therefore, calculated weights of deionised water and sodium hydroxide (99 wt% NaOH) from Merck and sodium aluminate (50.9 wt% Al_2O_3 + 31.2 wt% Na_2O + 17.9 wt% H_2O) from Sigma-Aldrich were mixed in a plastic bottle until they were completely dissolved. The required amount of Ludox AS-40 colloidal silica (40 wt% silica + 60 wt% water) from Sigma-Aldrich was then added to the solution and all were stirred using a magnetic stirrer for 12 h, after which the mixture was poured into a Teflon-lined stainless-steel autoclave and heated in an oven at 180 °C for crystallisation. After 24 h crystallisation time, the mixture was taken out of the oven, decanted and centrifuged several times while washing with deionised water after each centrifugation until the pH of the decanted solution dropped below 9. Finally, the samples were dried at 110 °C overnight until they were ready for characterisation. The optimum mordenite phase with the highest purity and crystallinity was optimised against some synthesis parameters while keeping the composition ratio of Na_2O–Al_2O_3–SiO_2–H_2O constant at the above molar gel composition. Three factors were optimised: the ageing time of the synthesis gel, the crystallisation time and the crystallisation temperature. It was found that the optimum synthesis conditions were an ageing time of 12 h, crystallisation time of 24 h and crystallisation temperature of 180 °C. This procedure was repeated more than 10 times and reproducible results were obtained, in terms both of the yields, crystallinity, bulk and framework Si/Al ratio and of the crystal morphology and size.

Post-synthesis modification techniques

Ion exchange

The procedure of ion exchange is carried out simply by immersing zeolite in a solution of the salt of the targeted cation at room temperature or a higher temperature if the exchange rate is to be improved [22], since the Brønsted acid site is generated after the modification of the parent zeolite in its sodium form by first exchanging the sodium ion with ammonium, then calcining the zeolite in the ammonium form at high temperatures to transform it into the hydrogen form. One gram of the parent commercial zeolites in their sodium form—320HOA and CBV10A—and in-house as-synthesised mordenite zeolite were dissolved in 100 ml of 0.5 M ammonium nitrate, (99 wt% purity, Sigma-Aldrich) and heated at 80 °C under reflux conditions and constant stirring for 1 h. This process was performed three times (multiple ion exchange) assuming a complete uptake of ammonium in the catalysts. The mixtures obtained were centrifuged, washed with deionised water until pH was 7, then dried at 100 °C overnight. After three ion exchanges, residual Na was determined by inductively coupled plasma (ICP) elemental analysis to be below 0.05 wt%.

Acid leaching

The treatment of zeolites by acid leaching leads to dealumination, which generally improves their efficiency in processes that result in rapid catalyst deactivation through coking, such as cracking. Acid leaching results in an increase in the Si/Al ratio through the removal of tetrahedral aluminium atoms from the framework and the subsequent introduction of mesopores into the structure, thus enhancing the thermal stability of the zeolite. It has the advantage of not forming EFAL inside the zeolite structure, as happens when Al is removed by steaming [23–26]. Acid attack was carried out by heating a suspension of in-house as-synthesised mordenite in its ammonium form (NH_4-MOR) in nitric acid or hydrochloric acid (BDH Chemicals) under reflux conditions and constant stirring under the acid leaching conditions listed in Table 2. The suspension was then filtered, washed with distilled water and dried at 110 °C overnight until it was ready for characterisation.

Table 2 Conditions of acid leaching by mineral acids

Condition	NH_4-MOR (wt)	Acid type	Acid conc. (molar)	Acid, vol. (ml)	Leaching time (h)	Temp (°C)
1	1	HCl	6	50	12	25
2	1	HCl	6	50	24	25
3	1	HCl	6	100	24	25
4	1	HCl	12.1 (conc)	100	24	120
5	1	HNO_3	10	100	16	25
6	1	HNO_3	6	50	24	25
7	1	HNO_3	15.8 (conc)	100	24	25
8	1	HNO_3	15.8 (conc)	50	24	25
9	1	HNO_3	15.8 (conc)	100	24	120
10	1	HNO_3	6	100	24	120
11	1	HNO_3	6	100	48	120
12	2	HNO_3	10	100	24	120
13	1	HNO_3	10	100	16.5	120
14	1	HNO_3	10	100	48	120
15	1	HNO_3	10	100	87	120
16	1	HNO_3	10	100	168	120

Table 3 Conditions of acid leaching by AHFS

Sample	AHFS, wt (g)	320USY, wt (g)	Mixing time (h)	Temp (°C)	Mixture volume (ml)
Optimising the mixture volume					
F1	0.6	2	1	25	25
F2	0.6	2	1	25	50
F3	0.6	2	1	25	75
F4	0.6	2	1	25	100
Optimising the mixing time					
F2	0.6	2	1	25	50
F5	0.6	2	3.5	25	50
F6	0.6	2	7	25	50
F7	0.6	2	14	25	50
Optimising the AHFS weight					
F8	0.3	2	1	25	50
F2	0.6	2	1	25	50
F9	1.2	2	1	25	50
Optimising the mixture temperature					
F2	0.6	2	1	25	50
F10	0.6	2	1	50	50
F11	0.6	2	1	75	50
F12	0.6	2	1	100	50

AHFS solution was used to remove EFAL from the USY samples: USY-A, USA-C, 320HOA and 350HOA. The 320HOA sample containing 2 wt% sodium with the ammonium form as the starting material was fine tuned using AHFS as a chelating compound to provide the best conditions to remove only the EFAL whilst maintaining the aluminium framework and original zeolite crystallinity. Thus, four main factors were considered when optimising the chemical treatments: the AHFS and USY mixture volume, the treatment duration time, the zeolite-to-AHFS weight ratio and the temperature of chemical treatment, as shown in Table 3. The zeolite-to-AHFS mass ratio was optimised in the range of 1.67–6.67. For AHFS treatments, 2 g of the steamed zeolite USY samples was placed in AHFS (99.999 % trace metals basis, Sigma-Aldrich) solution and the suspension was heated to the required temperature and stirred whilst being heated with refluxing. The suspension was filtered, washed by warm deionised water to remove any residual fluoride species until pH was 7, and then dried at 100 °C overnight. The solid obtained was subjected to three ion exchanges in ammonium nitrate solution in order to eliminate sodium and any residual fluoride species.

Steaming

Steaming or hydrothermal treatment dealumination involves calcining zeolites in the ammonium (or hydrogen) form, in the presence of steam at high temperatures. This results in the removal of aluminium atoms from the zeolite framework and their replacement by silicon atoms, thus increasing the Si/Al ratio of the framework. However, the aluminium atoms which have been removed remain in the zeolite as EFAL species, which can lead to enhanced acidity through the formation of Lewis acid sites [23]. The formation of EFAL species occurs through the migration of aluminium from the framework into the outer surface of crystals and is rate-controlled, which means that it can be accelerated or slowed down by changing the steaming temperature [27]. The in-house as-synthesised NH_4-MOR was dealuminated by steaming at 500, 600 and 700 °C for 12 h at each steaming temperature, using a fixed-bed laboratory microreactor. Deionised water was fed to the reactor using a high-performance liquid chromatography (HPLC) pump at a controlled flow rate of 0.125 ml min^{-1}. Nitrogen carrier gas (BOC, oxygen free) was dried by passage through a bed of molecular sieves and fed into the reactor at a controlled flow rate of 187.5 ml min^{-1} using a mass flow controller. The catalyst (1 g, bed particle size of 40 and 60 mesh) was located in the reactor between two beds of glass wool in the centre of the heated stable zone of the furnace. In a typical experiment, the NH_4-MOR was heated from ambient temperature to the required reactor temperature of 500 °C at a rate of 3 °C min^{-1} in a flow of air (50 ml min^{-1}) at ambient pressure for 2 h. When these conditions had stabilised, the water was introduced to the reactor. At the end of the required time period (12 h for steaming), the water feed was stopped and the catalyst was then cooled in a stream of dry nitrogen.

Metal loading

Metals from group VIII, such as platinum and nickel, are commonly loaded onto the zeolites used in commercial refining processes as bifunctional catalysts, having both a hydrogenation/dehydrogenation function provided by the metal and an acidic function provided by the zeolite acid sites [28, 29]. A common method for preparing bifunctional catalysts is a multi-step process consisting mainly of the deposition of a metal precursor compound over the support surface, either by impregnation, ion exchange or mechanical mixing; the drying and calcination of the catalysts; and the transformation of the precursor compound into the active metallic phase by reduction. It is well established that the metal dispersion and thus the metal/acid ratio depend markedly on the way the metal is introduced [30]. The catalysts listed in Table 1 were loaded with 1 wt% platinum via the ion-exchange technique using an amino complex of platinum, and were fully described in a previous work [29–31]. The CP814E catalyst was loaded with 1 wt% platinum by wet impregnation using tetraammine platinum (II) chloride, at room temperature with excess solution under constant stirring overnight, followed by evaporation at 70 °C and drying at 110 °C. A series of monometallic and bimetallic bifunctional catalysts was formed by loading 350HOA USY with different amounts of platinum, nickel and platinum–nickel by competitive ion exchange. The first series comprised monometallic bifunctional catalysts containing 0.5, 1, 1.5 and 2 wt% of platinum, the second was a series of monometallic bifunctional catalysts containing 0.5, 1, 1.5, 2, 10 and 20 wt% of nickel and the third series was of bimetallic bifunctional catalysts containing 1 wt% of platinum and 0.25, 0.5, 0.75 and 1 wt% of nickel. The monometallic catalysts were prepared using the ion-exchange method, while the method used for the bimetallic catalysts was simultaneous competitive ion exchange, which means that the platinum and nickel precursors, $Pt(NH_3)_4Cl_2H_2O$ and $Ni(NO_3)_26H_2O$ (Fischer Chemical, 98 % purity), respectively, were mixed with the zeolite at the liquid-to-solid ratio of 50 (1 g of zeolite to 50 ml of the metal complex solution), using ammonium hydroxide solution to adjust the pH of the suspension obtained at >9.

Composite catalysts

Two composite bifunctional catalysts consisting of beta and USA zeolites were prepared as follows: (1) the 1 wt% Pt/CP814E and 1 wt% Pt/USY-C catalysts were mixed physically at a mass ratio of 1:1; (2) CP814E and USY-C in their ammonium forms were mixed at a mass ratio of 1:1 and were loaded by 1 wt% Pt using the ion-exchange method.

Catalyst characterisation

Characterisation techniques used were ICP optical emission spectroscopy (ICP–OES), X-ray diffraction (XRD), ^{27}Al and ^{29}Si MAS-solid-state NMR, pyridine adsorption Fourier transform infrared spectroscopy, ammonia temperature programmed desorption (NH_3-TPD), scanning electron microscopy and energy dispersive X-ray analysis (SEM), transmission electron microscopy (TEM), X-ray photoelectron spectroscopy (XPS), hydrogen chemisorption of platinum, BET surface area analysis by nitrogen adsorption, and thermogravimetric analysis (TGA).

Catalyst loading and testing

It was fully described in a previous work [31].

Results and discussion

Deactivation behaviour of 320HOA USY zeolite

320HOA USY zeolite was supplied by Tosoh Corporation in the form of a white powder with particle sizes between 6 and 10 μm [15]. Figures 1 and 2 show the XRD pattern and SEM microgram of 320HOA USY, respectively, and Table 4 shows the in-house characterisation of 320HOA USY using various techniques.

USY zeolite structure was confirmed from the XRD pattern for a fresh sample with 2 wt% sodium content as supplied, 320HOA USY has a lower crystallinity but a

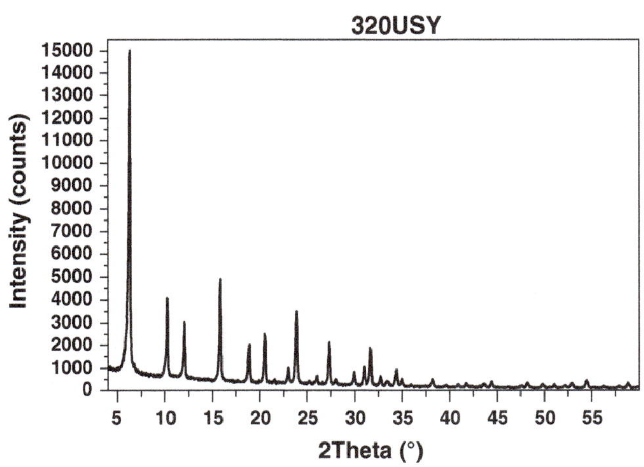

Fig. 1 XRD pattern of 320HOA USY

Fig. 2 SEM micrograph of 320HOA USY

higher unit cell size compared with CBV712 USY and CBV720 USY as explained in the previous work [31]. This can be related to the fact that 320HOA USY was produced via less steaming dealumination process and that no acid leaching was applied afterwards, which can be confirmed by the lower bulk Si/Al molar ratio. As a result, a reduced amount of unit cell size shrinking occurred. The rounded edged cubic crystallites, ranging in size between 1 and 3 µm, are combined with an amorphous phase which normally surrounds the zeolite Y crystals after the dealumination process, as is confirmed by the SEM images shown in Fig. 2. The ^{27}Al solid-state NMR for this USY catalyst, as shown in Fig. 3, indicated the presence of non-framework aluminium species as penta- and octa-coordinated aluminium, besides the tetrahedral aluminium framework. Four chemical environments were also obtained for Si, as shown in Fig. 4 for the ^{29}Si NMR solid state, which confirms that a shorter dealumination process was necessary to produce 320HOA USY. In addition, the Si/Al molar ratio for the framework was calculated to be 6.3, compared with 8.34 and 31.2 for CBV712 and CBV720 USY, respectively. Moreover, the ^{27}Al NMR peaks shapes and their corresponding chemical shifts were totally different from those for CBV712 and CBV720 USY, which may be explained by the different dealumination conditions that were applied to create these catalysts in the different structures. It can be concluded that most USY catalysts

Table 4 In-house characterisation of 320HOA USY

Elemental analysis (bulk)	Before ion exchange				After ion exchange with Pt precursor as tetraammine platinum (II) chloride $(Pt(NH_3)_4Cl_2).H_2O$			
	Si/Al mole ratio		2.90		Si/Al mole ratio	2.90	Pt wt%	0.83
XPS[a]	Si	Al	Si/Al	Pt	No Pt on the surface			
atomic conc %	24.42	10.89	2.24	0				
XRD	Crystallinity				Related to in-house as-synthesised Y			
	63.68							
NMR	Si/Al mole ratio of the framework				Four chemical environments of Si-NMR: Si(0Al), Si(1Al), Si(2Al) and Si(3Al)			
					Lewis acid sites existed: $Al(OSi)_4$, $(Al(H_2O)_5)^{3+}$ and $(Al(H_2O)_6)^{3+}$			
	6.30							
Ammonia TPD	NH_3 desorption through TPD/mmol g^{-1}				Ammonia adsorption calorimetry followed by temperature programmed desorption			
	0.30 ± 0.01							

Hydrogen chemisorption[b]	Metal dispersion	Metallic surface area (m^2 g^{-1} metal)		Metallic surface area (m^2 g^{-1} sample)	Crystallite size (nm)	
	14.89	36.77		0.36	7.61	
BET[b]	Surface area (m^2 g^{-1})	Pore volume (cm^3 g^{-1})		Pore size (Å)	Shows both micro- and mesoporosity	
	735.87	0.39		34.52		
TGA for coke content in spent catalyst (mass loss wt% in air) at different reaction temperatures	210 °C	Top	0.11	250 °C	Top	0.35
		Mid	0.14		Mid	0.31
		Bot	0.10		Bot	0.30
	230 °C	Top	0.24	270 °C	Top	0.10
		Mid	0.20		Mid	0.26
		Bot	0.13		Bot	0.15

[a] LPD Laboratory Services Ltd.

[b] MCA Services

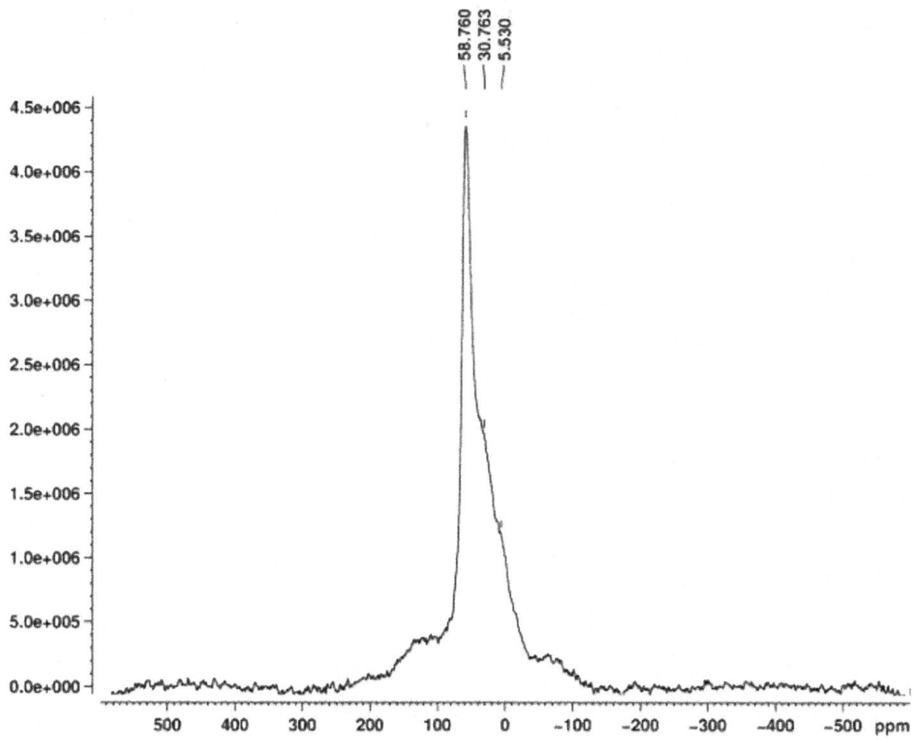

Fig. 3 ^{27}Al NMR spectra of 320HOA USY

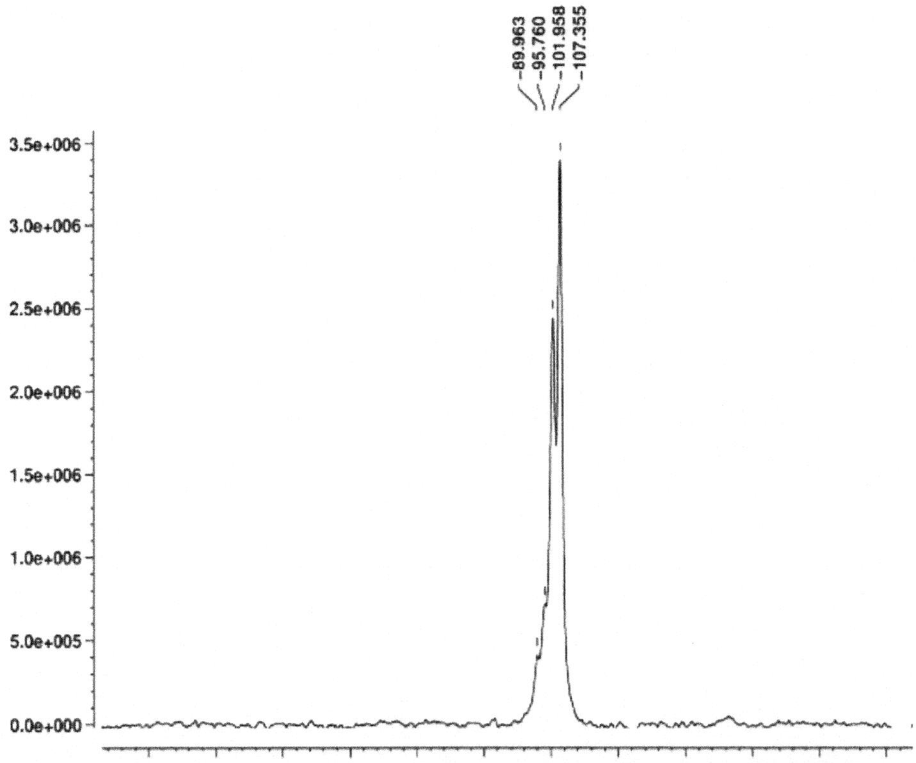

Fig. 4 ^{29}Si NMR spectra of 320HOA USY

have non-framework aluminium species; the nature and crystallographic locations within the zeolite cavities of USY catalysts are different from other catalysts, due to the dealumination conditions that have been applied to create them. Similar observations can be made for the ^{27}Al NMR peak shapes and the corresponding chemical shifts for USY-A compared with 320HOA USY as was described in [31]. They were all produced via steaming with no acid leaching, which is clearly reflected in their catalytic performance and deactivation behaviour.

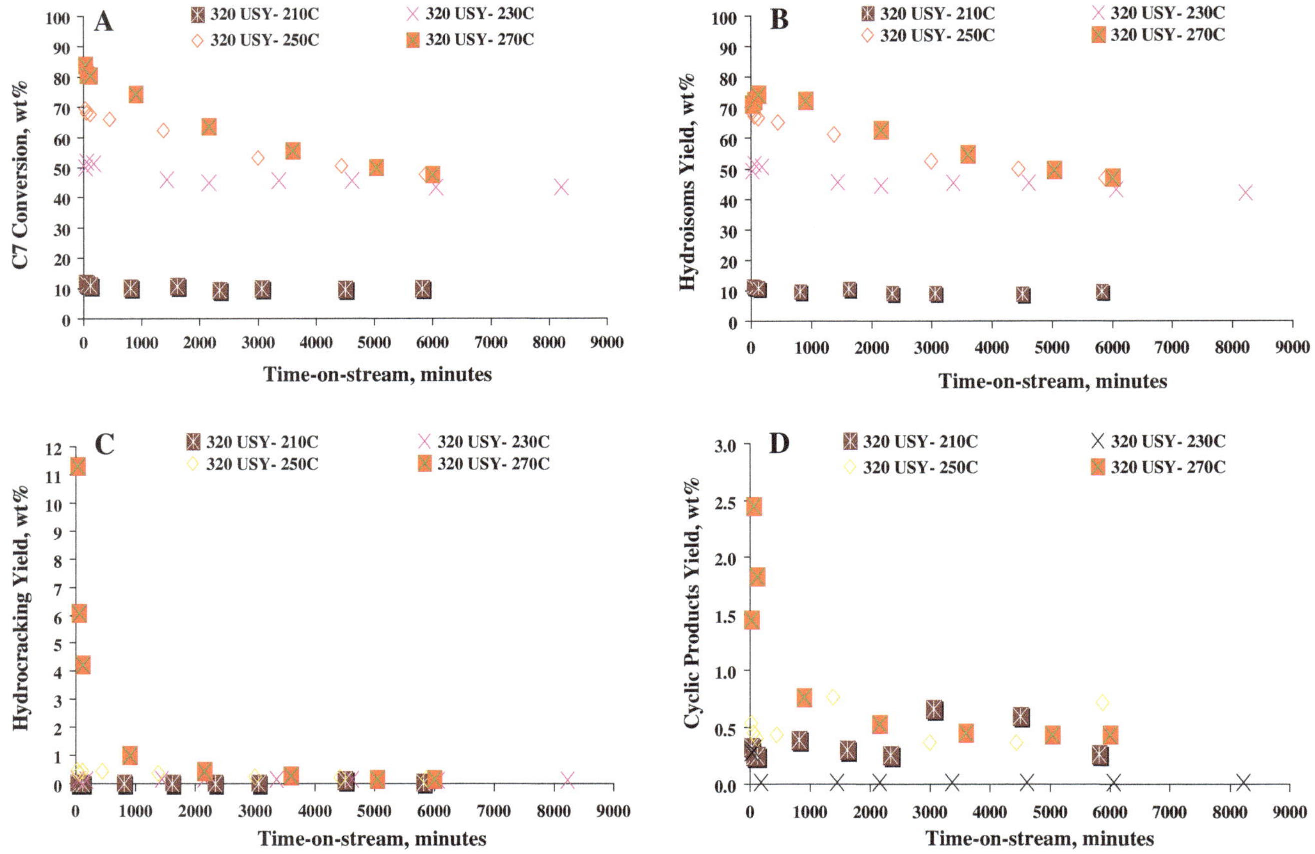

Fig. 5 Overall conversion (**a**), total yield of C7 isomers (mono + di + tribranched isomers) (**b**), total hydrocracking yield (**c**), total yield of the cyclic products (**d**) as a function of TOS for 320HOA USY at various reaction temperatures. WHSV = 5.13 h^{-1}; H_2/C7 (mol/mol) = 9; total pressure = 1 atm; temperature = 210, 230, 250 and 270 °C; 1 wt% Pt loading

The effect of TOS for the C7 conversion, hydroisomerisation, hydrocracking, and cyclic product yields at different reaction temperatures 210–270 °C are shown in Fig. 5a–d. Fresh catalyst was used for each run at each different temperature. Initially, the overall conversions were 12, 50, 70, and 84 wt% after 30 min following introduction of the feed to the reactor at the temperatures of 210, 230, 250, and 270 °C, respectively. At 210 °C, the catalyst lost about 1–2 wt% of its activity before levelling off and reaching a constant value after 60 h on stream, whereas the deactivation rate at 230 °C for the new activated catalyst was slightly faster, as can be seen in Fig. 5a. The overall conversion was 50 wt% after 30 min from the point at which the feed was introduced to the reactor, which then decreased with the TOS to 46 wt% after the first 24 h, and then decreased slowly within the deactivation rate range of 2–3 wt% every 24 h until the reaction terminated after about 6 days (140 h). A moderate drop in the catalyst activity occurred during the first few hours at the higher temperatures of 250 and 270 °C, after which a constant deactivating value was reached, ranging from 3 to 5 wt% every 24 h, until the reaction terminated. The initial selectivity towards the hydroisomerisation products was 98.5 and 84 % after 30 min following introduction of the feed to the reactor, which then increased throughout the TOS to 99 and 92 % after 120 min at 250 and 270 °C, respectively. The maximum isomer yield obtained was around 71 wt% at 270 °C and around 68 wt% at 250 °C, and selectivity towards cracked products was fairly high at the higher temperatures during the first few hours of the deactivation reaction. The highest selectivity for cracked products was 13.5 % 30 min after introducing the feed to the reactor at 270 °C, decreasing rapidly throughout the TOS and finally reaching 5 wt% after 120 min. The yield of cyclic products at the temperatures of 250 and 270 °C was also higher than at the lower temperatures.

It has been found that after the deactivation runs at temperatures of 210–270 °C, the hard coke content is lower than that found for the previously studied catalysts: CBV712USY, CBV720USY, and USY-A [31], which can be linked to the acidity properties of each catalyst. Thus, it can be seen that 320HOA USY has a lower acidity than the above catalysts resulting in a lower cracking activity and thus a lower coke formation. However, the negative effect of the presence of extra-framework Al species (EFAL) could be expected to be lower in 320HOA USY due to the pore

Table 5 The C7 conversion (wt%), product yield (wt%), and product selectivity (%) at different reaction conditions using the pressure rig for 1wt% Pt/320HOA USY

	Conversion (wt%)	Hydroisoms yield (wt%)	Hydrocrack yield (wt%)	Cyclic yield (wt%)	Hydroisoms selectivity (%)	Hydrocrack selectivity (%)	Cyclic selectivity (%)
WHSV (h^{-1})	T = 210 °C, P = 1 bar						
2.57	28.45	28.16	0.01	0.29	98.97	0.02	1.01
5.13	16.89	16.64	0.00	0.24	98.56	0.00	1.44
10.26	7.99	7.73	0.00	0.26	96.76	0.00	3.23
WHSV (h^{-1})	T = 210 °C, P = 8 bar						
2.57	3.54	3.25	0.00	0.29	91.73	0.00	8.27
5.13	4.27	3.98	0.00	0.29	93.14	0.00	6.86
10.26	0.52	0.27	0.00	0.25	52.70	0.00	47.30
WHSV (h^{-1})	T = 210 °C, P = 15 bar						
2.57	0.46	0.24	0.00	0.22	51.84	0.00	48.16
5.13	0.77	0.54	0.00	0.23	69.70	0.00	30.17
10.26	3.97	3.72	0.00	0.26	93.51	0.00	6.49
Pressure (bar)	T = 210 °C, WHSV = 5.13 h^{-1}						
1	16.89	16.64	0.00	0.24	98.56	0.00	1.44
8	4.27	3.98	0.00	0.29	93.14	0.00	6.86
15	0.77	0.54	0.00	0.23	69.70	0.00	30.17
WHSV (h^{-1})	T = 230 °C, P = 1 bar						
2.57	46.38	45.89	0.16	0.33	98.94	0.35	0.71
5.13	45.47	45.06	0.08	0.33	99.10	0.17	0.72
10.26	26.11	25.36	0.03	0.28	97.16	0.11	1.06
WHSV (h^{-1})	T = 230 °C, P = 8 bar						
2.57	32.35	32.00	0.04	0.31	98.91	0.13	0.97
5.13	19.90	19.56	0.03	0.31	98.29	0.16	1.55
10.26	9.08	8.80	0.00	0.28	96.94	0.00	3.05
WHSV (h^{-1})	T = 230 °C, P = 15 bar						
2.57	16.95	16.60	0.02	0.33	97.94	0.09	1.97
5.13	10.92	10.64	0.00	0.28	97.43	0.00	2.55
10.26	6.28	6.02	0.00	0.26	95.81	0.00	4.17
Pressure (bar)	T = 230 °C, WHSV = 5.13 h^{-1}						
1	45.47	45.06	0.08	0.33	99.10	0.17	0.72
8	19.90	19.56	0.03	0.31	98.29	0.16	1.55
15	10.92	10.64	0.00	0.28	97.43	0.00	2.55
WHSV (h^{-1})	T = 250 °C, P = 1 bar						
2.57	78.83	76.38	1.82	0.62	96.90	2.31	0.79
5.13	68.83	67.52	0.85	0.47	98.09	1.23	0.68
10.26	39.93	38.96	0.55	0.43	97.55	1.36	1.08
WHSV (h^{-1})	T = 250 °C, P = 8 bar						
2.57	70.07	68.91	0.83	0.32	98.35	1.19	0.46
5.13	52.62	51.99	0.66	0.31	98.81	1.25	0.58
10.26	30.51	30.05	0.19	0.27	98.50	0.61	0.89
WHSV (h^{-1})	T = 250 °C, P = 15 bar						
2.57	52.05	51.45	0.30	0.31	98.83	0.58	0.59
5.13	42.85	42.32	0.25	0.29	98.76	0.58	0.68
10.26	25.38	24.94	0.15	0.29	98.28	0.58	1.14

Table 5 continued

	Conversion (wt%)	Hydroisoms yield (wt%)	Hydrocrack yield (wt%)	Cyclic yield (wt%)	Hydroisoms selectivity (%)	Hydrocrack selectivity (%)	Cyclic selectivity (%)
Pressure (bar)	$T = 250$ °C, WHSV $= 5.13$ h^{-1}						
1	68.83	67.52	0.85	0.47	98.09	1.23	0.68
8	52.62	51.99	0.66	0.31	98.81	1.25	0.58
15	42.85	42.32	0.25	0.29	98.76	0.58	0.68
WHSV (h^{-1})	$T = 270$ °C, $P = 1$ bar						
2.57	78.02	71.93	4.18	1.91	92.19	5.36	2.45
5.13	72.53	69.68	1.88	0.97	96.07	2.59	1.34
10.26	41.78	40.09	0.96	0.73	95.94	2.31	1.75
WHSV (h^{-1})	$T = 270$ °C, $P = 8$ bar						
2.57	66.87	63.21	3.27	0.39	94.53	4.89	0.58
5.13	54.19	50.28	3.65	0.26	92.79	6.74	0.48
10.26	30.19	28.25	1.67	0.27	93.57	5.54	0.88
WHSV (h^{-1})	$T = 270$ °C, $P = 15$ bar						
2.57	58.01	54.98	2.71	0.32	94.78	4.68	0.54
5.13	43.24	40.59	2.37	0.29	93.86	5.47	0.66
10.26	29.63	27.47	1.90	0.26	92.69	6.42	0.89
Pressure (bar)	$T = 270$ °C, WHSV $= 5.13$ h^{-1}						
1	72.53	69.68	1.88	0.97	96.07	2.59	1.34
8	54.19	50.28	3.65	0.26	92.79	6.74	0.48
15	43.24	40.59	2.37	0.29	93.86	5.47	0.66

blockage effect, especially in the polymeric species [24], such that the coke content can be trapped within the pores since these will be smaller than those found for USY-A.

Effects of reaction pressure and contact time for 1 wt% Pt/320HOA USY

The effect of increasing the hydrogen partial pressure, and hence the overall reaction pressure and variation in contact time, representing the time taken by reactants to pass through the catalyst bed during a reaction, has been studied in this work at a range of temperatures from 210 to 270 °C for the hydroisomerisation of n-C7 over 1 wt% platinum-loaded 320HOA USY. A stainless-steel higher pressure unit was used, and the catalyst (1 wt% Pt/320HOA USY) was tested at temperatures ranging from 210 to 270 °C, WHSV of 2.57, 5.13, and 10.26 h^{-1}, and pressures of 1, 8 and 15 bar. At each temperature, products of the reaction were collected at the point where steady-state conditions were achieved; catalysts were calcined and then re-reduced before proceeding to the next temperature. Unfortunately, the reaction rig could not operate overnight at high pressures due to the necessary safety precautions and therefore it was not possible to fully study deactivation behaviours at these pressures.

Table 5 shows the results from the n-C7 conversion (wt%), product yield (wt%), and product selectivity (%) at the different reaction conditions using the pressure rig for the hydroisomerisation of n-C7 over 1 wt% Pt/320HOA USY. In general, it has been found that an increase in pressure results in a decrease in overall conversion and a slight decrease in selectivity to isomers with a noticeable increased selectivity to the cyclic products, particularly at the lower temperatures of 210 and 230 °C. There was no obvious cracking activity at the lower temperatures of 210 and 230 °C, and a constant WHSV was observed whilst changing reaction pressures. In contrast, at the highest temperature of 270 °C and WHSV of 5.13 h^{-1}, increased selectivity towards the cracked products was observed with increased reaction pressures, with comparable results being reported by Chao et al. [32] who investigated the effect of reaction pressure over the range 1–41.3 bar on mordenite catalysts with different Si/Al ratios and the zeolite Beta. In addition, they showed that pressure reduced the cracking yield for all catalysts as well as improved the mordenite catalysts' stability. Additionally, an increase in hydrogen pressure has been found to reduce the hydroisomerisation reaction rate [33], where an increase in the hydrogenation activity at higher hydrogen pressure results in the hydrogenation of intermediate olefins, due to a shorter intermediate olefin residence time inside the catalyst [34]. This minimises the cracking activity, since pressure reduces overall conversion but improves hydroisomerisation selectivity compared with that of cracking, which to some extent, was not the case for catalyst 1 wt% Pt/320HOA

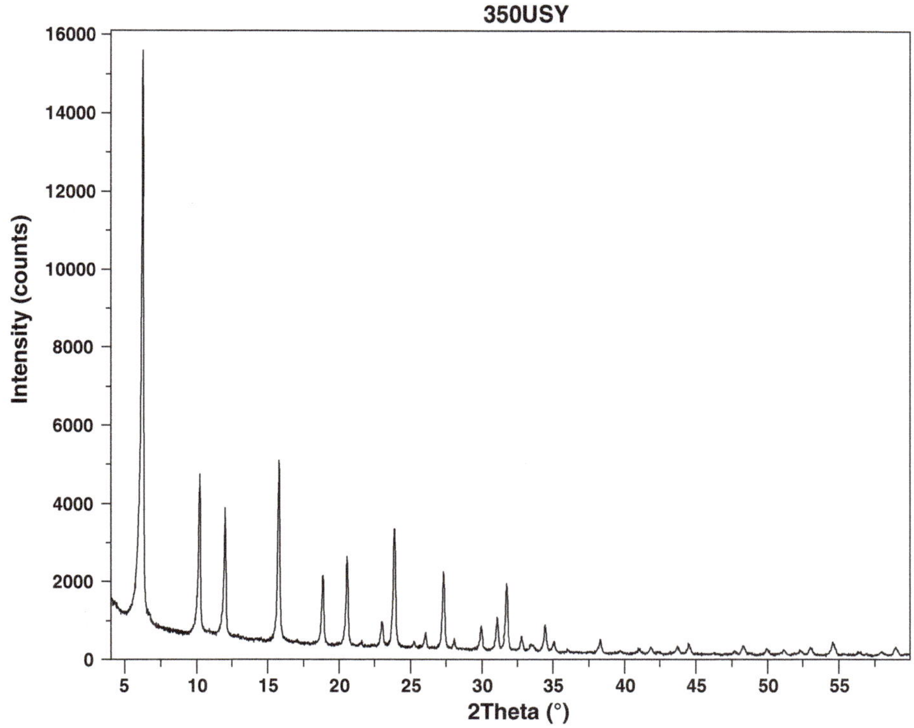

Fig. 6 XRD pattern for 350HOA USY

USY, whose small average pore diameter most likely limited the movement of larger molecules at high pressures, and therefore promoted cracking.

Table 5 illustrates that the conversion increased as the contact time increased (WHSV $\leq$ 10.26, 5.13, 2.57 h^{-1}) at constant temperature and pressure. However, increasing contact time between the reaction feed (n-C7) and the catalyst appears to cause a reduction in cyclic product yield, giving rise to an increase in cracking and hydroisomerisation selectivity.

Wang et al. [35] found that increasing the contact time results in an increased overall conversion of n-C7 at a constant temperature of 220 °C and in a reduced selectivity in hydroisomerisation versus cracking. Similar observations were made by Chica and Corma [36] when they tested the hydroisomerisation of nC_5, nC_6 and nC_7 over platinum-loaded zeolite supports. However, the decrease in selectivity to isomers was opposed by an increase in the overall yield of isomers at short contact times. It was also shown that with increasing contact times, the formation of multi-branched isomers also increases, leading to a lower mono/multi-branched isomer ratio. This behaviour could result from an initial formation of mono-branched isomers and their subsequent transformation into multi-branched isomers.

Deactivation behaviour of 350HOA USY zeolite

350HOA USY zeolite was supplied by Tosoh Corporation in the form of a white powder with particle sizes between 6 and 10 μm. The Figs. 6 and 7, respectively, show the XRD pattern and SEM microgram for 350HOA USY, and Table 6 the in-house characterisation for 350HOA USY using different techniques.

Fig. 7 SEM micrograph for 350HOA USY

Figures 8 and 9 show solid-state NMR spectra for ^{27}Al and ^{29}Si, respectively. ^{27}Al MAS NMR analysis revealed that the catalyst had EFAL species even though it was a dealuminated and partially acid-leached zeolite. Information regarding the Si atoms coordinated (via oxygen) with neighbouring T atoms (Si and Al) is provided by ^{29}Si MAS NMR, indicating two different chemical configurations of SiO_4, which were assigned to (4Si,0Al) and SiO_4 (3Si,1Al).

This confirms that 350HOA was produced by a dealumination process and acid leached to achieve a high Si/Al

Table 6 Some in-house characterisations of 350HOA USY

Elemental analysis (bulk)	Before ion exchange				After ion exchange with Pt precursor as tetraammine platinum (II) chloride ($(Pt(NH_3)_4Cl_2).H_2O$)			
	Si/Al mole ratio	5.55			Si/Al mole ratio	5.55	Pt wt%	0.90
XPS[a]	Si	Al	Si/Al	Pt	Pt is likely to be metallic			
Atomic Conc %	28.32	5.69	4.98	0.07				
XRD	Crystallinity %				Related to in-house as-synthesised Y			
	66							
NMR	Si/Al mole ratio for the framework				Two chemical environments for Si-NMR: Si(0Al) and Si(1Al)			
	12.38				Lewis acid sites are present			
Ammonia TPD	NH_3 desorption through TPD/ mmol g^{-1}				Ammonia adsorption calorimetry followed by temperature programmed desorption			
	0.76 ± 0.01							
BET[b]	Surface area (m^2 g^{-1})	Pore volume (cm^3 g^{-1})		Pore size (Å)	Shows both micro and mesoporosity			
	872.97	0.47		26.10				
TGA for coke content in spent catalyst (mass loss wt% in air) after different reaction temperatures	210 °C	Top	0.10	250 °C	Top	0.26		
		Mid	0.13		Mid	0.37		
		Bot	0.13		Bot	0.28		
	230 °C	Top	0.52	270 °C	Top	1.20		
		Mid	0.76		Mid	1.12		
		Bot	0.71		Bot	1.29		

LPD Laboratory Services Ltd.

[b] MCA Services

mole ratio; thus a high surface area was obtained, as there were fewer EFAL species. This would result in a mesoporous channel network, due to the larger pore volume. Moreover, it should be noted that the dealumination process required to produce 350HOA USY was less severe than for the CBV720USY catalyst [31], which has a higher Si/Al mole ratio in the framework than 350HOA USY. Nevertheless, both catalysts have mesoporous systems, due to larger pore volumes and surface areas, but less acidity and thus lower catalytic activity compared to just-dealuminated catalysts.

The effects of TOS on the C7 conversion, hydroisomerisation, hydrocracking, and cyclic product yields at different reaction temperatures 210–270 °C are shown in Fig. 10a–d, respectively. Fresh catalyst was used for each run at the different reaction temperatures. Initially, the overall conversion was 17, 56, 75, and 84 wt% after 30 min from the point when the feed was introduced to the reactor at the temperatures of 210, 230, 250, and 270 °C, respectively. At 210 °C, the catalyst lost about 1–2 wt% of its activity and then reached a pseudo-stable state for the initial 60 min on stream. The deactivation rate at 230 °C for the new activated catalyst was slightly faster, with the overall conversion being 56 wt% after 30 min after which it decreased with the TOS to 51 wt% after 27 h, and then decreased slowly within the deactivation rate range of 2–3 wt% every 24 h until the reaction finally terminated after about 6 days (140 h). Rapid initial deactivation occurred during the first few hours at the higher temperatures of 250 and 270 °C, followed by a constant deactivating value being reached, ranging from 3 to 5 wt%, every 24 h until the reaction terminated.

The initial selectivity towards the hydroisomerisation products was 98 and 85 % after the initial 30 min, and then increased throughout the TOS to 99 and 94 % after 120 min at 250 and 270 °C, respectively. Thus, the selectivity towards cracked products was fairly high at higher temperatures during the first few hours of the deactivation reaction, with the highest cracked product selectivity being 14 % 30 min after introducing the feed to the reactor at 270 °C; however, this decreased rapidly throughout the TOS to 4 % after 120 min. Additionally, the yield of cyclic products at the temperatures of 210 and 230 °C was higher than that at the higher temperatures.

The TOS results provide evidence that the higher activity of 350HOA USY compared to CBV720 USY [31] is related to the number of acid sites. Furthermore, moderately slow deactivation for both catalysts was observed, reaching the pseudo-stable state in the initial few minutes at the lower temperatures of 210 and 230 °C. The n_{Pt}/n_A ratio for 350HOA USY was calculated as 0.06, which is in the suggested range of $0.03 \leq n_{Pt}/n_A \leq 0.17$ for a well-

Fig. 8 ^{27}Al NMR spectra for 350HOA USY

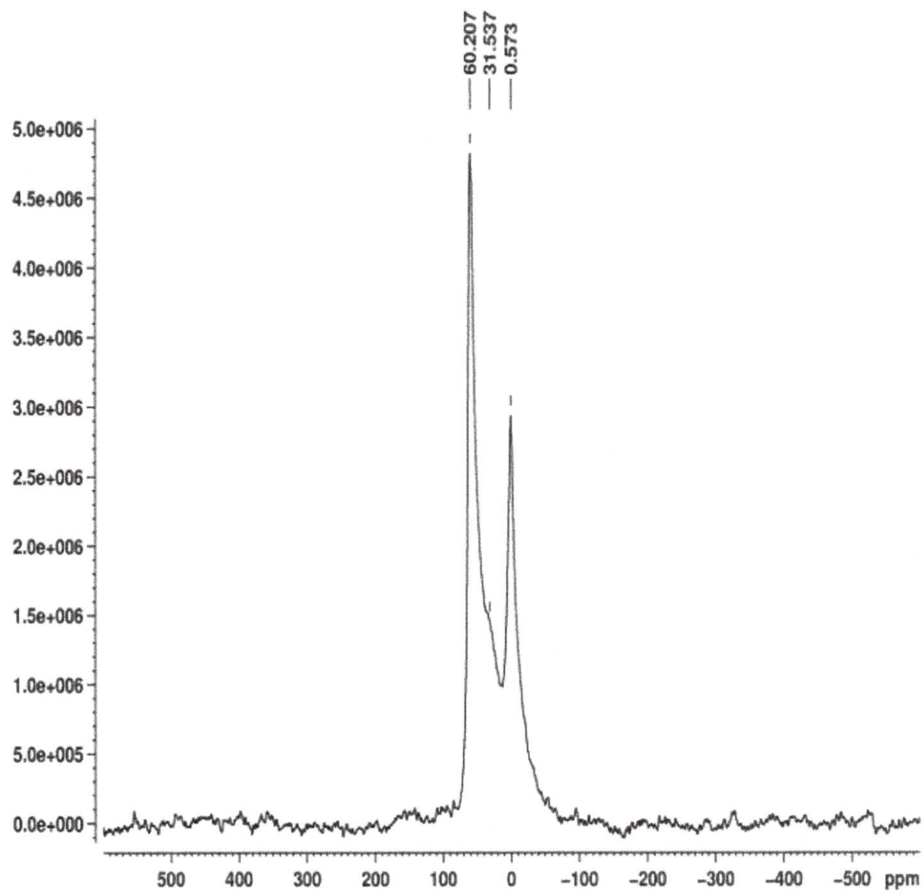

balanced bifunctional catalyst, in conjunction with 0.12 for CBV720 USY, which allowed deactivation to occur at reasonably low temperatures with lower coke content. Although the mesoporosity texture was similar for both catalysts, the average coke content was higher for 350HOA USY, which may be due to its higher acidity, as the framework Si/Al was lower, resulting in cracking activity at temperatures of 250 and 270 °C.

Extraction of EFAL in USY zeolites by AHFS post-synthetic modification

Ammonium hexafluorosilicate solution was used to remove the EFAL from the USY samples: USY-A, USA-C, 320HOA, and from the acid-leached 350HOA, due to the samples having different silicon-to-aluminium ratios, and thus different numbers of aluminium atoms as non-framework on their crystal unit cells. As with dealumination, after the process the aluminium removed from the framework remains on the crystal and migrates to the surface in an extension that is dependent on the partial pressure of the steam used. This can be confirmed by Si/Al ratio results on the above USYs surfaces by XPS analysis as shown in Tables 4 and 6 and it can be seen that the Si/Al ratios on the surfaces are much lower for zeolites that have only been dealuminated, such as USY-A, USY-C, and 320HOA than for the acid-leached zeolites such as CBV712, CBV720 and 350HOA. The resulting AHFS modified zeolite samples were characterised by XRD, MAS NMR, ammonia TPD, ICP, XPS and BET, with the effects of the treatments on the total amount and type of EFAL, the relationship between amount and nature of acid sites, and the deactivation behaviour discussed below.

Optimisation of chelating extraction parameters

The USY sample 320HOA containing 2wt% sodium with the ammonium form as the starting material has been fine tuned using a chelate compound (ammonium hexafluorosilicate $(NH_4)_2SiF_6$) to provide the best condition to remove only the non-framework aluminium whilst maintaining the aluminium framework and original zeolite crystallinity—as these are unaffected by this chelating agent. Four main factors were considered when optimising the chemical treatments: (1) the AHFS and USY mixture volume, (2) the treatment duration time, (3) the zeolite to AHFS weight ratio, and (4) the temperature used during the chemical treatment. Table 7 lists the resulting modified sample characteristics for each factor and illustrates, for the AHFS modified samples, that the increase in mixture volume

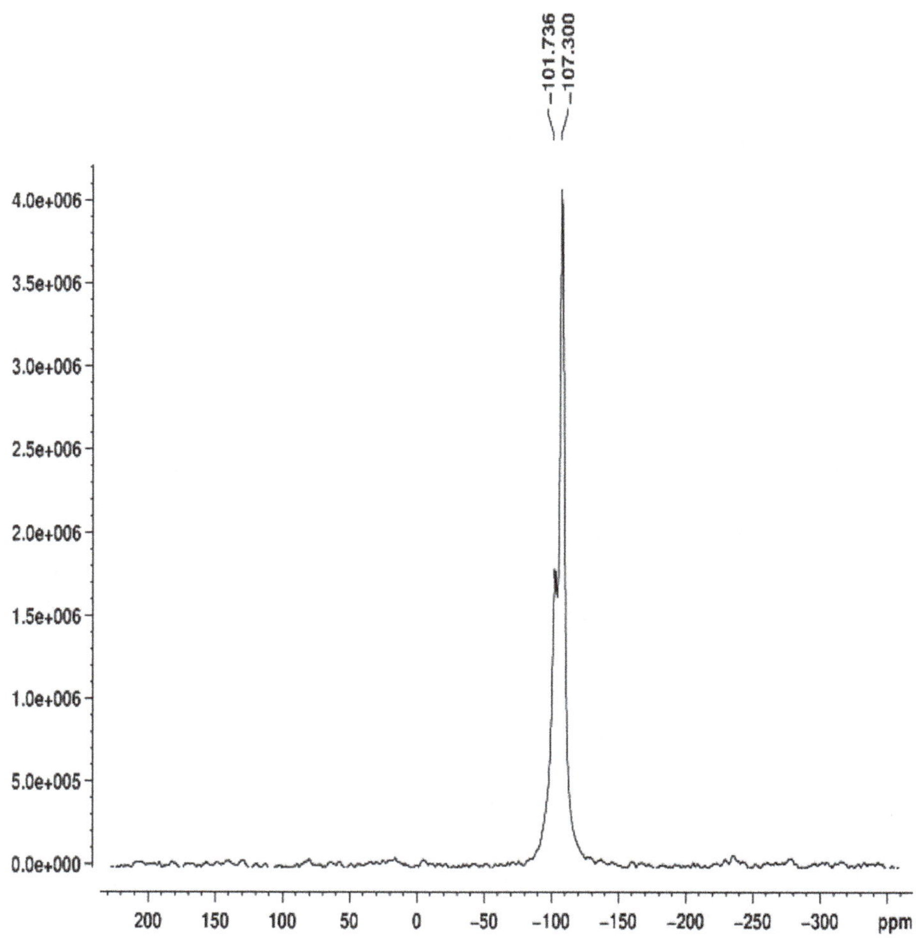

Fig. 9 ^{29}Si NMR spectra for 350HOA USY

caused the total aluminium content in the bulk to decrease from 9.73 wt% before the modification to between 4.83 and 5.51 wt%. Additionally, both the Si/Al mole ratio in the framework and the crystallinity were maintained in the ranges of 6.45–6.75 mol ratio and 57–68 %, respectively. There was no significant effect on the mixing time factors, such that the Si/Al mole ratio in the framework and the crystallinity were maintained in the ranges of 6.45–6.75 and 57–66 %, respectively. The zeolite to AHFS mass ratio was optimised in the range of 6.67–1.67, so the framework structure had been attacked by the AHFS at a ratio of 1.67 mol, leading to a Si/Al mole ratio in the framework which increased to 11.10.

A temperature range of 25–100 °C was used during the chemical treatment and it was found that the resulting modified samples had the total aluminium content in the range of 4.79–5.56 wt%, and crystallinity in the range of 57–64 %. As well as this, the framework was unaffected as the temperature rose, leaving the Si/Al in the framework in the range of 6.43–6.53 mol ratio. It can be seen that the total Si content in the modified samples is higher than that detected for the original sample 320HOA before the acid chelating treatment, which can explained by the existence of the silicon fluoride containing species on the outer surfaces of the modified zeolites, an effect that has been reported in the literature [27, 37]. The detected residual fluoride species was removed from the structure by applying an extra ammonium ion-exchange step.

Table 8 shows the unit cell for the AHFS modified samples, the starting material sample (320HOA), the total aluminium number Al_{total} (based on the elemental analysis), the framework aluminium number Al_F (measured using ^{29}Si MAS NMR $(Si/Al)_F$, and on the basis of 192 T atoms (Si or Al) per unit cell with the composition of $H_x(AlO_2)_x(SiO_2)_{192-x}$) and the EFAL number Al_{nF}, (the difference between the Al_{total} and Al_F) [26, 27, 38, 39]. To some extent, the value for $Al_{F/u.c}$ was greater than the value for the total $Al_{total/u.c}$ as shown for the samples designated as F9, F10, and F22 in the table, so this difference reflects the errors in analyses for some AHFS modified samples which may have resulted from the existence of silicon fluoride containing species during the chemical treatment. The Al_{nF} extraction efficiency percentage was calculated for each resulting sample by dividing the difference between Al_{nF} for the modified sample and Al_{nF} for the starting material, which is equal to 23 atoms for 320HOA.

It is clear from Table 8 that the most effective extraction of EFAL by AHFS chelating agent was for the samples F4,

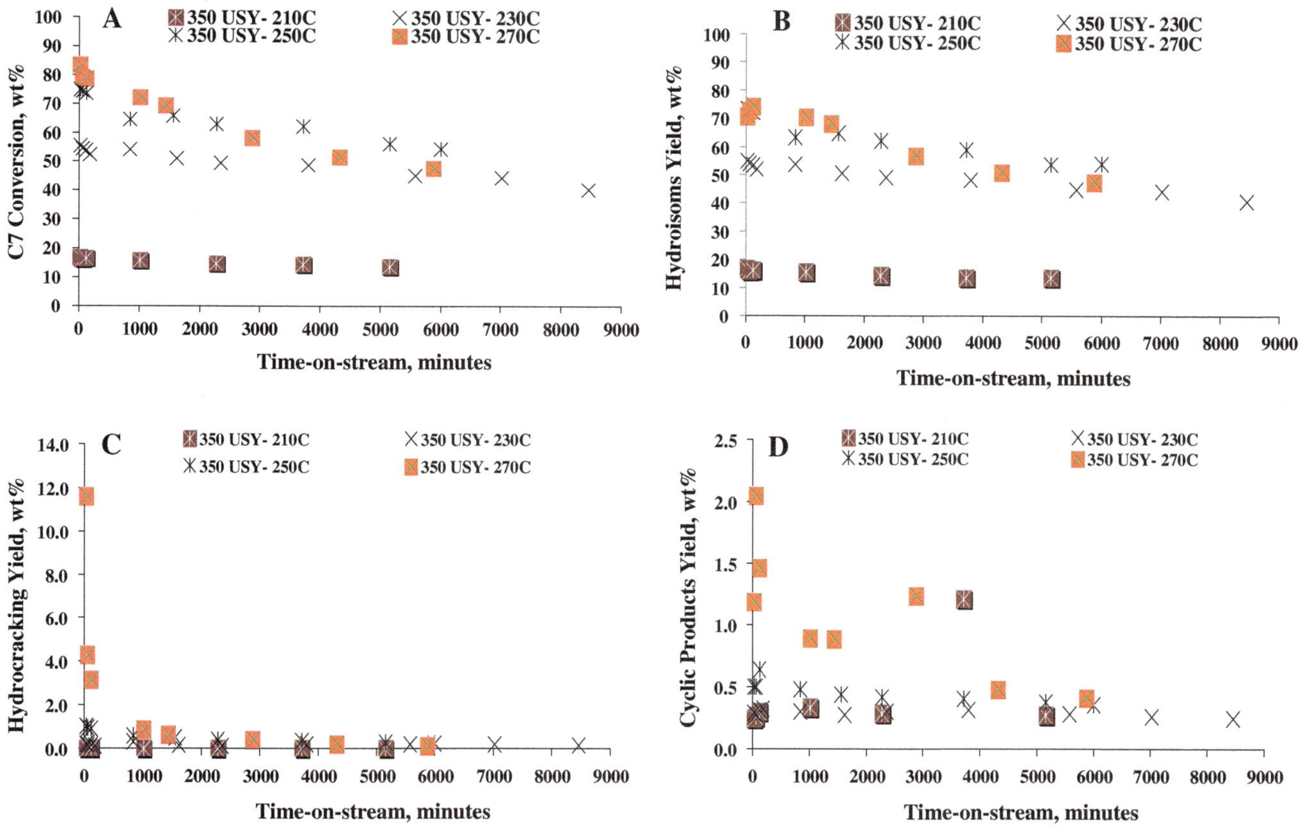

Fig. 10 Overall conversion (**a**), total yield of C7 isomers (mono + di + tribranched isomers) (**b**), total hydrocracking yield (**c**), and total yield of the cyclic products (**d**) as a function of TOS for 350HOA USY at various reaction temperatures. WHSV = 5.13 h^{-1}; H_2/C7 (molar) = 9; total pressure = 1 atm; temperature = 210, 230, 250 and 270 °C; 1 wt% Pt loading

F5, F6, F7, and F11. The results for the starting material (320HOA) are arranged in the table according to the chemical treatment conditions applied, where the EFAL formed represents approximately 47 % of the total aluminium. Furthermore, the obtained $(Si/Al)_T$ as bulk (measured by the chemical analysis) is almost in agreement with $(Si/Al)_F$ measured by NMR analysis for those samples (F4, F5, F6, F7, and F11) which has 100 % extraction efficiency. The lowest extraction efficiency was found to be 46 % for the modified sample F8, which was prepared by the chemical treatment using the highest mass ratio (6.67 wt/wt) of zeolite to AHFS, which can be considered as the mildest condition, with moderate extraction efficiency observed for the modified samples F1, F2, F3, and F12 in which partial extraction occurred, leaving the framework aluminium unaffected, and maintaining the framework $(Si/Al)_F$ ratio as prior to modification. This resulted in reasonable crystallinity values in the range of 61–66 % compared with the original crystallinity of 64 % for the starting material.

Catalysts USY-A, USY-C, 320HOA USY, and 320HOA USY were chemically treated by AHFS to remove the EFAL as these have a detrimental effect on catalytic and transport properties, and consequently adversely affects the coke formation and deactivation behaviour. The EFAL was removed from the catalysts in their ammonium form using the chelating agent AHFS under the same chemical treatment conditions as sample F2. This involved 50 ml of a solution mixture, stirred for 1 h at room temperature, with a zeolite to AHFS mass ratio of 3.33 wt/wt. From Table 9, it can be seen that samples F20 and F21, the modified samples of catalysts USY-A and USY-C, respectively, still have EFAL which has not been extracted by AHFS, and that framework Al atoms were attacked. In contrast, F2 was partially extracted but was not attacked by the AHFS. It can be interpreted that F2 was produced by the AHFS treatment of 320HOA USY which in its ammonium form contains approximately 2 wt% sodium, and thus the presence of sodium inhibits aluminium extraction [27]. Table 9 shows the in-house characterisation for the original and modified catalysts using different techniques, and it can be seen that the $(Si/Al)_F$ increased for modified samples F20 and F21, confirming that the AHFS has dealuminated the framework Al atoms, and subsequently decreased their crystallinity. It was also observed that the zeolites prepared with AHFS retained less than 0.6 wt% fluorine based on the ion chromatography analysis, which was not seen by XPS analysis measurements for the

Table 7 The chemical treatment parameters of AHFS and the resulting samples characteristics

	The chemical treatment conditions					Elemental analysis ICP			MAS NMR	XRD
	AHFSwt (g)	320USYwt (g)	Mixing time (h)	Temp (°C)	Vol (ml)	Si (wt%)	Al (wt%)	Bulk (Si/Al) (mol/mol)	Framework (Si/Al) (mol/mol)	Crystallinity (%)
320 USY	–	–	–	–	–	29.36	9.73	2.9	6.41	63.68
Optimising the mixture volume										
F1	0.6	2	1	25	25	30.01	5.51	5.23	6.71	63.57
F2	0.6	2	1	25	50	30.71	5.22	5.65	6.45	57.27
F3	0.6	2	1	25	75	31.18	5.33	5.62	6.75	66.47
F4	0.6	2	1	25	100	32.37	4.83	6.44	6.23	68.46
Optimising the mixing time										
F2	0.6	2	1	25	50	30.71	5.22	5.65	6.45	57.27
F5	0.6	2	3.5	25	50	33.12	4.75	6.70	6.75	66.28
F6	0.6	2	7	25	50	33.73	4.81	6.74	6.60	62.91
F7	0.6	2	14	25	50	33.72	4.6	7.04	6.59	66.69
Optimising the AHFS weight										
F8	0.3	2	1	25	50	30.94	7.37	4.03	6.51	60.57
F2	0.6	2	1	25	50	30.71	5.22	5.65	6.45	57.27
F9	1.2	2	1	25	50	37.68	1.68	21.55	11.10	67.11
Optimising the mixture temperature										
F2	0.6	2	1	25	50	30.71	5.22	5.65	6.45	57.27
F10	0.6	2	1	50	50	33.37	4.58	7.00	6.51	63.53
F11	0.6	2	1	75	50	32.5	4.79	6.52	6.43	64.47
F12	0.6	2	1	100	50	32.16	5.56	5.56	6.53	61.13

samples surface. In addition, the XPS measurements revealed that the surface atomic concentration for the aluminium atoms decreased for the AHFS modified samples; thus, it can observed that the $(Si/Al)_F$ mole ratio measured by NMR, (Si/Al) measured by the elemental analysis, and Si/Al measured by XPS present a closer match to each other, particularly in the case of sample F2 as opposed to samples F21 and F22 in which the Al_F was attacked by the AHFS. In terms of the obtained texture of the samples treated with AHFS, some mesoporosity developed within the crystal as the pore volume increased and pore size decreased, thus the total surface area was consequently increased for all of the modified samples.

Figure 11a–d shows the effect of TOS on the n-C7 conversion, hydroisomerisation, hydrocracking and cyclic product yields at the reaction temperature of 230 °C for the samples F2, F20 and F21, which are the AHFS modified samples for the catalysts 320HOA USY, USY-A, and USY-C, respectively. Figure 11a shows that the initial overall conversion was 63 wt% after 30 min from introducing the feed to the reactor at the temperature of 230 °C for F21 compared to 50 wt% for 320HOA USY before the modification, and that the catalyst lost about 1–2 wt% of its activity before reaching a constant "level-off" value after 60 h on stream. The deactivation rate for the F2 catalyst was slightly faster than the parent catalyst along the TOS to be in the range of 3–4 wt% and then reached a constant "level-off" value until the reaction terminated after about 4 days (5,580 min).

Figure 11b shows that the initial selectivity towards the hydroisomerisation products was 98 % for F2 compared to 99 % for 320HOA USY before the modification and kept almost the same along the deactivation run. As demonstrated in Fig. 11c and d, the selectivity towards cracked and cyclic products was fairly high for F2 in comparison with 320HOA USY which may be due to the n_{Pt}/n_A ratio being disturbed as a consequence of increasing the number of acid sites for F2.

Figure 12a shows that the initial overall conversion was 60 wt% after 30 min from the point when the feed was introduced to the reactor at the temperature of 230 °C for F20 compared to 70 wt% for USY-A before the modification [31], which is then decreasing with the TOS to 43 wt% after 780 min and decreasing slowly within the deactivation rate ranging between 1 and 3 wt% every 24 h until the reaction terminated after almost 4 days. In comparison with USY-A, which was completely deactivated in the initial few minutes, F20 deactivated much more steadily with TOS, which may be explained by the effect of pore blockage by the polymeric EFAL species being overcome even in the presence of some EFAL due to further dealumination occurring during the acid leaching

Table 8 Al content of AHFS modified samples and extraction efficiency

	Al_{total}/u.c	Al_F/u.c	Al_{nF}/u.c	Al_{nF} %	Extraction efficiency (%)
320USY	49	26	23	47	–
F1	31	25	6	19	75
F2	29	26	3	11	87
F3	29	25	4	15	82
F4	26	26	0	0	100
F5	25	25	0	0	100
F6	25	25	0	0	100
F7	24	24	0	0	100
F8	38	26	13	33	46
F9	9	16	*	*	–
F10	24	26	*	*	–
F11	26	26	0	0	100
F12	29	26	4	13	84
USY-A	50	24	26	52	**
F20	24	18	6	25	**
USY-C	51	22	29	57	**
F21	24	15	9	38	**
350USY	29	14	15	52	**
F22	5	13	*	*	–

* This difference reflects the errors in the analyses. ** Al_F attacked by AHFS

via the AHFS chelating agent. Figure 12b illustrates that the initial selectivity towards the hydroisomerisation products was 97 % for F20 compared to 92 % for USY-A before the modification and was maintained fairly steadily along the deactivation run. As shown in Fig. 12c, d, the selectivity towards cracked and cyclic products was fairly high for F20 compared with USY-A, which may be due to the n_{Pt}/n_A ratio being disturbed as a consequence of increasing the number of the acid sites (n_A) for F20, and also due to the presence of some EFAL species.

From the Fig. 13a, the initial overall conversion was 36 wt% after 30 min from introducing the feed to the reactor at the temperatures of 230 °C for F21 compared to 46 wt% for USY-C before the modification, and then decreasing with the TOS to 32 wt% after 1,000 min compared to 43 wt% after the same period of time, and decreasing slowly within the deactivation rate ranging between 1 and 3 wt% every 24 h until the reaction was terminated after almost 4 days. Figure 13b shows that the initial selectivity towards the hydroisomerisation products was 99 % for F21, which was the same as USY-C before the modification, and kept almost same along the deactivation run. As shown in Fig. 13c and d, the selectivity towards cracked and cyclic products was fairly similar for F21 compared with USY-C.

The influence of nickel on the stability of Pt/350HOA USY in the hydroisomerisation of n-C7

The catalytic properties of a bifunctional catalyst depend on the balance between its acidity and its hydrogenating functions. With respect to its selectivity and stability, increasing the proportion of the hydrogenation function at constant acidity leads to an improved catalyst performance, and for this reason, a series of monometallic and bimetallic bifunctional catalysts was formed by loading the H-USY (350HOA USY, with a global Si/Al = 5 mol/mol, supplied by Tosoh Corporation) with different amounts of platinum, nickel and platinum–nickel by competitive ion exchange, then investigating the effects of nickel as a second metal to Pt on the stability of the Pt/350HOA USY catalyst in the hydroisomerisation of n-C7 at 230 °C and 1 atm using H_2/C7 = 9 and WHSV = 5.13 h^{-1}. The first series comprised monometallic bifunctional catalysts containing 0.5, 1, 1.5 and 2 wt% of platinum supported on H-USY (350HOA USY), the second was a series of monometallic bifunctional catalysts containing 0.5, 1, 1.5, 2, 10 and 20 wt% of nickel, also supported on 350HOA USY, and the third series was of bimetallic bifunctional catalysts containing 1 wt% of platinum and 0.25, 0.5, 0.75 and 1 wt% of nickel, again supported on 350HOA USY. The monometallic catalysts were prepared using the ion-exchange method, while the method used for the bimetallic catalysts was simultaneous competitive ion exchange, which means that the platinum and nickel precursors, $Pt(NH_3)_4Cl_2H_2O$ and $Ni(NO_3)_26H_2O$, respectively, were mixed together with the zeolite, at the liquid-to-solid ratio of 50 (1 g of zeolite to 50 ml of the metal complex solution), using ammonium hydroxide solution to adjust the pH of the suspension obtained at >9 in order to enhance the uptake and homogenous distribution of the metals within the zeolite grain.

Table 10 shows some characteristics of the mono- and bimetallic bifunctional catalysts so formulated using characterisation techniques such as ICP, XRD, NMR and TGA. Some results were also obtained using XPS and TEM. The quantitative uptake of the loaded metals was verified by elemental analysis using inductively coupled plasma spectroscopy for the hydrofluoric acid digested zeolite samples. It was observed that the uptake of metals by the zeolite support was less than the calculated value. It has been reported that the USY zeolite in its ammonium form had greater ion-exchange efficiency than the zeolite H-USY, due to the greater reduction in the pH of the reaction medium in the latter case [40]. The pH was adjusted for this reason, even though the desired molar ratio for the bimetallic catalyst Pt–Ni/USY was obtained. Using X-ray diffraction, it was found that all the diffraction patterns of the metal-loaded catalysts were identical to that of the parent sample of 350HOA USY. Moreover, no

Table 9 In-house characterisation of original catalysts and those modified by AHFS

	AHFSwt (g)	USYwt (g)	Mixing time (h)	Temp (°C)	Vol (ml)	Si (wt%)	Al (wt%)	F (wt%)	Pt (wt%)	Bulk Si/Al, molar ratio	Framework Si/Al, molar ratio	Crystal (%)	NH_3 TPD (mmol g^{-1})	(Si/Al), Atomic Conc %(XPS)	Pt, Atomic conc % (XPS)	Surface area (m^2/g)	Pore volume (cm^3/g)	Pore size (Å)
320USY	–	–	–	–	–	29.36	9.73	–	0.83	2.90	6.41	63.68	0.3	2.24	0	735.87	0.39	34.52
F2	0.6	2	1	25	50	30.71	5.22	0.56	0.94	5.65	6.45	57.27	1.05	5.32	0	818.98	0.45	31.43
USY-A	–	–	–	–	–	26.75	9.71	–	1.01	2.87	7.11	99.16	0.86	2.22	0.03	755.67	0.42	27.99
F20	0.6	2	1	25	50	33.57	4.53	0.03	0.87	7.12	9.51	70.17	1.11	10.23	0.03	774.25	0.42	26.66
USY-C	–	–	–	–	–	28.51	9.80	–	0.99	2.79	7.84	98.35	0.60	1.25	0.03	519.16	0.37	41.84
F21	0.6	2	1	25	50	35.84	4.81	0.05	0.86	7.16	11.93	63.38	1.50	3.36	0	720.30	0.43	28.78
350USY	–	–	–	–	–	34.04	5.89	–	0.90	5.55	12.38	66.39	0.76	4.98	0.07	872.97	0.47	26.10
F22	0.6	2	1	25	50	39.39	1.06	0.05	0.84	36.19	13.69	72.58	0.49	16.16	0	917.30	0.46	22.12

diffraction peaks were assigned to Pt or Ni, which may indicate high dispersion of these metals. The crystallinity results obtained were in the range of 60–78, versus 66 % for the parent material. The TEM images for the 1 wt% Pt–1 wt% Ni, 2 wt% Pt and spent 2 wt% Pt/350USY catalysts are shown in Fig. 14. The black dots appearing on the support matrix are assumed to be Pt or Pt–Ni particles. The average size of the visible metal particles was found to be in the range of 3–7 nm for the Pt–Ni bimetallic catalyst and in the range of 3–30 nm for the monometallic samples, as measured manually and averaged [41], indicating that the presence of nickel and platinum in the bimetallic catalysts enhanced greatly the formation of metal particles and reduced their sizes. It was also found that the presence of platinum improved the nickel reduction in comparison to the presence of nickel only [40]. As in the faujasite zeolite structure (zeolite Y), the cavities had a maximum diameter of 1.3 nm, which means that the observed metal particles were located outside the zeolite cavities. Nevertheless, these metallic particles could have been located within the mesopore channels which were created during the hydrothermal dealumination of the Y zeolite, or at the outer surface of the catalyst grains, as has been reported [40]. It has also been suggested that there are likely to be smaller metal particles inside the cavities which are not detectable by TEM [42, 43]. It was observed that the average metallic particle size in the case of the 2 wt% Pt/350USY was unchanged by the deactivation reaction, which may indicate that no metal sintering occurred; however, this conclusion cannot be assured, as TEM was able to study only limited regions of the samples and consequently the results should be interpreted as mainly local [43].

The Figs. 15, 16, and 17 show the effect of TOS on the n-C7 conversion, hydroisomerisation, hydrocracking and cyclic product yields at the reaction temperature of 230 °C for the first, second and third series of monometallic and bimetallic bifunctional catalysts containing platinum and nickel, supported on H-USY (350HOA USY). Figure 15a indicates that the initial overall conversion values were 47, 56, 57, and 51 wt% for the monometallic catalysts containing Pt of 0.5, 1, 1.5, and 2 wt%, respectively, after 30 min from introducing the feed to the reactor at the temperature of 230 °C. All the catalysts underwent small decreases in conversion with TOS up to 12 h, after which there was a significant decrease in conversion for the catalyst containing 0.5 wt% Pt. The catalyst containing 2 wt% Pt was found to have the smaller drop in conversion, while that containing 0.5 wt% Pt underwent greatest fall in activity during the TOS. The higher sustainability of the catalyst containing 2 wt% Pt can be explained by the formation of catalytically active metallic particles and better balance between metal and acid sites, as shown in Table 10.

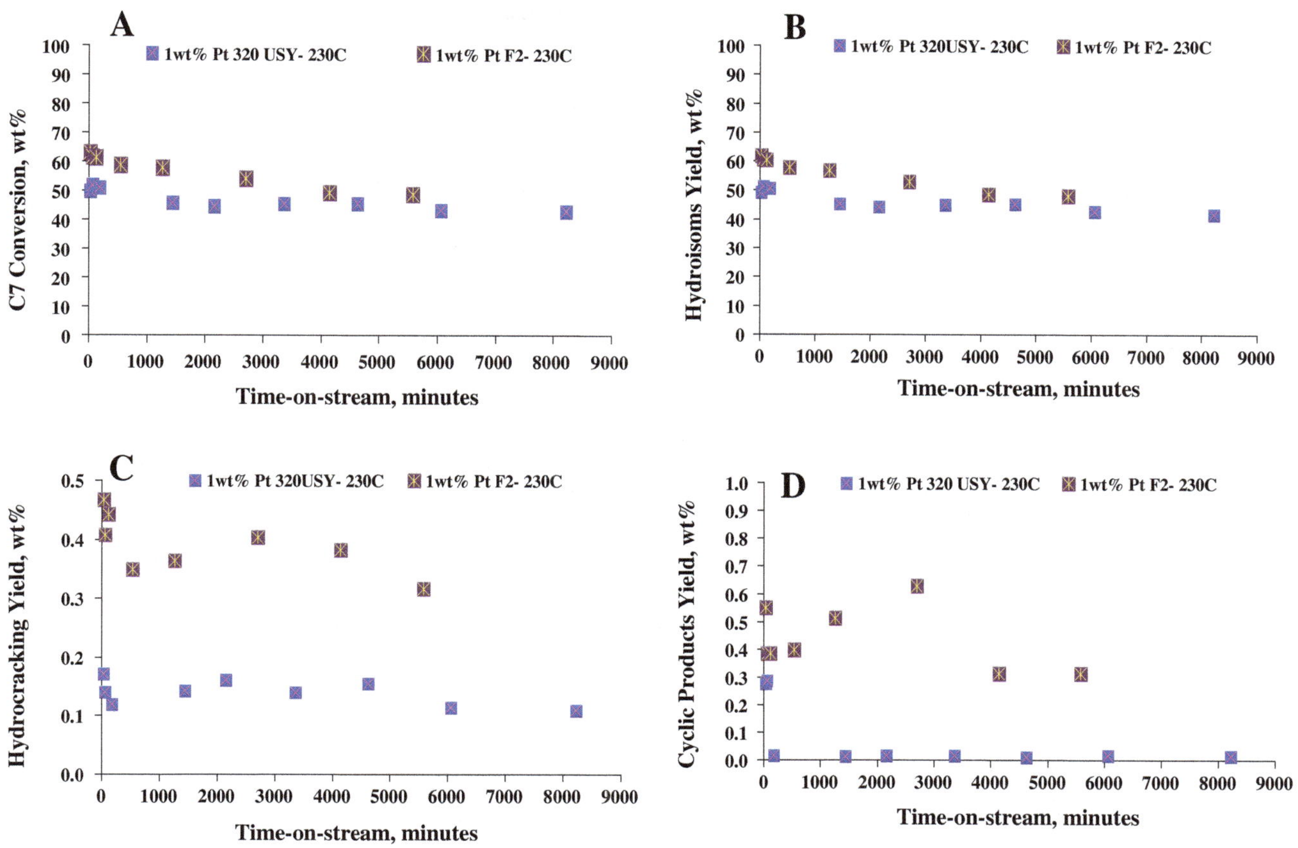

Fig. 11 Overall conversion (**a**), total yield of C7 isomers (mono + di + tribranched isomers) (**b**), total hydrocracking yield (**c**), and total yield of the cyclic products (**d**) as a function of TOS for F2 catalyst at various reaction temperatures. WHSV = 5.13 h^{-1}; H_2/C7 (mol/mol) = 9; total pressure = 1 atm; temperature = 230 °C; 1 wt% Pt loading

Figure 15b shows that the initial selectivity towards the hydroisomerisation products ranged between 98 and 99 % for all the catalysts, remaining almost unchanged throughout the deactivation run. Furthermore, as shown in the Fig. 15c and d, the selectivity towards cracked and cyclic products respectively was fairly in the range of 1 and 2 % for all the catalysts even if their n_{Pt}/n_A ratio is quite different. Nevertheless, n_{Pt}/n_A ratio was found to be within the optimum balance limits of 0.03 and 0.17 ($0.03 \leq n_{Pt}/n_A \leq 0.17$) for the most catalysts, which may reflect the great dehydrogenation–hydrogenation function capacity of platinum even at the lower contents [40, 41, 44]. The average hard coke contents were found to be in the range of 0.1–0.2 wt% for all the catalysts except that containing 1 wt% platinum, for which the coke content was about 0.7 wt%, which may due to the effect of the longer TOS when compared with the other catalysts.

Figure 16a shows that 30 min after introducing the feed into the reactor at 230 °C, the initial overall conversion rates were 52 and 59 wt%, respectively, for the monometallic catalysts containing 10 and 20 wt% Ni. After 180 min on stream, there was a large drop in activity to 30 and 45 wt%, then a sustained and rapid deactivation rate until a pseudo-stable state was reached at activity levels of 10 and 16 wt%, respectively, persisting until the reactions terminated. Deactivation was also found to be very rapid for the monometallic catalysts containing Ni at 0.5, 1, 1.5 and 2 wt%. This behaviour may be explained in terms of the poor hydrogenating capacity of Ni compared to Pt [44], making it more difficult for the Ni atoms to be fully reduced, so that not enough Ni metallic sites were available to sustain the reaction, since the hydroisomerisation reaction includes a dehydrogenation step [40, 42]. Therefore, selectivity to form the C7 isomers was lower than for the catalysts containing Pt, while higher cracked and cyclic product yields were obtained, which means that higher coke content was formed, possibly blocking and poisoning the active sites of the catalysts during the deactivation reaction. In addition, the low stability of catalysts with higher nickel content may be explained by the tendency of Ni to polymerise olefins and hence to form coke [42].

Figure 17a shows that the initial overall conversion rates were 66, 72, 66 and 67 wt% for the bimetallic bifunctional catalysts containing 1 wt% Pt and 0.25, 0.5, 0.75 and 1 wt% Ni, respectively, 30 min after introducing the feed into the reactor. All of these catalysts showed small decreases in conversion with TOS up to 12 h, after which there was a continuous moderate decrease in conversion

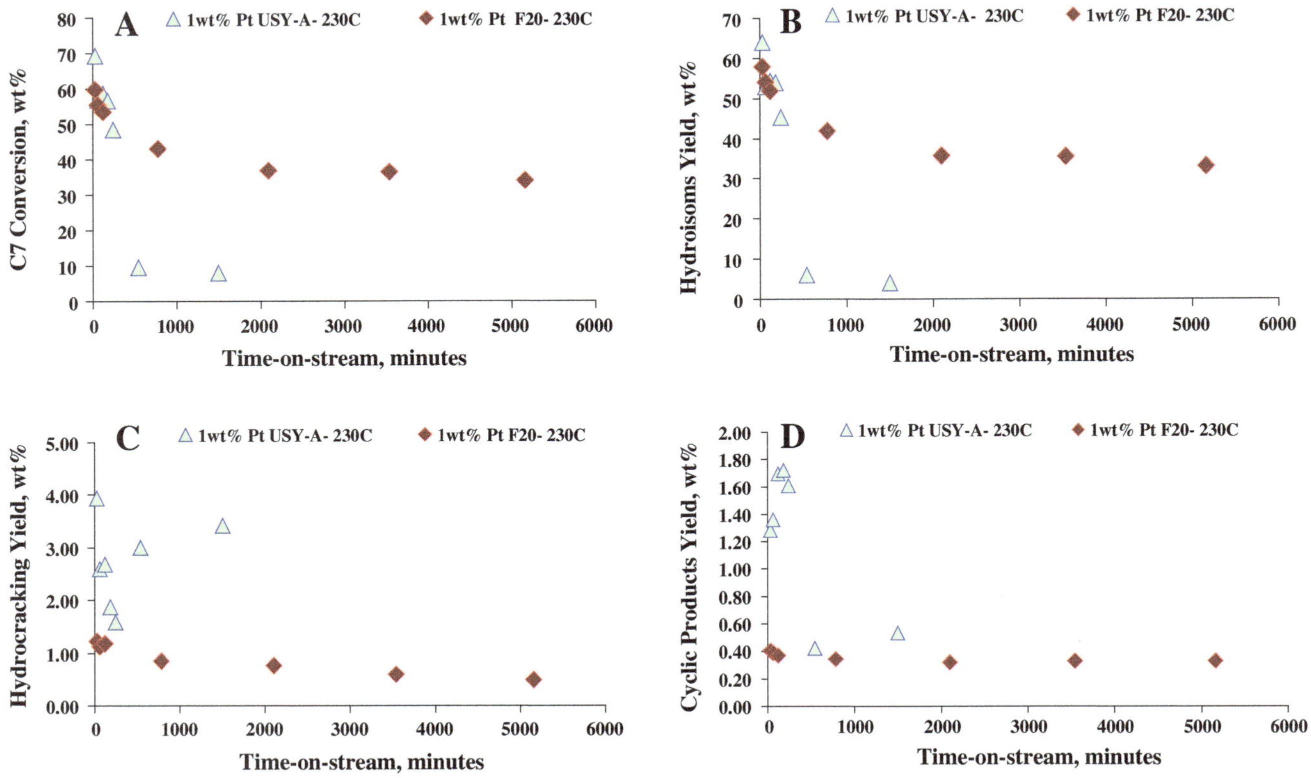

Fig. 12 Overall conversion (**a**), total yield of C7 isomers (mono + di + tribranched isomers) (**b**), total hydrocracking yield (**c**), and total yield of the cyclic products (**d**) as a function of TOS for F20 catalyst at various reaction temperatures. WHSV = 5.13 h^{-1}; H_2/C7 (mol/mol) = 9; total pressure = 1 atm; temperature = 230 °C; 1 wt% Pt loading

until the reactions terminated. Figure 17b indicates that initial selectivity towards the hydroisomerisation products ranged between 98 and 99 % for all of the catalysts, remaining almost constant throughout the deactivation run. Furthermore, as shown in the Fig. 17c and d, selectivity towards cracked and cyclic products respectively was broadly in the range of 1–2 % for all of the catalysts. On the other hand, the maximum hard coke content was found in the bimetallic catalyst containing 1 wt% Pt and 1 wt% Ni, as shown in the Table 10.

It can be concluded that a small addition of Pt motivated a substantial increase in hydrogenating activity when compared with the catalysts containing only Ni, as reported in the literature [30]. It also appears that the presence of platinum enhanced the reduction of the nickel cations, increasing the dispersion of the two metals as the nickel particles provided support for the Pt atoms, indicating that the Pt atoms were inserted into the nickel crystallite [40, 42]. On the other hand, it was found that increasing Ni loadings inhibited both the activity and selectivity of the catalyst, which may be interpreted as indicating that the formation of larger bimetallic Ni–Pt particles at higher Ni loadings resulted in some Ni content not being reduced, thus contributing to pore blockage of the catalyst and reducing the availability of catalyst acid sites accordingly [44].

Effect of platinum loading method on CP814E beta

Bifunctional zeolite catalysts typically have two types of active centres, the hydrogenation/dehydrogenation sites provided by the metal (such as platinum) for the formation of olefinic intermediates, and the acidic site for isomerisation and cracking. In the case of Pt-loaded zeolite catalysts in the ammonium form, the method most commonly used to introduce platinum into the pores and obtain a homogeneous filling in the crystallites of the zeolite is cationic exchange, using platinum tetraammine $[Pt(NH_3)_4]^{2+}$ complexes in an aqueous solution in the presence of a competing cation, generally an ammonium ion, NH_4^+. A second method to introduce platinum in the pores is incipient wetness impregnation by which the platinum tetraammine will randomly be deposited in the zeolite pores whereas only one platinum tetraammine complex can be associated with an aluminium site, and thus an atomic dispersion of the complexes inside the pores is expected with the ion-exchange method. Therefore, more dispersed platinum and consequently smaller particles using the ion-exchange method would be produced than via the impregnation method [27, 29, 45]. The purpose of this study was to examine the effects of platinum loading of beta zeolite by ion-exchange and impregnation methods, examining the

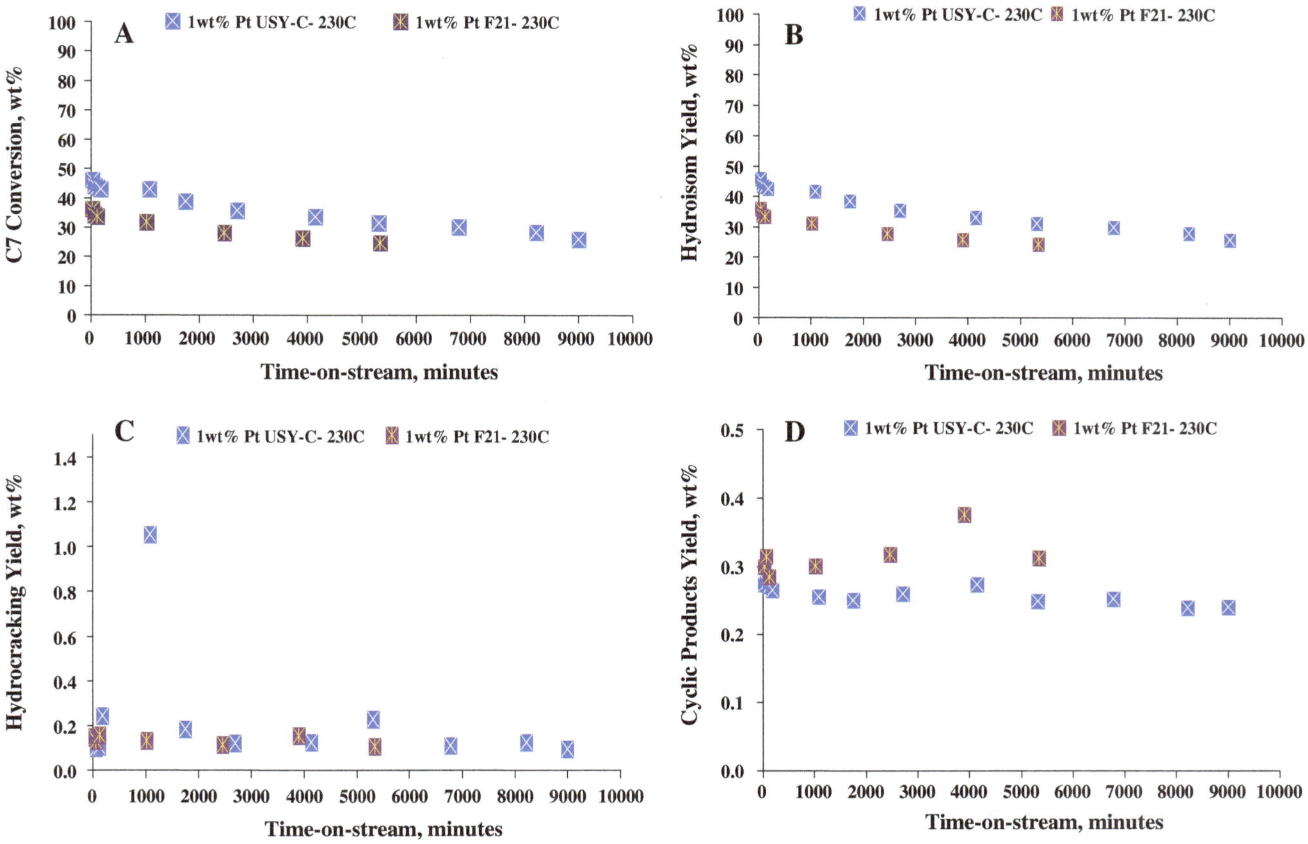

Fig. 13 Overall conversion (**a**), total yield of C7 isomers (mono + di + tribranched isomers) (**b**), total hydrocracking yield (**c**), and total yield of the cyclic products (**d**) as a function of TOS for F21 catalyst at various reaction temperatures. WHSV = 5.13 h^{-1}; H_2/C7 (mol/mol) = 9; total pressure = 1 atm; temperature = 230 °C; 1 wt% Pt loading

TOS stability of 1wt% Pt-loaded CP814 beta, where its deactivation behaviour was studied in the previous work [31], in the hydroisomerisation of n-C7 at constant reaction conditions and the effect of pH on the ion-exchange process.

Figure 18a–d shows the effects of TOS for the C7 conversion, hydroisomerisation, hydrocracking, and cyclic product yields for the catalyst with 1 wt% Pt/CP814E in which the platinum was loaded by one of three procedures: (1) preparation of the ion-exchange system using an aqueous solution of tetraammine platinum (II) chloride complex mixed with the ammonium-beta zeolite CP814E and the addition of the competing cation NH_4^+ using ammonia solution at pH > 9, (2) preparation of the ion-exchange system using an aqueous solution of tetraammine platinum (II) chloride complex mixed with the ammonium-beta zeolite CP814E without adding ammonia solution such that the pH = 4–5, and (3) preparation by wet impregnation using an aqueous solution of tetraammine platinum (II) chloride complex mixed with the ammonium-beta zeolite CP814E without adding ammonia solution, such that the pH = 4–5. As Fig. 18a shows, initially the overall conversion of C7 for the three catalysts was 77, 88, and 90 wt% after 30 min from the point when the feed was introduced to the reactor at 230 °C, prepared by the three procedures described above. The catalysts lost about 1–2 wt% of their activity before reaching a pseudo-stable state for the initial 60 min on stream, and then slightly decreased with a further increase in time until the reactions were finally terminated. On the other hand, the selectivity towards hydroisomerisation products was 99, 85, and 85 % after 30 min and then increased with time to 99, 89, and 89 % for the three catalysts, respectively. Furthermore, the greatest selectivity towards cracked products was 15 % and slightly decreased with TOS to approximately 9 % for the catalysts which were prepared by the ion-exchange method with pH = 4–5 and by the wet impregnation method. The selectivity towards the cyclic products was very low for the three catalysts along the deactivation runs.

The variation in C1/C3 mass ratio with TOS for the three catalysts at constant reaction conditions is shown in Fig. 19, with both the acid and metal functions being initially active and deactivated rapidly as indicated by the C1/C3 ratio. This could be explained by a change occurring in the balance between the two functions with TOS due to coke depositions for both functions. Furthermore, in all the

Table 10 Different metal loadings wt% on 350HOA USY

	Al (wt%)	Si (wt%)	Bulk (Si/Al) (mol/mol)	Framework (Si/Al) (mol/mol)	Crystallinity (%)	Actual Ni (wt%)	Actual Pt (wt%)	Theoretical Ni (wt%)	Theoretical Pt (wt%)	Ni/Pt (mol/mol)	n_{Pt}/n_A	Hard coke (wt%)
350HOA USY	5.89	34.04	5.55	12.38	66	–	–	–	–	–	–	–
Pt/350HOA USY												
0.5 %Pt	–	–	–	–	–	–	0.36	–	0.50	–	0.02	0.19
1 %Pt	5.89	34.04	5.55	12.38	66	–	0.90	–	1.00	–	0.06	0.66
1.5 % Pt	–	–	–	–	–	–	1.17	–	1.50	–	0.08	0.13
2 % Pt	–	–	–	–	–	–	1.46	–	2.00	–	0.10	0.14
Pt-Ni/350HOA USY												
1 % Pt–0.25 %Ni	5.29	27.50	5.00	–	60	0.22	0.90	0.25	1.00	0.81	0.11	0.47
1 % Pt–0.5 % Ni	5.29	27.60	5.01	–	63	0.44	0.91	0.50	1.00	1.61	0.16	0.62
1 % Pt–0.75 %	5.31	27.65	5.00	–	61	0.68	0.92	0.75	1.00	2.46	0.21	0.43
1 % Pt–1 % Ni	5.31	27.97	5.06	11.57	60	0.92	0.91	1.00	1.00	3.36	0.27	1.01
Ni/350HOA USY												
0.5 % Ni	–	–	–	–	–	0.44	–	0.50	–	–	0.10	–
1 % Ni	–	–	–	–	–	0.88	–	1.00	–	–	0.20	–
1.5 % Ni	–	–	–	–	–	1.28	–	1.50	–	–	0.29	–
2 % Ni	–	–	–	11.03	–	1.86	–	2.00	–	–	0.42	–
10 % Ni	4.74	27.74	5.62	10.86	73	7.48	–	10.00	–	–	1.68	–
20 % Ni	4.28	24.83	5.57	10.58	–	11.96	–	20.00	–	–	2.68	–

three catalysts, the metal function deactivated faster than the acid function; however, this was not severe enough to cause a drop in the n-C7 conversion. The worst metal/acid balance was found for the catalysts that were prepared by ion exchange with the pH = 4–5 and by the wet impregnation method, as indicated by the lower values of C1/C3, and as such higher cracking activity was observed for those two catalysts. Moreover, the two catalysts demonstrated poor metal dispersion resulting in some of the metal being deposited on the outside of the crystals and not within the zeolite pores, as compared to the catalyst that was prepared by ion exchange with the pH > 9.

Effect of CP814E composites

The zeolite-based composite materials consisting of two distinct phases with entirely different pore structure and acidity have been used extensively both as catalysts, and to minimise the formation of coking, thus increasing the lifetime of the catalyst. Beta zeolite can also be used as an additive to FCC cracking catalysts because it favours reactions that improve the octane number of gasoline since the feedstocks consist of bulky poly-nuclear molecules which cannot diffuse inside the zeolite pores. Composite catalysts made of a zeolitic and a non-zeolitic active oxide/poly-oxide as a matrix can be used to crack the bulky feed molecules into smaller molecules which can diffuse inside the zeolite pores providing a two-phase system of the composite that operates synergistically. This synergism is a result of the open structure of the matrix that allows easy intra-particle diffusion and favours the primary cracking of the bulky feed molecules, as a result of the balance between the acidic sites and the metallic sites due to the dispersion of Pt throughout the matrix [27, 46, 47].

In this work, TOS experiments were carried out on the hydroisomerisation of n-C7 using hybrid catalysts consisting of CP814E and USY-C at identical reaction conditions, and thus the relation between zeolite pore structure and its resistance to coke deactivation is examined for the composite catalysts formed. Two composite bifunctional catalysts were prepared as follows: (1) the catalysts 1 wt% Pt/CP814E and 1 wt% Pt/USY-C were mixed physically at mass ratio of 1:1. (2) CP814E and USY-C in their ammonium forms were mixed at mass ratio of 1:1, and were loaded by 1 wt% Pt using the ion-exchange method.

The effect of TOS on the C7 conversion, hydroisomerisation, hydrocracking, and cyclic product yields at reaction temperature of 230 °C are shown in the Fig. 20a–d,

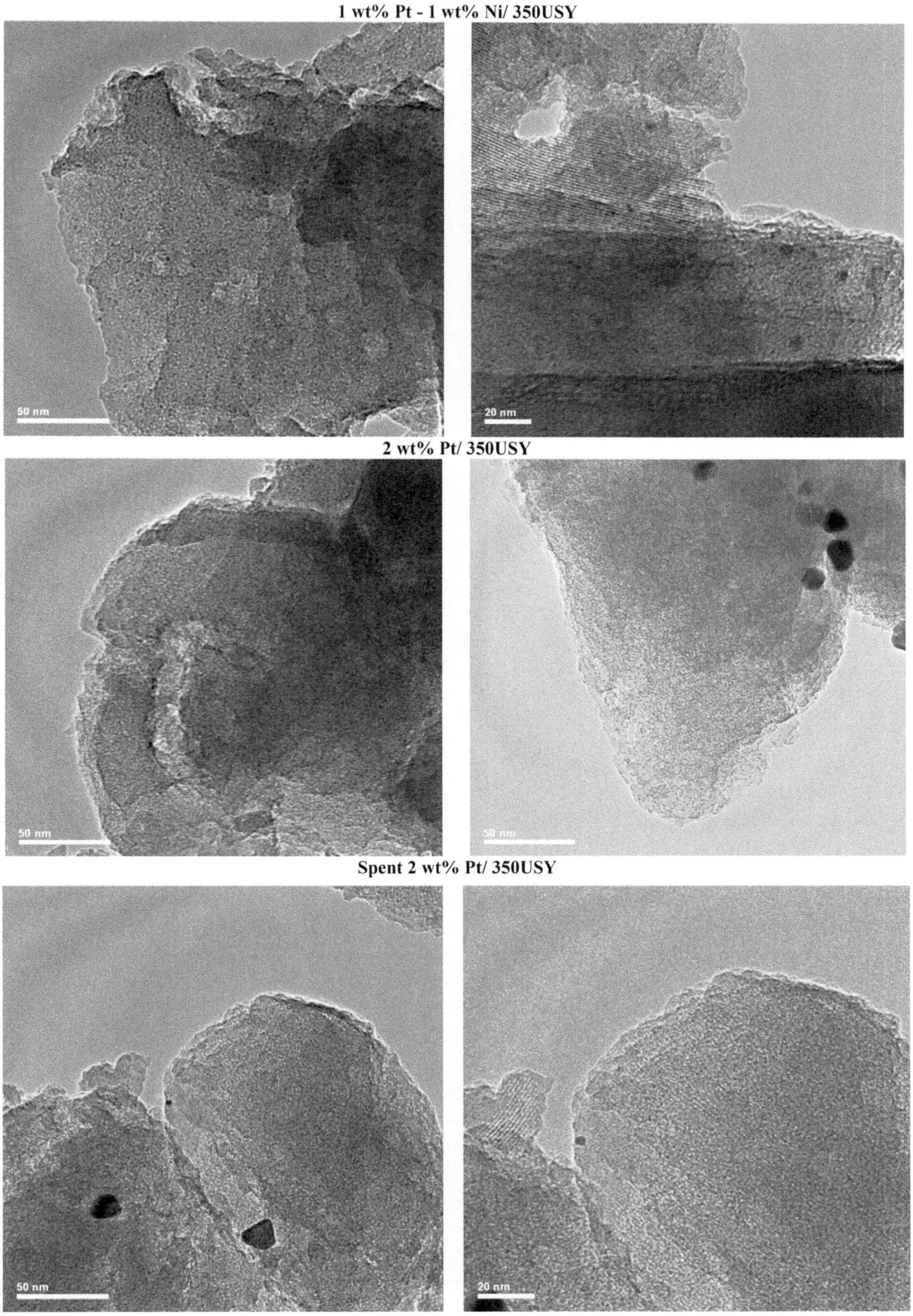

Fig. 14 TEM images of 1 wt% Pt–1wt% Ni, 2 wt% Pt and spent 2 wt% Pt/350USY

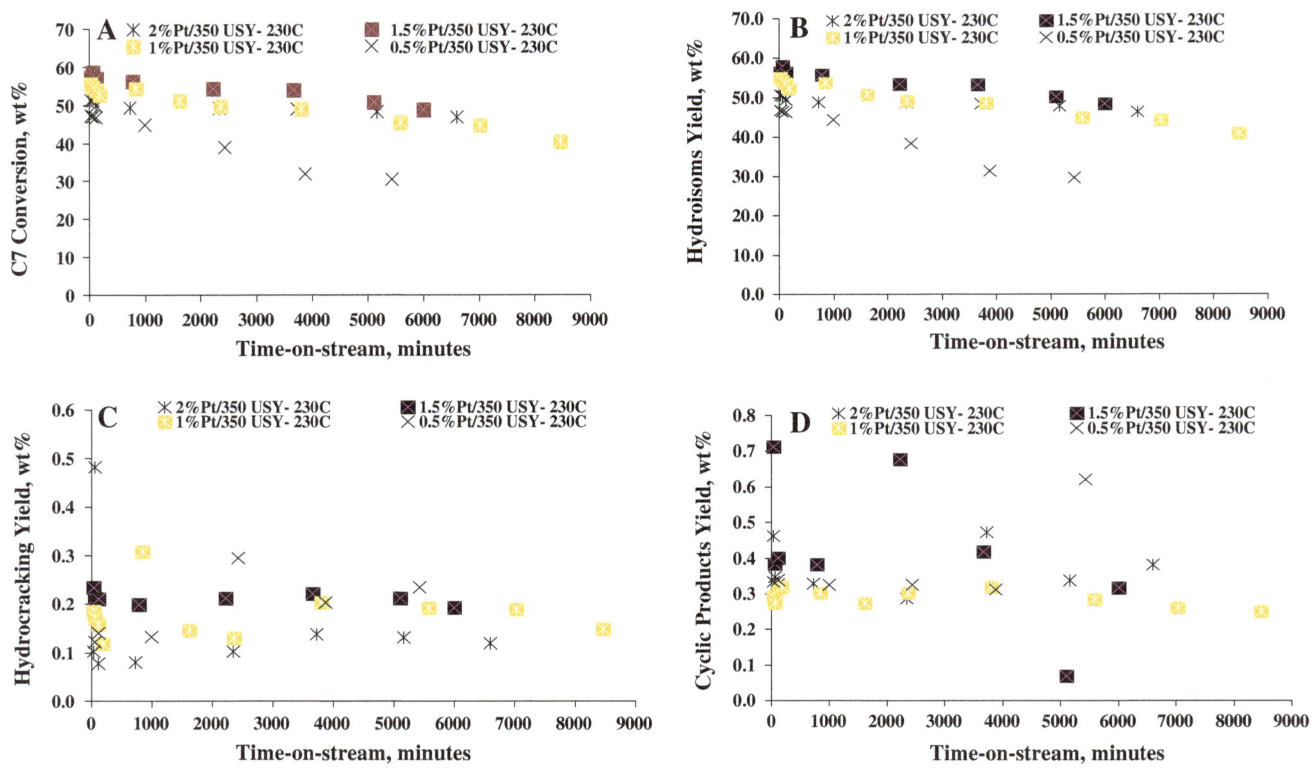

Fig. 15 Overall conversion (**a**), total yield of C7 isomers (mono + di + tribranched isomers) (**b**), total hydrocracking yield (**c**), and total yield of the cyclic products (**d**) as a function of TOS for the monometallic catalysts containing Pt at various reaction temperatures. WHSV = 5.13 h^{-1}; H_2/C7 (mol/mol) = 9; total pressure = 1 atm; temperature = 230 °C; 1 wt% Pt loading

respectively, for the catalysts: 1 wt% Pt/CP814E, 1 wt% Pt/USY-C, 1 wt% Pt/USY-CP814E physically mixed, and 1 wt% Pt/USY-CP814E physically mixed and loaded by Pt together by the ion-exchange method with ammonia added to adjust the pH to be more than 9. As Fig. 20a shows, initially the overall conversion of C7 for the 1 wt% Pt/USY-CP814E physically mixed was 66 wt%, which is in between that of the corresponding individual catalysts, compared with 77 and 46 wt% for the catalysts 1 wt% Pt/CP814E and 1 wt% Pt/USY-C, respectively. The conversion then decreased abruptly at the initial stages of the reaction, and finally reached a relatively steady value, after which the conversion decreased steeply with the TOS until the reaction was terminated after 183 h. Thus, this composite catalyst demonstrated a better TOS activity than 1 wt% Pt/USY-C catalyst, but slightly worse than 1 wt% Pt/CP814E. Even its selectivity toward the C7 isomers and cracked products was very similar to both catalysts along the deactivation run.

On the other hand, the composite catalyst 1 wt% Pt/USY-CP814E, physically mixed and loaded with Pt using ion exchange and subsequently deactivated as with the 1 wt% Pt/CP814E catalyst, offered slightly higher activity towards the cracked products along the deactivation run, with the initial conversion of 78 wt% decreasing slightly with TOS to 67 wt% with the reaction terminating after 83 h. The selectivity towards the C7 isomers and cracked products was 96 and 3 %, respectively, and the composite catalyst demonstrated a better TOS activity than the 1 wt% Pt/USY-C catalyst, and a similar deactivation pattern to the 1 wt% Pt/CP814E even though the selectivity towards the cracked products was higher. This indicates that the 1 wt% Pt/USY-C catalyst was the dominant component in the formulated composite catalyst of 1 wt% Pt/USY-CP814E and provided a better TOS stability and higher activity. On the other hand, the 1 wt% Pt/CP814E catalyst was the dominant component for the composite catalyst 1 wt% Pt/USY-CP814E loaded with Pt using ion exchange. Figure 21 shows the deposited hard coke contents over the top, middle, and bottom zones of the individual and composite catalysts. It can be seen that the lowest hard coke content was found for the 1 wt% Pt/USY-C and the highest for the 1 wt% Pt/CP814E catalyst, with the composite catalyst having a hard coke content part way between that of the corresponding individual catalysts. The total acidity for ammonia TPD for the 1 wt% Pt/CP814E, 1 wt% Pt/USY-C, and 1 wt% Pt/USY-CP814E physically mixed was found to be 0.63, 0.60 and 1.88 mmol g^{-1}, respectively. Thus, it can be seen that even the composite catalyst has the highest acidity which is almost double compared with the corresponding individual catalysts. It has moderate hard coke content and fairly good TOS catalytic stability.

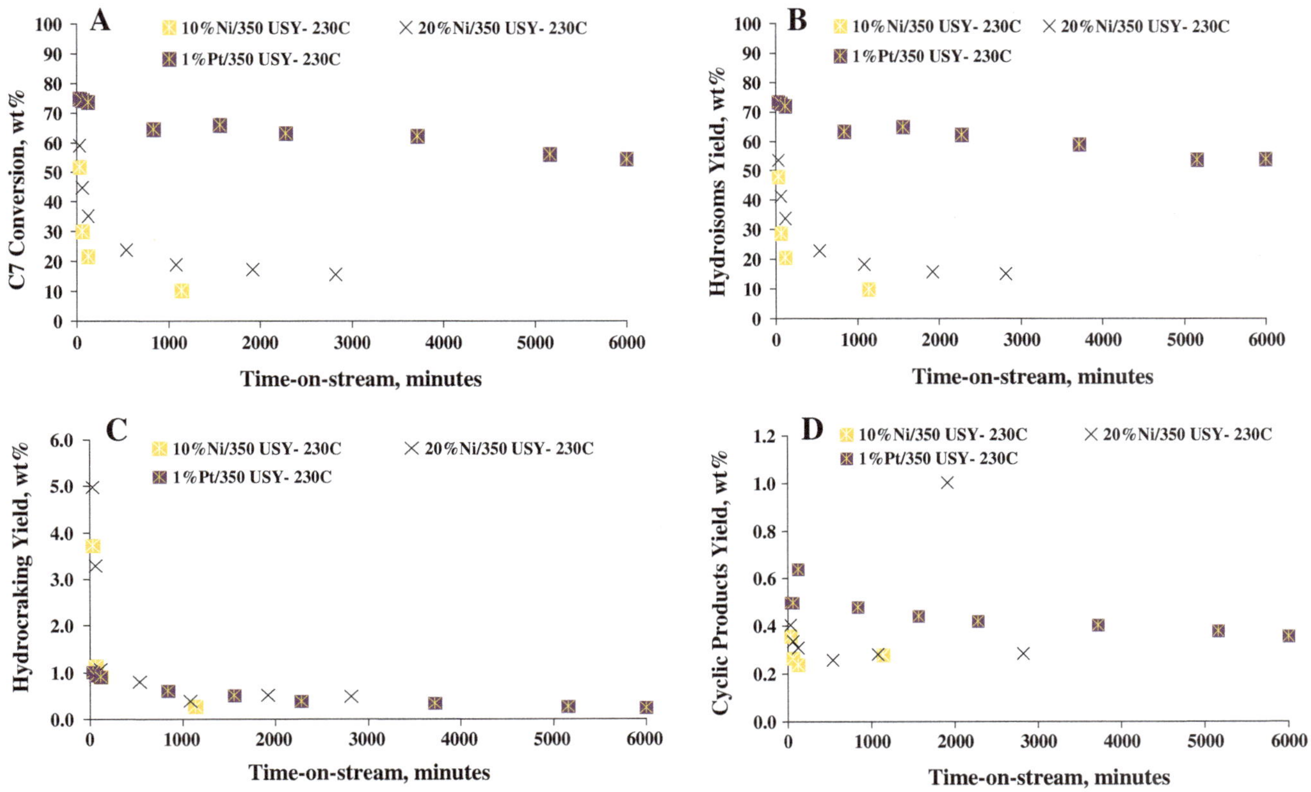

Fig. 16 Overall conversion (**a**), total yield of C7 isomers (mono + di + tribranched isomers) (**b**), total hydrocracking yield (**c**), and total yield of the cyclic products (**d**) as a function of TOS for the monometallic catalysts containing Ni at different reaction temperatures. WHSV = 5.13 h^{-1}; H_2/C7 (mol/mol) = 9; total pressure = 1 atm; temperature = 230 °C; 1 wt% Pt loading

Main characteristics of in-house mordenite zeolites

Mordenite zeolite was synthesised using the molar gel composition of (6 Na_2O: Al_2O_3: 30 SiO_2: 780 H_2O), which is the synthesis recipe of mordenite obtained from a verified synthesis handbook [21]. The mordenite phase with the highest purity and crystallinity was optimised against certain synthesis parameters while keeping this composition ratio constant. Three factors were optimised: the ageing time of the synthesis gel, the crystallisation time and the crystallisation temperature. Table 11 shows the phase crystallinity, the bulk Si/Al ratio and the framework Si/Al ratio of the mordenite obtained by optimising these three factors. It can be seen that the optimum synthesis conditions were when the ageing time was 12 h, crystallisation time was 24 h and crystallisation temperature was 180 °C. Under these conditions, the mordenite zeolite phase obtained had a crystallinity of 82 %, bulk Si/Al ratio of 8.44 (mol/mol) and framework Si/Al ratio of 9.02 (mol/mol). This procedure was repeated more than 10 times and reproducible results were obtained in terms of the yield, crystallinity, bulk and framework Si/Al ratio, as well as of the crystal morphology and size.

Figures 22 and 23 show the SEM microgram and XRD pattern of this in-house mordenite catalyst. The former shows that the catalyst contained large crystallites of 15–20 μm whose elliptical shape was different from that of commercial catalysts, and it has been previously reported that synthetic mordenite presents several morphologies, as a function of the crystallisation conditions [48]. The XRD patterns show a pure phase of mordenite consistent with that found previously [49]. However, it had a lower crystallinity than CBV21A [31], which also has a higher bulk Si/Al ratio. Figure 24a and b shows respectively the ^{27}Al and ^{29}Si solid-state NMR spectra of this in-house mordenite catalyst. It can be seen that the catalyst had only one ^{27}Al peak at about 56 ppm, which was assigned to tetrahedrally coordinated aluminium, and there were no additional EFAL peaks. In the ^{29}Si MAS NMR spectrum, two peaks at 112 and 106 ppm were assigned to Si(0Al) and Si(1Al) configurations, respectively.

Post-synthesis modification of in-house mordenite zeolite

Post-synthesis treatment by steam or acid attack, often referred to as dealumination, can remove aluminium from the zeolite framework. Steaming, which is performed at temperatures above 450 °C, extracts aluminium atoms from the framework and deposits them on the external

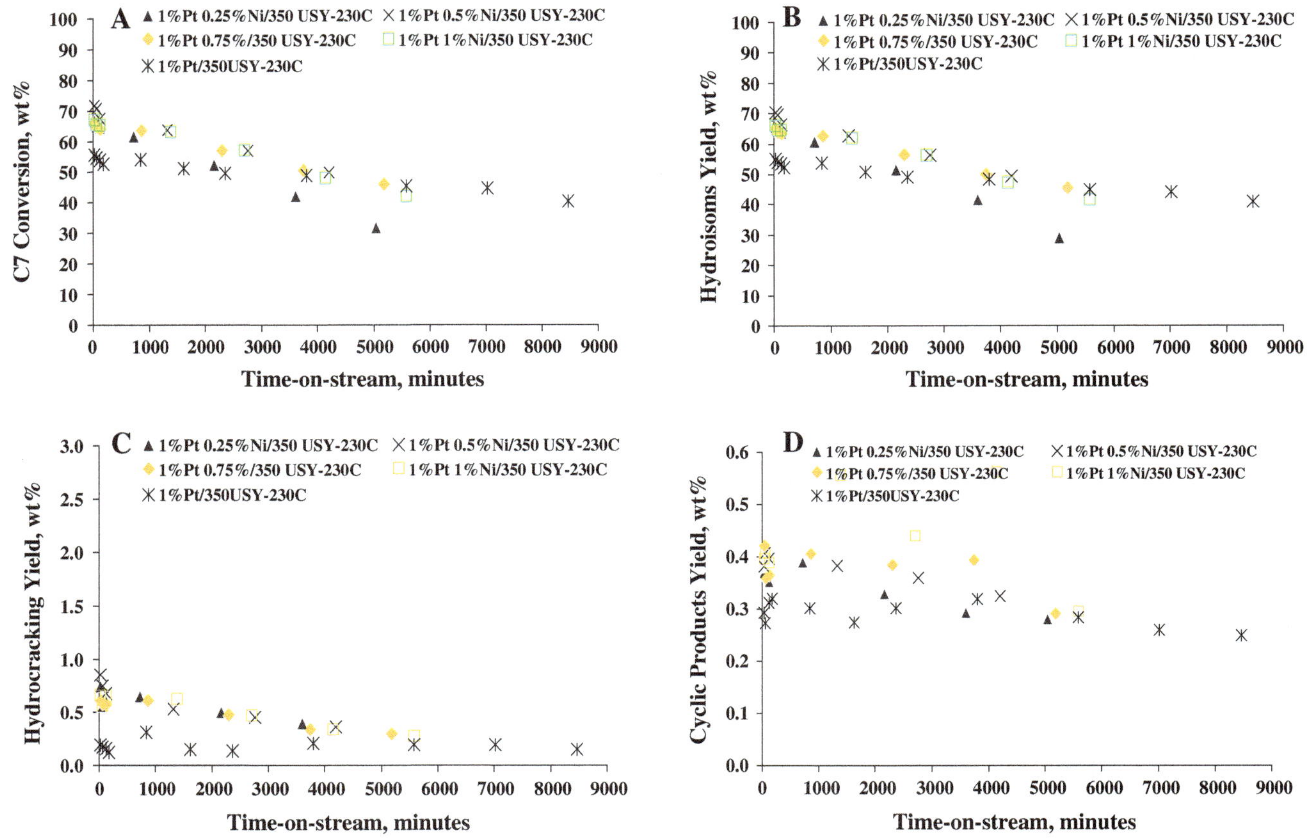

Fig. 17 Overall conversion (**a**), total yield of C7 isomers (mono + di + tribranched isomers) (**b**), total hydrocracking yield (**c**), and total yield of the cyclic products (**d**) as a function of TOS for the bimetallic catalysts containing Pt–Ni at different reaction temperatures. WHSV = 5.13 h^{-1}; H_2/C7 (mol/mol) = 9; total pressure = 1 atm; temperature = 230 °C; 1 wt% Pt loading

surface of the zeolite. An additional acid treatment can also be employed to completely remove aluminium atoms from the zeolite pore system. Increasing the framework Si/Al ratio by reducing the aluminium content in this way can result in the isolation of the remaining aluminium atoms, which in turn increases the strength of the acid sites and decreases the total number of active sites. This serves to improve both thermal and hydrothermal stability, while also generating mesopores, thus making the crystallite more accessible and reducing the severity of the diffusion limitations. Shortening the diffusion path length is of paramount importance for increasing the activity and stability of mordenite, as it is a one-dimensional, non-mesoporous zeolite. A small number of blockages can deactivate large parts of the mordenite structure [18, 50–52].

Two series of dealuminated mordenite catalysts were prepared by acid leaching and steaming applied to the in-house synthesised mordenite in its ammonium form. Table 12 shows the different acid leaching conditions that were applied to produce different catalysts with differing Si/Al ratios (acidity) based on the severity of acid leaching. In addition, three dealuminated mordenite catalysts were produced via the steaming dealumination technique at temperatures of 500, 600 and 700 °C. Table 13 shows the main characteristics of the acid-leached and steamed catalysts.

From Tables 5 and 6, it can be seen that the highest Si/Al molar ratios obtained for the framework and bulk were 19.13 and 42.76 (mol/mol), respectively, by acid leaching using nitric acid at a concentration of 10 molar, a temperature of 120 °C and a leaching time of 87 h (3.6 days), as in condition 15 in Table 12. A similar Si/Al ratio for the framework was obtained by increasing the acid leaching time to 168 h (7 days) at the same mixing temperature, as in condition 16, the most severe that was applied. It is generally expected that the bulk Si/Al molar ratio should remain the same even after steaming modification and that the steamed mordenite should have the same bulk Si/Al molar ratio as synthesised mordenite, but the framework Si/Al molar ratio was higher and increased as the steaming temperature increased from 500 to 700 °C, as shown in Table 13, which means that the framework aluminium atoms were dealuminated by steam to become EFAL.

Figures 25 and 26 show the SEM micrograms of steamed mordenites and the XRD patterns of acid-leached and steamed mordenites, respectively. From the SEM images, it can be seen that the steamed catalysts contained

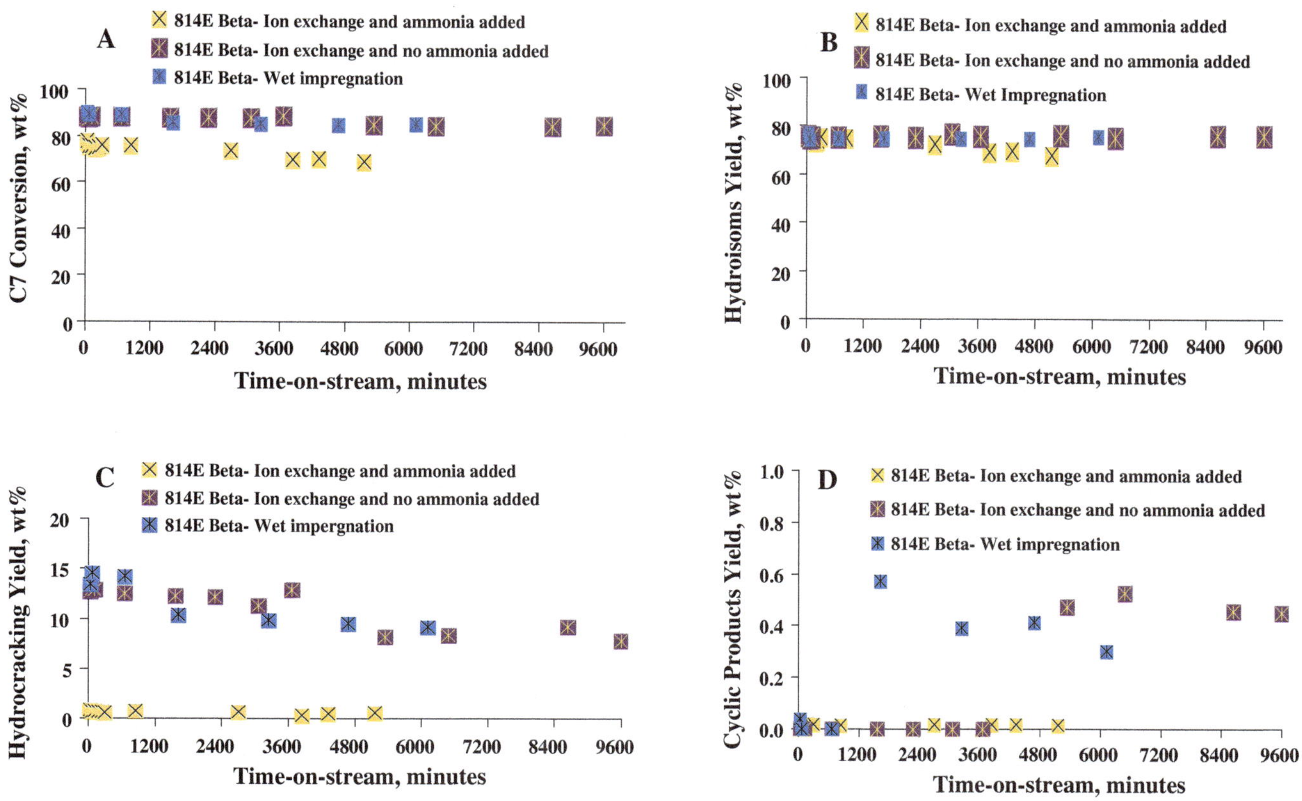

Fig. 18 Overall conversion (**a**), total yield of C7 isomers (mono + di + tribranched isomers) (**b**), total hydrocracking yield (**c**), and total yield of the cyclic products (**d**) as a function of TOS for CP814E. WHSV = 5.13 h^{-1}; H_2/C7 (mol/mol) = 9; total pressure = 1 atm; temperature = 230 °C; 1 wt% Pt loaded by different methods

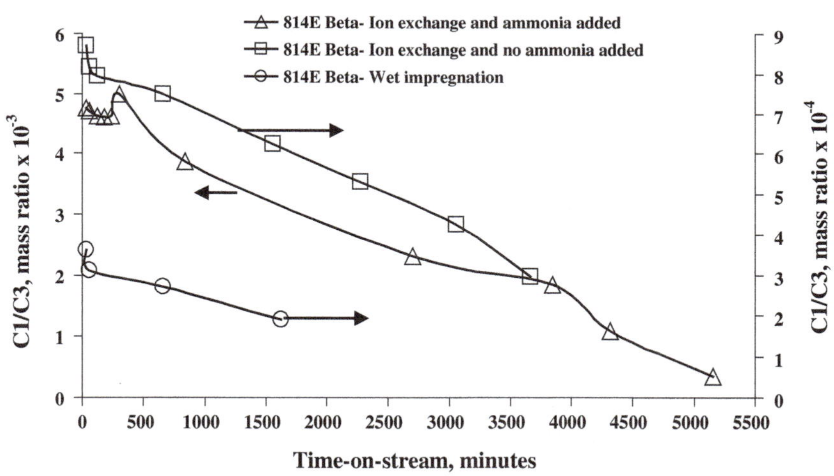

Fig. 19 Variation of the C1/C3 mass ratio as a function of TOS of CP814E

large and small crystallites of different sizes with some smaller debris and agglomerates whereas the acid-leached catalyst kept almost its original large crystal shape with some holes on the crystal surface. The XRD patterns confirm the pure mordenite phase for the steamed and acid-leached samples, with less overall crystallinity than as-synthesised mordenite.

Figures 27 and 28 show the ^{27}Al and ^{29}Si solid-state NMR spectra of acid-leached and steamed samples. It can be seen that the acid-leached catalysts had two ^{27}Al peaks, at about 56 and 0 ppm, assigned to a tetrahedrally coordinated aluminium atom in the framework and a smaller quantity of octahedrally coordinated aluminium as EFAL. Conversely, the steamed samples showed a very large quantity of EFAL at unique resonance spectra for octahedral and tetrahedral aluminium atoms at about 1 and 3 ppm, respectively, alongside in-framework aluminium in tetrahedral coordination. The quantity of EFAL also

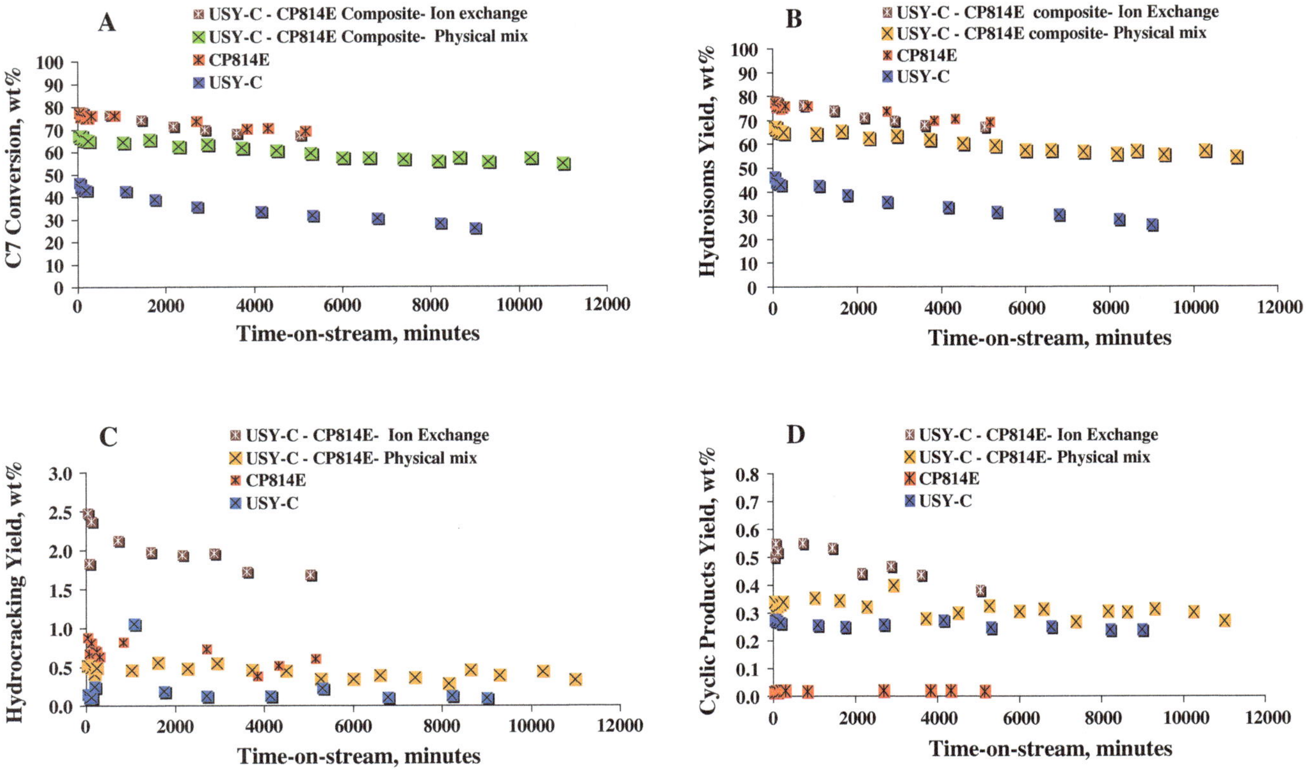

Fig. 20 Overall conversion (**a**), total yield of C7 isomers (mono + di + tribranched isomers) (**b**), total hydrocracking yield (**c**), and total yield of the cyclic products (**d**) as a function of TOS for USY and CP814E composites. WHSV = 5.13 h^{-1}; H_2/C7 (mol/mol) = 9; total pressure = 1 atm; temperature = 230 °C; 1 wt% Pt loading

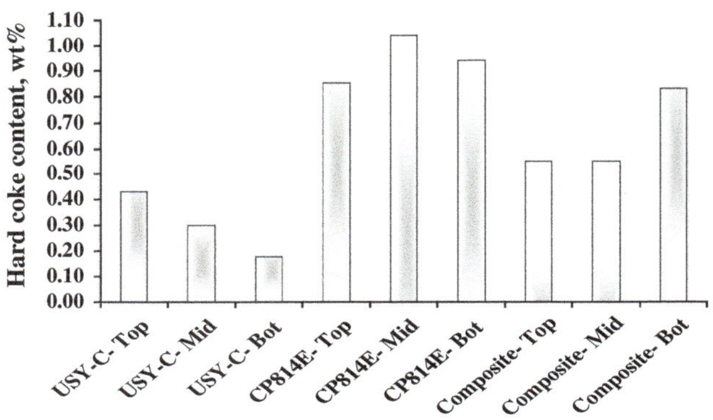

Fig. 21 The hard coke content over the aged CP814E, USY-C, and USY-CP814E physically mixed

increased as the steaming temperature increased. In the ^{29}Si MAS NMR spectra, two peaks at about 114 and 107 ppm were assigned respectively to Si(0Al) and Si(1Al) configurations, and these chemical shifts differ slightly from the sample produced by steaming. In addition, silicon atoms are different crystallographically as indicated by the different peak shape, depending upon the steaming temperature that was employed. However, the strong increase in the Si(0Al) signal intensity as the steaming temperature increased indicates that the replacement of Al by Si in the framework was more effective [20].

Table 13 and Fig. 29 show that the dealumination treatments by acid leaching and steaming resulted in an increase in mesoporosity, as illustrated with N_2-physisorption as a sharp increase in adsorption at high relative pressures, indicating the formation of large mesopores and a corresponding increase in the external surface area. Additionally, an average pore size of 27 Å was observed in the treated samples, compared with 18 Å for as-synthesised mordenite.

TOS behaviour during C7 hydroisomerisation conversion was examined for the following commercial and in-house mordenite catalysts: CBV10A, CBV21A, 640HOA, 660HOA, 690HOA [31], acid-leached mordenite and in-house mordenite steamed at 600 °C. All of the above catalysts were loaded with 1 wt% Pt by the ion-exchange method. Three very acidic commercial catalysts—1 wt% Pt/CBV10A, 1 wt% Pt/640HOA and 1 wt% Pt/660HOA—and the acid-leached mordenite deactivated very rapidly during the initial few seconds, due to a rapid coking that may have poisoned the acid sites of these catalysts and changed the effective acid site density needed for an ideal bifunctional conversion. As a consequence, no TOS results were reported for these catalysts. Figure 30a–d shows the effects of TOS at a reaction temperature of 230 °C on C7 conversion, hydroisomerisation, hydrocracking and cyclic

Table 11 In-house mordenite synthesis parameters and the resulting sample characteristics

Ageing time (h)	Crystallisation time (h)	Crystallisation temp (°C)	XRD crystallinity (%)	ICP Si/Al (mol/mol)	ICP Na (wt%)	NMR Si/Al (mol/mol)
1	1 day	180	43	9.50	3.63	8.03
3			64	9.70	3.99	9.13
4			75	8.36	3.26	8.97
6			77	8.13	3.15	8.97
8			75	7.83	3.08	8.98
10			75	7.92	3.07	8.98
12			82	8.44	3.14	9.02
15			80	7.99	3.23	8.97
24			49	8.92	3.09	8.62
12	12	180	Amorphous	–	–	–
	20		76	–	–	8.71
	22		75	–	–	8.89
	23		71	–	–	8.92
	24		82	8.44	3.14	9.02
	25		76	–	–	8.63
	26		72	–	–	8.63
	28		70	–	–	8.65
	36		Amorphous	–	–	–
12	1 day	170	Amorphous	–	–	–
		175	32	–	–	7.67
		180	82	8.44	3.14	9.02
		185	72	–	–	8.80
		190	71	–	–	8.63

Fig. 22 SEM micrograph of in-house and as-synthesised mordenite

product yields respectively for the following catalysts: 1 wt% Pt/CBV21A, 1 wt% Pt/690HOA and in-house mordenite steamed at 600 °C.

Initially, the overall conversions were 12, 41 and 52 wt% after 60 min from the point when the feed was introduced to the reactor at 230 °C, for the 1 wt% Pt/CBV21A, 1 wt% Pt/690HOA and steamed mordenite catalysts, respectively. A rapid deactivation occurred for the 1 wt% Pt/CBV21A catalyst with high cracking activity, resulting in selectivity towards the cracked products at

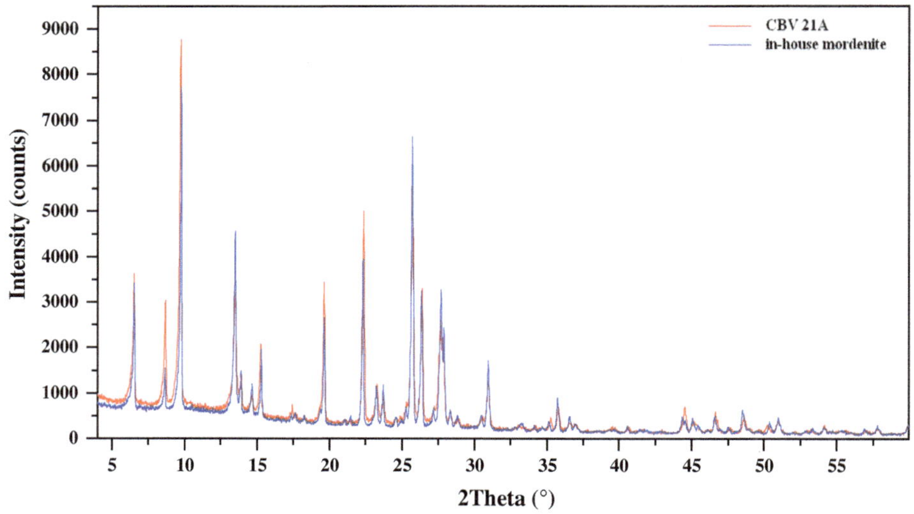

Fig. 23 XRD patterns of in-house and as-synthesised mordenite

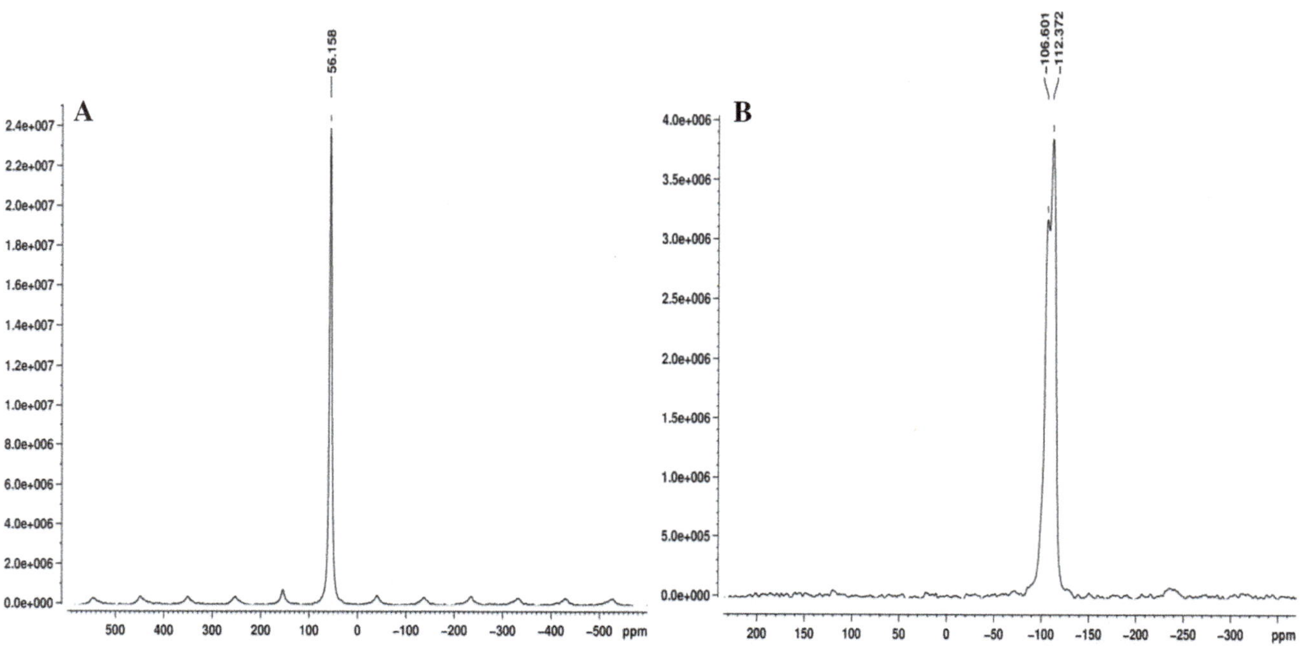

Fig. 24 ^{27}Al (**a**) and ^{29}Si (**b**) MAS NMR spectra of in-house and as-synthesised mordenite

24 %, compared with 71 % for the isomer products, and thereafter the catalyst was almost deactivated after 240 min on stream. On the other hand, greater TOS stability was shown by the 1 wt% Pt/690HOA and steamed mordenite catalysts. However, a shorter coking rate was observed for the steamed mordenite, so it preserved its initial activity for almost 24 h, after which it deactivated at the rate of 8–10 wt% every day before achieving a pseudo-stable state at the end of the reaction. In contrast, the 1 wt% Pt/690HOA catalyst deactivated continuously from the point when the feed was introduced to the reactor at the slower rate of 3–5 wt% every day, before achieving a pseudo-stable state at the end of the reaction. In addition, selectivity towards C7 isomers and cyclic products for both catalysts was in the range of 97–98 and 2–3 %, respectively.

Faster deactivation and the mono-dimensional pore system of 1 wt% Pt/CBV21, coupled with a higher density of strong acid sites and as-synthesised zeolite, may have caused the lower activity, which may also be attributable to platinum crystallites blocking the pore openings, as previously reported [35, 53]. There are side pockets in the pore topology of mordenite that match the 12MR and 8MR crossing. The higher number of Brønsted acid sites in the cavities also increases the sensitivity of the 8 MR pores to coking and they can easily fill up and become blocked by carbonaceous deposits [54]. The average lifetime of the carbocations on the surface can be lengthened by strong

Table 12 Acid leaching conditions applied and the main characteristics of samples obtained

Condition	NH_4-MOR (wt)	Acid type	Acid conc. (molar)	Acid, Vol. (ml)	Leaching time (h)	Temp (°C)	XRD crystallinity (%)	NMR Si/Al (mol/mol)	ICP Si/Al (mol/mol)
1	1	HCl	6	50	12	25	71	9.53	–
2	1	HCl	6	50	24	25	73	9.63	–
3	1	HCl	6	100	24	25	74	9.69	–
4	1	HCl	Conc	100	24	120	70	15.32	–
5	1	HNO_3	10	100	16	25	73	9.24	–
6	1	HNO_3	6	50	24	25	70	9.32	–
7	1	HNO_3	Conc	100	24	25	70	9.36	–
8	1	HNO_3	Conc	50	24	25	72	9.39	–
9	1	HNO_3	Conc	100	24	120	78	15.30	–
10	1	HNO_3	6	100	24	120	73	16.09	–
11	1	HNO_3	6	100	48	120	75	16.78	–
12	2	HNO_3	10	100	24	120	70	17.10	–
13	1	HNO_3	10	100	16.5	120	77	17.88	–
14	1	HNO_3	10	100	48	120	74	18.75	–
15	1	HNO_3	10	100	87	120	74	19.13	42.76
16	1	HNO_3	10	100	168	120	76	19.25	–

Table 13 The main characteristics of acid-leached and steamed mordenite samples

Bulk (Si/Al) mol/mol	Framework (Si/Al) (mol/mol)	Crystallinity (%)	Bulk Pt (wt%)	Acidity (mmol g^{-1})	Surface area (m^2 g^{-1})	Pore volume (cm^3 g^{-1})	Pore size (Å)
As-synthesised mordenite							
8.44	9.02	82	–	1.24	406.95	0.17	18.02
Acid-leached mordenite							
42.76	19.13	74	–	0.48	478.85	0.23	26.57
Mordenite steamed at 500 °C							
8.97	20.38	66	–	0.30	419.52	0.21	26.97
Mordenite steamed at 600 °C							
9.06	28.84	64	0.58	0.22	415.00	0.20	26.12
Mordenite steamed at 700 °C							
8.89	∞	67	–	–	404.93	0.20	26.68

acidity and narrow pore diameter, while the diffusion of the branched products can be slowed down. Each of these factors will help to crack the tertiary carbocations that have formed during isomerisation prior to desorption and will assist in the re-adsorption and cracking of the branched paraffins before they leave the pores and enter the gas stream [55].

There are various possible explanations for the slower deactivation of 1 wt% Pt/690HOA and steamed mordenite. First, it may be due to the medium acid strength of the catalysts and their reduced acid site density. Alternatively, it may be due to the balance between acid and metal active sites. It could also be ascribed to the existence of mesopores and to the slight increase in the pore volume, which may be beneficial to the diffusion of bulky moieties. Finally, it may be due to post-synthesised mordenite having smaller crystallites. The last two characteristics can be seen as reducing the resistance of the product molecules to formation and diffusion, resulting in less cracking and deactivation of the catalyst caused either by blockages at the site or by plugging of channels. The method of dealumination applied to the final lattice Si/Al ratio influences catalyst activity. For example, steam dealumination leaves the lattice with aluminium removed as EFAL, whereas samples that are practically free of EFAL can result from acid leaching, so EFAL may have a negative effect by blocking pores and preventing the Pt from being well dispersed during the ion-exchange process. This may explain the deactivation of the 1 wt% Pt/690HOA steamed mordenite catalysts.

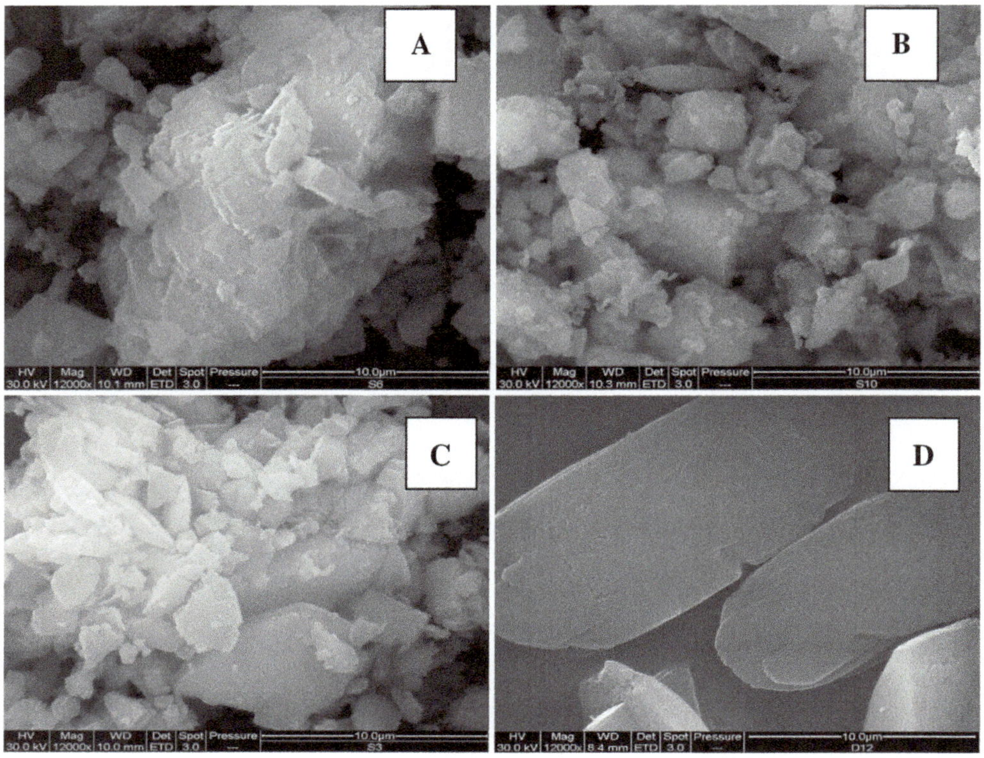

Fig. 25 SEM micrographs of mordenites steamed at 500 °C (**a**), 600 °C (**b**), 700 °C (**c**) and acid leached (**d**)

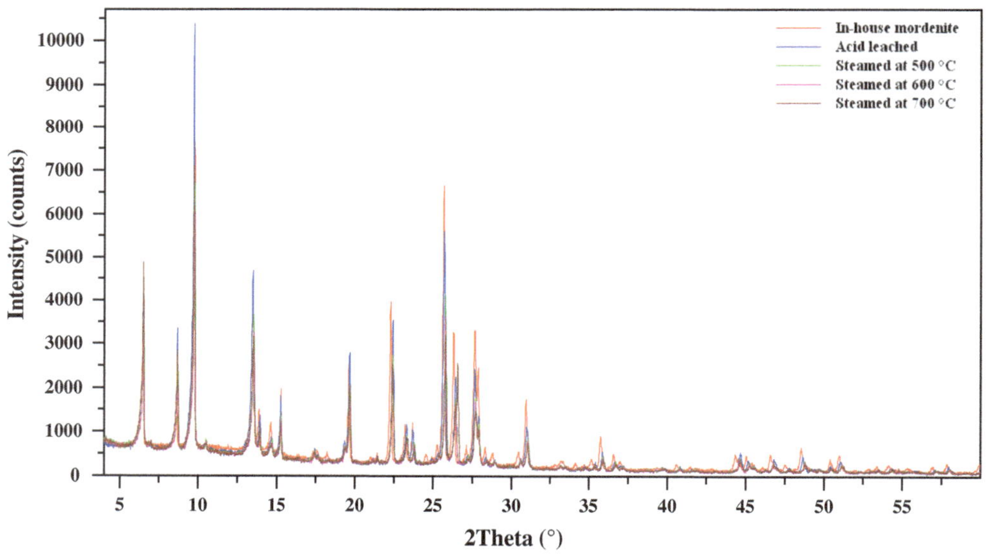

Fig. 26 XRD patterns of acid-leached and steamed mordenites

Deactivation behaviour of two selected commercial mesoporous aluminosilicates and a γ-alumina-based catalyst

TOS stability during the hydroisomerisation of n-C7 under identical reaction conditions was examined using two selected commercial mesoporous aluminosilicates—Al-SBA-15 and Al-MCM-48—and a γ-alumina catalyst, which were loaded with 1 wt% platinum by the ion-exchange technique and tested using the atmospheric pressure glass fixed-bed flow reactor system. The two mesoporous aluminosilicates, Al-SBA-15 and Al-MCM-48, containing 8 and 3 wt% aluminium, respectively, were supplied by Claytec (USA) [20] as white powders in their proton form. Their main surface texture features and the methods by which the alumination and protonation were

Fig. 27 ^{27}Al NMR of acid-leached and steamed samples

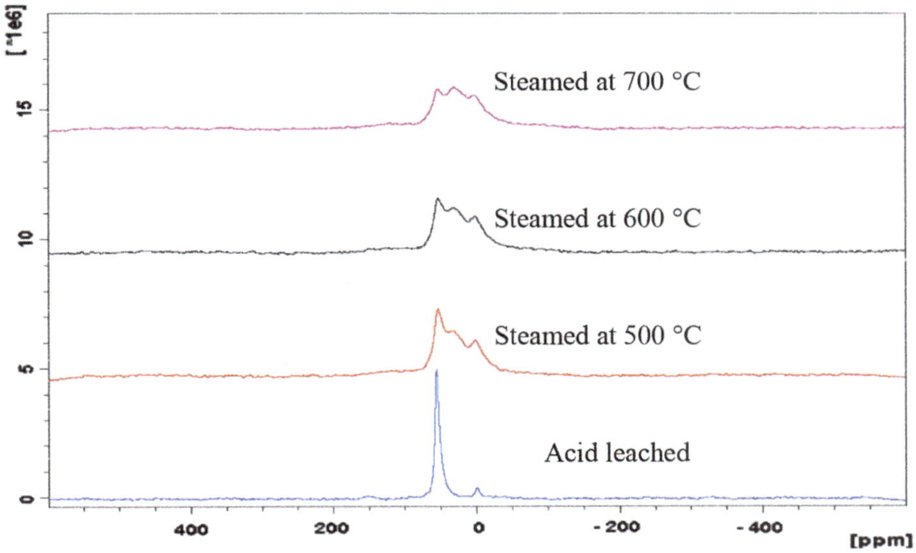

Fig. 28 ^{29}Si NMR of acid-leached and steamed samples

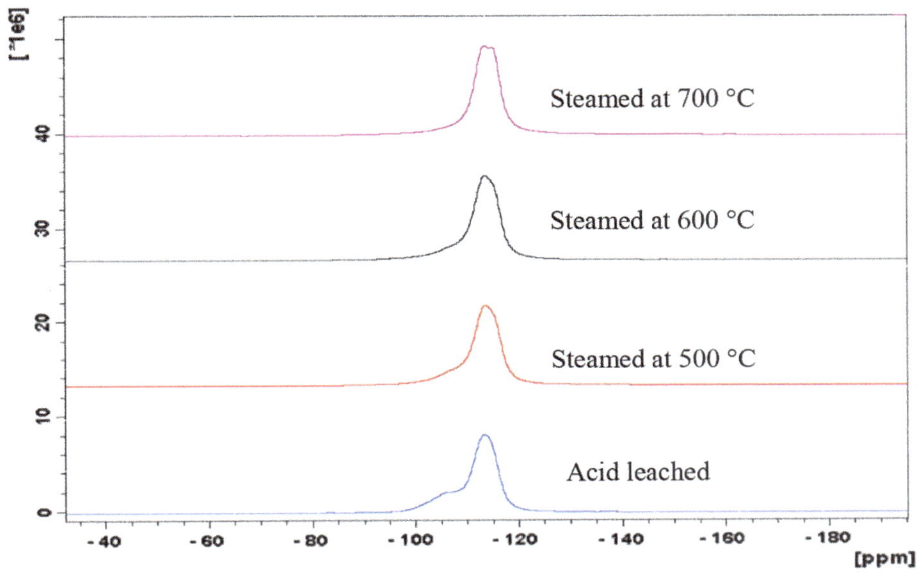

preformed, as specified by the supplier, are shown in Table 14.

Figure 31 shows the SEM image of Al-SBA-15 and Al-MCM-48 catalysts. It can be observed that both samples consist of non-uniform and irregular-shaped particles varying in size from 0.5 to 2 µm. XRD patterns of the Al-SBA-15 and Al-MCM-48 catalysts are shown in Fig. 32. The Al-SBA-15 sample shows well-resolved patterns with a prominent diffraction peak (100) and two additional diffraction peaks indexed to (110) and (200) reflections, which match well with the pattern reported for SBA-15 [56], indicating that the Al-SBA-15 product prepared by post-alumination had a well-ordered hexagonal meso-structure, whereas the XRD pattern of the Al-MCM-48 sample was identical to the one reported by Beck et al. [57]. Four diffraction peaks were also clearly observed and indexed to (211), (220), (420) and (332), indicating that the Al-MCM-48 product prepared by post-alumination had a well-ordered cubic mesostructure. Table 15 shows some in-house characterisations of the Al-SBA-15, Al-MCM-48 and γ-alumina catalysts using various different techniques.

Solid-state MAS NMR spectroscopy of ^{27}Al indicated the presence of tetrahedrally coordinated aluminium atoms in the Al-SBA-15 (A) and Al-MCM-48 lattices. It also revealed extra-framework octahedral and pentahedral Al in the pores of both molecular sieves, which may have been formed during hydrothermal crystallisation and/or by dea-lumination during calcination [58], as shown in Fig. 33. Therefore, it can be confirmed that not all Al atoms were inserted in the frameworks of SBA-15 and MCM-48.

In the ^{29}Si MAS NMR spectra, two peaks at 110 and 102 ppm were assigned to Si(0Al) and Si(1Al)

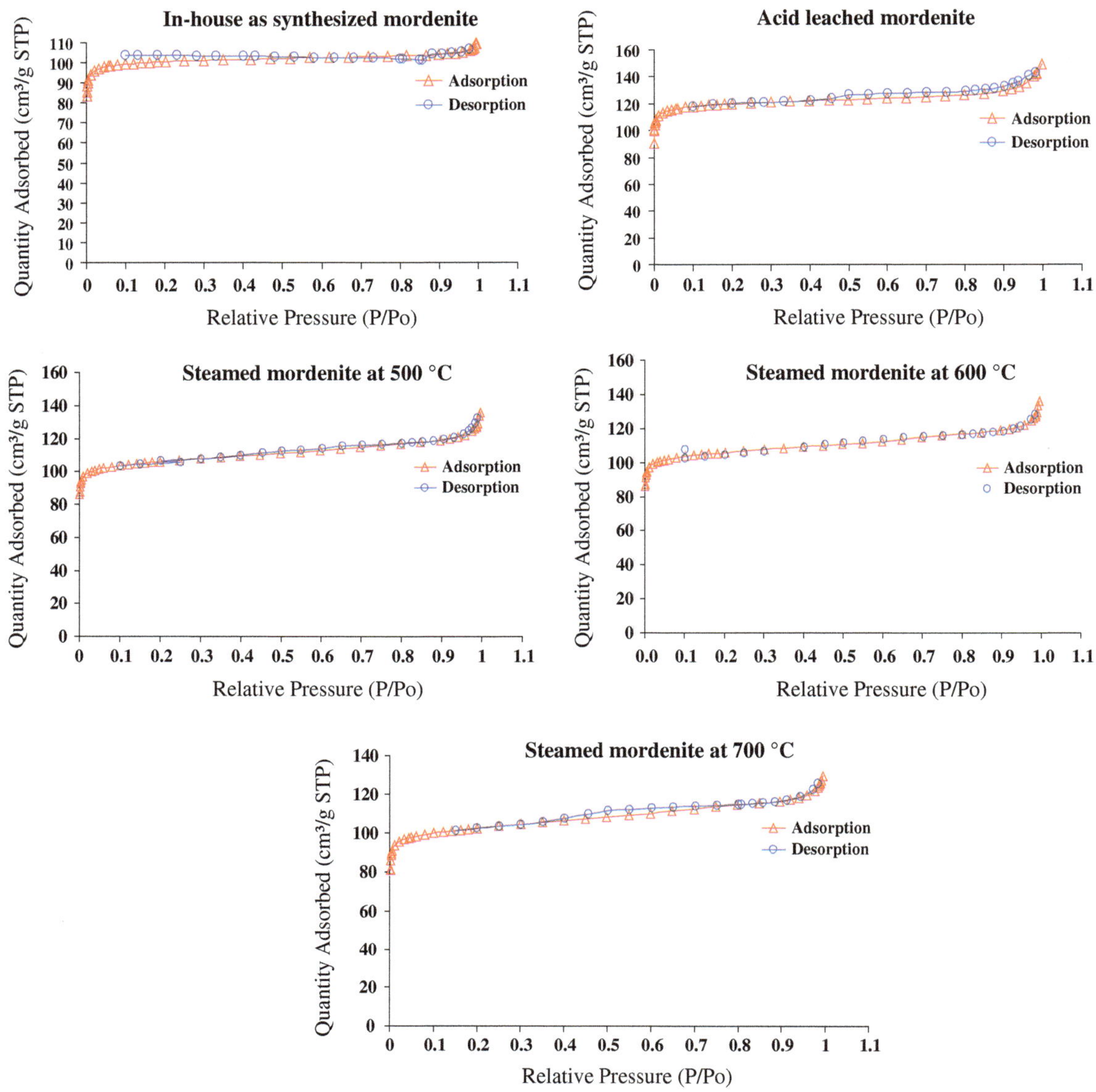

Fig. 29 N_2 adsorption and desorption isotherms of acid-leached and steamed samples

configurations respectively for Al-SBA-15, and three peaks at 110, 101, and 92 ppm were assigned to Si(0Al), Si(1Al) and Si(2Al) configurations respectively for Al-MCM-48, as shown in Fig. 34. This confirmed that some Si atoms were substituted by Al atoms to generate the mesoporous aluminosilicate structure for both catalysts.

The Al-SBA-15 and Al-MCM-48 catalysts were also characterised using nitrogen adsorption at −196 °C, as shown in Fig. 35, to confirm that these materials had retained their mesostructures after the post-alumination process to substitute the silicon with aluminium. Figure 35 shows the nitrogen adsorption isotherms for both catalysts, which are type IV according to the IUPAC classification [59]. The isotherm of the mesoporous molecular sieve exhibits a sharp inflection characteristic of capillary condensation within uniform pores, where the P/P_0 position of the inflection point is related to the diameter of the pore. The sharp rise near 0.4 P/P_0 corresponds to condensation in the primary mesopores. As the size of the primary mesopores increases, the capillary condensation step shifts to higher relative pressures and hysteresis becomes more pronounced. The existence of a hysteresis loop in the isotherms indicates the presence of mesopores and its shape is related to their shape: roughly, a vertical loop indicates cylindrical mesopores, whereas a horizontal one indicates inkbottle-shaped mesopores [51].

As shown in Tables 14 and 15 alongside Fig. 35, the two commercial catalysts exhibited high pore size and pore volume and retained their mesostructures, even after the post-alumination process. Table 14 shows that Al-SBA-15 had a much lower surface area than expected for mesoporous materials (>1,000 $m^2\ g^{-1}$), which can be explained

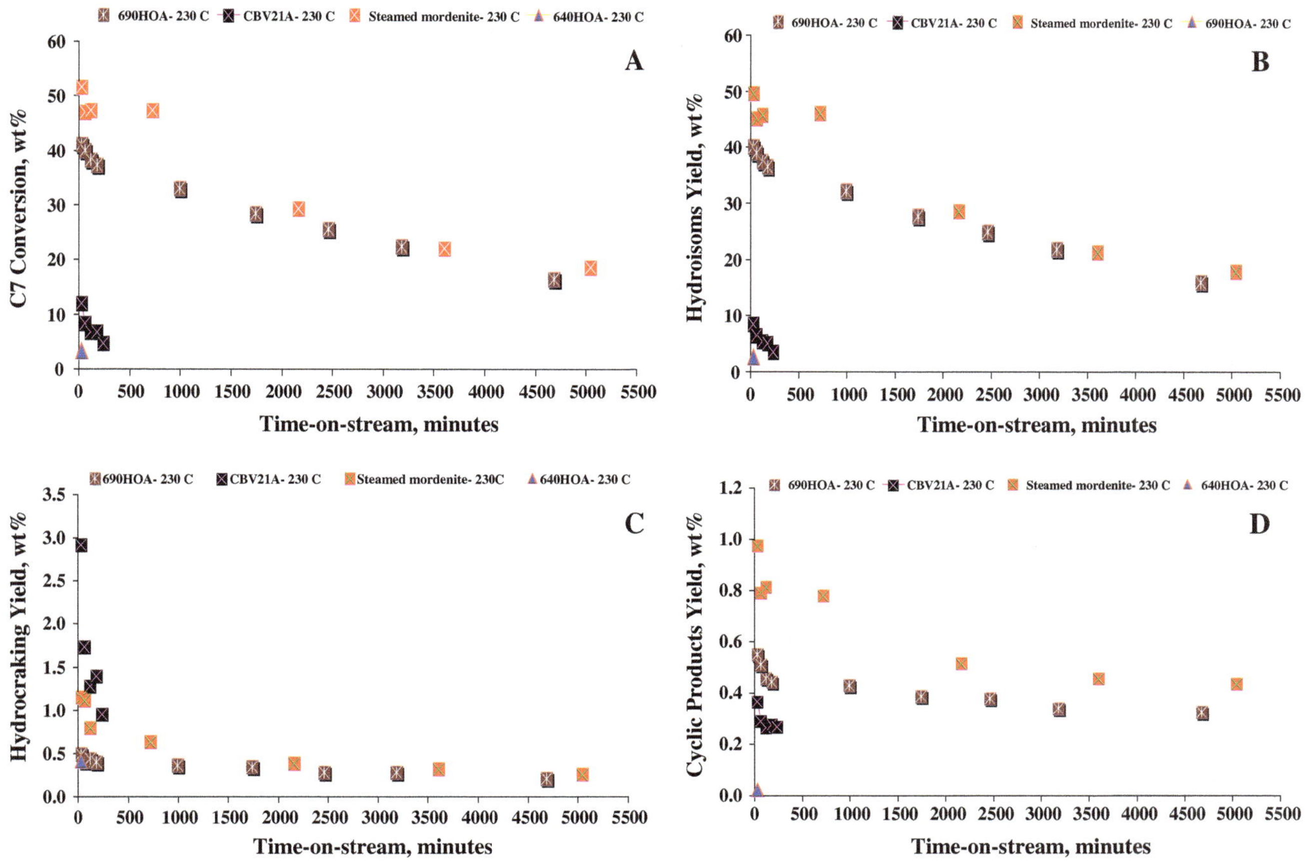

Fig. 30 Overall conversion (**a**), total yield of C7 isomers (mono + di + tribranched isomers) (**b**), total hydrocracking yield (**c**) and total yield of cyclic products (**d**) as a function of TOS for commercial and in-house steamed mordenite catalysts. WHSV = 5.13 h^{-1}; H_2/C7 (mol/mol) = 9; total pressure = 1 atm; temperature = 230 °C; 1 wt% Pt loading

in two ways. First, the presence of EFAL on the mesostructure may partially block the mesopores and subsequently hinder nitrogen adsorption. The existence of EFAL was revealed by NMR, as shown in Fig. 33a. Alternatively, the Al-SBA-15 mesostructure may have partially collapsed, so some residual amorphous material was generated after the post-alumination process. Despite the appearance of the lower surface area, the Al-SBA-15 catalyst retained its general mesostructure, as revealed by the XRD and nitrogen adsorption isotherms shown in Figs. 32 and 35 respectively, while its large pore size and pore volume were comparable with those of Al-MCM-48.

The deactivation behaviour of the 1 wt% Pt/Al-SBA-15 and 1 wt% Pt/Al-MCM-48 catalysts was investigated during the catalytic conversion of n-C7 at reaction temperatures of 210, 230, 250 and 270 °C and at atmospheric pressure, using the glass fixed-bed flow reactor system loaded with 1 g of each catalyst, as described in "Results and discussion" section. The constant mole ratio of hydrogen to n-C7 was 9 and WHSV was 5.13 h^{-1}. At temperatures below 300 °C, very low n-C7 conversion was found, and so 370 °C was chosen as the optimum temperature for studying the activity decay of these catalysts.

Figure 36a–d illustrates the effects of TOS on the C7 conversion, hydroisomerisation, hydrocracking and cyclic product yields at reaction temperature of 370 °C for the two catalysts loaded with 1 wt% Pt by the ion-exchange method. The γ-alumina support supplied by Alfa-Aesar (Johnson–Matthey) was also loaded with 1 wt% Pt using the wet impregnation method and tested under the same reaction conditions.

Initially, the respective overall conversions for 1 wt% Pt/Al-SBA-15, 1 wt% Pt/Al-MCM-48 and 1 wt% Pt/γ-Al_2O_3 were 51, 39 and 24 wt%, 30 min after the introduction of feed to the reactor at 370 °C. The 1 wt% Pt/Al-SBA-15 catalyst deactivated slightly with TOS, reaching a conversion of 25 wt% (<26 wt% of the initial catalytic activity) after about 2 days, then remained in a pseudo-stable state until the reaction terminated after 4 days. The 1 wt% Pt/Al-MCM-48 started with less activity than the 1 wt% Pt/Al-SBA-15, but with greater TOS stability, deactivating slightly to reach a conversion of 30 wt% (<9 wt% of the initial catalytic activity) after 2 days, then maintained a pseudo-stable state until the reaction terminated after 3 days. Deactivation was rapid for the 1 wt% Pt/γ-Al_2O_3 catalyst, with high cyclisation activity under the conditions of the investigation, resulting in selectivity towards the cyclic products at 55 %, compared with 40 % for the isomer products; thereafter, the catalyst was almost deactivated after 180 min on stream. The 1 wt% Pt/Al-SBA-15 and 1 wt% Pt/Al-MCM-48 catalysts also exhibited fairly high cyclisation selectivity, of 20 and 27 wt%, respectively, decreasing with TOS to 7 and 9 wt%, respectively, by the end of the reactions. The cracking selectivity of the 1 wt% Pt/Al-SBA-15, 1 wt% Pt/Al-MCM-48 and 1 wt% Pt/γ-Al_2O_3 catalysts was 1, 3 and 10 wt%, respectively. The maximum isomer yields for 1 wt% Pt/Al-SBA-15, 1 wt% Pt/Al-MCM-48 and 1 wt% Pt/γ-Al_2O_3 were found to be 40, 27 and 10 wt%, respectively, decreasing with TOS in the same way as conversion.

As the major reactions were hydroisomerisation and cyclisation, negligible cracking was observed for the

Table 14 Main surface texture features and the alumination and protonation methods of Al-SBA-15 and Al-MCM-48 as supplied by Claytec (USA) [20]

Sample	BET surface area ($m^2\ g^{-1}$)	Total pore volume @ $P/P_0 = 0.98$ ($cm^3\ g^{-1}$)	BJH pore size (nm)	Alumination and protonation methods
8 wt% Al-SBA-15	408	0.75	8.6	Calcination of the as-made pure silica mesophase at 600 °C to remove surfactant, followed by grafting reaction of the pure silica mesophase with aluminium i-propoxide to achieve alumination, followed by another calcination at 550 °C to form the protonated derivative
3 wt% Al-MCM-48	1,217	1.4	2.4	Sodium aluminate was used for Al incorporation during framework assembly, followed by ammonium ion exchange in the presence of surfactant, then calcination at 600 °C to form the protonated derivative

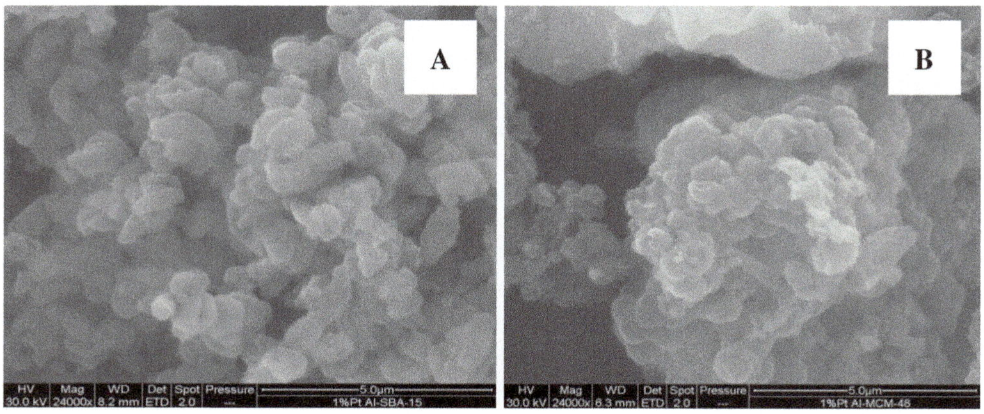

Fig. 31 SEM micrographs of Al-SBA-15 (**a**) and Al-MCM-48 (**b**)

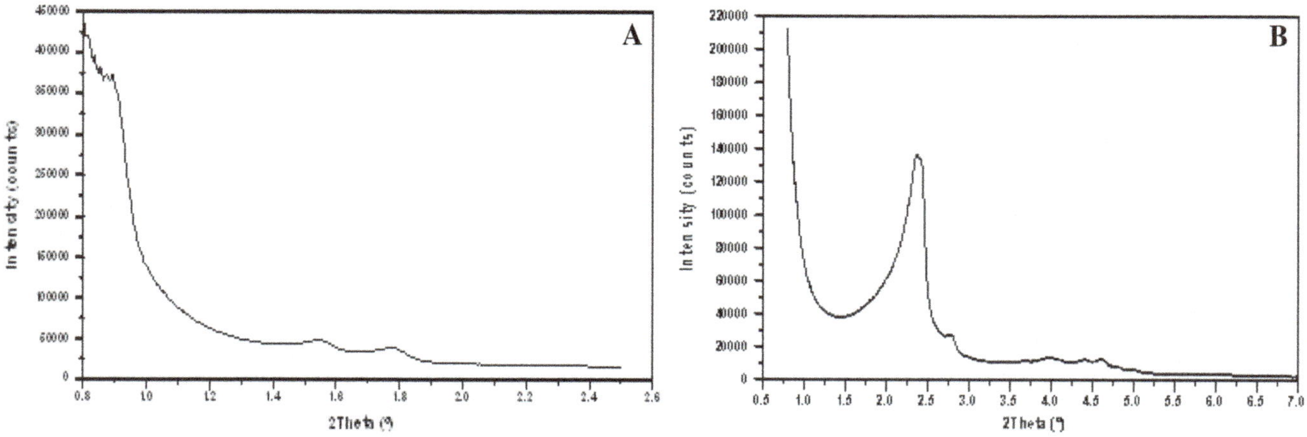

Fig. 32 XRD patterns of Al-SBA-15 (**a**) and Al-MCM-48 (**b**)

Table 15 Some characterisations of Al-SBA-15 and Al-MCM-48 catalysts

Bulk (Si/Al) (mol/mol)	Framework (Si/Al) (mol/mol)	Bulk Pt (wt%)	Si/Al atomic conc % (XPS)	Pt atomic conc % (XPS)	Acidity (mmol g^{-1})	Surface area (m^2 g^{-1})	Pore volume (cm^3 g^{-1})	Pore size (Å)
Al-SBA-15								
11.05	6.39	0.95	10.62	0.04	0.29	–	–	–
Al-MCM-48								
32.65	9.33	0.95	–	–	–	1,035	1.33	34.36
γ-alumina								
–	–	–	–	–	0.19	211	0.59	90.49

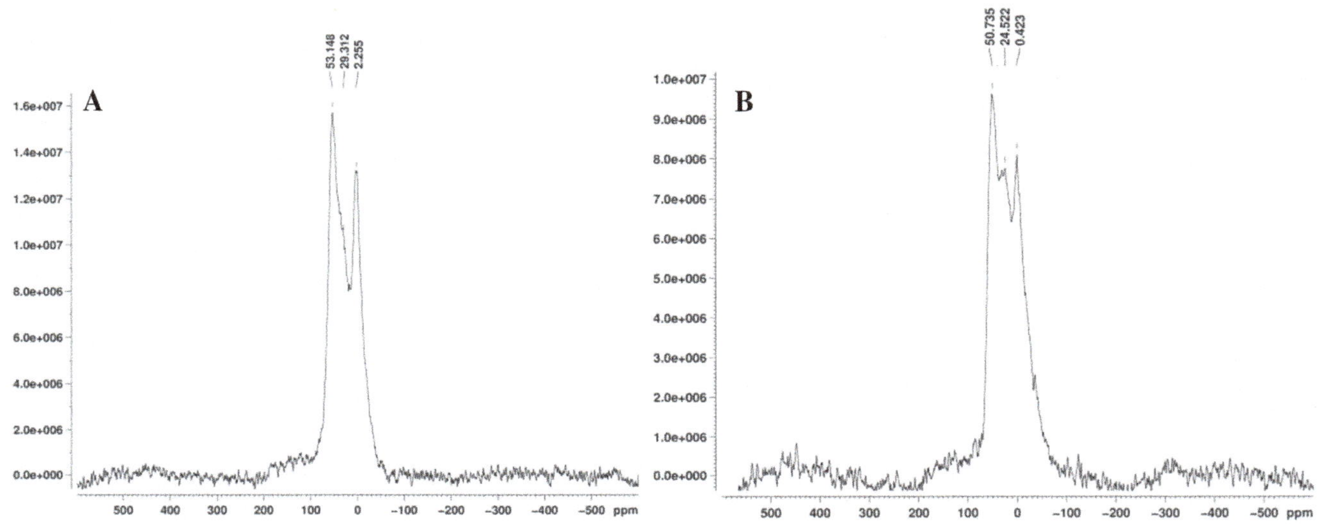

Fig. 33 ^{27}Al MAS NMR spectra of Al-SBA-15 (**a**) and Al-MCM-48 (**b**)

1 wt% Pt/Al-SBA-15 and 1 wt% Pt/Al-MCM-48. Small amounts of alkenes were also detected in the products. Although the deactivation rate was significant during the first few minutes, the activity of the catalyst became reasonably stable after about 2 days. A certain amount of initial deactivation has been reported by other workers during the isomerisation of *n*-alkanes over Pt catalysts, which is expected under atmospheric pressure conditions [60]. However, it was shown that a great quantity of oligomers (cyclic products) is produced from propene at low temperatures on MCM-41-type catalysts and that the pore size of the mesoporous material can make a difference when the aim is to oligomerise larger olefins or to produce large lubricant molecules in general [12].

It is reported that the very high surface area of ordered mesoporous materials makes it easier to obtain highly dispersed noble metal catalysts [12, 61]. However, because these catalysts have fewer acid sites, as shown in Table 15, the turnover frequency (TOF) of n-C7 hydroisomerisation at 370 °C over these sites would be expected to be low, which means in turn that there would be a relatively low level of interaction between the low acid sites in the large pore walls of these catalyst and the iso-heptane molecules generated by dehydrogenation on the platinum sites. Thus, these alkenes will be oligomerated to produce the cyclic products. It can be concluded that the acid and metal sites are not sufficiently well balanced to produce ideal bifunctional catalysts for n-C7 conversion. The oligomers formed may be attributed to the formation of carbonaceous species on the catalyst surface, which contribute to the poisoning and/or blocking of the active centres. Thus, it is clear that good isomerisation activity is obtained when the hydrogenation activity of the metal (Pt) and the Brønsted acid activity of the bifunctional catalyst are in balance. In comparison to the 1 wt% Pt/γ-Al_2O_3 catalyst, the 1 wt% Pt/Al-SBA-15 and 1 wt% Pt/Al-MCM-48 produced smaller quantities of cracked products, probably due to their effectively greater Pt dispersion.

The maximum coke content for the 1 wt% Pt/Al-SBA-15 catalyst was found to be 0.49, 0.72, and 0.70 wt% in the top, middle and bottom reactor zones, respectively, while the minimum coke content for the 1 wt% Pt/γ-Al_2O_3 catalyst was found to be 0.36 wt%, as revealed by TGA analysis. It can be clearly seen that the middle and bottom

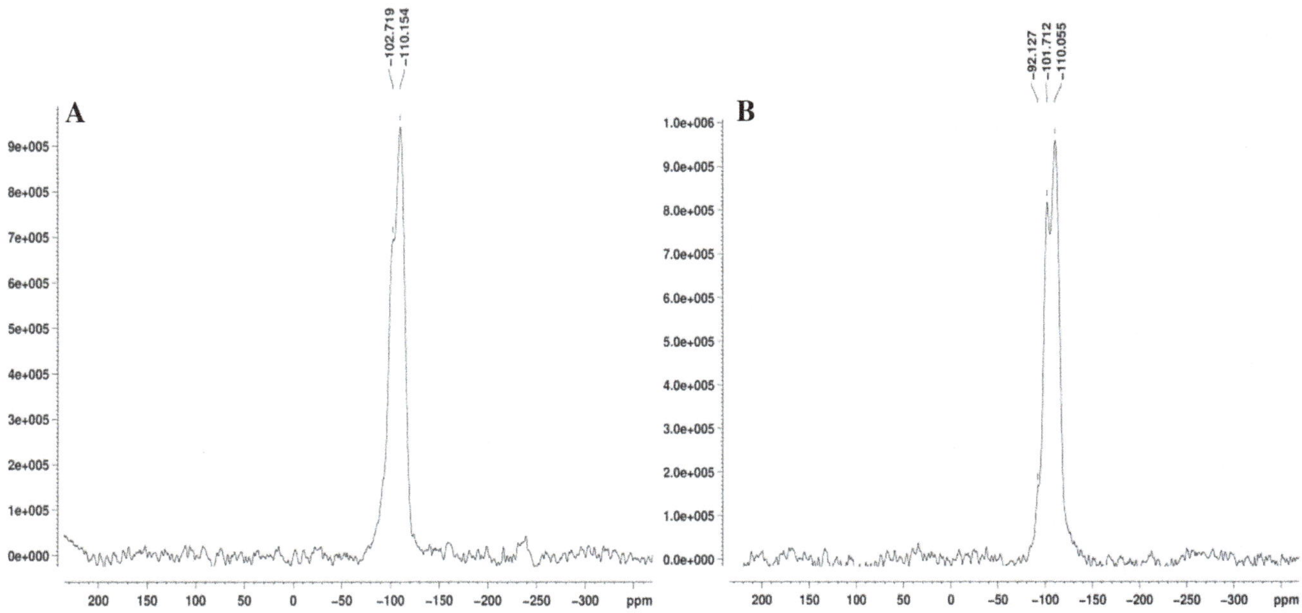

Fig. 34 ^{29}Si MAS NMR spectra of Al-SBA-15 (**a**) and Al-MCM-48 (**b**)

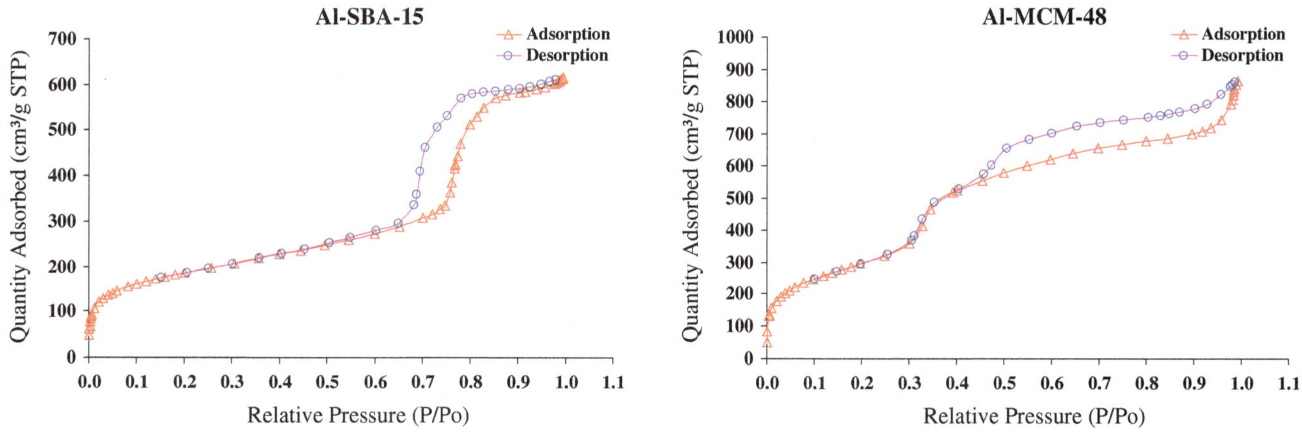

Fig. 35 N_2 adsorption and desorption isotherms of Al-SBA-15 and Al-MCM-48

zones of 1 wt% Pt/Al-SBA-15 had the highest hard coke content in the catalyst bed, whereas there was slightly less coke in the top zone. This suggests that the middle and bottom zones of the catalyst bed contributed the most activity, where the reaction took place. In addition, the lower coke content in the top zone is to some extent indicative of the low acidity of this catalyst and the fact that no diffusion limitations were encountered with it, since such limitations would result in coking and would cause the top section of the bed to be coke rich.

Conclusions

There are many conclusions which can be drawn from this investigation of the deactivation of a range of commercial bifunctional nanoporous-based catalysts during the hydroisomerisation of normal heptane (n-C7) under identical reaction conditions. The main cause of deactivation of acid zeolite catalysts is generally coking, as coke compounds become trapped in the zeolite pores and can poison the acid sites or block their access:

- It has been shown that many different parameters control the stability of USY zeolites in the hydroisomerisation of n-C7. Not only reaction conditions but also catalyst characteristics play the crucial roles.
- For most of the catalysts studied, total conversion increased dramatically as reaction temperature increased in the range of 210–270 °C. However, the total selectivity of C7 isomers at lower temperatures was very high throughout the deactivation reaction at

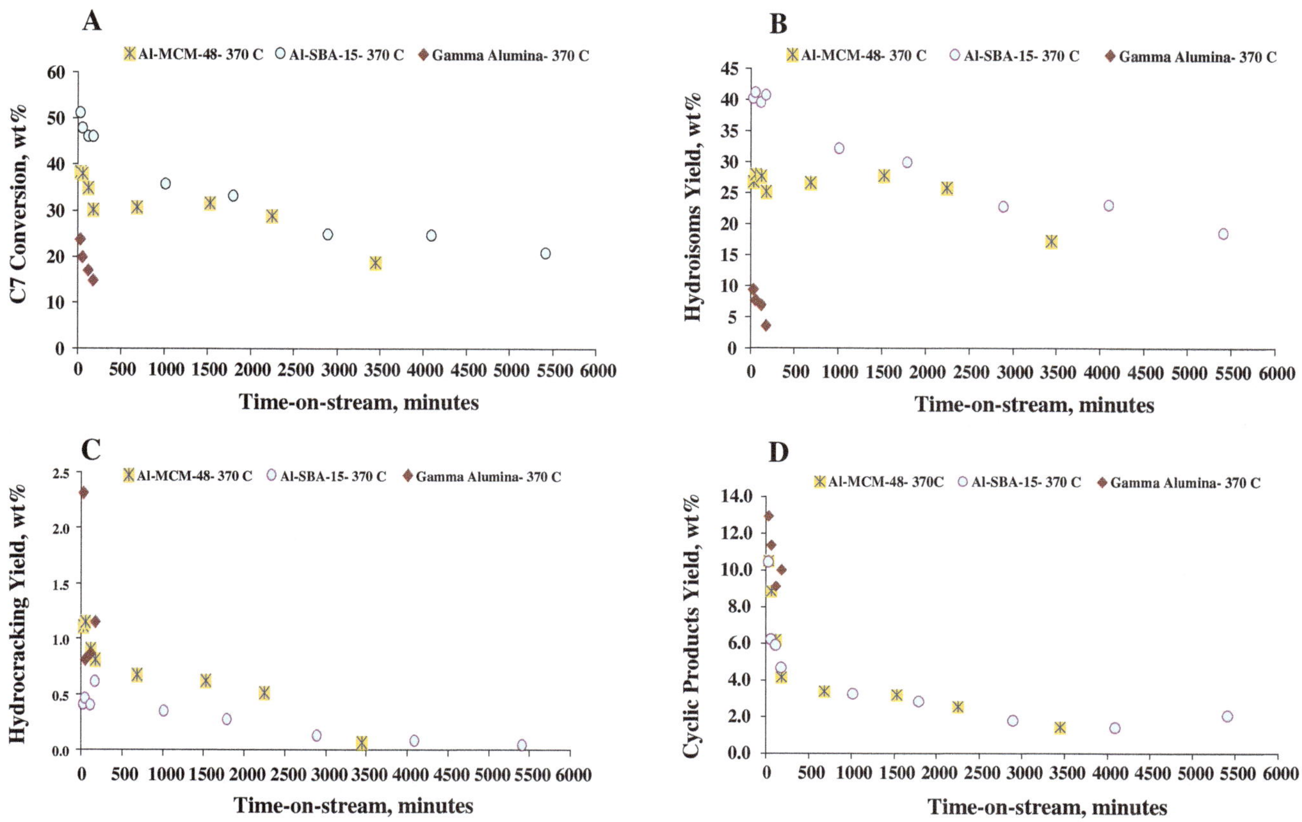

Fig. 36 Overall conversion (**a**), total yield of C7 isomers (mono + di + tribranched isomers) (**b**), total hydrocracking yield (**c**) and total yield of the cyclic products (**d**) as a function of TOS for the Al-SBA-15, Al-MCM-48 and γ-alumina catalysts. WHSV = 5.13 h^{-1}; H_2/C7 (mol/mol) = 9; total pressure = 1 atm; temperature = 370 °C; 1 wt% Pt loading

210 and 230 °C, taking advantage of the thermodynamic equilibrium.

- The selectivity of cracked products increased at the higher temperatures of 250 and 270 °C during the first few hours of the deactivation reaction and then decreased rapidly throughout the TOS. The selectivity of cyclic products was slightly higher at the lower temperatures of 210 and 230 °C.
- Increasing pressure resulted in decreased overall conversion and slightly decreased selectivity to isomers with noticeable increased selectivity to the cyclic products, particularly at the lower temperatures of 210 and 230 °C.
- There was no obvious cracking activity at the lower temperatures of 210 and 230 °C, and constant WHSV as the reaction pressure changed, and in contrast there was increased selectivity towards the cracked product as the reaction pressure increased at the highest temperature of 270 °C.
- The total conversion increased as the contact time increased at constant temperature and pressure. However, increasing the contact time between the n-C7 feed and the catalyst seems to cause an increase in cracking selectivity.
- Zeolite Y commercial catalysts that were generated by steaming at different levels of severity behaved very differently to each other and to acid-leached commercial catalysts.
- USY zeolite pore structures which can inhibit coke formation in conjunction with careful dealumination to decrease the number of acid sites are necessary for the development of time-stable reforming zeolite-based catalysts.
- There is a clear correlation between the obtained yield of the cracked products, which relates to the existence of the stronger acid sites and to the hard coke. Maximum hard coke values were found at 270 °C.
- The balance between the number of Pt atoms and the number of acid sites plays an important role in determining the activity, selectivity and stability of the bifunctional catalyst. The greater the value of n_{Pt}/n_A, the slower the coke formation and the higher the catalyst stability.
- Lower platinum dispersion may have resulted in the formation of a considerably greater quantity of coke.
- Lower acidity, hence less cracking activity, resulted in the catalyst being more stable against deactivation by coking, which is the case for zeolites with high Si/Al ratios.

- High surface area and larger pore volume, due to the more severe dealumination process and acid leaching, reduced the hard coke content, thus slowing the deactivation rate.
- The AHFS acid-leached zeolite had a larger pore volume (smaller pore size and larger surface area), due to the alleviation of the obstruction by EFAL of the mesopore system, thus slowing the formation of coke and providing good catalytic stability.
- The higher Pt loading, up to 2 wt%, improved catalytic stability, due to a better balance and closeness of the catalytic functions.
- Pt–Ni bimetallic catalysts were better than monometallic catalysts containing only Ni because they had higher activity, better selectivity for C7 isomers and more stability. The presence of platinum in the bimetallic catalysts improved the formation of metal particles to a great extent and reduced their sizes, while enhancing the reduction of nickel.
- The best method to introduce the platinum into the beta zeolite pores was found to be the ion-exchange method at pH > 9 which achieved a better metal/acid balance, lower cracking activity, and relatively good metal dispersion.
- The metal functions deactivated faster than the acid functions for all the catalysts loaded with platinum, irrespective of the method used. However, this was not sufficiently severe to cause a drop in n-C7 conversion due to the unique channel system and pore structure of beta zeolite.
- The TOS behaviour of USY zeolites can be enhanced by forming composite catalysts of USY and beta zeolite, providing a unique pore structure and acidity to improve their TOS stability.
- The metal function deactivated faster than the acid function in all the composite catalysts but did not cause a severe drop in n-C7 conversion, indicating that the composite zeolite pore structure plays a very important role in the conversion, and an appropriate combination of the pore structure and acidity showed superior TOS behaviour.
- Although the composite catalyst offers a highest acidity compared to the corresponding individual catalysts; coking is moderate and as such the TOS stability is reasonable.
- Synthesis parameters greatly affect mordenite morphology in terms of the size and aspect ratio of the mordenite crystals, so a range of mordenite morphologies were obtained for a variety of catalysts, both in-house and commercial. Each catalyst had different diffusion and porosity properties; therefore, performance differed among them.
- The framework aluminium was removed to extra-framework positions as a result of steam dealumination as well as giving the final in-framework Si/Al ratio. On the other hand, acid leaching resulted in samples that were practically free of EFAL, which in turn led to the formation of aluminium gradients along the zeolite crystallites. A more uniform aluminium distribution was achieved by steam treatment.
- The main factors controlling catalyst stability and selectivity were pore topology, acid strength, acid site density and location of the acid sites. The crucial feature for coke formation was the pore topology of the catalyst.
- The synthesis method of mesoporous materials and the post-alumination process applied significantly affect the number and strength of acid sites and thereby their catalytic activity, selectivity and stability.
- As the mesoporous molecular sieves present very high total surface area and large pore size, this may depress the TOF, causing the olefins (alkenes) produced during the dehydrogenation of alkanes on the metal sites to have less interaction with the acid sites on which they can be hydrogenated to give their relevant isomers. Thus, the alkenes will be oligomerated to produce the cyclic products, which may be attributed to the formation of carbonaceous species on the catalyst surface, contributing to the poisoning and/or blocking of the active centres.

Acknowledgments I would like to thank Dr. Patrick Hill (Manchester University) for his help in carrying out the SEM analysis, Dr. Vladimir Zholobenk (Keele University) who kindly provided the FTIR technique, Dr. Christopher Muryn (Manchester University) for his help in performing the XRD analyses, Professor Michael Anderson and Dr. Barbara Gore (Manchester University), who kindly provided the NMR technique, Mr. Martin Jennings (Manchester University) for his help with the elemental analysis and Mr. David Gordon (Manchester University) for his cooperation in making the glass reactor.

References

1. Guisnet M, Ribeiro FR (2011) Deactivation and regeneration of zeolite catalysts. Imperial College Press, London
2. Bartholomew CH (2003) Catalyst deactivation and regeneration. In: Kirk-Othmer encyclopedia of chemical technology. Wiley, London
3. Zhang W, Smirniotis PG (1999) On the exceptional time-on-stream stability of HZSM-12 zeolite: relation between zeolite pore structure and activity. Catal Lett 60:223–228

4. Stöcker M (2005) Gas phase catalysis by zeolites. Microporous Mesoporous Mater 82:257–292
5. Gopal S, Smirniotis PG (2004) Factors affecting isomer yield for *n*-heptane hydroisomerization over as-synthesized and dealuminated zeolite catalysts loaded with platinum. J Catal 225:278–287
6. Kuznetsov PN (2003) Study of *n*-octane hydrocracking and hydroisomerization over Pt/HY zeolites using the reactors of different configurations. J Catal 218:12–23
7. Villegas JI, Kumar N, Heikkila T, Smieskova A, Hudec P, Salmi T, Murzin DY (2005) A highly stable and selective Pt-modified mordenite catalyst for the skeletal isomerization of n butane. Appl Catal 284:223–230
8. McDaniel CW, Maher PK (1968) New ultrastable form of faujasite. In: Molecular sieves. London: Soc Chem Ind, pp 86–194
9. Van Laak ANC, Sagala SL, Zečević J, Friedrich H, de Jongh PE, de Jong KP (2010) Mesoporous mordenites obtained by sequential acid and alkaline treatments—catalysts for cumene production with enhanced accessibility. J Catal 276:170–180
10. Moushey DL (2006) Formation of mesoporosity in zeolite and mesoporous molecular sieve structures through use of carbon as a secondary templating agent. Dissertation, University of Cincinnati, USA
11. Logar NZ, Kaucic V (2006) Nanoporous materials: from catalysis and hydrogen storage to wastewater treatment. Acta Chim Slov 53:117–135
12. Corma A (1997) From microporous to mesoporous molecular sieve materials and their use in catalysis. Chem Rev 97:2373–2419
13. Pinnavaia TJ, Liu Y (2002) Aluminosilicate mesostructures with improved acidity and hydrothermal stability. J Mater Chem 12:3179–3190
14. Zeolyst International, http://www.zeolyst.com/our-products/standard-zeolite-powders/zeolite-y.aspx. Accessed 29 May 2011
15. Tosoh Corporation, http://www.tosoh.com/Products/tcdzeo.htm#hzc. Accessed 26 June 2011
16. Rawlence DJ, Earl, GJ (1993) Materials having novel structure factors for catalysis of hydrocarbon transformation and development of selective processes. BRITE-EURAM 4633, 5th Meeting Report, Elf Solaize
17. Abudawood R, Alotaibi F, Garforth A (2011) Hydroisomerization of *n*-heptane over Pt-loaded USY zeolites. Effect of steaming, dealumination and the resulting structure on catalytic properties. Ind Eng Chem Res 50:9918–9924
18. Abudawood R (2010) Hydroisomerization of alkanes over metal-loaded zeolite catalysts. Dissertation, University of Manchester
19. Alotaibi F (2011) Comparative study of the time-on-stream stability of bifunctional nanoporous-based catalysts in *n*-heptane hydroisomerisation. Dissertation, University of Manchester
20. Claytec, Inc., http://claytecinc.com/products. Accessed 29 May 2011
21. Robson H, Lillerud KP (2001) Verified Syntheses of zeolitic materials, 2nd edn. Elsevier, Amsterdam
22. Weitkamp J, Puppe L (1999) Catalysis and zeolites. Springer, Berlin
23. Wang QL, Torrealba M, Giannetto G, Guisnet M, Perot G, Cahoreau M, Caisso J (1990) Dealumination of Y zeolite with ammonium hexafluorosilicate: a SIM-XPS study of the aluminium distribution. Zeolites 10:703–706
24. Wang QL, Giannetto G, Torreaaba M, Perot G, Kappenstein C, Guisnet M (1991) Dealumination of zeolites: II. Kinetic study of the dealumination by hydrothermal treatment of a NH4NaY zeolite. J Catal 130:459–470
25. Wang QL, Giannetto G, Guisnet M (1991) Dealumination of zeolites: III. Effect of extra-framework aluminum species on the activity, selectivity, and stability of Y zeolites in *n*-heptane cracking. J Catal 130:471–482
26. Remy MJ, Stanica D, Poncelet G, Feijen EJP, Grobet PJ, Martens JA, Jacobs PA (1996) Dealuminated H–Y zeolites: relation between physicochemical properties and catalytic activity in heptane and decane isomerization. J Phys Chem 100:12440–12447
27. Van Bekkum H, Flanigen EM, Jacobs PA, Jansen JC (2001) Introduction to zeolite science and practice, 2nd edn. Elsevier, Amsterdam, pp 261–288
28. Roussel ML, emberton J, Guisnet M, Cseri T, Benazzi E (2003) Mechanisms of *n*-decane hydrocracking on a sulfided NiW on silica–alumina catalyst. J Catal 218: 427–437
29. Guerin M, Kappenstein C, Alvarez F, Giannetto G, Guisnet M (1988) Preparation of PtHY catalysts: influence on the catalytic properties of the complexes used as platinum precursors. Appl Catal 45:325–333
30. Martins A, Silva JM, Ribeiro FR, Ribeiro MF (2006) Hydroisomerization of *n*-hexane over Pt–Ni/HBEA using catalysts prepared by different methods. Catal Lett 109(1–2):83–87
31. Alotaibi F, Abudawood R, Al-Megren H, Al-Kinany H, Garforth A (2013) The time-on-stream stability of some selected bifunctional nanoporous-based catalysts in *n*-heptane hydroisomerisation. J Appl Petrochem Res (submitted)
32. Chao K, Lin C, Lin C, Wu H, Tseng C, Chen S (2000) *n*-Heptane hydroconversion on platinum-loaded mordenite and beta zeolites: the effect of reaction pressure. Appl Catal A 203:211–220
33. Tran MT, Gnep NS, Szabo G, Guisnet M (1998) Isomerization of *n*-butane over H-mordenites under nitrogen and hydrogen: influence of the acid site density. J Catal 174:185–190
34. Guisnet M, Fouche V (1991) Isomerization of *n*-hexane on platinum dealuminated mordenite catalysts: I. Influence of the silicon-to-aluminium ratio of the zeolite. Appl Catal 71:283–293
35. Wang ZB, Kamo A, Yoneda T, Komatsu T, Yashima T (1997) Isomerization of *n*-heptane over Pt-loaded zeolite β catalysts. Appl Catal 159:119–132
36. Chica A, Corma A (1999) Hydroisomerization of pentane, hexane, and heptane for improving the octane number of gasoline. J Catal 187:167–176
37. Lónyi F, Lunsford JH (1992) The development of strong acidity in hexafluorosilicate-modified Y-type zeolites. J Catal 136:566–577
38. Cruz JM, Corma A, Fornes V (1989) Framework and extra-framework aluminum distribution in $(NH_4)_2F_6Si$-dealuminated Y-zeolites—relevance to cracking catalysts. Appl Catal 50:287–293
39. Gola A, Rebours B, Milazzo E, Lynch J, Benazzi E, Lacombe S, Delevoye L, Fernandez C (2000) Effect of leaching agent in the dealumination of stabilized Y zeolites. Microporous Mesoporous Mater 40:73–83
40. Jordao MH, Valencia E, Vicemario S, Dilson C (1999) Characterization of Pt/HUSY and Pt-Ni/HUSY catalysts by transmission electron microscopy. Mate Res 2:219–223
41. Karthikeyan D, Lingappan N, Sivasankar B (2008) Hydroisomerization of *n*-octane over bifunctional Ni-Pd/HY zeolite catalysts. Ind Eng Chem Res 47:6538–6546
42. Barsi FV, Cardoso D (2009) Bimetallic Pt–Ni catalysts supported on USY zeolite for *n*-hexane isomerization. Braz J Chem Eng 26–2:353–360
43. Roldán R, Beale AM, Sánchez-Sánchez M, Romero-Salguero FJ, Jiménez-Sanchidrián C, Gómez JP, Sankar G (2008) Effect of the impregnation order on the nature of metal particles of bi-functional Pt/Pd-supported zeolite beta materials and on their catalytic activity for the hydroisomerization of alkanes. J Catal 254:12–26
44. Eswaramoorthi I, Lingappan N (2003) Ni–Pt/H-Y zeolite catalysts for hydroisomerization of *n*-hexane and *n*-heptane. Catal Lett 87(3–4):133–142
45. Kampers FWH, Engelen CWR, van Hoof JHC, Konigsberger DC (1990) Influence of preparation method on the metal cluster size

of platinum/ZSM-5 catalysts as studied with EXAFS. J Phys Chem 94(23):8574–8578
46. Changlei J, Bo M, Xiwen Z, Fengxiang L, Zhizhi Z, Bo Q (2008) Catalytic performance of Pt/HY-β in *n*-octane hydroisomerization. Petrol Sci 6:299–305
47. Smirniotis PG, Davydov L, Ruckenstein E (1999) Composite zeolite-based catalysts and sorbents. Catal Rev 41:43–113
48. Hamidi F, Bengueddach A, Di Renzo F, Fajula F (2003) Control of crystal size and morphology of mordenite. Catal Lett 87:149–152
49. Treacy MMJ, Higgins JB (2001) Collection of simulated XRD powder patterns for zeolites. Elsevier, Amsterdam
50. Janssen AH, Koster AJ, de Jong KP (2001) Three-dimensional transmission electron microscopy observations of mesopores in dealuminated zeolite Y. Angew Chem Int Ed 40:1102–1104
51. Van Donk S, Janssen AH, Bitter JH, de Jong KP (2003) Generation, characterization, and impact of mesopores in zeolite catalysts. Catal Rev 45:297–319
52. Janssen AH, Koster AJ, de Jong KP (2002) On the shape of the mesopores in zeolite Y: a three-dimensional transmission electron microscopy study combined with texture analysis. J Phys Chem 106:11905–11909
53. Lee JK, Rhee HK (1997) Characteristics of Pt/H-beta and Pt/H-mordenite catalysts for the isomerization of *n*-hexane. Catal Today 38:235–242
54. Van Donk S (2002) Adsorption, diffusion and reaction studies of hydrocarbons on zeolite catalysts. PhD Dissertation, Utrecht University, Netherlands
55. Chica A, Corma A, Miguel PJ (2001) Isomerization of C5–C7 *n*-alkanes on unidirectional large pore zeolites: activity, selectivity and adsorption features. Catal Today 65:101–110
56. Nie C, Huang L, Zhao D, Li Q (2001) Performance of Pt/Al-SBA-15 catalysts in hydroisomerization of *n*-dodecane. Catal Lett 71:117–125
57. Beck JS, Vartuli JC, Roth WJ, Leonowicz ME, Kresge CT, Schmitt KD, Chu CTW, Olson DH, Sheppard EW (1992) A new family of mesoporous molecular sieves prepared with liquid crystal templates. J Am Chem Soc 114:10834–10843
58. Eimer GA, Pierella LB, Monti GA, Anunziata OA (2002) Synthesis and characterization of Al-MCM-41 and Al-MCM-48 mesoporous materials. Catal Lett 78:65–75
59. Gregg SJ, Sing KSW (1982) Adsorption, surface area and porosity. Academic Press, London
60. Chaudhari K, Das TK, Chandwadkar AJ, Sivasanker S (1999) Mesoporous aluminosilicate of the MCM-41 type: its catalytic activity in *n*-hexane isomerization. J Catal 186:81–90
61. Chaudhari K (2000) Synthesis, characterization and catalytic properties of mesoporous molecular sieves. PhD Thesis, Catalysis Division, National Chemical Laboratory, Pune

14

Characterization of polyethylene synthesized by zirconium single site catalysts

Abdulhamid A. Alsaygh · Jehan Al-hamidi · Fares D. Alsewailem · Ibrahim M. Al-Najjar · Vladimir L. Kuznetsov

Abstract New non-metallocene, bridged-zirconium center catalysts were prepared by Schiff base condensations reaction of two equivalents of appropriate 2-hydroxybenzaldehyde with one equivalent of 4,4′-methylene bis(2-methylcyclohexylamine), and subsequent metathesis reaction with $ZrCl_4(THF)_2$, and tested for ethylene polymerization reaction at different conditions. Catalyst productivities were found to be high, in the range of 29.5×10^3 kg/mole·Zr·h to 2.3×10^3 kg/mole·Zr·h, dependents on the polymerization condition. The molecular weight of the product, i.e. polyethylene, varied from 397,000 to 988,000. ^{13}C NMR study of the polymers indicate that fraction of branches and end groups percentage range from 0.01 to 0.55. X-ray diffraction analysis showed that polyethylene samples had highly crystalline structure within the orthorhombic space group (*Pnam*) and carbon–carbon interatomic distance of 0.154061 nm and the C–C–C angle of 112.192°.

Keywords Polyethylene · Zirconium · Single site catalysts

A. A. Alsaygh (✉) · F. D. Alsewailem · I. M. Al-Najjar
Petroleum and Petrochemicals Research Institute, King Abdulaziz City for Science and Technology (KACST), P.O. Box 6086, Riyadh 11442, Saudi Arabia
e-mail: aalsaygh@kacst.edu.sa

J. Al-hamidi
Chemistry Department, College of Science, Princess Nora Bint Abdulrahman University, Riyadh, Saudi Arabia

V. L. Kuznetsov
Department of Chemistry, University of Oxford, South Parks Road, Oxford OX1 3QR, UK

Introduction

The area of olefin polymerization catalysis has endured a remarkable increase in research outcome over the past 20 years, with many academic and industrial research centers involving in the design of new organometallic precatalysts for the controlled synthesis of polyolefin products. During that period, studies on the Group 4 metallocenes afforded much needed insight into the nature of the activated species. During the first half of the 1990s, interest grew in developing new generation "non-metallocene" catalysts, partly to avoid the growing patent minefield in Group 4 cyclopentadienyl systems, but also because late transition metal systems also offered the commercially attractive possibility of being able to incorporate polar comonomers into polyolefin materials, to give modified surface properties at low levels of comonomer incorporation, to change the bulk properties of the polyolefin at higher levels of incorporation.

The potential for this approach was realized with the discovery in the late 1990s of highly active ethylene polymerization catalysts based on iron, a metal with no previous track record in olefin polymerization. Another significant development over the past years or so has been the introduction of systems capable of catalyzing the living polymerization of olefinic monomers. The absence of chain-transfer or chain-termination processes allows access to polyolefinic materials with very high molecular weight (>600,000) and narrow molecular weight distributions (typically <1.1), block copolymers, and polymers with novel topologies [1, 2]. Ultrahigh molecular weight polyethylene is an important material for production of high performance fibers where the polymer chains are extended and aligned as in the crystalline material. The performance of polymers is improved as the chain length and crystallinity are improved [3].

Polyethylene is the simplest possible polymer; however, the simplicity of the molecular formula $(–CH_2–)_n$ does not reflect the real complexity of the material and its properties. Polyethylene is generally synthesized as a mixture of crystalline and amorphous components. The physical properties of the polyethylene are usually described using at least a two-phase model consisting of crystalline blocks in an amorphous matrix. [4] The crystalline phase ranges from around 50 to >98 % and its concentration generally increases with increasing molecular weight and decreases with increasing chain branching. As the temperature of a polymer is raised, the percentage of crystallinity diminishes and above the melting point the structure is entirely amorphous, and consists of a viscous liquid.

Polycrystalline polyethylene forms chained molecules that are distributed on lattice points with a rigorous periodicity, thus forming a long-range order crystalline structure. For the first time the basic structural features of polyethylene were determined, by means of powder diffraction, in 1939 by Bunn [5] who found that polyethylene crystallizes in the orthorhombic structure (space group *Pnam*) with two polymeric chains per unit cell. The orthorhombic unit cell, which is identical to the crystal structure of straight-chain alkanes, contains four CH_2 groups. Paraffin's are identical with polyethylene in regard to their molecular structure except for their molecular length; they crystallize in the same crystal form as the polyethylene analogues.

^{13}C NMR and X-ray powder diffraction (XRD) techniques are widely used in polymer science. Solution and melt ^{13}C NMR has been established as a useful tool for a quantitative determination of branch content of short- and long-chain branched polyethylenes while XRD can reveal important information on the morphology, structure and the degree of crystallinity of polythene [6–8]. In the present paper we report the synthesis of high molecular weight polyethylene using Zr-based single site catalyst and results of the use of melt-state ^{13}C MAS NMR and X-ray diffraction techniques for the determination of branch content of long-chain polyethylene's and their crystal structure.

Experimental

All operations involving air sensitive reagents and materials during catalysts precursor synthesis were carried out under nitrogen atmosphere using standard Schlenk, vacuum or dry box techniques. Solvents were dried and distilled under nitrogen prior to use of by passing through a column of activated alumina and a supported copper (Q5) catalyst. Imino-ligands were synthesized by condensation of two equivalents of appropriate 2-hydroxybenzaldehyde with one equivalent of 4, 4′-methylene bis(2-methylcyclohexylamine), and subsequent metathesis with $ZrCl_4(THF)_2$ [9].

The polymerization reactions were performed in stainless steel 1 L reactor, Polyclave Buchi glass uster. Solvent (Toluene) and MAO were loaded in the reactor in a glove box and removed after sealed. The reactor then was perched with ethylene. After reaching the reaction temperatures and pressure, catalyst was injected into the reactor. The polymerization product was transferred to 1 L flask with 100 mL methanol (5 % HCl). Polymer was isolated by filtration and washed with acidic methanol and dried under vacuum [10].

^{13}C MAS NMR measurements of melted polymers were performed using Varian Unity Inova 300 NMR Spectrometer operated at 75 MHz frequency and 2.58 kHz spin rate. Powdered polyethylene samples were placed into 7.5 mm ceramic rotors and heated to 170 °C during measurements. ^{13}C spectra were recorded with acquisition time of 100.0 ms, 90° pulse angle, 5 s recycle delay and 2,070 scans per spectra. Branch content was determined by numerical peak integration of branching ^{13}C NMR signals.

X-ray powder diffraction patterns of polymers were recorded using a Philips PW3040/60 X'pert PRO Console diffractometer equipped with a focus graphite monochromator and two Soller slits, 1 mm divergence slit and a 0.2 receiving slit. The spectra were recorded using $CuK_{\alpha1}$ (1.5406 Å) incident radiation; the tube voltage was 45 mV and tube current was 40 mA. The scan diffraction angle varied from 5° to 100° 2θ angular range and the samples were scanned at a rate of 0.008°/min in 2θ; the collection time was 3 h. The spectra were recorded in air at 25 °C using sample holders with a 1 mm deep depression filled with polymer powders. The sample holder was rotated at a rate of 0.5 Hz to enhance scattering signal.

Results and discussions

The ethylene polymerization activity of the zirconium dihalide complexes was investigated using MAO as co-catalyst. The activity of imino- of bridged-zirconium catalysts was found to be 29.5×10^3 to 2.3×10^3 kg $mmol^{-1}$ h^{-1} Zr^{-1}, which is considerably higher than the activities seen from (alpha-dimines) $ZrCl_2$ mononuclear catalysts [11]. The molecular weight of the polyethylene product is also significantly larger ranging between 397,000 and 988,000 g/mol (Table 1).

Results of ^{13}C MAS NMR of a series of molten polymers performed at 170 °C are shown in the last column of the Table 1. The total fraction of branches and end groups of main chains was determined by calculating the ratio between the area of peak at around 14.2 ppm (corresponding to CH_3 group shown as **1** in the scheme below) to

Table 1 Polymerization conditions, molecular weight and fraction of branches and end groups in several polyethylene samples

Sample	Polymerization conditions	Molecular wt (g/mol)	Fraction of branches and end groups (%)
1	40 Bar–24 °C–10 MAO	610,000	~0.02
2	10 Bar–24 °C–20 MAO	397,000	0.22
3	10 Bar–60 °C–10 MAO	–	0.21
4	10 Bar–24 °C–10 MAO	397,000	0.47
5	30 Bar–24 °C–10 MAO	696,000	0.04
6	20 Bar–24 °C–10 MAO	959,000	~0.01
7	10 Bar–40 °C–10 MAO	–	0.08
8	10 Bar–24 °C–7 MAO	–	0.15
9	10 Bar–24 °C–15 MAO	517,000	0.35

the area of main carbon peak at 30 ppm, which corresponds to long-chain CH_2 groups (δ and further carbons in the scheme below). Peaks corresponding to carbons 2 and 3 (23 and 32.3 ppm, respectively) are visible in all spectra and their observed intensities are the same as those for terminal carbon labeled as 1.

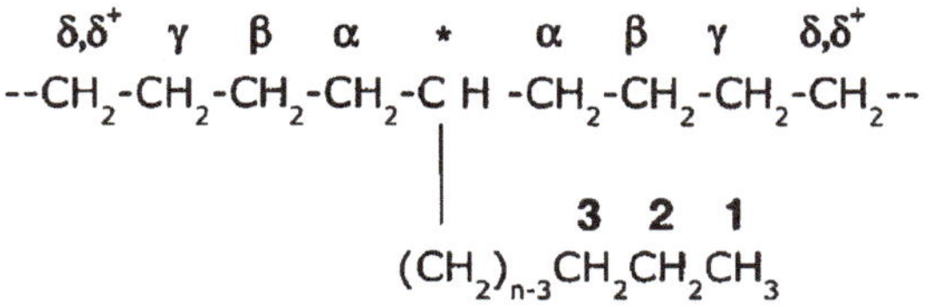

As seen from Table 1, the fraction of branches and end groups correlates well with molecular weight of polymers. Samples for tests 1, 5 and 6 have smallest fraction of branches and the largest molecular weight.

Typical results of X-ray diffraction measurements of polyethylene samples are presented in Fig. 1. All samples exhibit very similar diffraction patterns with sharp high-intensity peaks at 21.5° and 23.9° in 2θ and several lower intensity peaks in the range of 30°–55° and 74°–82° in 2θ. The spectra in the range of 90°–100° are not shown in Fig. 1 because they do not contain any well-defined diffraction peaks. For all the samples studied it was found that there are no appreciable changes in the position and the width of crystalline peaks.

The indexing of polyethylene XRD pattern is presented in Fig. 1 and agrees well with literature data [12]. All well defined peaks are indexed within the *Pnam* space group with the only exception of a weak shoulder at around 19° in 2θ which cannot be assigned to any *hkl* index within this space group. Some authors attribute the peaked feature of the maximum at about 19° in 2θ to the presence of a small amount of monoclinic polyethylene [13]. However, according to Fontana et al. [14] the monoclinic polyethylene phase can be formed only at high pressures (around 6 GPa) and the formation of this crystallographic modification during polyethylene synthesis (10–40 bar) is very unlikely.

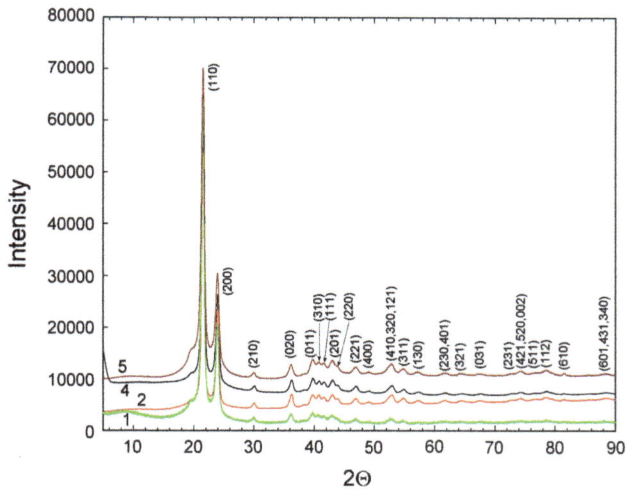

Fig. 1 XRD spectra of long-chain polyethylene samples 1, 2, 4 and 5 with all the diffraction peaks indexed within the *Pnam* space group

It should be noted that both crystalline orthorhombic polyethylene and amorphous polyethylene exhibit a diffraction peak at 19° in 2θ [15] and from our point of view, the weak shoulder at 19° corresponds to a broad peak of amorphous or semi-amorphous polyethylene phase.

Rietveld refinement of XRD spectra was performed using the structure refinement program GSAS [16], with the orthorhombic space group (*Pnam*) of crystalline polyethylene reported by Bunn [17] as the structural model. The peak shape was fitted with a convoluted pseudo-Voigt function to allow for broadening induced though finite crystallites. The preferred orientation of the films was not accounted for at this stage and it was assumed that the crystallite orientation in the polyethylene powders is completely random. The results of the Rietveld refinement of sample 5 are shown in Fig. 2 over the range of 5° to 100° in 2θ. Figure 3 displays refinement results over an enlarged area between 28° and 60° in 2θ. It is seen that there is a very good agreement between experimental and calculates intensities and positions of X-ray peaks with the exception of one peak at 43.5° in 2θ (Fig. 3).

Figures 4a and b depict the projections of packing of orthorhombic polyethylene chains along the *c* and *b* axes, respectively. The two planes containing the skeletal carbon chains within the unit cell are almost orthogonal to each other while all-*trans* chains are parallel, as seen in Fig. 4a. It has been shown that the setting angle of the molecular packing φ, i.e. the angle between the plane containing the ···C–C–C··· zigzag molecular chains and the *b* axis of the unit cell (see Fig. 3a), is in the range 41°–46° [5, 17]. Such differences may be attributed to the presence of an amorphous component that may significantly distort the diffraction profile in both peak position and intensity [18]. While the crystalline constituent gives sharp diffraction peaks, the amorphous phase produces three broad diffuse peaks at around 19°, 41° and 80° in 2Θ for CuK_α radiation.

The orthorhombic unit cell of branched polyethylene is expanded along axis *a* and, to a smaller degree, along axis

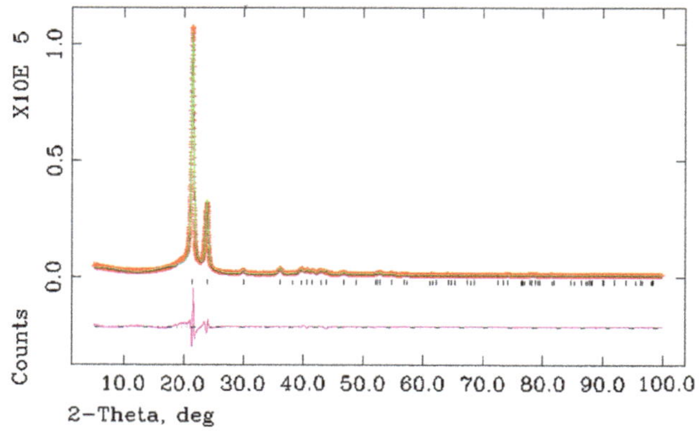

Fig. 2 Rietveld refinement of XRD spectrum of polyethylene sample 5. *Red crosses* experimental X-ray diffraction points, *green line* Rietveld refinement results, *pink line* difference line, *black ticks* calculated position of X-ray peaks

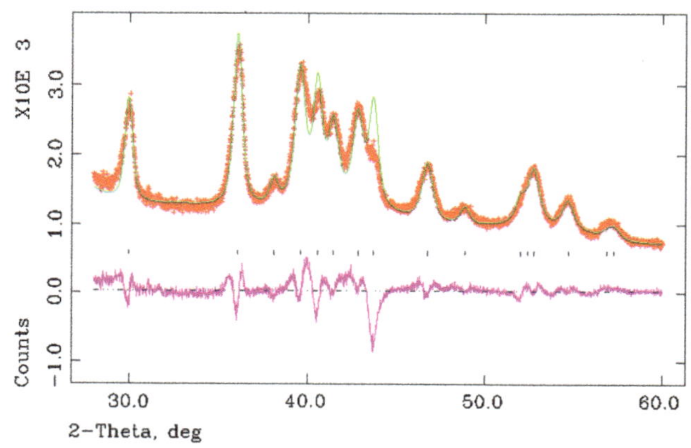

Fig. 3 Rietveld refinement of XRD spectrum of polyethylene sample 5. Enlarged area between 28° and 60° in 2θ is shown

b. At room temperature the unit cell parameters or linear polyethylene are $a = 0.74069$ nm, $b = 0.49491$ nm and $c = 0.25511$ nm [19]. The difference in *a* unit cell parameters might be due to a different degree of branching in polyethylene samples used by different authors.

With rising temperature the dimensions of the *a* and *b* axes change, while the *c* dimension does not change appreciably since it is determined by the strong intra-chain covalent C–C and C–H bonds unlike weak inter-chain van der Waals interactions which determine lattice parameters along the *a* and *b* axes. Due to a significant difference in the attractive bond forces between atoms parallel and perpendicular to the chain axis *c* the orthorhombic crystal phase of polyethylene, like most other crystalline polymers, has pronounced anisotropy of mechanical, optical and dielectric properties.

The simulated crystal structure of polyethylene sample 5 along and perpendicular to the *c* axis is presented in Figs. 5 and 6.

The calculated lattice parameters of sample 5 are as follows: $a = 0.74510 \pm 0.00003$ nm, $b = 0.49724 \pm 0.00002$ nm and $c = 0.25573 \pm 0.00004$ nm. The lattice parameters *a* and *b* are strongly affected by nonbonded interactions between the chains and, potentially, by the degree of branching. The *c* lattice parameter of polyethylene sample 5 which is mainly determined by the length of ···C–C–C··· zigzag chains is very close to the literature data for polyethylene samples [19] while *a* and *b* lattice parameters in sample 5 are significantly larger. On the basis of our results on ^{13}C MAS NMR investigation which shoved very small degree of branching ($\sim$0.01 %) in this sample it is unlikely that branching can explain larger values of both *a* and *b* lattice parameters in sample 5 comparing to those reported in the literature, for example $a = 0.74241(7)$ nm and $b = 0.49491(5)$ nm; $c = 0.25534(1)$ nm [13]. One of the possible reasons could be

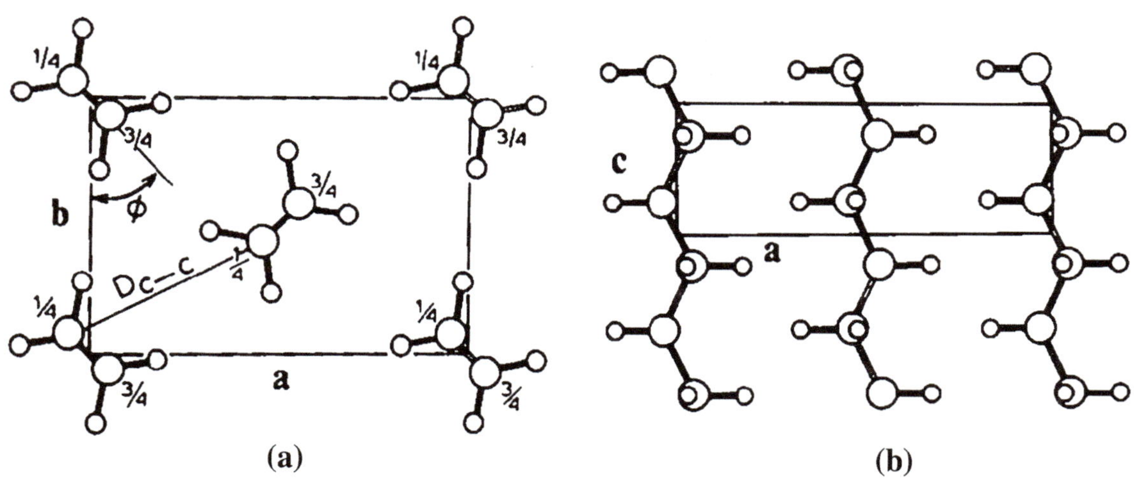

Fig. 4 Projection of polyethylene packing along the *c* axis **a** and along the *b* axis **b**. Polymer chain axes are parallel to axis *c*

Fig. 5 Simulated crystal structure of polyethylene sample 5 along the *c* axis

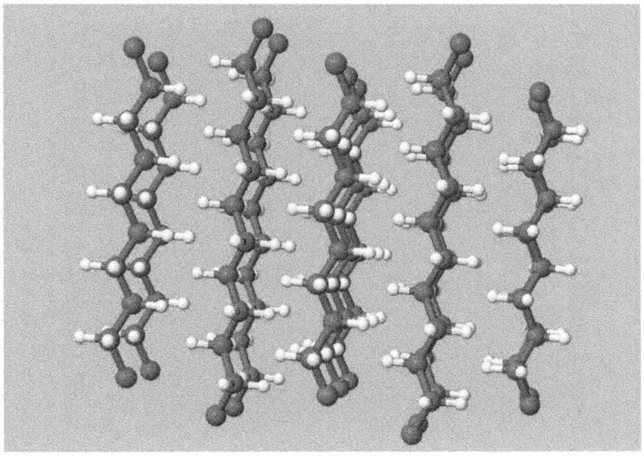

Fig. 6 Simulated crystal structure of polyethylene sample 5 perpendicular to axis *c*

the fact that literature X-ray diffraction data were obtained on re-melted polyethylene samples while we used as-synthesised polyethylene powder.

The calculated carbon–carbon interatomic distance in polyethylene sample 5 is 0.154061(1) nm and the C–C–C angle is 112.191(7)°. The calculated unit cell volume is 0.09474(7) nm^3 and calculated density of sample 5 is 0.9836 g/cm^3 which corresponds to a very high density polyethylene. Lattice parameters and densities of other samples in Table 1 are expected to be very similar to those of sample 5 as the positions of XRD diffraction peaks are almost identical.

Conclusion

In the polymerization of ethylene, the activity of the catalyst and the molecular weight of the polymer are highly dependent on the nature of the metal and the sizes of the group (R″) attached to the imino nitrogen atoms and the ortho constituent R. Developed Zr-based single site catalysts and optimized reaction condition enabled the synthesis of very high density polyethylene samples with very low branching content and high crystallinity.

Acknowledgments Authors would like to thank King Abdulaziz City for Science and Technology (KACST) for founding and supporting our collaborations with Oxford University. Solid-state NMR spectra were obtained at the EPSRC UK National Solid-state NMR Service at Durham.

References

1. Alsaygh A (2013) Distinct new catalysts for utilization of ethylene gas to polymer and oligomer. Appl Petrochem Res 3:101–105
2. Gibson VC, Spitzmesser SK (2003) Advances in non-metallocene olefin polymerization catalysis. Chem Rev 103:283–315
3. Crist B (1995) The ultimate strength and stiffness of polymers. Annu Rev Mater Sci 25:295–323
4. Young RJ, Lovell PA (1991) Introduction to polymers. Chapman, London
5. Bunn CW (1939) The crystal structure of long-chain normal paraffin hydrocarbons. Trans Faraday Soc 35:482–491
6. Hatfield GR, Killinge WE, Zeigler RC (1995) MeltlState ^{13}C MAS NMR determination of comonomer type and content in ethylene/α-Olefin copolymers. Anal Chem 67:3082–3085
7. Hay JN, Kemmish DJ, Langford JL, Rae AIM (1984) Polymer Commun 25:175
8. Ballirano P, Caminiti R, D'Ilario L, Martinelli A, Piozzi A, Maras A (1998) J Mater Sci 33:3519
9. Luo H-K, Schumann H (2005) New bi-nuclear and multi-nuclear α-diimine/nickel catalysts for ethylene polymerization. J Mol Catal A Chem 227(1–2):153–161 (And references therein)
10. Hsieh K-C, Chang C Jr, Lee M-T, Hu C-H, Hung C-H, Lee HM, Huang J-H, Wang M-H, Lee T-Y (2004) Hafnium chloride and hafnium methyl complexes bearing substituted pyrrolyl ligands: synthesis, characterization, and ethylene polymerization. Inorg Chim Acta 357:3517–3524
11. Britovsek GJP, Baugh SPD, Hoarau O, Gibson VC, Wass DF, White AJP, Williams DJ (2003) The role of bulky substituents in the polymerization of ethylene using late transition metal catalysts: a comparative study of nickel and iron catalyst systems. Inorg Chim Acta 345:279–291
12. Laridjani M, Leboucher P (2009) The structural dilemma of bulk polyethylene: an intermediary structure. PLoS ONE 4:e6228
13. Caminiti R, Pandolfi L, Ballirano P (2000) Structure of polyethylene from x-ray powder diffraction: influence of the amorphous fraction on data analysis. J Macromol Sci Phys B39:481–492
14. Fontana L, Vinh DQ, Santoro M, Scandolo S, Gorelli FA, Bini R, Hanfland M (2007) High-pressure crystalline polyethylene studied by x-ray diffraction and ab initio simulations. Phys Rev B 75:174112

15. VanderHart DL, Khoury F (1984) Polymer 25:1589
16. Larson AC, Von Dreele RB (2004) General structure analysis system (GSAS). Los Alamos National Laboratory Report LAUR, pp 86–748
17. Dorset DL, Moss B (1983) Polymer 24:291–294
18. Schwartz KB, Von Dreele RB (1997) Adv X-Ray Anal 39:515
19. Clark ES (2007) Chapter 38. Physical properties of polymers handbook. In: Mark JE (ed) Unit cell information on some important polymers, 2nd edn. Springer, New York

15

The effect of lanthanum addition on the catalytic activity of γ-alumina supported bimetallic Co–Mo carbides for dry methane reforming

Liam J. France · Xian Du · Naif Almuqati · Vladimir L. Kuznetsov · Yongxiang Zhao · Jiang Zheng · Tiancun Xiao · Abdulaziz Bagabas · Hamid Almegren · Peter P. Edwards

Abstract The effect of lanthanum addition to γ-alumina supported bimetallic carbides has been studied for the reaction of dry methane reforming using four different lanthanum loading levels of 1, 5, 10 and 15 wt% of lanthanum. It has been demonstrated that the addition of lanthanum to supported bimetallic carbides at low loading levels (1 wt%) results in smaller carbide crystallite sizes compared to catalysts containing either no lanthanum or higher lanthanum loading levels (5–15 wt%). Increased lanthanum loading results in increased carbon dioxide desorption at 500–700 °C. Reactions indicated that increased lanthanum loading resulted in significantly reduced product yields due to increased reverse water–gas shift activity. All materials exhibited degrees of sintering during the reaction. It was found that cobalt reacted with lanthanum species to form a $LaCoO_3$ phase. The 1 wt% catalyst possessed superior catalytic properties for dry methane reforming and was tested for 100 h. After an initial loss of activity, the catalyst appeared to stabilise, however, a decrease of ~3 % in the H_2:CO ratio, evidence of carbide crystallite growth and carbon deposition, indicated that a shift in the side reactions had occurred during the reaction.

L. J. France · X. Du · N. Almuqati · V. L. Kuznetsov ·
J. Zheng · T. Xiao · P. P. Edwards (✉)
Department of Chemistry, University of Oxford, South Parks Rd, Oxford OX1 3QR, UK
e-mail: Peter.Edwards@chem.ox.ac.uk

L. J. France
e-mail: liam.france@chem.ox.ac.uk

Y. Zhao
School of Chemistry and Chemical Engineering, Shanxi University, Taiyuan 030006, China

A. Bagabas · H. Almegren
Petrochemical Research Institute, King Abdulaziz City of Science and Technology (KASCT), King Abdullah Rd, Riyadh 11442, Saudi Arabia

Keywords Dry methane reforming · Bimetallic carbide · Lanthanum · Carbon dioxide

Introduction

Due to our current reliance on petrochemical feed stocks, carbon dioxide emissions will continue to soar while we are employing combustion-based processes. From the environmental perspective, the application of carbon capture and storage technologies would effectively negate carbon dioxide emissions, resulting in a stabilisation of manmade contributions. Although such technologies are more than viable for stationary sources of carbon dioxide emissions, there are still no viable routes for incorporation into vehicles equipped with combustion engines. However, storage of carbon dioxide is a waste of a free carbon-based resource, when it may be converted into fuels or their precursors. This could help to provide a solution to two major current issues; global warming and the imminent energy crisis.

One potential route that is currently being examined by the start-up company NewCO$_2$Fuel is the application of dry methane reforming (DMR) employing renewable energy sources for generating synthesis gas (syngas) [28]. Prior to this, DMR has only found limited industrial application. Most notably in the form of the SPARG and CALCOR processes [11]. The SPARG process combines methane combustion (MC), partial oxidation of methane (POM), steam reforming (SR) and DMR. By implementing such a process, MC not only yields the oxygenated feed streams used in both SR and DMR, but also supplies energy to the endothermic reactions resulting in lower external energy requirements. POM, SR and DMR produce syngas as the product [32]. However, commercially in SR, carbon monoxide is further reacted in the presence of water to yield hydrogen and carbon dioxide via the water gas shift reaction (WGS). SR and DMR are described by Eqs. 1 and 2, respectively.

Steam reforming

$$H_2O + CH_4 \rightarrow 3\,H_2 + CO \quad +206\ \text{kJ mol}^{-1} \tag{1}$$

Dry methane reforming

$$CO_2 + CH_4 \rightarrow 2\,H_2 + 2\,CO \quad +247\ \text{kJ mol}^{-1} \tag{2}$$

Both of these reforming processes are highly endothermic requiring significant external energy input to proceed. While they are similar in the respect that they are both endothermic, they could not be more different chemically. SR utilises water, which results in significant hydrogen yield giving a syngas ratio of 3:1. DMR utilises carbon dioxide, which results in a more endothermic process, however, it reforms two major greenhouse gases and yields (theoretically) a 1:1 syngas composition, which is known to be particularly useful for the synthesis of hydrocarbons [18].

Subsequently, much research has been conducted on the DMR process, in particular with cobalt and nickel, which are well known for their significant C–H bond activation ability [4]. While this is a mature reaction, many "new" classes of catalysts have been investigated in recent years to examine their performance in this promising area. In particular, group VI carbides have shown a particular affinity towards such processes [2]. This has been attributed to the hybridised electron configuration of the metal, which resembles that of the noble metals and has been shown previously to possess some properties similar to those exhibited by platinum [22].

In particular, monometallic molybdenum carbide, Mo_2C, has been tested as a catalyst for a number of reactions, including SMR [7], DMR [8, 9, 21, 26], POM [36] and hydrotreating [37]. The use of supported metal carbides was also considered in an attempt to further improve catalytic activity and stability of these materials. Studies undertaken by Brungs et al. [2] examined the use of a number of supports (Al_2O_3, SiO_2, TiO_2 and ZrO_2) for molybdenum carbide (Mo_2C) catalysts. The authors were able to determine a stability order for the supported materials which showed that alumina supported monometallic carbides exhibited the highest catalytic stability in DMR.

It has also been reported that the addition of a second metal, e.g. Co, has a significant effect on the catalytic activity of Mo carbides [1, 37]. Xiao et al. produced a series of CoMo bimetallic carbides with different Co:Mo ratios. It was observed that at low Co:Mo ratios, a homogeneous bimetallic carbide phase was formed. However, materials with higher Co:Mo ratios, e.g. $Co_{0.4}Mo_{0.6}C_x$ and $Co_{0.5}Mo_{0.5}C_x$, exhibited a number of impurity phases, primarily Co and Mo_2C [6, 37]. A small number of studies have been conducted on the use of supported bimetallic carbide systems for DMR, employing a rather diverse number of supports such as γ-alumina [6, 10, 19, 30], activated carbon [23, 27], MCM-41 [35] and SBA-15 [14].

Additive studies were reported by Darujati et al. specifically the use of ceria was explored as an additive to γ-alumina supported Mo_2C. It was found that although no obvious change in activation energy was observed for Mo_2C as a bulk or supported catalyst, the ceria-promoted material exhibited a higher activity, which was attributed to improved dispersion of the catalyst due to the support. However, ceria was found to act as an active site for strong CO_2 adsorption, which lowered the reaction order of CO_2 to zero. A redox cycle was also proposed which accounted for CO_2 and CO interactions with the Ce^{3+} and Ce^{4+} oxides [9]. It was determined that ceria had to be impregnated into the support and calcined prior to molybdenum impregnation. Additionally, Darujati et al. examined additives such as K and Zr. It was found that K significantly decreased surface molybdenum concentrations resulting in reduced activity and rapid deactivation. Zr was found to have no obvious effect on either methane conversion or deactivation rates [8].

Lanthanum-like cerium and their oxides are well known as catalyst additives and their use is well documented. In zeolitic materials, both are known to possess similar properties in that they enhance acid site density in conjunction with mediating carbon lay down during fluidised catalytic cracking [29] and [20]. Additionally, their application as a phase stabiliser for transitional support materials such as titania and alumina has been documented [12, 13]. It has been reported previously that simple perovskites of the type $LaMO_3$ (M = Ni/Co/Fe) have been reduced and then used as a catalyst in the DMR reaction. It was found that lanthanum oxide adsorbed carbon dioxide, which readily converted carbon generated by methane cracking (MCr) into carbon monoxide [33]. While the use of ceria as an additive for supported molybdenum carbide has been reported previously the use of the more abundant lanthanum and its oxides has not to the best of our knowledge been investigated previously on either monometallic or bimetallic carbides.

The present work examines the effect of La addition to γ-alumina supported bimetallic carbides in DMR, with particular emphasis on the stability of the catalysts during reactions in the presence of excess carbon dioxide at high reaction temperatures. Analysis of the catalysts prior to reaction was undertaken to understand the effects that the presence of La may have had on the formation of the carbide. Additionally, analysis of post-reaction catalysts were conducted to examine changes of the active phase during reaction and to assess the stability of the carbides in the presence of lanthanum during dry methane reforming.

Experimental

Catalyst preparation

γ-alumina high surface area catalyst support (Alfa Aesar, 250 $m^2 g^{-1}$) was ground and sieved to a particle size of 125–250 μm. Six different supports were generated in 5 g batches, these contained 0, 1, 5, 10 and 15 wt% La. These were produced by adding together lanthanum (III) nitrate hexahydrate (Sigma-Aldrich, Fluka ≥99.0 %) and citric acid (Alfa Aesar, Anhydrous ACS, 99.5 + %) in a 1.0:1.5 molar ratio, respectively. The resultant mixture was added in a required proportion to sieved alumina and mixed with manual grinding for 30 min prior to adding water and mixing to a stiff paste. For the purpose of comparison, the 0 wt% La sample also underwent mixing and adding of water. The supports were dried at 110 °C for 3 h and then heated in air at 10 °C min^{-1} to 850 °C and kept at this temperature for 3 h. Once cooled, the supports were ground and sieved to a particle size of 125–250 μm.

3 g of each support was added to 50 ml of a solution containing ammonium heptamolybdate tetrahydrate (Sigma-Aldrich, ACS, ≥99.99 %) cobalt (II) nitrate hexahydrate (Sigma-Aldrich, ACS, ≥98 %), nitric acid (Sigma-Aldrich, 70 %) and an excess of citric acid (Alfa Aesar, Anhydrous ACS, 99.5 + %) and distilled water. The mixture was stirred vigorously for 24 h and then heated at 120 °C for 2 h to remove excess water. The resultant solids were transferred to alumina crucibles, heated at 5 °C min^{-1} to 600 °C and kept at this temperature for 10 h. Once cooled, the precursor solids were removed and sieved to <250 μm. The target loading level of the oxide precursor was 40 wt% (~30 wt% carbide) in all instances with a Co:Mo molar ratio of 0.67.

The precursors (~2.5 g) were then reduced and carburised under a continuous flow of methane (BOC, 99.9 %) and hydrogen (BOC, 99.9 %) with a ratio of 20:80 volume %, respectively, and a flow rate of 40 ml min^{-1}, while the temperature was increased at 3 °C min^{-1} to 750 °C and held for 3 h. The samples were then allowed to cool to room temperature at 3 °C min^{-1} under argon (BOC 99.9 %). All catalysts were left in the quartz tube at room temperature for 24 h exposed to air prior to removal, subsequent characterisation and catalytic tests.

Catalytic tests

Catalytic reactions were undertaken in a stainless steel fixed bed plug flow micro reactor (Fig. 1) at atmospheric pressure. To reduce dead volume attributed to the reactor tube, carborundum (Fisher, General purpose grade, 24 mesh) was loaded above and below the catalyst bed separated with quartz wool wads (Fisher) to prevent the mixing of catalyst (0.5000 ± 0.0005 g) with the carborundum. The catalyst was heated at 5 °C min^{-1} to 850 °C under 40 ml min^{-1} flow of dry nitrogen (BOC 99.999 %). Once the temperature had stabilised at 850 °C, the gas supply was switched to a mixture (Air products special gas mixture, 99.99 %) containing methane, carbon dioxide and nitrogen (33, 39 and 28 %, respectively) at 40 ml min^{-1}, giving a weighted GHSV of 4,800 ml hr^{-1} g^{-1}. Reactions were performed initially for 5 h to establish the most active catalyst, which was then tested for 100 h. Gas analysis was undertaken every 30 min for the former reactions and every hour for the latter. The composition of the reactor effluent was monitored for CH_4, CO_2, N_2, H_2 and CO using a Shimadzu GC 2014 Gas Chromatograph equipped with TCD and FID detectors employing a column-switching technique between a 1 m mole sieve 13× 60–80 mesh packed column and a 3 m Hayesep N 60–80 mesh packed column; argon (BOC 99.999 %) was used as the carrier gas. Post-reaction, the spent catalyst was cooled to room temperature under 40 ml min^{-1} of dry nitrogen (BOC, 99.999 %) overnight and then exposed to air prior to unloading.

Conversion of component x was defined as:

$$1-\left(\frac{[X_{\text{oulet}}]\times[\text{IS}_{\text{inlet}}]}{[X_{\text{inlet}}]\times[\text{IS}_{\text{outlet}}]}\right)\times 100 \tag{3}$$

where X is the concentration of either CH_4 or CO_2 in the feed, IS is the concentration of the internal standard (N_2).

The hydrogen and carbon monoxide yields were defined as:

$$\frac{H_{2\,\text{outlet}}}{2CH_{4\,\text{inlet}}} \tag{4}$$

and

$$\frac{CO_{\text{outlet}}}{2CO_{2\,\text{inlet}}} \tag{5}$$

The product ratio was defined as:

$$\frac{H_{2\,\text{outlet}}}{CO_{\text{outlet}}} \tag{6}$$

where all yields and the product ratio were calculated from N_2 corrected flow rates at STP (ml min^{-1}).

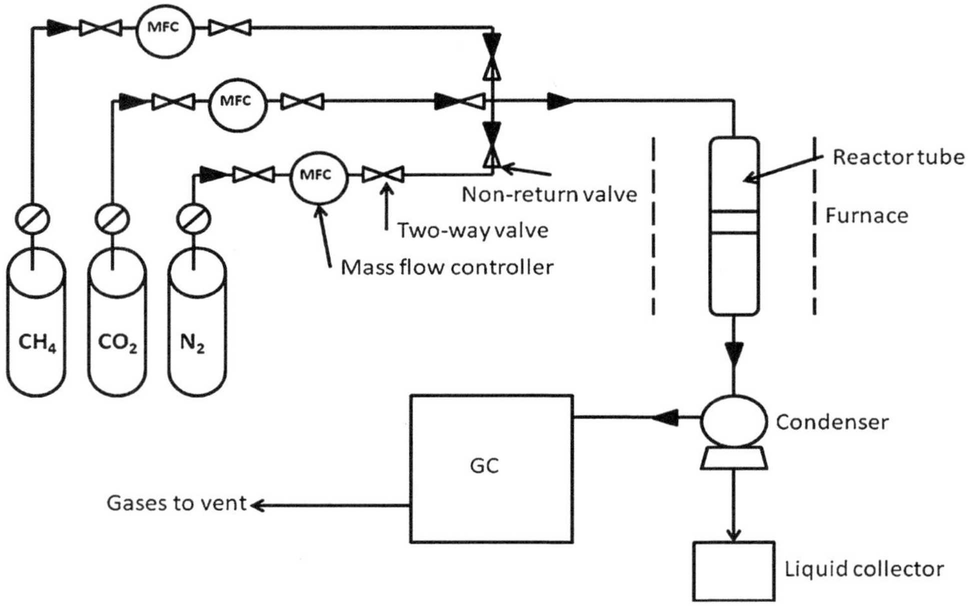

Fig. 1 A schematic representation of the dry methane reforming fixed bed plug flow reactor

The effect of the reactor tube, carborundum and quartz wool has been accounted for by undertaking a blank experiment. In all instances, this conversion amounted to <1 % of that demonstrated by the reactor tube loaded with catalysts. All 5 h data sets presented below represent an average of two experiments with a maximum difference between experiments of ±3 % for CO_2 conversion and CO yield data and ±2 % for CH_4 conversion and H_2 yield data. All experiments were carbon balanced and found to be close in the range 96–103 %.

X-ray diffraction (XRD)

Measurements were undertaken on a PANalytical X'Pert Pro diffractometer, utilising $CuK_{\alpha 1}$ radiation; the X-ray tube was operated at 40 kV and 45 mA. The scanning range was 5°–85° in 2θ, with a step size of 0.0020° and a scan speed of 0.0056° s^{-1}. Crystallite sizes of the carbidic phase were calculated by employing the Scherer equation and are only presented for comparative purposes.

Thermogravimetric analysis (TGA)

TGA data were collected on a Perkin Elmer STA6000. 4–7 mg of material was heated in an alumina crucible under a 30 ml min^{-1} flow of air from 50 to 950 °C at a ramp rate of 10 °C min^{-1}.

Nitrogen physisorption (BET method)

N_2 adsorption–desorption isotherms were measured at −196 °C using a Micromeritics ASAP2020 specific surface area and porosity analyser. The specific surface area was determined from the N_2 adsorption isotherms by application of the BET equation. Prior to analysis, the samples were degassed under vacuum at 150 °C for 5 h.

Carbon, hydrogen and nitrogen analysis (CHN)

The total weight percentage of carbon, hydrogen and nitrogen was determined for fresh and spent catalysts. This was performed via combustion analysis using an Exeter Analytical CE-440 analyser. Values given within are an average of three measurements and correspond to a maximum difference of 3 %.

Results and discussion

Synthesised catalyst characterisation

Synthesised catalysts were analysed via a variety of techniques to determine composition prior to and after reaction. Table 1 illustrates the carbon contents both prior to and post-reaction; initially it can be observed that all samples contained a significant fraction of carbon, catalysts containing 0, 1, 5, 10 and 15 wt% La had 10.5–12.3 wt% of carbon. In comparison, the bulk Mo_2C had ~6 wt% carbon present, hence, this was viewed as a large excess which may have resulted from longer than necessary carburisation times and come about through the methane decomposition reaction as the carbide phase formed. The bulk composition of the bimetallic carbide is predicted as being $Co_xMo_yC_z$ where $x = 2.29$, $y = 3.43$ and $z = 1$. Which equates to 8.48 wt% Co, 20.75 wt% Mo, 0.77 wt% C 0.70–10.50 wt% La and 69.30–59.50 wt% Al_2O_3. It was also

Table 1 Catalyst characteristics of the La containing supported bimetallic carbide catalysts prior to and post-reaction

Lanthanum doping level (wt%)	Carbon content prior to reaction (wt%)	BET surface area ($m^2\ g^{-1}$)	Carbide crystallite size prior to reaction (nm)	Carbon content post-reaction (wt%)	Carbide crystallite size post-reaction (nm)
0	11.28	85	12	0.73	23
1	10.64	87	5	0.67 (5 h) 5.15 (100 h)	20 (5 h) 52 (100 h)
5	12.32	84	11	0.69	20
10	10.50	78	11	0.65	30
15	12.25	66	13	0.75	20

found that in general, the surface area of the supported catalysts decreased with increased lanthanum content. This may have been an effect caused by the increase in lanthanum content and a decrease in alumina content, resulting in similar levels of coke across supports of reduced surface area and porosity.

XRD patterns of the supports (Fig. 2a) exhibited a small degree of variation in peak intensity which may be attributed to the dilution effect of the lanthanum additive. All catalysts exhibited a number of phases present (Fig. 2b). Unsurprisingly, the most dominant phase was that of carbide. It has been reported in a number of studies that the bimetallic carbide exhibits diffraction peaks shifted to higher diffraction angles compared to Mo_2C and were similar to those observed for the β-molybdenum carbide [35, 36]. The γ-alumina phase was also observed, in conjunction with amorphous carbon and Co. At the Co:Mo ratio employed (0.67), the carbide structure is known to undergo a degree of phase separation during synthesis which results in a mixture of mono and bimetallic carbides as well as metallic Co [6]. It can also be observed that when small quantities of lanthanum (1 wt%) were added, the growth of carbidic species was suppressed in comparison to that observed for the undoped sample, as seen from the intensity of the carbide phase peaks (Fig. 2b). As lanthanum addition increased this effect appeared to decrease, where at 10–15 wt% of lanthanum a significant increase in carbide diffraction peak intensity was observed. This is in line with the observations of Martinez et al. where they observed that catalysts possessing large amounts of lanthanum oxide addition resulted in reduced catalytic activity and stability. From this, they concluded that while lower loading levels of lanthanum oxide resulted in good dispersion and hence an improvement in the active metal dispersion on the catalyst. High loadings resulted in a reversal of this effect, where active metal dispersion was

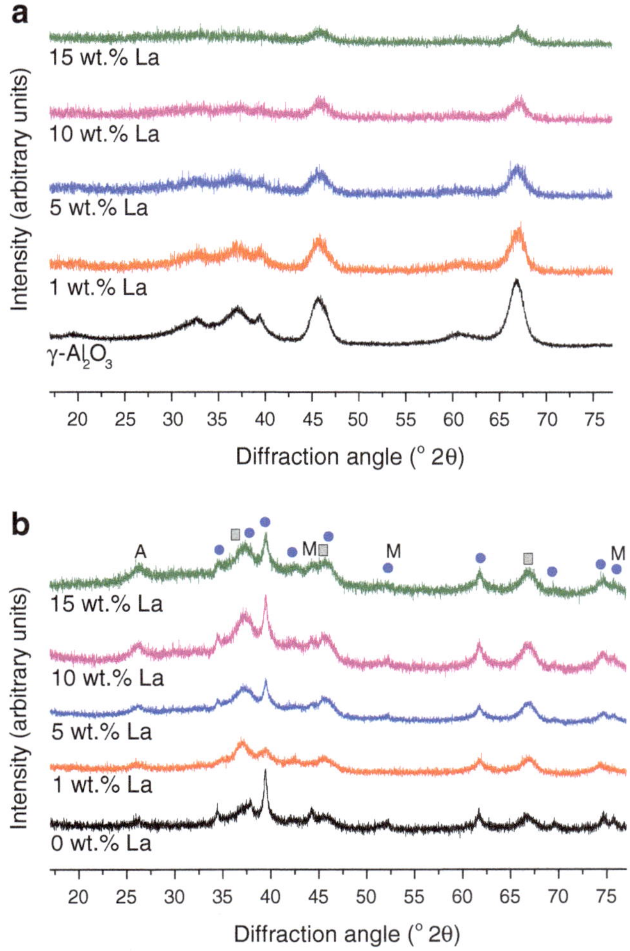

Fig. 2 **a** XRD patterns of supports containing La additive. **b** XRD patterns of the La-containing alumina supported bimetallic carbide (40 wt%) catalysts prior to reaction. *A* carbon, *filled circle* carbide, *filled square* γ-alumina and *M* cobalt

found to be worse than that employed on the catalyst with no lanthanum oxide addition [25]. Interestingly, in our study lanthanum oxide diffraction peaks were not observed on any of the supported samples even up to 15 wt% of lanthanum, which suggested that lanthanum derived species were highly dispersed on the support. This observation is in good agreement with work previously undertaken by [13] who reported that a La:Al ratio of 0.15 (~9.3 wt% of La) resulted in high dispersion of the La phase as La^{3+}. These dispersion effects were observed to be present at higher La:Al ratios (0.25–0.50), however, it was found that excess lanthanum species may have become partially incorporated in the alumina framework giving rise to species of the type $LaAlO_3$ resulting in reduced support surface area.

TGA was used to examine the reaction of the supported bimetallic carbides under an oxidising atmosphere. It can be seen in Fig. 3 that all catalysts exhibited a series of mass changes over the range 200–650 °C. Monometallic carbides

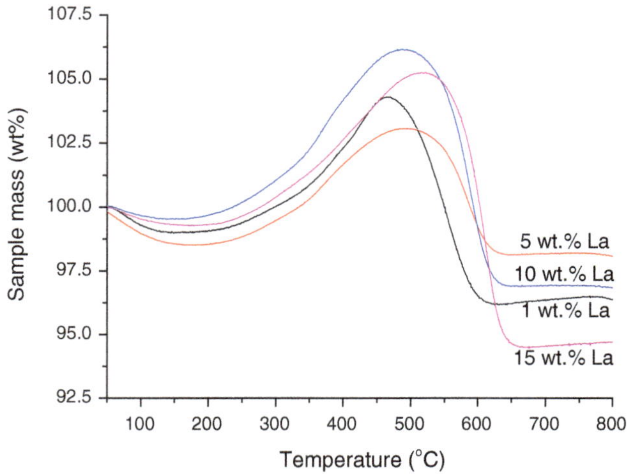

Fig. 3 TGA of the La-containing alumina supported bimetallic carbide catalysts prior to reaction

are known to be highly reactive in the presence of oxygen and readily undergo a transformation to their respective oxide phases [2]. However, in this instance, this does not appear to be a rapid transformation for any of the catalysts. In all instances, these events occur over a prolonged period from 200 °C until a maximum mass increase was observed between 475 °C (1 wt% La) and 550 °C (15 wt% la). Presumably, this is due to the formation of a protective oxide overlayer on the carbidic species, inhibiting rapid oxidation [30]. Additionally cobalt, the second metal employed in these systems, is known to undergo a convoluted oxidation process over a wide temperature range, which may also significantly contribute to the elongation of the oxidative events [34]. After this process, a significant mass loss is observed which coincides with the removal of carbonaceous species deposited on the catalyst during the carburisation process, these temperatures are similar to those observed by Jiang et al. [16] over spent nickel catalyst supported on aerogels. Interestingly, the increased oxidation temperatures coincided with increased lanthanum content. This could suggest that increased La enhanced the carbidic and metallic species oxidative resistance. This effect appears to have affected the combustion process associated with deposited carbon species. It may be inferred that the combustion of the deposited carbon is a catalytic process and that the observed differences in temperature are associated with the formation of active metal oxide species [24].

Catalytic tests

To further examine the effects that lanthanum addition has on the bimetallic carbide, reactions were undertaken employing an excess of carbon dioxide at 850 °C for a duration of 300 min. It can be observed in Fig. 4a that all lanthanum dispersed catalysts exhibited a methane conversion of above 97.5 % which was well in line with calculated equilibrium values (>99 %) [3]. However, for the undoped sample, a lower conversion indicates that lanthanum may have acted as a promoter for methane decomposition. It was demonstrated by Jiang et al. [17] that the addition of lanthanum oxide to nickel-containing catalysts resulted in enhanced methane conversion, especially at higher temperatures. It was apparent that after 300 min very little change was observed for the methane conversion. This suggested that conversion of methane occurred readily over all materials regardless of carbon dioxide conversion. To further verify whether this was due to dry methane reforming, methane decomposition or a combination of both, the carbon dioxide conversion was analysed to establish the presence of such potential side reactions.

The samples containing 1 and 10 wt% of lanthanum exhibited similar CO_2 conversions comparable to those observed for the undoped sample (Fig. 4b). Interestingly, this runs in parallel to the carbon content of each catalyst and more importantly to amount of carbon remaining post-reaction (Table 1). We may assume that due to the carbon balances observed, the surface carbon prior to reaction is removed before sampling of the product gases occurs at 30 min. Additionally, it is proposed that this occurs in part due to the larger amount of CO_2 employed in this process in comparison to methane. The catalyst with 1 wt% La exhibited a higher degree of stability during the 5 h reaction, whereas the undoped and the 10 wt% La catalysts exhibited a small decrease in conversion during the 300 min reactions. Both the 5 and 15 wt% La exhibited very similar initial carbon dioxide conversion values at ~94 %, however, the 15 wt% La deactivated during the course of the reaction to ~89 %. A slower rate of deactivation was observed for the 5 wt% La catalyst which displayed only a decrease of 3 %.

The yields generated by each catalyst were analysed to assess the effects that lanthanum addition had on side reactions commonly found in dry methane reforming, most notably hydrogen-consuming reactions, for example reverse water–gas shift (RWGS). It can be observed in Fig. 4c, d that the sample containing 1 wt% La exhibited the highest yield of hydrogen (~96 %) and carbon monoxide (~82 %). While higher loadings show an effect which favours significant side reactions that result in reduced yields of hydrogen and carbon monoxide. The final assessment of the catalysts was the analysis of the product ratios (Fig. 4e), which in a near ideal scenario (Eq. 2) would be close to 1:1. However, this is normally difficult to achieve due to the ongoing side processes, for example MCr and RWGS. The materials that resulted in ratios closest to the optimal were the 0 and 1 wt% La-doped catalysts, which produced a fairly stable ratio of approximately 0.98–1.02 (based upon duplicate data sets). All of

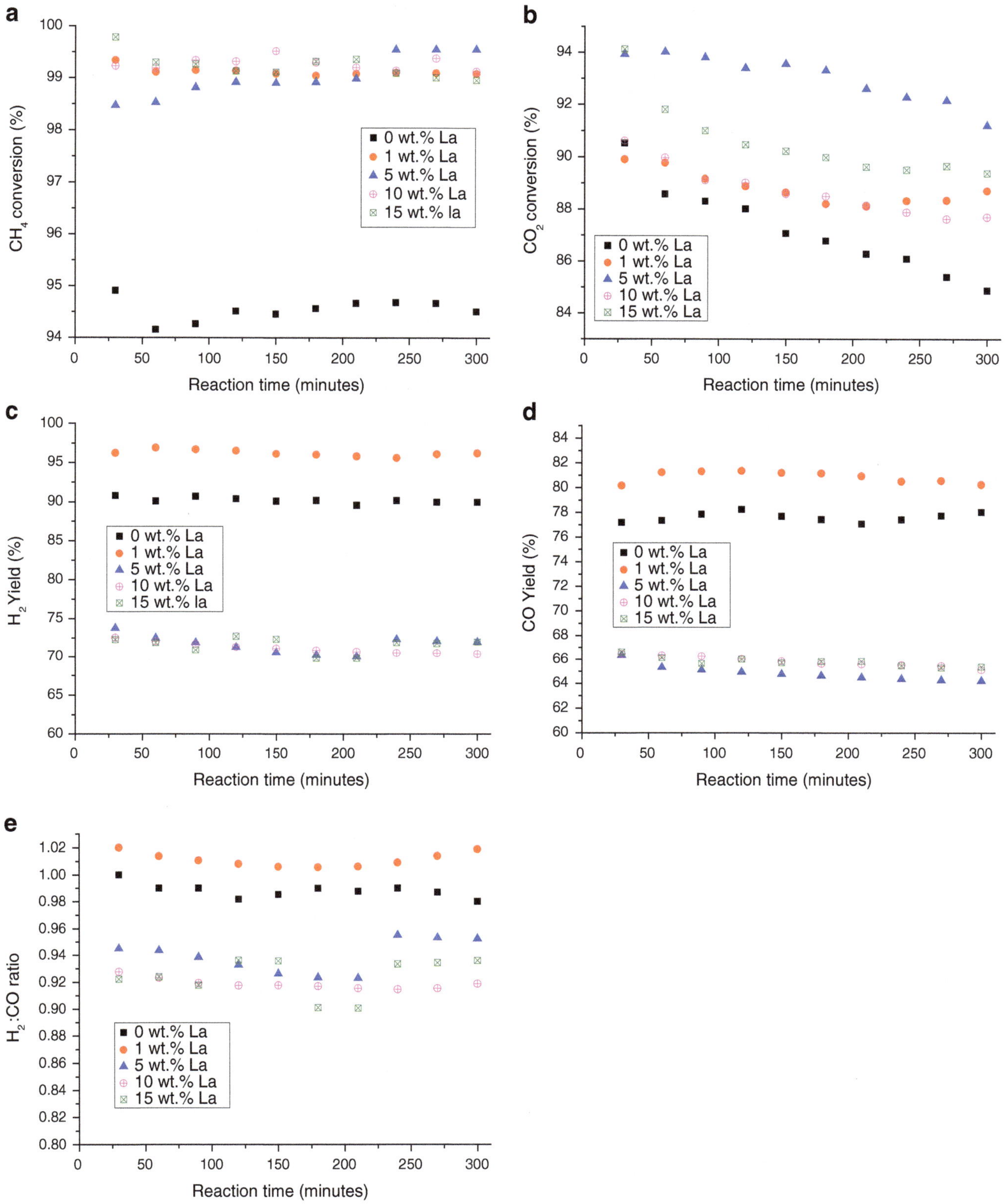

Fig. 4 Dry methane reforming reaction data for La containing supported bimetallic carbides. **a** methane conversion, **b** carbon dioxide conversion, **c** hydrogen yield, **d** carbon monoxide yield and **e** H_2:CO ratio

the other H_2/CO ratios were in the range of 0.90–0.96. It is evident at higher loadings of lanthanum that carbon monoxide is the favoured product.

These observed differences in reactivity may stem from the presence of new diffraction peaks observed in post-reaction XRD patterns of spent catalysts (Fig. 5). It was

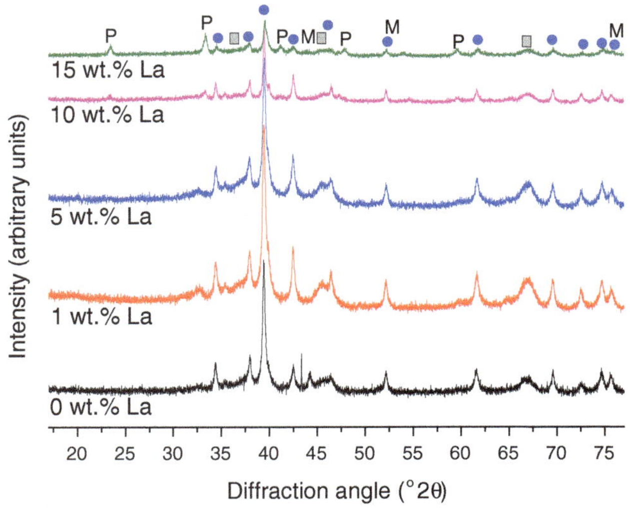

Fig. 5 Post-reaction XRD patterns of the La-containing alumina supported bimetallic carbide catalysts. *P* $LaCoO_3$, *filled circle* carbide, *filled square* γ-alumina and *M* = cobalt

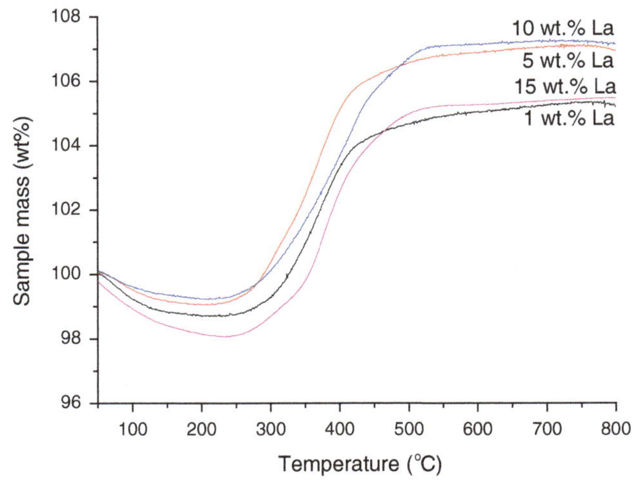

Fig. 6 Post-reaction TGA of the La-containing alumina supported bimetallic carbide catalysts

found that a new phase was formed during the reaction and matched the perovskite $LaCoO_3$. Recently, Chawla et al. [5] demonstrated that lanthanum cobaltite possesses a degree of catalytic activity towards dry methane reforming, they also found that the working catalyst exhibited a high degree of RWGS activity as indicated by a H_2:CO ratio lower than 1. This may partially explain why all loading levels above 1 wt% La exhibit a high degree of RWGS activity.

Post-reaction characterisation

Catalysts were cooled to room temperature under flowing nitrogen and once cooled, allowed to passivate in air prior to removal from the reactor. Elemental analysis (Table 1) was conducted on the post-reaction catalysts and in all instances, the carbon content of the material had decreased significantly, indicating removal of the excess amorphous carbon via direct reaction with carbon dioxide (reverse Boudouard (RB) reaction $\Delta H^{o}_{298} = 172$ kJ mol^{-1}). Similar levels of residual carbon were observed over the spent catalysts, which in conjunction with post-reaction XRD patterns (Fig. 5) indicated the presence of carbide phases in post-reaction catalysts. It should also be noted that although not presented in Table 1, all post-reaction samples displayed a very small quantity of hydrogen (0.03 wt%), which may be in the form of adsorbed hydrocarbons [31]. XRD patterns also indicated that the carbide phase peak intensity had increased substantially on catalysts that contained lower lanthanum content. Previously, lanthanum has been demonstrated to act as a dispersion agent on alumina providing improved dispersion of supported phases [13]. Hence, loss of lanthanum through the formation of a perovskite phase [15] would reduce the amount of lanthanum and alter this dispersion effect which may have resulted in the significant sintering during reaction (Table 1).

TGA thermographs (Fig. 6) exhibit a number of phenomena in post-reaction catalysts. Initially, a small mass loss was observed at a similar level to that found prior to reaction, suggesting that this may be associated with either water loss and/or buoyancy as a result of the up flow system. After this initial loss, an increase in mass was observed indicating the presence of carbidic and metallic species being present over the spent catalyst which oxidised during TGA experiments. In all instances, these mass increases were found to begin within the range 200–225 °C which was similar to those observed prior to reaction. At maximum oxidation, it became apparent that there were differences in the degree of oxidation of the materials. Additionally, the remaining residual carbon observed via elemental analysis did not appear to be associated with graphitic coke. Based upon the findings from XRD, the carbidic phases still appear to be the major components on the support. This is further verified by the theoretical carbon content of the carbides. Bulk bimetallic carbides synthesised by Xiao et al. [37] exhibited a carbon content of ~2.56 wt% at the same ratio employed within this study. The approximate loading level of the supported carbides in this instance is 30 wt%. Based upon this, the theoretical carbon content of the catalyst should be 0.77 wt%. Hence, it is observed that post-reaction carbon contents of all materials are similar to those expected theoretically.

100 hour catalytic test

Based upon the observations made during the 5 h test reactions (low activity towards side reactions), the 1 wt%

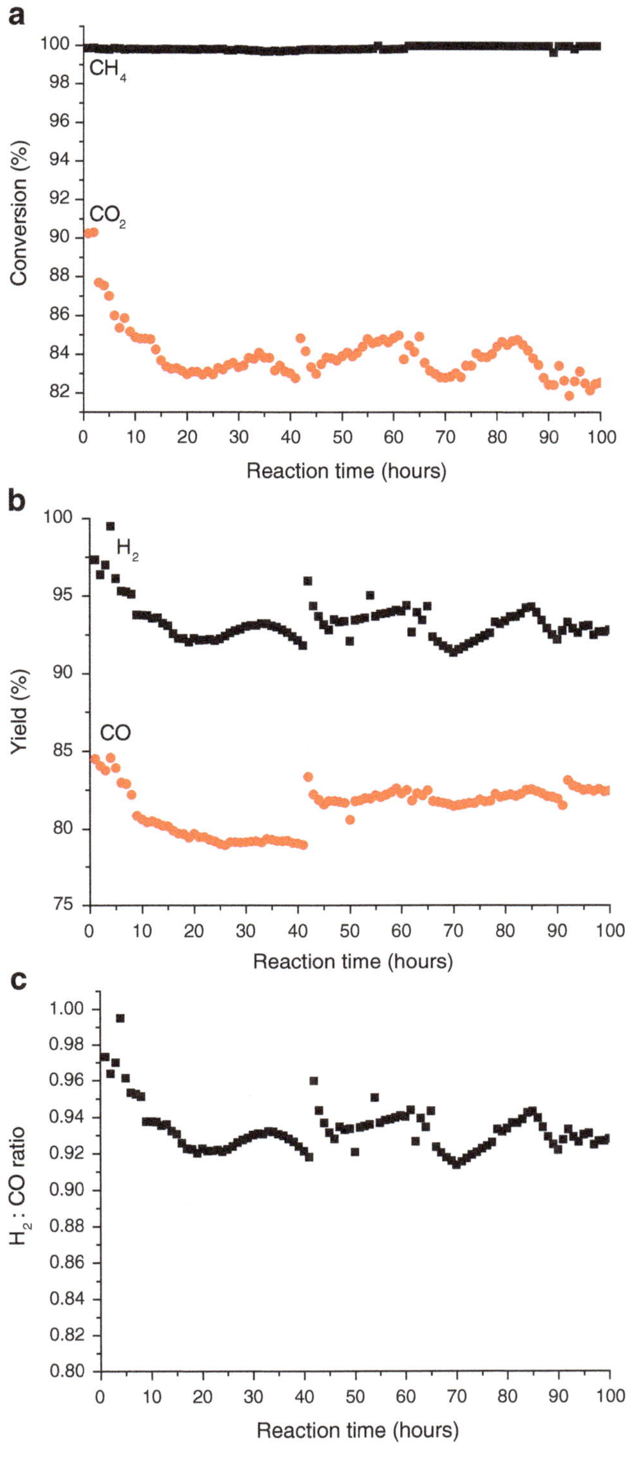

Fig. 7 100 hour reaction data for the 1 wt% La-containing alumina supported bimetallic carbide catalyst. **a** methane and carbon dioxide conversion, **b** hydrogen and carbon monoxide yields and **c** H_2:CO ratio

La doped supported bimetallic carbide catalyst was tested in an extended time on stream study for 100 h. It was found that during the course of the reaction, a drop in carbon dioxide conversion was observed, however, after ~12 h, this appeared to stabilise at ~84 % with minor fluctuations of ±1 % (Fig. 7a). In contrast to this, the methane conversion (~99 %) remained unchanged for 100 h. Analysis of the yields (Fig. 7b) revealed a gradual decline in the product quantity. The yield of carbon monoxide was found to decrease from ~85 to 82 % during the course of the reaction. This trend was mirrored by the hydrogen yield where it decreased from ~99 to 92 %.

During the reaction, the carbon dioxide feed failed at 41 h, subsequently all feeds except nitrogen were switched immediately to prevent carbon deposition occurring and the reactor was cooled to room temperature. The reactor was brought back online in the same manner as initial start-up and measurements were resumed. It can be observed that at ~42 h, an increase in yields (Fig. 7b) was apparent, suggesting that the catalyst had been partially reactivated through the reheating cycle under the inert gas. However, soon after this, the catalyst re-established similar conversion and yield levels to those observed prior to reactor shutdown. A small increase in the carbon monoxide yield may be attributable to carbon laid down on the catalyst as lanthanum species (which are known to help prevent carbon deposition [33]) were converted to the perovskite. Presumably with no corresponding increase in CO_2 conversion this can be associated with the RB reaction. The product ratio (Fig. 7c) exhibited a general decline from 0.99 to 0.96 across the course of the reaction, indicating an increase in the carbon monoxide and/or a subsequent decrease in the hydrogen content of the produced syngas.

Post-reaction characterisation

In the previous section, three major effects were observed during the reaction, these being; decreasing CO_2 conversion, decreasing H_2 yield and an increasing CO yield during the latter stages of the reaction. To explain these observations, the spent catalyst was characterised after reaction and compared to the pre-reaction and the spent catalyst obtained post 5 h. After 100 h, the catalyst exhibited a significant degree of sintering as evidenced by the change in the calculated crystallite sizes (Fig. 8; Table 1). Very little change in the cobalt metal and perovskite diffraction peaks was observed at 100 h time on stream in comparison to that observed for 5 h, suggesting that these changes occurred early on in the reaction. It can be observed in Fig. 9 that the 100 h time on stream catalyst exhibited an almost identical TGA thermograph to the 5 h time on stream catalyst after 600 °C. Prior to this, two additional features are observed in the form of an inflection point at 375 °C which may be associated with delayed oxidation of sintered particles (evidenced previously via XRD) in conjunction with the loss of amorphous carbon.

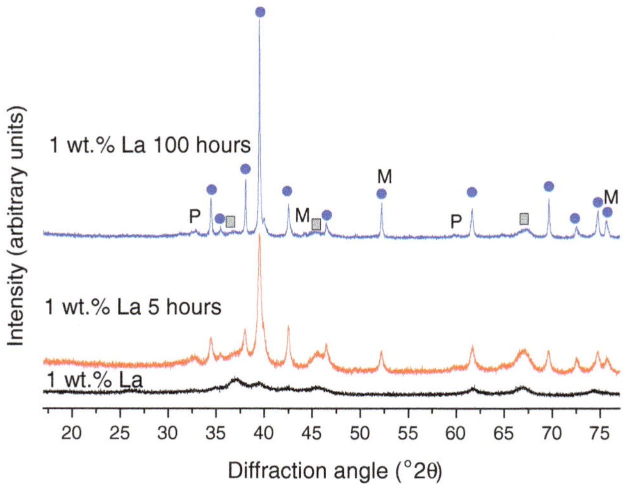

Fig. 8 XRD patterns of the 1 wt% La-containing alumina supported bimetallic carbide catalysts prior to and post-reaction at 5 and 100 h TOS. *P* = $LaCoO_3$, *filled circle* carbide, *filled square* γ-alumina and *M* = cobalt

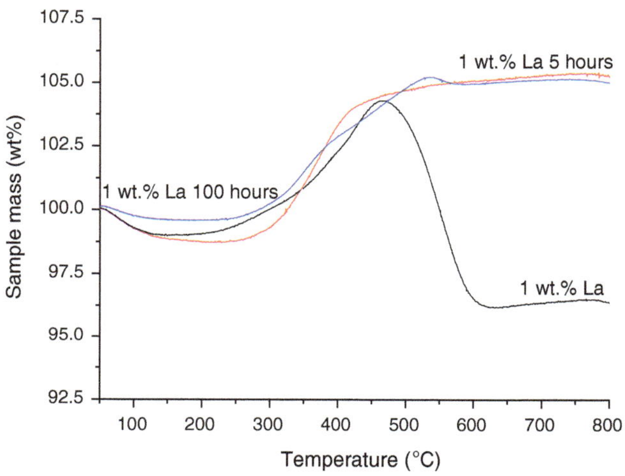

Fig. 9 TGA of the 1 wt% La-containing alumina supported bimetallic carbide catalysts prior to and post-reaction at 5 and 100 h time on stream

The second feature is also apparent at ~525 °C associated with subsequent oxidation of graphitic carbon [16].

Summary

Addition of lanthanum exhibited a number of beneficial effects at low loading levels, such as reduced carbide crystallite size and improved methane conversion in conjunction with improved catalyst stability. Higher loading levels resulted in similar carbide crystallite sizes as those observed over the catalyst containing no lanthanum addition. This increase in lanthanum content had a negative effect upon the reaction resulting in significant enhancement of side reactions in conjunction with increased carbon dioxide conversion. As an interesting aside during these reactions, it was apparent that a new phase was formed ($LaCoO_3$), which previously has never been successfully synthesised directly over γ-alumina. It has been documented that such phases possess activity for both dry methane reforming as well as reverse water gas shift, perhaps indicating, why larger lanthanum loadings had such an adverse effect upon the reaction.

At longer reaction times, the 1 wt% catalyst exhibited an initial loss of carbon dioxide conversion and hydrogen and carbon monoxide yields. Indicating that after 5 h, a significant change in the material occurred, causing a more pronounced change in the reaction than that observed previously during the 5 h runs. However, after this period the catalyst exhibited significant stability, with only a small increase occurring in the generation of carbon monoxide and subsequently a small (3 %) reduction in the H_2:CO ratio during the 100 h examined. Based upon the observations made post-reaction, it appears that the loss of activity may be related to sintering of the active phase. The change observed in the carbon monoxide yield is believed to a shift in side reactions from reverse water gas shift to the reverse Boudouard reaction, which would account for this enhancement effect.

Conclusions

Increased addition of lanthanum resulted in increased particle size of the carbide phase over the alumina support. This also occurred in conjunction with an increase in CO_2 adsorption over the catalyst with increasing lanthanum content. Several changes were observed in the catalytic activity of the catalysts. Upon addition of lanthanum an increase of 4–5 % methane conversion was observed, indicating the presence of lanthanum, enhanced reactions associated with methane. Further to this, materials that contained significantly more carbon (3, 5 and 15 wt%) possessed a higher conversion of carbon dioxide, however, only the materials containing 0, 1 and 3 wt% exhibited stability over the 5 h reaction. Catalysts containing 3-15 wt% lanthanum possessed low product yields due to a number of ongoing side reactions such as the reverse water–gas shift reaction. However, the 0 and 1 wt% catalysts exhibited high product yields indicating minimised side reactions. Post-reaction data suggested that these differences in activity may be attributed to the formation of the $LaCoO_3$ perovskite structure for samples containing higher lanthanum loading levels. For materials containing 1 wt% La, a significant sintering effect was observed, which during the course of the short time on stream study had very little effect on the catalyst stability. Longer reaction times for the 1 wt% La-containing catalyst exhibited a decrease in conversion up to 12 h time on stream, which may have been caused by significant sintering of the active carbide phase and hence a reduction in the effective catalyst surface area. However, after 12 h, the catalyst remained relatively stable exhibiting a small increase in carbon monoxide yield, which was as a result of carbon deposition on the catalyst due to the loss of the lanthanum phase through the generation of the $LaCoO_3$ phase.

Acknowledgments The authors would like to thank KACST for their kind generosity in funding the KOPRC lab, a post-doctoral research position and supporting the carbon dioxide activation project at the University of Oxford. We would like to thank Professor Xiaoqin Qin from Shanxi University for kindly undertaking gas adsorption measurements. We would also like to acknowledge Dr Justin Hargreaves from the University of Glasgow for his useful advice and fruitful discussions on binary and ternary interstitial compounds.

References

1. Al-Megren H, Gonzalez-Cortes S, Xiao T, Green M (2007) A comparative study of the catalytic performance of Co–Mo and Co(Ni)-W carbide catalysts in the hydrodenitrogenation (HDN) reaction of pyridine. Appl Catal A 329:36
2. Brungs AJ, York APE, Claridge JB, Marquez-Alvarez C, Green MLH (2000) Dry reforming of methane to synthesis gas over supported molybdenum carbide catalysts. Catal Lett 70:117
3. Budiman AW, Song S-H, Chang T-S, Shin C-H, Choi M-J (2012) Dry reforming of methane over cobalt catalysts: a literature review of catalyst development. Catal Surv Asia 16:183
4. Burghgraef H, Jansen APJ, van Santen RA (1995) Methane activation and dehydrogenation on nickel and cobalt: a computational study. Surf Sci 324:345
5. Chawla S, George M, Patel F, Patel S (2013) Production of synthesis gas by carbon dioxide reforming of methane over nickel based perovskite catalysts. Procedia Eng 51:461
6. Cheng J, Huang W (2010) Effect of cobalt (nickel) content on the catalytic performance of molybdenum carbides in dry-methane reforming. Fuel Process Technol 91:185
7. Christofoletti T, Assaf JM, Assaf EM (2005) Methane reforming on supported and non-supported molybdenum carbides. Chem Eng J 106(2):97
8. Darujati ARS, Thomson WJ (2005) Stability of supported and promoted-molybdenum carbide catalysts in dry-methane reforming. Appl Catal A 296(2):139
9. Darujati ARS, Thomson WJ (2006) Kinetic study of a ceria-promoted Mo2C/γ-Al2O3 catalyst in dry-methane reforming. Chem Eng Sci 61(13):4309
10. Diaz B, Sawhill SJ, Bale DH, Main R, Phillips DC, Korlann S, Self R, Bussel ME (2003) Hydrodesulfurization over supported monometallic, bimetallic and promoted carbide and nitride catalysts. Catal Today 86:191
11. Er-rbib H, Bouallou C, Werkoff F (2012) Dry reforming of methane—review of feasibility studies. Chem Eng Trans 29:163
12. Francisco MSP, Mastelaro VR (2002) Inhibition of the anatase-rutile phase transformation with addition of CeO_2 to $CuO\text{-}TiO_2$ system: raman spectroscopy, X-ray diffraction and textural studies. Chem Mater 14:2514
13. Haack L, de Vries J, Otto K, Chattha M (1992) Characterization of lanthanum modified γ-alumina by X-ray photoelectron spectroscopy and carbon dioxide adsorption. Appl Catal A 82:199
14. Huang T, Huang W, Huang J, Ji P (1868) Methane reforming reaction with carbon dioxide over SBA-15 supported Ni-M0 bimetallic catalysts. Fuel Process Technol 2011:92
15. Ito T, Zhang Q, Saito F (2004) Synthesis of perovskite-type lanthanum cobalt oxide nanoparticles by means of mechanochemical treatment. Powder Technol 143–144:170
16. Jiang Z, Liao X, Zhao Y (2013) Comparative study of the dry reforming of methane on fluidised aerogel and xerogel Ni/Al_2O_3 catalysts. Appl Petrochem Res 3:91
17. Jiang P, Shang Y, Cheng T, Bi Y, Shi K, Wei S, Xu G, Zhen K (2003) Methane decomposition over Ni alpha-Al_2O_3 promoted by La_2O_3 and CeO_2. J Nat Gas Chem 12:183
18. Kim S, Lee S, Song K (2007) Performance of Ni catalyst supported on La-hexaaluminate in CO_2 reforming of CH_4. Korean J Chem Eng 24(3):477
19. Korlann S, Diaz B, Bussell ME (2002) Synthesis of bulk and alumina supported bimetallic carbide and nitride catalysts. Chem Mater 14:4049
20. Kumar R, Wu-Cheng C, Sutovich KJ, Ziebarth MS, Yuying S (2014) High matrix surface area catalytic cracking catalyst stabilized with magnesium and silica, Patent US 2014/0021098
21. LaMont DC, Thomson WJ (2004) The influence of mass transfer conditions on the stability of molybdenum carbide for dry methane reforming. Appl Catal A 274(1–2):173
22. Levy RB, Boudouard M (1973) Platinum-like behaviour of tungsten carbide in surface catalysis. Science 181:547
23. Liang C, Ma W, Feng Z, Li C (1833) Activated carbon supported bimetallic CoMo carbides synthesized by carbothermal hydrogen reduction. Carbon 2003:41
24. Lojewska J, Kolodziej A, Zak J (2007) Promoted cobalt oxide catalyst on the metallic structured reactor filling for VOC combustion as an alternative to noble metal catalysts. Pol J Chem Technol 1:15
25. Martinez R, Romero E, Garcia L, Bilbao R (2003) The effect of lanthanum on Ni-Al catalyst for catalytic steam gasification of pine sawdust. Fuel Process Technol 85:201
26. Pritchard ML, McCauley RL, Gallaher BN, Thomson WJ (2004) The effect of sulfur and oxygen on the catalytic activity of molybdenum carbide during dry methane reforming. Appl Catal A 275(1–2):213
27. Puello-Polo E, Gutierrez-Alejandre A, Gonzalez G, Brito JL (2010) Relationship between sulfidation and HDS catalytic activity of activated carbon supported Mo, Fe–Mo, Co–Mo and Ni–Mo carbides. Catal Lett 135:212
28. Reuters (2013) Israel's new motor fuels strategy leans on gas, www.trust.org/item/20131224115423-3lf0m Accessed 21 Feb 2014
29. Sanchez-Castillo MA, Madon RJ, Dumesic JA (2005) Role of rare earth cations in Y zeolite for hydrocarbon cracking. J Phys Chem B 109:2164
30. Schwartz V, Oyama ST, Chen JG (2000) Supported bimetallic Nb-Mo carbide: synthesis, characterization and reactivity. J Phys Chem B 104:8800
31. Silverwood IP, Hamilton NG, Laycock CJ, Staniforth JZ, Ormerod RM, Frost CD, Parker SF, Lennon D (2010) Quantification of surface species present on a nickel/alumina methane reforming catalyst. Phys Chem Chem Phys 12:3102
32. Song C (2001) Tri-reforming: a new process for reducing CO_2 emissions. Chem Innov 31:21

33. Sutthiumporn K, Maneerung T, Kathiraser Y, Kawi S (2012) CO2 dry-reforming of methane over La0.8Sr0.2Ni0.8M0.2O3 perovskite (M = Bi Co, Cr, Cu, Fe): roles of lattice oxygen on C–H activation and carbon suppression. Int J Hydrogen Energy 37:11195
34. Tompkins HG, Augis JA (1981) The oxidation of cobalt in air from room temperature to 467 °C. Oxid Met 16:355
35. Wang XH, Zhang MH, Li W, Tao KY (2008) Synthesis and characterization of cobalt-molybdenum bimetallic carbide catalysts. Catal Today 131:111
36. Xiao T, Wang HT, York APE, Green MLH (2002) Effect of sulfur on the performance of molybdenum carbide catalysts for the partial oxidation of methane to synthesis gas. Catal Lett 83(3–4):241
37. Xiao T, York APE, Al-Megren H, Williams CV, Wang HT, Green MLH (2001) Preparation and characterisation of bimetallic cobalt and molybdenum carbides. J Catal 202(1):100

Removing of hydrocarbon contaminated soil via air flushing enhanced by surfactant

Th. Abdel-Moghny · Ramadan S. A. Mohamed · E. El-Sayed · Shoukry Mohammed Aly · Moustafa Gamal Snousy

Abstract Surfactants enhanced air sparging actually acts to displace the organic contaminant entrapped in soil pores. In this work, a comparison study was carried out between two air-flushing modes, namely, continuous air flushing and pulsed air flushing, which was conducted to remediate soil contaminated with waste-lubricant oil. Therefore, coarse sand was artificially polluted and mixed well with waste-lubricant oil at different concentrations of 10, 25 and 50 wt% to give the soil an oil blend. Then a laboratory glass column was established and backed with contaminated soil to study the effect of flow rate, pollutant and surfactant concentrations on the removal of waste-lubricant oil from soil. The contaminated soil was washed with pure water and flushed with both air-flushing modes at a pressure of 2 kPa and flow rate of 6 L min^{-1}. After that fixed 300 mL nonionic surfactant solutions ($NPEO_{9.3}$) at concentrations of 3, 5 and 7 wt%, were poured individually along with air injection at the same pressure and flow rate. The treated soil was washed several times with pure water to eliminate the residual surfactant solutions. It was found that water washing and air injection remove 27 % of oil; however, air injection along with surfactant solutions increased the oil removal efficiency up to 90 %. Moreover, both air-flushing modes succeeded in removing the pollutant with majority to pulsed air mode over continuous mode; therefore, pulsed air flushing was applied for 25 and 50 wt% waste lubricant oils in presence of 3 wt% nonionic surfactant.

Keywords Air injection · Soil washing · Surfactant flushing · Waste-lubricant oils

Th. Abdel-Moghny (✉)
Applications Department, Egyptian Petroleum Research Institute, Ahmed El-Zomer, Nasr City, Box. No. 11727, Cairo, Egypt
e-mail: thanaa_h@yahoo.com

R. S. A. Mohamed · E. El-Sayed · M. G. Snousy
Geology Department, Faculty of Science, El-Minia University, El-Minia, Egypt
e-mail: esamelsayed@yahoo.com

M. G. Snousy
e-mail: moustafa_gamal93@yahoo.com

S. M. Aly
Executive General Manager Collection and Treatment of Used Oil Petrotreat Co, Cairo, Egypt
e-mail: shokrymorsy@yahoo.com

Introduction

Somasundaran et al. [19] investigated the feasibility of using flotation process to remove non-volatile hydrophobic compounds (paraffin oil) from artificially contaminated soil (particle size 0.075–0.83 mm) using sodium dodecyl sulfate (SDS). Their study indicated that soil washing through flotation with 0.1 %-mass SDS solution is effective in reducing the amount of oil in soil comparing with the *ex situ* soil washing method, which showed 50 % less oil removal at concentration of 0.5 %-mass. Limited literature has shown that significant amount of petroleum oil may be removed from contaminated soil by flotation process [2, 19, 23, 25].

Soil contamination with spillage lubricant oils generally results from leaking underground storage tanks, pipelines and accidental spills [4, 16]. Air-flushing technique involves introducing forced air into the artificially contaminated soil to encourage removal of contaminants. Air flushing is a cost-effective, time-efficient system for the remediation of volatile organic compounds (VOCs), particularly dissolved petroleum hydrocarbon and/or biodegradable contaminants [14, 15, 18]. Two mechanisms suggested for air flushing to reduce the dissolved organic compounds in subsurface area: (1) physical stripping (volatilization) as air moves through the aquifer and (2) aerobic biodegradation of VOCs through increased oxygen supply [7].

Surfactants enhance organic contaminant act by two mechanisms. First, surfactants reduce the interfacial tension between water and contaminants that slow the mobility of the organic components, thereby, surfactants can be able to transfer the hydrophobic organic compounds (HOCs) to the mobile phase [1]. Secondly, surfactants are capable of forming aggregates known as micelles, thus solubilizing HOCs. Numerous studies have indicated that surfactants can enhance recoveries of non-aqueous phase liquids [6, 17].

Urum et al. [22], study removal of crude oil from soil using air flushing assisted stirred tank reactors. Two surfactants (rhamnolipid and SDS) were tested and the effects of different parameters (i.e., temperature, surfactant concentrations, washing time, volume/mass ratio) were investigated under varying washing modes namely, stirring only, air flushing only and the combination of stirring and air flushing.

Surfactant-enhanced air flushing was conducted to remove perchloroethene (PCE) sources from laboratory flow chambers packed with sand. The resident water was supplemented with an anionic surfactant, (SDBS), to reduce the water's surface tension, and then sparged with nitrogen gas at a constant flow rate of 0.12 min^{-1} [9, 10].

The present work aimed to applying air flushing as remedial technology for removing the waste-lubricant oil from soil in presence of nonionic surfactant polyethylene glycol nonylphenyl ether ($NPEO_{9.3}$). In this respect three different concentrations (3, 5 and wt 7 %) poured individually into laboratory glass column packed with contaminated sand. Two air injection modes continuous (direct injection) and pulsed (on/off interval mode) were applied and the effect of surfactant, pollutant concentrations, washing time and pressure had been studied. The results discussed based on application of surfactant enhancing air flushing techniques as the most suitable and cheapest remedial technology for removal of oil polluted soil.

Materials and methods

Materials

1. Coarse sand with diameter range from 0.5 to 1 mm with porosity equal 25 % was used as porous medium.
2. Commercial grade nonionic surfactant polyethylene glycol nonylphenyl ether ($NPEO_{9.3}$) was purchased from Egyptian market and used as received, and its physical properties are given in as following; number of moles E.O is 9.3, HLB is 13.0, average molecular weight is 629.37 g mol^{-1}, surface tension, dynes/cm (0.01 % aq., 25 °C) is 32, specific gravity, g ml^{-1} is 1.05–1.07, and its biodegradability is 90.
3. Waste-lubricant oils were collected from different factories and companies and were used as pollutant. The physical properties of the used lubricating oils are as the following, the specific gravity at 20 °C is 0.875, flash point (close cup Pensesky Martin) is 140 °C, water and sediment is 1.28 (vol%), water content [Dean and Stark method is 0.79 (vol%)], Viscosity at 37.8 °C is 209.235cst, ash content is 0.714 wt%, asphaltene content is 4.995 wt%.
4. n-Hexane supplied by Aldrich Chemical Co. and used without further treatment.

Determination of surface tensions and CMC

Surface tensions of surfactant solutions were determined with Wilhelmy plate Krüss K100 Tensiometer instrument that operates based on DuNouy principle. The critical micelle concentration (CMC) was determined by measuring the surface tension versus surfactant concentration.

Solubilization of lubricant oils in surfactant solutions

The ability of surfactant solutions to solubilize the waste-lubricant oil was investigated by mixing 30 ml surfactant solutions at concentrations ranging from 0.01 to 0.05 % with 0.5 ml waste-lubricant oils. Then the solutions were continuously shaken at 25 °C for 48 h. The sample vials were allowed to settle at 24 h for phase separation. After that, the aqueous phase was withdrawn and extracted with chloroform to determine the concentration of solubilized lubricant oil using ultraviolet spectroscopy.

Soil contamination

According to Urum et al. [21], a fixed mass of 1 kg of coarse sand (0.5–1 mm) was artificially polluted and mixed well with waste-lubricant oils at different concentrations of 10, 25 and 50 wt%, to give soil oil blend.

Experimental set-up and procedure

As shown in Fig. 1, a cylindrical Plexiglas column having dimension (105 × 5 × 4.5 cm), was packed with 31 cm height of 10 wt% contaminated sandy soil. The outlet end

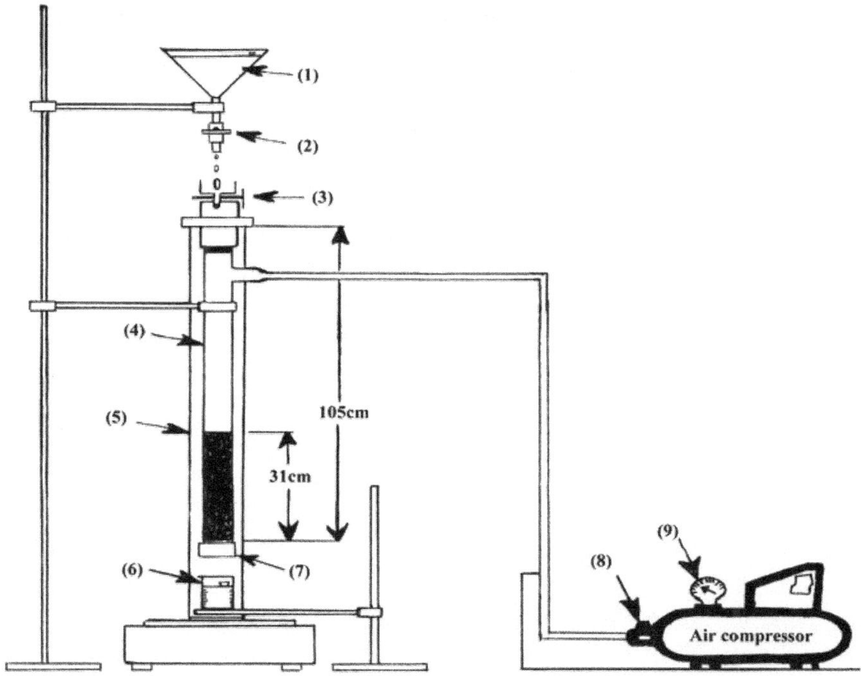

Fig. 1 The laboratory experimental model

(1) Washing solutions, (2) Valve, (3) Input column valve, (4) Column, (5) Column support, (6) Effluent beaker, (7) Wire mesh, (8) On/off control button, (9) Gauge.

of the column was fitted with a fine wire mesh screens (50 μm diameter) to prevent soil wash out. The column was cautiously packed with coarse sand (diameter range 0.5–1 mm) to insure better distribution of injected air during air flushing and preventing the sand particles from clogging the orifice where air was introduced. Then the column was filled with nonionic surfactant solutions ($NPEO_{9.3}$), prior to packing sand to avoid air bubble entrapment. Two air injection modes were applied in this study, the first one is continuous (direct injection), the second is pulsed (on/off interval mode), both modes are carry out to increasing contaminant mass removal. Air was injected at the side arm of the column into the coarse sand saturated with surfactant solution using opening air compressor system, as described by Kommalapati [11] Air delivery was controlled with a low-pressure regulator and flow meter. The flow rates ranged between 6 L min^{-1}. The air inlet stream was directed to create turbulent air current and to provide better distribution of the injected air with similar and strong focusing at all points of soil surface. The procedure used for soil packing was reproduced with 25 and 50 wt% contaminated sandy soil without substantial variation in the characteristics of the packed soil.

Soil remediation by surfactant flushing

The experiment was conducted to study the pollutant removal with nonionic surfactant solutions, and then study the effect of water washing cycles for eliminate the residual oil entrapped in soil. In this respect, a fixed 300 mL of ($NPEO_{9.3}$) at 3, 5 and 7 wt% concentrations were poured into cylindrical plexiglas column containing 10 % contaminated soil. Then a compressed air at a pressure of 2 bar (2 kg/cm^2) and flow rate 6 L min^{-1} was introduced through the side arm of the column. The pressure was monitored using a pressure gauge as shown in Fig. 1. All experiments were carried out at 25 ± 1 °C. Experiments were also conducted in downflow (gravity-stable). During the downflow flushing experiments, surfactant solutions were poured individually into the top, and the oily waste was recovered from the bottom of the soil column. To rinse off the washed soil, the remaining waste oil was improved by different water cycles. Then the effluent was collected in 50 ml centrifuge tubes, for both surfactant and water runs, and the oil remaining in soil was determined.

Analysis of effluent oily waste

The collected effluent distinct into two phases: oil phase and an aqueous phase. The oil phase was recovered by centrifuging the effluent at 10,000 rpm for 10 min and the supernatant aqueous phase was pipette out. The soil and oily waste mixture was separated and weighted. Finally, the oils remaining in soil was extracted by shaken laterally 1 g of soil for 5 min with 10 cm^3 of *n*-hexane, the washing with *n*- hexane was continued until nearly all the oily waste was removed from the soil. The *n*-hexane/oils extract was collected into one volumetric flask up to 50 cm^3 with *n*-hexane and oil percentage was determined using ultra violet spectroscopy.

Determination of oil removing percentage

The oils removed from soil using single air flushing in absence and presence of surfactant was determined and the results are given in Tables 1 and 4.

Results and discussion

Waste-lubricant oils includes crankcase (engine) oil, brake fluid, automatic transmission fluid, power steering fluid, liquid and semi-solid gear, chain, and ball bearing lubricants, and hydraulic fluid. Waste-lubricant oils are not considered hazardous waste unless it is mixed with a hazardous waste such as a chlorinated solvent. The problem of soil and groundwater pollution was widely recognized in recent years. The flushing air creates air bubbles where the oil will adhere.

Surface tension and CMC of surfactant

The surface tension of nonionic surfactant solutions was determined using 0.01 % of (polyethylene glycol nonylphenyl ether) in pure distille water @ 25 °C. It is found that the surface tension of surfactant solutions is 32 m N m^{-1}, and its CMC have spans between 0.13 and 0.2 g L^{-1} or (0.2–0.31 mmol L^{-1}).

Apparent solubility

The surfactant concentrations used in current study were 3 %, 5 % and 7 wt%, that equivalent to 15, 25 and 35 times more than its CMC. It was found that the apparent waste-lubricant oil solubility increased linearly with increasing surfactant concentrations above their CMC. The increased solubility may increase the potential removal of trapped waste-lubricant oil droplets. At low concentrations in aqueous solution, single molecules are present. The use of surfactants enhances the solubility of oil significantly by partitioning it into the hydrophobic cores of surfactant micelles. However, beyond a CMC, the surfactant molecules will aggregate, form micelles and reduce the thermodynamic energy in the system. Thereby, surfactants act to reduce the free energy of the system by replacing the bulk molecules of higher energy at an interface.

Air flushing technology and its application

With stress emphasis on air injection mode its recognized that there are two air injection mode well-known, the first one is continuous (direct injection), the second is pulsed (on/off interval mode), both modes are carry out to increasing contaminant mass removal.

Effect of continuous air flushing on removed oil

The compressed air has been injected into the saturated sand column by opening air compressor as illustrated in Fig. 1; in this respect, 1 kg of coarse sand contaminated with 10 % waste oil was subjected to applying continuous air flushing via separate two steps as the following: (1) flushing the contaminated soil with air only for 2,700 s as a blank, (2) flushing the contaminated soil with different surfactant solutions enhanced with air injection followed by washing several times with pure water to eliminate the residual surfactant solutions. The removal rate of 10 wt% waste lubricant oils using continuous air injection was given in Table 1. The results showed that the quantities of waste oil repel out from soil due to action of air flushing only is 27 wt%, and the consuming time during such process is 2,700 s. The flushing time applying and selected depending on no effluent was passing. After that the action of surfactant solutions enhanced with air injection were apply on the same model and the results are recorded in Table 2. The results revealed that the values of removed oil are raised up to 93 %, 96 %, and 97 wt% after flushing with 3 %, 5 % and 7 wt% surfactant solutions at 600, 200, and 90 s, respectively. This can be explained by, during injection of air flushing enhanced by surfactant solutions, the waste lubricant oils entrapped in soil pores were migrate because the interfacial tension between waste lubricant oils and the surfactant solution became low enough to force the blobs or ganglia of waste lubricant oils to move away. Sequentially, it can rapidly remove a large fraction of pollutant mass entrapped in the soil. It is also

Table 1 Removed oil % @ 10 wt% polluted soil using continuous air flushing

Condition	C_i (mg/g) is the initial soil concentration of pollutant	C_A (mg/g) is the final soil concentration of pollutant after air injection.	Removed oil % $= \frac{(C_i - C_A)}{(C_i)} \times 100$
After injection for 45 min	0.0001	$7.29E^{-05}$	27

Table 2 Removed oil % @ 10 wt% polluted soil using continuous air flushing with pollutant, different surfactant solution, and water

Air pulsing No.	Air flushing and pollutant (blank)		Air flushing, pollutant and 3 % surfactant (NPEox)		Air flushing, pollutant and and 5 % surfactant (NPEox)		Air flushing, pollutant and 7 % surfactant (NPEox)		Air flushing, pollutant, and water	
	Flushing time (s)	% of removed oil	Flushing time (s)	% of removed oil	Flushing time (s)	% of removed oil	Flushing time (s)	% of removed oil	Flushing time (s)	% of removed oil
First	2,700	27	600	93	200	96	90	97	30	30
Second			78	94	35	95	16	98		
Third			54	96	18	97	13	98		
Fourth			30	97	12	98	12	99		

shown in Table 2, that the water washing cycles in presence of air injection are succeeded for removing the residual oil, this means that during water washing, the percentage of oil removing increased significantly, and then became nearly constant at the end of the process. Therefore, the use of water in the remediation of highly contaminated soil as polishing treatment solution was shown of great interest. Our results proved the performance of water in the remediation of hydrocarbon-contaminated soil as found by Fernandez et al. [3].

Effect of flushing time and surfactant concentrations on removed oil

Figure 2 exhibits the effect of continuous air flushing time for remediation 10 % polluted soil in absence and presence of surfactant solutions, it is clear a linear relationship between removing oil with surfactant concentrations and inverse relationship with flushing time. This means that the efficiency of removed oil increases as the surfactant concentrations increased. The drastic decrease in time is attributed to decrees of pollutant concentration, which results from cleaning or opening of air channels due to increasing of surfactant concentrations from 3 % to 7 wt%.

Surfactant concentrations used in current study are above its CMC. A good relationship between surfactant concentrations and percentage of removing oil is shown in Fig. 3. It is clear that the solubility of waste-lubricant oil being dependent on surfactant concentrations also surfactants reduce the interfacial tension between water and contaminants that slows the mobility of the organic components. In addition, surfactants are capable of forming aggregates known as micelles, thus solubilizing waste-lubricant oil. Therefore, in order to overcome the difficulty of removal of non-volatile constituents like lubricating oil, the author advise by adding nonionic surfactant solution to enhance its recovery by solubility or desorption and subsequently accelerate the solubilization of the contaminated soil in presence of air flushing.

Effect of pulsing air flushing and time on removed oil

In this part the treatment of contaminated soil with different concentration 10 %, 25 % and 50 wt% (100, 250 and 500 g/kg) was carried out by pulsing air flushing in presence of surfactant solution followed by six water washing cycles. This step was performed at (28.37 cf/min) to evaluate the effectiveness of surfactant concentration versus different pollutant concentrations. The results of this treatment process are given in Table 3. The results reveal that in the case of 10 wt% pollutant the removed oil are super grade and exceed to 93 wt% within 40 s.

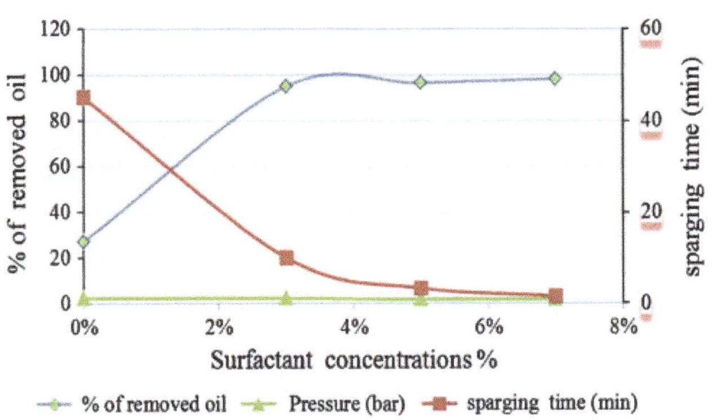

Fig. 2 Effect of continuous air flushing for remediation of 10 % polluted soil in absence and presence of surfactant

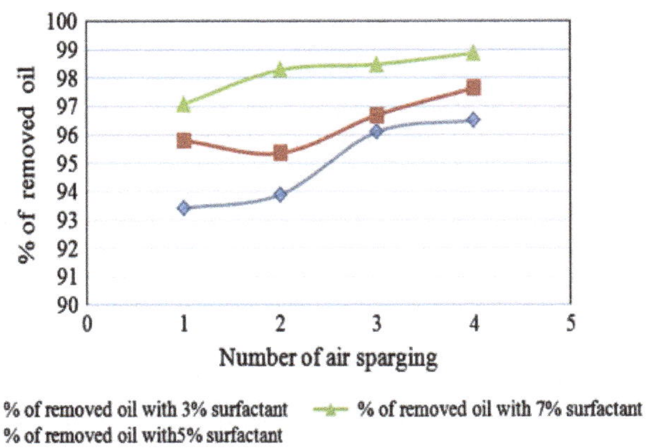

Fig. 3 Effect of air flushing for remediation 10 wt% polluted soil in presence of different surfactant concentrations

Table 3 Effect of time on removed oil% @ 10 wt%, and 25 wt% and 50 wt% polluted soil using pulsing air flushing @ 3 wt% surfactant

Air pulsing No.	10 wt% contaminated soil		25 wt% contaminated soil		50 wt% contaminated soil	
	% of removed oil	Flushing time (s)	% of removed oil	Flushing time (s)	% of removed oil	Flushing time (s)
First	93	40	63	173	11	2,700
Second	94	26	69	30	7	2,100
Third	97	20	77	29	7	2,000
Fourth	98	19	82	26	7	1,970
Fifth	99	19	84	23	7	1,970
Sixth	99	19	86	22	6	1,964
Seventh	99	19	91	19	3	1,962

The injection of air sparging (by pulsed mode) in presence of 3 wt.% surfactant help us to understand airflow pattern in the case of 10 wt% of pollutant concentration. Therefore, the movement and distribution of air through the aquifer in laboratory models have been studied to investigate airflow patterns in saturated soils at 25 wt% pollutant up to 50 wt%.

In the case of 25 % pollutant concentration Table 3, the efficiency of removed oil after first air flushing is 63 wt% and increased sharply tells to 91 wt% at seventh air flushing, this can attributed to air injection is temporarily turned off the water flows into the air channels and mixed with entrapped oil and formed emulsion. This mixing occurs in both microscale and macroscale air channels in sparging zones.

In the case of highly saturated soil up to 50 wt% pollutant concentration the date given in Table 3 reveals that, only 7 wt% of pollutant is removed under 2 kPa @ 2,100 s, after that the pressure was decreased and fixed at 1.5 kPa. This can attribute to that the oil exists as free phase. Free phase or mobile oil exists when the saturation is high enough to form pore-to-pore connections over a large area producing a continuous fluid. Furthermore, to recover the free phase of the waste lubricant oils in the treatment process it must be disposed of in accordance with applicable requirements. As well as it is necessary to use free-product recovery methods, which include each of vacuum or pumps to capture as much of the free flowing oil as possible.

The sum of all results illustrated that for a well-defined plume, a pump and treat system can be considered as an effective first line of defense in preventing further migration and in removing the bulk of free products. However, if the results of these studies are applied at relatively high airflow rates, the contaminant removal efficiency would be increased but the costs of air injection would also be significantly increased.

Effect of pulsing air flushing on removed oil

The treatment of contaminated soil was carried out using two steps: (1) injection by single air as marker, (2) washing with surfactant solution followed by washing with many water cycles. The removed oil using pulsing air injection is shown in Table 4.

The results reveal that in the case of the 10 % pollutant, the percentage of oil removal is super grade and exceeded 96 wt%; this means that the waste lubricant oil entrapped in soil pores migrated by the action of surfactant solution and forced the blobs or ganglia to move away.

In the case of the 25 wt% pollutant concentration the removed oil is nearly 79 wt%; this means that the residual waste lubricant oil existing and entrapped in soil pores drained off leaving behind some amount of liquid trapped by capillary forces that holds a liquid to a solid surface.

In case of 50 wt% pollutant concentration the oil removed is not satisfactory and reached 7 wt% only, this can be attributed to the high degree of saturation, and the residual waste lubricant oil can exist as free phase or as mobile non-aqueous phase liquid. This means that the free phase of non-aqueous phase liquid exists when the saturation is high enough to form pore-to-pore connections over a large area, producing a continuous fluid capable of flowing under an imposed gradient or its own gravitational potential. Furthermore, to recover the free phase of the non-aqueous phase liquid in the treatment process unit it must be disposed of accordance with applicable requirements, as well as its necessary to use free-product recovery methods which include each of vacuum trucks or pumps to capture as much of the free-fluid oil as possible. Generally, to solve the problem of how to remediate the high pollutant concentration, first, it must be capture the free phase by mechanical treatment as first remediation defense, followed by chemical treatment.

The advantage of pulsed over continuous air flushing

In this section a comparison study was carried out between pulsed and continuous air flushing mode conducted for remediation of soil contaminated with lubricant oil. Such comparison was carried out to determine which air flushing mode will consumed lower energy. Both of continuous

Table 4 Removed oil % @ 10, and 25 and 50 wt% polluted soil using pulsing air flushing @ 3 wt% surfactant

Number of air flushing	C_i (g/kg) Initial pollutant concentration in soil	C_A (g/kg) Final pollutant concentration in soil after 45 min water washing	C_f (g/kg) Final pollutant concentration in soil (at the end of the surfactant treatment)	$C_A - C_f$ Different between pollutant concentrations	$(C_A - C_f)/C_i$ Different between pollutant concentrations over initial concentration	$\frac{(C_A - C_f)}{C_i} \times 100$ % of removed oil
The percentage of removal of 10 % polluted oil from soil using 3 % surfactant						
First	100	99.8	6.39	93.41	0.9341	93
Second	100	99.8	5.91	93.89	0.9389	94
Third	100	99.8	3.11	96.69	0.9669	97
Fourth	100	99.8	2.16	97.64	0.9764	98
Fifth	100	99.8	1.2	98.6	0.986	99
Sixth	100	99.8	0.49	99.31	0.9931	99
Seventh	100	99.8	0.39	99.41	0.9941	99
The percentage of removal of 25 % polluted oil from soil using 3 % surfactant washing						
First	250	237.77	81.5	156.27	0.62508	63
Second	250	237.77	65	172.77	0.69108	69
Third	250	237.77	45.6	192.17	0.76868	77
Fourth	250	237.77	33.2	204.57	0.81828	82
Fifth	250	237.77	27.8	209.97	0.83988	84
Sixth	250	237.77	22.4	215.37	0.86148	86
Seventh	250	237.77	11	226.77	0.90708	91
The percentage of removal of 50 % polluted oil from soil using 3 % surfactant						
First	500	499.76	447	52.76	0.10552	11
Second	500	499.76	463	36.76	0.07352	7
Third	500	499.76	467	32.76	0.06552	7
Fourth	500	499.76	470	29.76	0.05952	6
Fifth	500	499.76	466	33.76	0.06752	7
Sixth	500	499.76	468	31.76	0.06352	6
Seventh	500	499.76	483	16.76	0.03352	3

(direct injection) and pulsed air flushing modes (on/off intervals) were applied to remediation course soil contaminated with 10 wt% waste lubricant oils in absence and presence of 3 wt% surfactant concentration. The results found in Tables 2 and 3 shows that both air flushing modes are succeeded to remove the same percentage from pollutant (93 wt%) at different time 600 and 40 s for continuous and pulsed air flushing, respectively, this means the performance of using the pulsed air flushing mode over the continuous mode for soil remediation. Therefore, the pulsed air flushing mode was applied with coarse soil contaminated with 25 and 50 wt% waste lubricant oils in presence of 3 wt% nonionic surfactant.

Several mechanisms have been postulated to explain why pulsed operation improves contaminant removal by air flushing. Induced surfactant solution flow and surfactant solution mixing might be the two dominating mechanisms.

First, as air is introduced into contaminated soil, air displaces flushing solution in the largest pores and creates temporary flushing solution flow around flushing point [12].When an air-flushing system achieves steady state, preferential airflow pathways consisting of the largest network of pores are formed in the soil, and the induced local flushing solution flow ceases [20]. Pulsing air injection frequently creates non-steady-state conditions and induces flushing solution circulation as the air channels form and collapse during each cycle. The induced flushing solution flow created by the pulsed air flushing substantially enhances the contaminant and oxygen mass transfer in the soil [5]. Second, contaminants in the immediate vicinity of air channels can be removed within few time of the start of flushing, but contaminated present at a greater distance from the air channels is less treated because of the limited mass transport (i.e., diffusion) [8].

The advantage of pulsed over continuous air flushing is that the pulsed can be used with shallow soil contaminated with saturated hydrocarbons. In addition, an optimum pulsing frequency can be determined based on the observed time for the hydrocarbon removal and airflow rate until reach steady state. Yang et al. [24] evaluated the field performance of pulsed air flushing in a short-term pilot test and during long-term system operation. Based on their successful results for short-term pilot test, the air flushing system was set to operate long term under pulsed conditions at the selected optimum pulsing frequency. This innovation resulted in higher reduction rates of dissolved benzene, toluene, ethylbenzene, and xylenes (BTEX) than those observed during the continuous operation. Performance monitoring of the air flushing system at 2, 8, and 12 months of pulsed operation indicated an increase in the hydrocarbon removal rate by a factor of up to three as compared to continuous operation, resulting in cost savings from shorter treatment time and less energy usage.

Master's thesis by Lambert [13], evidence that pulsed air flushing, otherwise known as pulsed bioflushing, is more effective than continuous air flushing because pulsing enhances treatment by inducing groundwater movement and mixing.

The cost of surfactant and air flushing operating mode

One liter commercial grade nonionic surfactant nonylphenol ethoxylate ($NPEO_9$) that purchased from Egyptian market is approximately equal 3US$, and 1,000 ml of 3 % surfactant solution can remediate 100 kg soil contaminated by 10 % waste lubricant oils. i.e., one tone soil polluted with 10 % hydrocarbons need 30US$ for complete remediation. Meanwhile, the amount of energy of the compressors used is 1.6 kWh/ton at 220 V and DC power supplier, where the energy cost is 1 U$/ton.

Conclusion

The removal of poured waste lubricant oils from soil using air flushing with and without surfactant were carried out using laboratory scale model. The effects of surfactant concentrations, water washing cycles and washing time were also studied. The results discussed based on application of surfactant enhancing air-sparging technique as the most suitable and cheapest remedial technology for removal of oil-polluted soil. Reducing the surface tension was found to promote airflow through the preferential air channel. These observations support the use of surfactant to improve air flushing of contaminated zones. This technique, in fact, was found to accelerate the remediation process for NAPL-contaminated soils. Flow time was measured at constant pressure (2 kPa) the evaluated surfactant concentration varied from 3, 5, and 7 wt%, to ensure the micellar solubilization process.

Acknowledgments This work was supported by a grant from Egyptian Petroleum Research Institute, El-Minia University, Faculty of Science, and Geology Department and Engineering Petrotreat Co.

Studied soil reach

The research work carried out throughout this study at Water laboratory Geology Department, El-Minia University, Faculty of Science, El-Minia- Egypt, and divided into two main parts: the first part is a laboratory study started since April 2009 ending July 2009, in this respect a simulated lab model is designed and packed with artificial polluted soils submerged with nonionic surfactant in presence of air sparging as mentioned in experimental part, where the second part are started in August 2009 ending January 2010 by applying the results of the first part on some rails polluted areas located in El-Minia Government City, the results did not mentioned in this articles, as well as did not published or sending for publication tell now.

References

1. Cheah PS, Reible D, Valsaraj KT, Constant D, Walsh W, Thibodeaux LJ (1998) Simulation of soil washing with surfactants. J Hazard Mater 59:107–122
2. Chou C-C, Ososkov V, Zhang L, Somasundaran P, (1998) Removal of vonvolatile hydrophobic compounds from artificially and naturally contaminated soils by column flotation. J. Soil Contam 7:559–557
3. Fernandez Perez F, Luque de Castro MD (2000) Micelles training for improvement of continuous subcriticalwater extraction of polycyclic aromatic hydrocarbons in soil prior to high performance liquid chromatography fluorescence detection. J Chromatogr A 902:357–367
4. Gallego JLR, Loredo J, Llamas JF, Va′zquez F, Sa′nchez, (2001) J Biodegradation 12:325–335
5. Heron G, Gierke JS, Faulkner B, Mravik S, Wood L, Enfield CG (2002) Pulsed air sparging in aquifers contaminated with dense nonaqueous phase liquids. Ground Water Monit. Remed. 22(4): 73–82
6. ITRC (2003) Interstate Technology & Regulatory Council: Technical and regulatory guidance for surfactant/cosolvent flushing of dnapl source zones. DNAPL-3. www.itrcweb.org/Documents/DNAPLs-3.pdf
7. Johnson PC, Johnson RL, McWhorter DB, Hinchee RE, Goodman I (1993) An overview of in situ air sparging. Gr Water Monit Remediat 13(4):127–135
8. Johnston CD, Rayner JL, Briegel D (1999) In situ air sparging—just how efficient is it in remediating groundwater contaminated by dissolved petroleum hydrocarbons. In: Proceedings of Contaminated Site Remediation Conference, Fremantle, Western Australia, March, pp 697–704
9. Kim H, Soh HE Annable MD Kim D-J (2004) Surfactant-enhanced air sparging in saturated sand. J Environ Sci Tech. 38(4):1170–1175

10. Kim H, Annable MD, Rao PSC, Cho J (2009) Laboratory evaluation of surfactant-enhanced air sparging for perchloroethene source mass depletion from sand. J Environ Sci Health Part A 44(4):406–413
11. Kommalapati RR (1995) remediation of contaminated soils using a plant-based surfactant, PhD Dissertation, Louisiana State University, Baton Rouge, LA
12. Kueper BH, Frind EO (1988) An overview of immiscible fingering in porous media. J Contam Hydrol 2(2):95–110
13. Lambert J (2008) Pulsed Biosparging of the E10 Gasoline Source in the Borden Aquifer Master's thesis, University of Waterloo, p 164
14. Maqsood I, Huang GH, Huang YF (2004) A groundwater monitoring design through site characterization, numerical simulation and statistical analysis—a North American case study. J Environ Inform 3(1):1–23
15. Marley MC, Hazebrouck DJ, Walsh MT (1992) The application of in situ air sparging as an innovative soil and groundwater remediation technology. Gr Water Monit Remediat
16. Molina-Barahona L, Rodriguez-Va′ ZR, Herna′ ndez-Velasco M, Vega-Jarquı′n C, Zapata-Pe′ rez O, Mendoza-Cantu′ A, Albores A (2004) Appl Soil Ecol 27:165–175
17. Mulligan CN, Yong RN, Gibbs BF (2001) Surfactant-enhanced remediation of contaminated soil. A review. Eng Geol 60:371–380
18. Reddy KR, Adams JA (1996) In situ air sparging: a new approach for groundwater remediation. Geotech News 14(14):27–32
19. Somasundaran P, Zhang L, Zheng J, Ososkov V, Chou C-C (1997) Removal of nonvolatile hydrophobic compounds from soils by flotation I. Laboratory investigation using a mechanically agitated machine. Adv Environ Res 1:157–165
20. Tomlinson DW, Thomson NR, Johnson RL, Redman JD (2003) Air distribution in the Borden aquifer during in situ air sparging. J Contam Hydrol 67:113–132
21. Urum K Pekdemir T, Çopur M (2003) Process Safety Environ Protect (Solid Waste Manage.) 81(3):203–209
22. Urum K, Pekdemir T, Ross D, Grigson S (2005) Crude oil contaminated soil washing in air sparging assisted stirred tank reactor using biosurfactants. Chemosphere 60:334–343
23. Varadaraj R (1995) Decontamination of hydrocarbon containing substrates (LAW 100). Patent Number 5,417,864, USA
24. Yang X, Beckmann D, Fiorenza S (2005) Field study of pulsed air sparging for remediation of petroleum hydrocarbon contaminated soil and groundwater. J Environ Sci Technol 39(18):7279–7286
25. Zhang L, Somasundaran P, Ososkov V, Chou CC (2001) Flotation of hydrophobic contaminants from soil. J Coll Surf A Phys Chem Eng Aspects 77:235–246

17

Role of metal substitution in lanthanum zirconate pyrochlores ($La_2Zr_2O_7$) for dry (CO_2) reforming of methane (DRM)

Devendra Pakhare · Daniel Haynes · Dushyant Shekhawat · James Spivey

Abstract Dry (CO_2) reforming of methane (DRM) produces syngas with H_2/CO ratio of ≤ 1. In this work, we report the use of pyrochlores ($A_2B_2O_7$) as catalysts for this reaction. Here, we examine lanthanum zirconate pyrochlores, LZ, with lanthanum and zirconium occupying the A and B sites, respectively. Three catalysts are tested: LZ and two pyrochlores in which the B-site has been isomorphically substituted with (a) Ru (2 wt%) (designated LRuZ) and (b) Pt (3.78 wt%) (designated LPtZ). These levels of substitution by weight correspond to identical atomic levels of substitution at the B-site. The effect of isomorphic substitution of Ru and Pt on the structure and activity of the pyrochlores in DRM is studied in this work. ICP-OES, XRD, H_2 and CH_4 TPR were used to characterize the structure of the catalysts. XRD results confirmed the formation of the $La_2Zr_2O_7$ phase in the bulk of the pyrochlores. H_2 TPR showed higher reduction temperatures for LPtZ compared to LRuZ, suggesting a less reducible, stable Pt in the pyrochlore structure compared to Ru. Quantitative analysis showed that LRuZ consumed 0.534 mg H_2/g_{cat}, whereas LPtZ consumed only 0.161 mg H_2/g_{cat} corresponding to much greater extent of reduction for LRuZ compared to LPtZ. CO formation during CH_4 TPR showed that both materials have reactive lattice oxygen which helps in reducing carbon formation during DRM. Temperature programmed oxidation studies of the catalyst immediately after CH_4 TPR showed greater carbon formation over LPtZ (1.67 g_c/g_{cat}) compared to LRuZ (1.17 g_c/g_{cat}) suggesting that Pt activates CH_4 to a greater extent compared to Ru. Temperature programmed DRM surface reaction showed that the light off temperature for LRuZ was about 45 °C lower than that of LPtZ. These are novel results and constitute the first report of which we are aware for Pt and Ru substituted pyrochlores for DRM.

D. Pakhare · J. Spivey (✉)
Department of Chemical Engineering, Louisiana State University, Baton Rouge, LA 70803, USA
e-mail: jjspivey@lsu.edu

D. Haynes · D. Shekhawat
National Energy Technology Laboratory,
U.S. Department of Energy,
3610 Collins Ferry Rd., Morgantown, WV 26507, USA

Keywords Dry reforming · Lanthanum zirconate · Pyrochlores · Lattice oxygen · Isomorphic substitution · Perovskite

Introduction

Dry reforming of methane (DRM) produces commercially important syngas that can be used for chemical processes like Fischer–Tropsch or synthesis of various chemicals [5, 6]. One of the problems associated with DRM is catalyst deactivation due to carbon deposition [3, 5, 8]. In addition to carbon deposition, DRM is highly endothermic reaction ($\Delta H_{298K} = 59.1$ kcal mol^{-1}) and requires temperatures approaching 900–1,000 °C to reach high equilibrium syngas yields. These temperatures lead to sintering of conventional supported metal catalysts. Thus, there is a need to develop catalyst that will resist deactivation due to sintering and carbon deposition [2, 11, 20, 28]. Pyrochlores are ternary metal oxides based on the fluorite structure with general formula of $A_2B_2O_7$ [25]. Large cations like rare earth elements occupy the A-site, whereas the B-site is occupied by a transition metal with a smaller atomic radius [16]. The pyrochlore structure allows limited isomorphic substitution of catalytically active metals like Ru, Rh, Pt at the B-site.

The most preferred metal for the DRM reaction is Ni due to its low cost and availability [8, 10, 17–19]; however, this metal undergoes severe deactivation due to carbon deposition. Noble metals like Ru, Rh, and Pt have greater resistance to carbon deposition, but are not economical [4, 7, 9, 21, 22]. In our previous work, we have studied Ru (on B-site) and Sr (on A-site) substituted pyrochlores for DRM [12]. In the present work, we report a study of Ru and Pt substituted (at the B-site) lanthanum zirconate pyrochlores. Despite their well-known thermal stability and reports of their use for other purposes such as thermal barrier coatings, we are aware of only one study of these materials for DRM, i.e., by Ashcroft et.al [1]. In that paper, Ashcroft et al. studied $Eu_2Ir_2O_7$ pyrochlores and observed that above 342 °C under DRM conditions, these materials completely decompose to Eu_2O_3 and Ir metal. However, in one of our previous works, we have used Ru and Sr substituted pyrochlores which have proven to be extremely thermally stable at 850 °C [12].

Here, we perform ICP-OES, XRD, and TPR using H_2 and CH_4 as reducing gases and study the effect of substitution of Ru and Pt (separately) at the B-site of the pyrochlore structure. Temperature programmed surface reaction (TPSR) was conducted to study the changes in the catalyst light-off temperatures as a result of change in the active metal substituted on the B-site.

Experimental section

Catalyst synthesis

The LZ, LRuZ, and LPtZ pyrochlores were synthesized by modified Pechini method using salts of lanthanum nitrate hexahydrate [$La(NO_3)_3.6H_2O$], zirconium oxynitrate [$ZrO(NO_3)_2.xH_2O$], and either ruthenium chloride ($RuCl_3$) or hydro-chloroplatinic acid [$H_2PtCl_6{\cdot}(H_2O)_6$] salts as precursors, and anhydrous citric acid ($C_6H_8O_7$) and ethylene glycol ($C_2H_6O_2$) as complexing and polymerizing agents, respectively. Stoichiometric amounts of metallic precursors were separately dissolved in 50 ml of deionized water, and then combined into a larger beaker with continuous stirring. An aqueous citric acid (CA) solution was then added to the stirring solution in the beaker. The molar ratio between CA and total metal ions was kept at 1.2. The solution was heated to 70 °C while stirring on a hot plate before ethylene glycol (EG) was added. The molar ratio of EG to CA was 1:1. The solution was stirred under isothermal conditions for several hours until a clear viscous gel was obtained. To promote polyesterification reaction between CA and EG, the gel was further heated in a heating mantle at 130 °C. This resulted in an amorphous polyester-type resin. The solid material was then collected and calcined at 1,000 °C for 8 h to form the mixed metal oxide material.

Inductively coupled plasma-optical emission spectroscopy (ICP-OES)

The samples were quantitatively weighed to the nearest 0.00001 g and then transferred to a Teflon digestion sample holder. Four mL of concentrated trace metal grade hydrochloric acid was added to the sample and was loosely covered for 16 h. The sample was then heated in a sand bath for 1 h at 95 °C. After cooling, the samples were sealed and digested in a CEM Mars 5 microwave. After digestion, the Teflon sample holders were heated on a sand bath at ~95 °C until acid was nearly evaporated. Then, 5 mL of freshly prepared aqua regia was added to each sample and loosely covered overnight. The microwave digestion was repeated with holding at 200 °C for 20 min. The samples were then cooled and sat overnight. The so prepared samples were then analyzed on a Perkin Elmer 2000 DV Inductively coupled plasma optical emission spectrometer at wavelengths of 240.272 nm for Ru, 333.749 nm for La, 257.139 nm for Zr, and 214.423 nm for Pt.

X-ray diffraction (XRD)

Crystalline nature of the freshly calcined pyrochlores was studied using X-Ray Diffraction spectra obtained from a Bruker/Siemens D5000 system. The system was composed of a ceramic X-ray tube with Cu Kα radiation operating at a wave length of 1.54184 Å. The X-ray tube operated at a voltage of 40 kV with a current of 30 mA. Angle of divergence slit for the incident X-ray beam was set to 1° and the antiscatter slit was set at 0.5°. For analysis, 2θ was varied from 10° to 80° at a sweep rate of 0.02°/min.

Temperature programmed reduction (TPR)

Temperature programmed reduction by H_2

Temperature programmed reduction of the catalysts was conducted in an AMI 200 (Altamira instruments, Pittsburgh, PA, USA) fixed bed reactor. For each run, 50 mg of catalyst was loaded into the 6.5 inch (length) by 6 mm (outer diameter) by 4 mm (inner diameter) quartz tube. The catalyst was packed in place using quartz wool plugs. Thermocouple that ran axially through the tube touching the quartz wool was used to monitor the bed temperature continuously. TPR was performed on LZ, LPtZ and LRuZ pyrochlores. Before starting TPR, samples were oxidized in flowing O_2/He at 30 mL/min from ambient (ca. 35° C) to 950 °C at a rate of 5 °C/min. This was done to oxidize the metals on the surface and in the bulk which could then consume H_2 during the TPR. The sample was then cooled to room temperature in flowing He at 30 mL/min. This also ensured oxidation and desorption of any adsorbed species on the catalyst surface. Next, the sample was exposed to mixture of 10 % H_2/Ar flowing at 30 mL/min with the temperature ramped from 40 to 950 °C at 5 °C/min. The temperature was held at 950 °C for 30 min and then cooled to room temperature.

The amount of metal reduced in pyrochlores was quantified. This estimation of the extent of metal reduction in the catalyst was done using silver oxide (Ag_2O) standards. Known quantity of Ag_2O was reduced under similar conditions as the catalysts. From stoichiometric calculations, the amount of hydrogen required for reduction of the given amount of Ag_2O was determined. Repeating TPR of the Ag_2O standard gave a calibration curve relating the area under the TPR profile of the catalyst to the hydrogen consumption of that catalyst during the reduction reaction. The area under the TPR peak of the catalyst was related to the H_2 uptake by that catalyst during TPR.

Temperature programmed reduction by CH_4

This characterization method was used to study the oxygen conductivity of the pyrochlore structure. For this about 50 mg of catalyst was dried at 500 °C in flowing He at 30 mL/min and cooled to room temperature in flowing He. The catalyst bed was then subjected to reduction by 10 % CH_4/He flowing at 30 mL/min and the temperature was ramped from 40 to 950 °C at the ramp rate of 5 °C/min. The conditions were maintained isothermal at 950 °C for 30 min. The product stream during the CH_4 TPR was analyzed using the mass spectrometer (MS) that was hooked up with the reactor outlet. The MS was calibrated for known concentration of CO and CO_2 and the concentration of CO and CO_2 in the product line was determined.

Temperature programmed oxidation (TPO)

After subjecting the catalysts to TPR by CH_4, a TPO was conducted for studying quantitatively and qualitatively the oxygen conductivity of the pyrochlores. For conducting the TPO, the catalyst was cooled to room temperature (ca. 35 °C) in flowing He at 20 mL/min. Then, it was oxidized in flowing 5 % O_2/He at 30 mL/min from room temperature to 950 °C and the ramp rate was 5 °C/min. The conditions were maintained isothermal at 950° C for 30 min. The CO and CO_2 emitted during the TPO were tracked using the mass spectrometer (MS) hooked up to the reactor outlet. The amount of CO and CO_2 formed during TPR by CH_4 and TPO was calculated using the calibration gases of 5 % CO/He and 10 % CO_2/He.

Temperature programmed surface reaction (TPSR)

For carrying out TPSR experiment, 50 mg of catalyst was loaded into the U-tube quartz reactor and was dried at 300 °C in flowing He. After treatment with He, it was cooled to room temperature (ca. 25 °C) and the reactant gases were allowed to flow over the catalysts. The temperature was then programmed to rise from 25 to 900 °C with the ramp rate of 5 °C/min and the gas hourly space velocity (GHSV) was maintained at 48,000 mL g_{cat}^{-1} h^{-1}. The MS signal for the products from the reactor was recorded.

Results and discussions

Inductively coupled plasma-optical emission spectroscopy

Table 1 shows the bulk concentration of the metals in the pyrochlores determined using ICP-OES. These results show similar atomic level substitution for Ru and Pt at the B-site.

X-ray diffraction (XRD) study of fresh catalyst

XRD pattern of the freshly calcined catalysts (see Fig. 1) showed the formation of the cubic unit cell crystalline $La_2Zr_2O_7$ phase. The triangle marked peaks denote the angle of diffraction of the X-rays from the $La_2Zr_2O_7$ phase of the catalyst. The diffraction angles attributed to the $La_2Zr_2O_7$ phase observed in this work are in accordance with the ones observed in the literature [13, 14]. As

Table 1 ICP-OES results for LZ, LPtZ, and LRuZ pyrochlores

Catalyst\metal	La		Zr		Pt or Ru	
	Atom (%)	ICP (wt%)	Atom (%)	ICP (wt%)	Atom (%)	ICP (wt%)
LZ	32.8	45.6	47.5	43.4	–	–
LPtZ	33.1	45.8	34.4	31.4	1.86	3.64
LRuZ	33.9	47.2	34.7	31.7	1.6	1.63

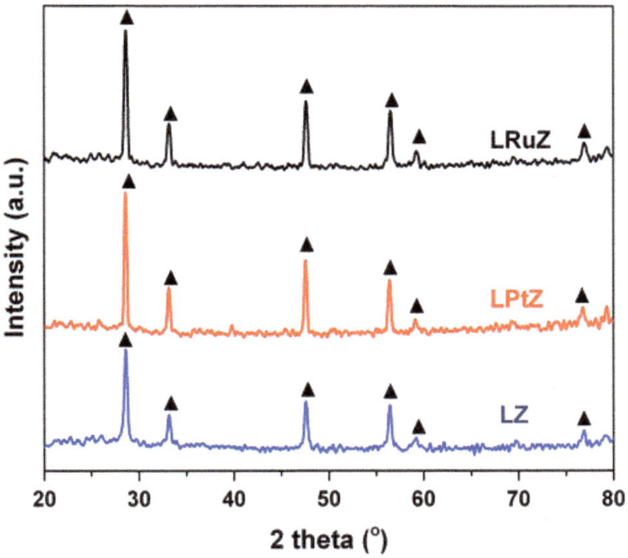

Fig. 1 XRD pattern of the freshly calcined LZ, LPtZ, and LRuZ pyrochlores. *Filled triangles* $La_2Zr_2O_7$)

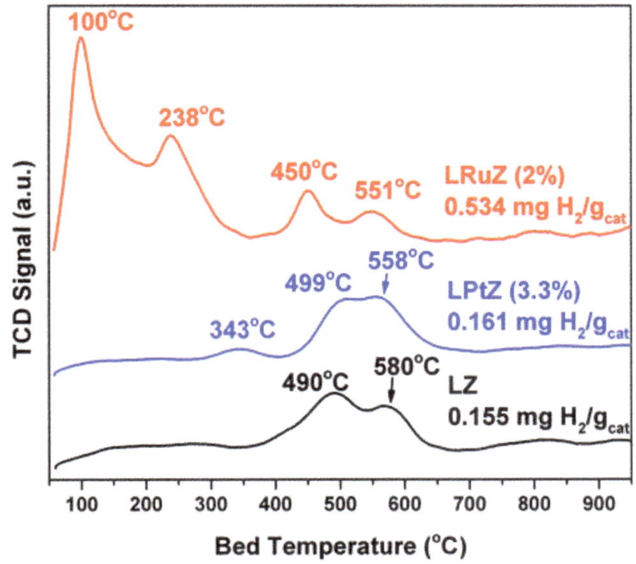

Fig. 2 Temperature programmed reduction of freshly calcined LZ, LPtZ, and LRuZ pyrochlores

expected, the substitution of Pt and Ru on the B-site did not affect the pyrochlore crystalline structure.

Haynes et al. have studied pyrochlores with Sr and Ru substitution on A and B sites, respectively, for other fuel reforming reactions [14–16]. They observed that when substitution of Sr and Ru was beyond a certain limit, a $SrZrO_3$ perovskite phase was formed, which diffracted X-rays at the angle of about 32°. This is in agreement with the fact that there is a limit to the extent of substitution at either A or B sites, beyond which the excess of the substituted metal separates out as a perovskite phase [13]. There was no formation of any such perovskite phase observed in Fig. 1, suggesting that the extent of substitution of Pt and Ru is low enough to avoid the formation of any separate perovskite phase. Comparing the diffraction pattern for LZ with LPtZ and LRuZ shows that there was no shift in the diffraction angle as a result of Pt (3.3 wt%) or Ru (2 wt%) substitution on the Zr^{+4} site. The substitution of Pt and Ru atoms on the B-site caused no apparent change in the lattice constant of the lanthanum zirconate.

Temperature programmed reduction (TPR)

Temperature programmed reduction (TPR) by H_2

The H_2 TPR profile for LZ, LPtZ, and LRuZ is shown in Fig. 2. LZ pyrochlore shows reduction at 490 °C and 580 °C and the total consumption of H_2 during reduction amounts to 0.155 mg H_2/g_{cat}. This suggests that only 0.6 % of the total lanthanum zirconate is reduced during TPR.

Reduction of LPtZ shows a small peak at 343 °C and two peaks at 499 and 558 °C. The 343 °C peak could be due to the reduction of either PtO_x in the pyrochlore lattice to Pt metal or of Zr at the Pt–Zr interface [24]. Souza et.al [24] observed similar reduction peaks for Pt/ZrO_2 catalysts and attributed it to the Pt-assisted reduction of Zr, possibly by hydrogen spillover, where Pt activates and homolytically dissociates H_2 which then reduces the Zr at the Pt–Zr interface. A similar phenomenon could be occurring over LPtZ. However, it is important to consider the fact that the class of catalysts used in the work by Souza and in this work is different.

The reduction peaks at 499 and 558 °C for LPtZ can be attributed to the reduction of lanthanum zirconate as seen in the profile of LZ. The total hydrogen consumption by LPtZ is 0.161 mg H_2/g_{cat} which is very close to that consumed by LZ, i.e., 0.155 mg H_2/g_{cat}. This indicates that the presence of Pt does not significantly increase the reducibility of the pyrochlore itself. Also, if all of Pt (assuming PtO_2 form) in the pyrochlore was reduced during TPR, then the hydrogen consumed by Pt would be 1.3 mg H_2/g_{cat} which is much greater than the total hydrogen consumption of LPtZ (0.161 mg H_2/g_{cat}). In pyrochlores, most of the metal is substituted in the bulk of the pyrochlore and is inaccessible to H_2 during TPR experiment. The extent of reduction of LPtZ during TPR is only 0.64 % of the total metal content, showing that the reduction is limited to the surface atoms, as expected.

TPR of LRuZ, however, shows multiple reduction peaks at 100, 238, 450, and 551 °C. The presence of a low temperature reduction peak at 100 °C suggests that at least a small amount of Ru at the pyrochlore surface is not strongly bound to the pyrochlore structure. Though not detectable by XRD, these small clusters of Ru exhibit a reduction behavior similar to that for RuO_2/Al_2O_3 [16]. This surface Ru appears to be formed by the oxygen pretreatment (to 950 °C; "Temperature programmed reduction by H_2"), during which Ru in the pyrochlore is destabilized from the B-site and diffuses to the surface. The peak at 238 °C is assigned to the reduction of Ru substituted in the pyrochlore structure. In one of our earlier work with Ru (at B-site) and Sr (at A-site) substituted pyrochlores, we observe reduction of Ru at 280 °C [12]. The reduction of Ru in the LRuZ is about 40 °C lower than that reported in the earlier study, possibly due to a slightly weaker interaction of Ru in the pyrochlore structure. The similarity of the reduction peaks at 450 and 551 °C for LRuZ and those at 490 and 580 °C for LZ in Fig. 2 suggests that both are attributed to reduction of the lanthanum zirconate pyrochlore. Presumably, the slightly lower peaks for LRuZ suggest that the reduction of LZ is enhanced somewhat by Ru via hydrogen spillover. The amount of hydrogen consumed by LRuZ is 0.534 mg H_2/g_{cat} which is much greater than that by LPtZ or LZ, largely due to reducibility of the Ru that appears to be on the pyrochlore surface rather than in the structure. If all the Ru in LRuZ reduces, then the hydrogen consumption by Ru (assuming RuO_2 form) alone would be 1.4 mg H_2/g_{cat} which is far greater than the total hydrogen consumption of LRuZ of 0.534 mg H_2/g_{cat}. This suggests that most of the Ru is still substituted in the pyrochlore structure, which is inaccessible during TPR. The total metal (La, Zr, and Ru together) reduced during LRuZ TPR is 2.12 % of the total metal content, again consistent with the premise that the reduction is limited to the surface of the pyrochlore.

Temperature programmed reduction (TPR) by CH_4

Lanthanum zirconate pyrochlores similar to those reported here have been used by Haynes et al. [14–16] for reforming of diesel-range alkanes. They observed that the lattice oxygen of these materials is reactive and helps in resisting carbon formation during reforming reactions. To study lattice oxygen conductivity on the pyrochlores of interest, TPR of these materials using CH_4 as a reducing agent was carried out. CH_4 is believed to activate on the catalyst to form surface carbon and hydrogen. This surface carbon then must be oxidized to limit the accumulation of carbon deposits. During DRM, CO_2 acts as an oxidizing agent and helps in limiting carbon formation over the catalyst. However, during CH_4 TPR, there is no CO_2 or O_2 in the feed or in the pretreatment of the catalyst. Thus, the only oxygen available for oxidation of surface carbon is the lattice oxygen.

Figure 3a shows the CO and CO_2 formation during the CH_4 TPR of LZ pyrochlore. The signal for CO_2 was very weak and no distinct peak could be identified. CO formation begins at about 560 °C and reaches a first peak at 605 °C. Other peaks at 720 and 920 °C were also observed for CO formation.

Substitution of Pt and Ru on the B-site decreases the temperature at which the lattice oxygen in the pyrochlore begins to react with methane (see Fig. 3b, c). For LPtZ, CH_4 TPR resulted in CO formation at 397, 650 and 950 °C. Substantial amounts of CO_2 were also formed during CH_4 TPR of LPtZ, unlike LZ. For LRuZ (Fig. 3c), the reduction temperatures were much lower compared to LPtZ and LZ. The first peak for CO formation was observed at 283 °C followed by a large peak at 449 °C and small peaks at 582 and 814 °C.

High temperature reduction peaks (720 and 920 °C) were observed for LZ and one similar reduction peak (950 °C) was also observed for LPtZ. This suggests that these high temperature peaks in LPtZ profile is due to the reaction of oxygen associated with the lanthanum zirconate part of LPtZ pyrochlore. The low temperature peaks (397 and 650 °C) are due to the reaction of oxygen associated with Pt in the pyrochlore structure. For LRuZ, the reduction peaks were at very low temperatures (283 and 449 °C) compared to LZ (605 and 720 °C) and LPtZ (397 and 650 °C). This is consistent with the H_2 TPR results, which suggested that Ru is not stable within the pyrochlore structure, and migrates from the B-site to the surface. The low temperature reduction peaks for LRuZ (Fig. 2c) could be due to the reaction of oxygen associated with Ru. H_2-TPR of LRuZ showed that Ru substitution increases the reducibility of lanthanum zirconate, causing it to reduce at temperatures (450 and 551 °C) lower than LZ (490 and 580 °C) (Fig. 2). A similar phenomenon was observed in CH_4-TPR of LRuZ where the reaction of oxygen associated with lanthanum zirconate was observed at 814 °C compared to 920 °C for LZ and 950 °C for LPtZ. Because H_2 is a stronger reducing agent than CH_4, it reduces the catalyst at much lower temperatures than CH_4. A comparison of Fig. 2 (H_2 TPR) and Fig. 3 (CH_4 TPR) shows this to be true for all the catalysts.

We are not aware of any reports in which CH_4-TPR has been used to study the lattice oxygen reactivity of Ru and Pt pyrochlores and in which CH_4-TPR and H_2-TPR are directly compared.

TPO of catalysts spent for CH_4 TPR

The surface carbon formed during CH_4 TPR was oxidized and the resultant CO/CO_2 profile (TPO) plots for LZ, LPtZ, and LRuZ are shown in Fig. 4a, b, and c, respectively. The major product during TPO was CO_2 and very small amounts of CO were formed at the same temperatures as CO_2. The oxidizable carbon formed on the surface were then quantified and normalized with respect to the weight of the catalyst.

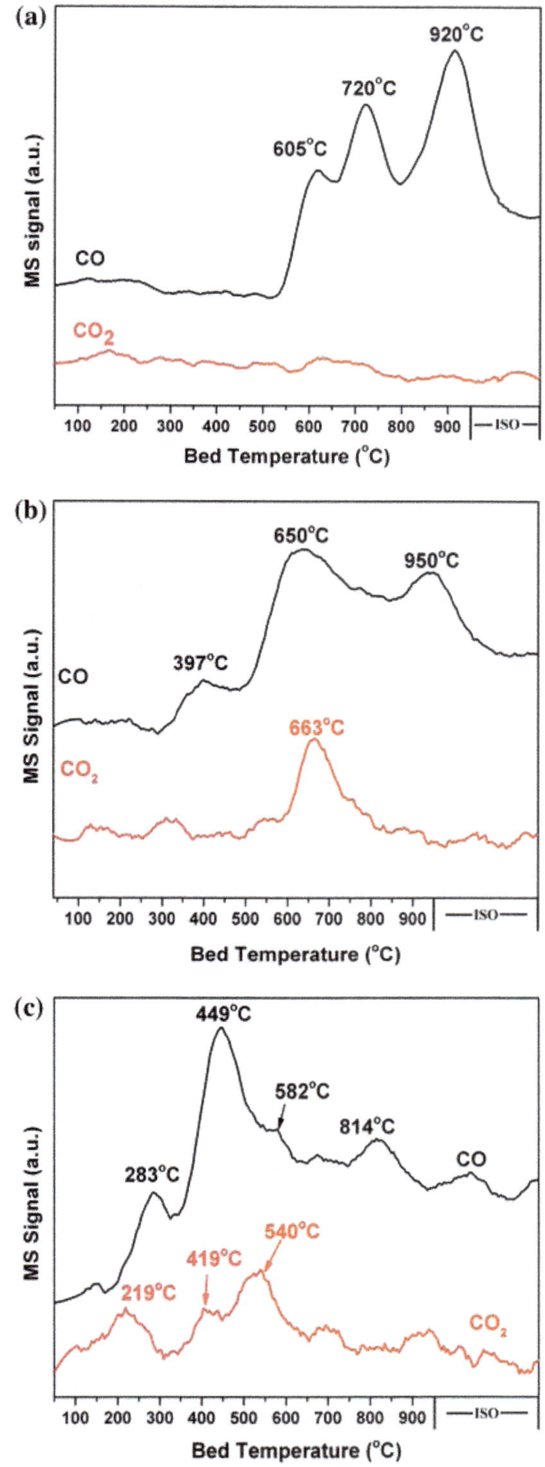

Fig. 3 TPR by CH_4 of freshly calcined **a** LZ, **b** LPtZ, and **c** LRuZ from 50–950 °C at the ramp rate of 5 °C/min. (ISO represents that the conditions were isothermal at 950 °C for 30 min for LZ and 60 min for LPtZ and LRuZ). The plots **a**, **b**, and **c** differ in *Y*-axis scale

TPO of the LZ pyrochlore shows a low temperature CO_2 peak at 340 °C and CO peak at 335 °C (see Fig. 4a). Verykios et al. [27] in their work on Rh/Al_2O_3 catalysts observed surface carbon which oxidized at 330 °C and called it (δ) phase of carbon. The peak at 340 °C could be due to the oxidation of this $C_{(\delta)}$ phase. Dehydrogenated carbon often oxidizes at higher temperatures compared to the partially hydrogenated surface carbon [23]. A similar dehydrogenated form of carbon could be responsible for the 600 °C TPO peak. The total amount of carbon formed over LZ during CH_4 TPR which was later oxidized to CO and CO_2 is 0.261 g_C/g_{cat}. The carbon formed over the LPtZ was qualitatively found to be more reactive, and presumably more hydrogenated, than the carbon formed on LZ pyrochlores. The low temperature peak was found to be at 260 °C (Fig. 4b). This carbon could be polymeric in nature deposited near the metal (Pt) site in the pyrochlore structure [23]. The high temperature peaks at 503 and 569 °C could be due to the oxidation of more dehydrogenated type of carbon deposited away from the Pt site [16, 23]. The total amount of carbon formed over LPtZ is about 1.67 gc/ g_{cat}, which is much greater than that compared to LZ. This suggests that Pt is the active site for activation of CH_4 during DRM.

The TPO of the LRuZ pyrochlores after CH_4 TPR suggests that the carbon formed was qualitatively different from that observed over LZ and LPtZ pyrochlores. The peak at 516 °C could be due to the oxidation of the same dehydrogenated polymeric carbon found on the LPtZ deposited away from the active metal [16, 23]. A high temperature peak was observed at 776 °C, which was not observed over LZ or LPtZ. This carbon could be graphitic in nature and is extremely stable, requiring a high temperature for oxidation. In one of our earlier studies with Ru and Sr substituted pyrochlores, we observed similar graphitic carbon oxidizing at 790 °C [12]. This carbon could be considered to be progressively formed during CH_4 TPR, first starting with hydrogenated polymeric carbon that is further dehydrogenated with increasing temperature.

Table 2 shows the total amount of carbon formed over the three catalysts. Carbon on LRuZ is 1.17 g_C/g_{cat}, somewhat less than for LPtZ (1.67 g_C/g_{cat}).

Recall from the CH_4 TPR spectra in Fig. 3b, c that LRuZ has more reactive lattice oxygen than LPtZ. This is likely the reason for smaller amount of carbon formation over LRuZ. Also, the greater amount of carbon on LPtZ would suggest greater amount of CH_4 is activated over Pt. This would mean that Pt is more active towards CH_4 activation as compared to Ru in the pyrochlore structure. A significant difference is observed in the reactivity of carbon formed over LRuZ and LPtZ. The reason could be due to the difference in the activation mechanism of CH_4. Activation of CH_4 occurs by either formate formation or by direct decomposition to form $C_{(s)}$ and $H_{2(g)}$ over the catalyst [26]. As the carbon formed over LPtZ is observed to be more reactive than that over LRuZ; Pt could be activating CH_4 via formate formation. Direct decomposition would

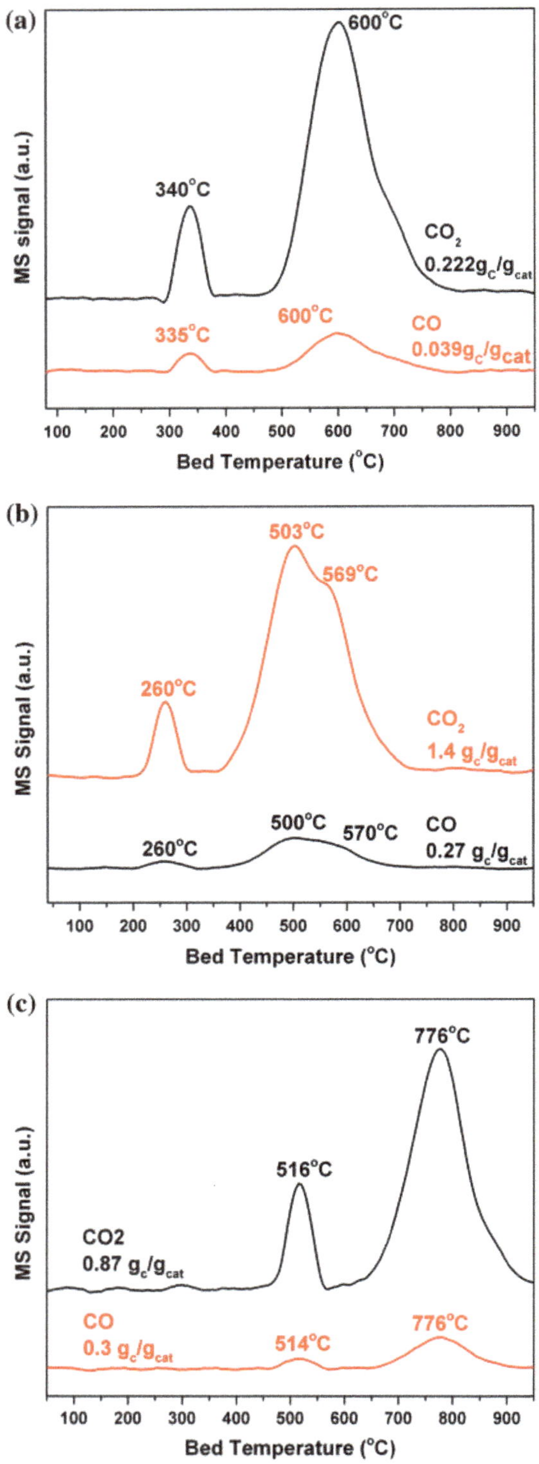

Fig. 4 Temperature programmed oxidation of the catalysts spent for CH_4 TPR **a** LZ, **b** L2RhZ, and **c** L5RhZ. Oxidized from 50–950 °C at the ramp rate of 5 °C/min. The plot of each catalyst differs in *Y*-axis scale

result in a dehydrogenated form of carbon which could become more graphitic and less reactive, resulting in a higher TPO peak. Thus, the TPO results suggest that Ru could be activating CH_4 via direct decomposition.

Table 2 Summary of carbon formed; as computed from the TPO of the catalysts spent during TPR by CH_4

Catalyst	Total amount of carbon formed during TPR by CH_4 (g_C/g_{cat})
LZ	0.261
LPtZ	1.67
LRuZ	1.17

Temperature programmed surface reaction (TPSR)

TPSR of LZ, LPtZ and LRuZ is shown in Fig. 5a, b, and c, respectively. This experiment was performed to determine the catalyst light-off temperature and to study the effect of metal substitution on the B-site of the pyrochlore. TPSR plot shows that LZ does not convert CH_4 or CO_2 to any significant extent at any temperature. The initial drop in CO_2 signal is due to the experimental error, as it is observed over all catalysts. Lack of any activity on LZ could be attributed to the absence of any active metal in the pyrochlore structure that is needed to activate CH_4.

It appears that LPtZ (see Fig. 5b) activates in two stages. It starts to form CO and H_2 at about 460 °C, but the rate of product formation increases more rapidly after 615 °C. These catalysts were not reduced during pre-treatment before TPSR. During TPSR, CH_4 appears to start in situ reduction of Pt, creating more active sites for DRM with time [1]. Once formed, H_2 and CO continue to reduce Pt to a greater extent resulting in faster creation of active sites, autocatalytically increasing reaction rate with temperature. This suggests that metallic Pt in the pyrochlore is the active site for DRM.

Figure 5c shows that LRuZ has a single light-off temperature of 415 °C, which is about 45 °C lower than for LPtZ. As a result of simultaneous occurrence of RWGS, small amount of water was also formed over both the catalysts. The temperature for water formation corresponds to the light-off temperature of each catalyst. In case of LPtZ, water formation began at about 460 °C, but the rate increased at 615 °C. However for LRuZ, water formation started at about 415 °C and was almost constant up to 900 °C.

Conclusion

XRD confirmed the formation of the $La_2Zr_2O_7$ phase in the freshly calcined pyrochlores. H_2 TPR showed that the reduction profile for LPtZ was very similar to that of LZ, suggesting that Pt is strongly bound within the pyrochlore structure. A somewhat lower reduction peak for LPtZ compared to LZ suggests that hydrogen spill-over from Pt

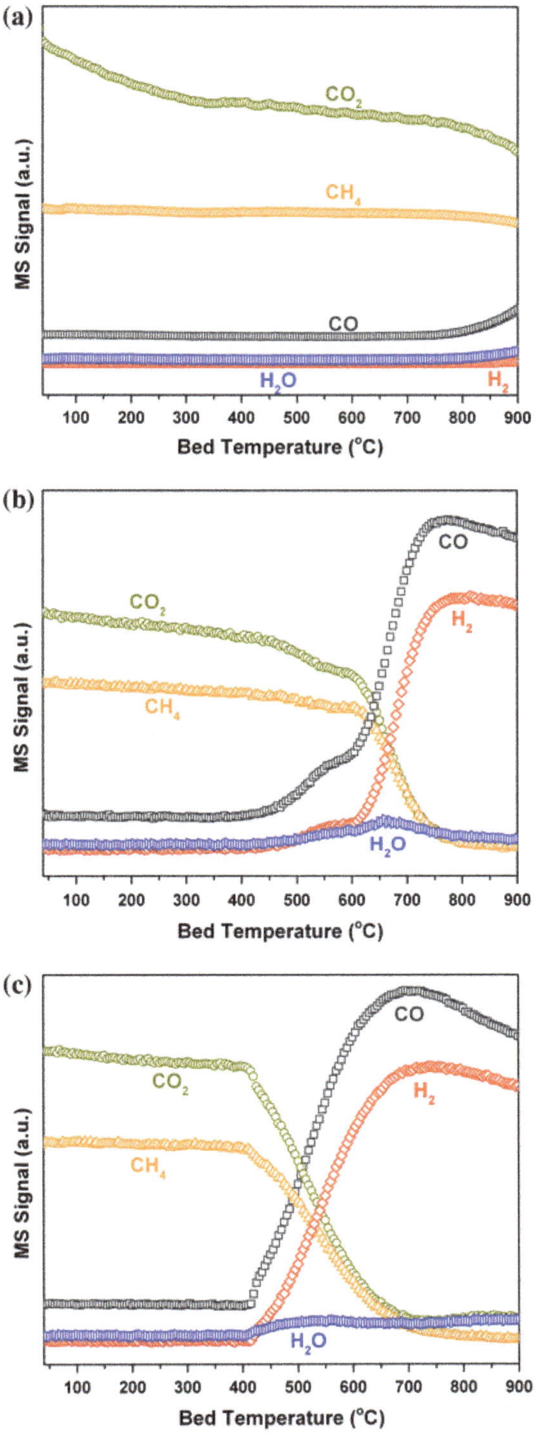

Fig. 5 TPSR plots for **a** LZ, **b** LPtZ, and **c** LRuZ from 40 to 900 °C at pressure of 1 atm and gas hourly space velocity of 48,000 mL h^{-1} g_{cat}^{-1}

reduces Zr ions at a lower temperature than when Pt is not present. For LRuZ, very low TPR peaks were observed suggesting that surface Ru was likely formed during a 950 °C oxygen pretreatment and was reducible at temperatures expected for conventional supported Ru clusters. CH_4 TPR suggested that all the pyrochlores, with or without Pt or Ru substitution, have mobile lattice oxygen that is reactive towards carbon formed over the catalyst during CH_4 TPR.

Ru substitution also assists in reduction of lanthanum zirconate. Quantitative and qualitative analysis of carbon formed over LRuZ and LPtZ during CH_4 TPR suggests a difference in the mechanism of CH_4 activation. Ru appears to activate CH_4 via direct decomposition resulting in the formation of surface carbon and hydrogen. Pt seems to activate CH_4 via formyl or formate group formation. Thus, the carbon formed over LPtZ is more hydrogenated and oxidizes at a lower temperature compared to that formed over LRuZ.

TPSR experiments suggest that the light-off temperature of LRuZ is about 45 °C lower compared to LPtZ. Greater surface coverage of Ru metal due to migration of Ru to surface as a result of unstable substitution on the B-site could be the reason for lower light-off temperature of LRuZ compared to LPtZ.

Acknowledgments This material is based upon work supported as part of the Center for Atomic Level Catalyst Design, an Energy Frontier Research Center funded by the U.S. Department of Energy, Office of Science, Office of Basic Energy Sciences under Award Number DE-SC0001058. We thank Ms. Kim Hutchison at North Carolina State University, and Ms. Wanda LeBlanc for helping in getting the ICP, and XRD data, respectively.

References

1. Ashcroft AT, Cheetham AK, Green MLH, Vernon PDF (1991) Partial oxidation of methane to synthesis gas using carbon-dioxide. Nature 352(6332):225–226.
2. Bitter JH, Seshan K, Lercher JA (2000) On the contribution of X-ray absorption spectroscopy to explore structure and activity relations of Pt/ZrO_2 catalysts for CO_2/CH_4 reforming. Top Catal 10(3–4):295–305.
3. Bradford MCJ, Vannice MA (1999) CO_2 reforming of CH_4 over supported Ru catalysts. J Catal 183(1):69–75.
4. Carrara C, Munera J, Lombardo EA, Cornaglia LM (2008) Kinetic and stability studies of Ru/La(2)O(3) used in the dry reforming of methane. Top Catal 51(1–4):98–106.
5. Choudhary VR, Mondal KC, Mamman AS, Joshi UA (2005) Carbon-free dry reforming of methane to syngas over $NdCoO_3$ perovskite-type mixed metal oxide catalyst. Catal Lett 100(3–4): 271–276.
6. Daza CE, Gallego J, Moreno JA, Mondragon F, Moreno S, Molina R (2008) CO_2 reforming of methane over Ni/Mg/Al/Ce mixed oxides. Catal Today 133:357–366.
7. Erdohelyi A, Cserenyi J, Solymosi F (1993) Activation of CH_4 and its reaction with CO_2 over supported RH catalysts. J Catal 141(1):287–299.

8. Fan M-S, Abdullah AZ, Bhatia S (2009) Catalytic technology for carbon dioxide reforming of methane to synthesis gas. Chemcatchem 1(2):192–208.
9. Ferreira-Aparicio P, Marquez-Alvarez C, Rodriguez-Ramos I, Schuurman Y, Guerrero-Ruiz A, Mirodatos C (1999) A transient kinetic study of the carbon dioxide reforming of methane over supported Ru catalysts. J Catal 184(1):202–212.
10. Gallego GS, Batiot-Dupeyrat C, Barrault J, Florez E, Mondragon F (2008) Dry reforming of methane over LaNi1-yByO3 ± delta (B = Mg, Co) perovskites used as catalyst precursor. Appl Catal A Gen 334(1–2):251–258.
11. Garcia-Dieguez M, Pieta IS, Herrera MC, Larrubia MA, Malpartida I, Alemany LJ (2010) Transient study of the dry reforming of methane over Pt supported on different gamma-$Al_{(2)}O_{(3)}$. Catal Today 149(3–4):380–387.
12. Gaur S, Pakhare D, Wu HY, Haynes DJ, Spivey JJ (2012) CO_2 reforming of CH_4 over Ru-substituted pyrochlore catalysts: effects of temperature and reactant feed ratio. Energy Fuels 26(4):1989–1998.
13. Hayakawa I, Kamizono H (1993) Durability of an $La_2Zr_2O_7$ waste form containing various amounts of simulated HLW elements. J Nucl Mater 202(1–2):163–168.
14. Haynes DJ, Berry DA, Shekhawat D, Spivey JJ (2008) Catalytic partial oxidation of *n*-tetradecane using pyrochlores: effect of Rh and Sr substitution. Catal Today 136(3–4):206–213.
15. Haynes DJ, Berry DA, Shekhawat D, Spivey JJ (2009) Catalytic partial oxidation of *n*-tetradecane using Rh and Sr substituted pyrochlores: effects of sulfur. Catal Today 145(1–2):121–126.
16. Haynes DJ, Campos A, Berry DA, Shekhawat D, Roy A, Spivey JJ (2010) Catalytic partial oxidation of a diesel surrogate fuel using an Ru-substituted pyrochlore. Catal Today 155(1–2):84–91.
17. Hou ZY, Yokota O, Tanaka T, Yashima T (2003) Characterization of Ca-promoted Ni/alpha-Al_2O_3 catalyst for CH_4 reforming with CO_2. Appl Catal A Gen 253(2):381–387.
18. Hou ZY, Yokota O, Tanaka T, Yashima T (2003) Investigation of CH_4 reforming with CO_2 on meso-porous Al_2O_3-supported Ni catalyst. Catal Lett 89(1–2):121–127.
19. Hu YH, Ruckenstein E (1996) An optimum NiO content in the CO_2 reforming of CH_4 with NiO/MgO solid solution catalysts. Catal Lett 36(3–4):145–149
20. Luo JZ, Yu ZL, Ng CF, Au CT (2000) CO_2/CH_4 reforming over Ni-La_2O_3/5A: an investigation on carbon deposition and reaction steps. J Catal 194(2):198–210.
21. O'Connor AM, Schuurman Y, Ross JRH, Mirodatos C (2006) Transient studies of carbon dioxide reforming of methane over Pt/ZrO_2 and Pt/Al_2O_3. Catal Today 115(1–4):191–198.
22. Sadykov VA, Gubanova EL, Sazonova NN, Pokrovskaya SA, Chumakova NA, Mezentseva NV, Bobin AS, Gulyaev RV, Ishchenko AV, Krieger TA, Mirodatos C (2011) Dry reforming of methane over Pt/PrCeZrO catalyst: kinetic and mechanistic features by transient studies and their modeling. Catal Today 171(1):140–149.
23. Shamsi A, Baltrus JR, Spivey JJ (2005) Characterization of coke deposited on Pt/alumina catalyst during reforming of liquid hydrocarbons. Appl Catal A Gen 293:145–152.
24. Souza M, Aranda DAG, Schmal M (2001) Reforming of methane with carbon dioxide over Pt/ZrO_2/Al_2O_3 catalysts. J Catal 204(2):498–511.
25. Subramanian MA, Aravamudan G, Rao GVS (1983) Oxide pyrochlores: a review. Prog Solid State Chem 15(2):55–143.
26. vanKeulen ANJ, Seshan K, Hoebink J, Ross JRH (1997) TAP investigations of the CO_2 reforming of CH_4 over Pt/ZrO_2. J Catal 166(2):306–314.
27. Zhang ZL, Tsipouriari VA, Efstathiou AM, Verykios XE (1996) Reforming of methane with carbon dioxide to synthesis gas over supported rhodium catalysts.1. Effects of support and metal crystallite size on reaction activity and deactivation characteristics. J Catal 158 (1):51–63.
28. Zhang ZL, Verykios XE (1996) Mechanistic aspects of carbon dioxide reforming of methane to synthesis gas over Ni catalysts. Catal Lett 38(3–4):175–179.

18

Framework flexibility and rational design of new zeolites for catalysis

Asel Sartbaeva · Stephen A. Wells

Abstract Zeolite materials have proved very useful as chemical catalysts and the search for new zeolite structures with novel channel and pore shapes is ongoing. We discuss a geometric feature of zeolite frameworks, the flexibility window, which may provide a criterion to identify hypothetical structures which can be synthesised as zeolites. In recent research using data on zeolite frameworks under compression, we show strong links between this geometric feature and the physics of zeolite frameworks.

Keywords Zeolite · Framework · Flexibility window · Geometric simulation

Introduction

The aluminosilicate zeolites have provided the chemical industry with a number of very useful catalytic materials. The corner-linked tetrahedral framework of a zeolite defines a system of channels with well-defined geometry, accessible to small molecules. As a result, a zeolite can display very high surface area, shape selectivity, and chemical activity provided by solid-acid sites in the framework. This is a near-perfect combination of properties for a catalyst. Petrochemical refining is perhaps their most important application—a majority of the world's petrol supply now passes through the pores of zeolites with the **FAU** and **MFI** frameworks.

A. Sartbaeva (✉)
Department of Chemistry, Inorganic Chemistry Laboratory, University of Oxford, South Parks Road, Oxford OX1 3QR, UK
e-mail: asel.sartbaeva@chem.ox.ac.uk

S. A. Wells
Department of Physics and Centre for Scientific Computing, University of Warwick, Gibbet Hill Road, Coventry CV4 7AL, UK

Zeolite catalysts would undoubtedly be even more widely used if a wider variety of framework structures were available. At present, however, only just over 200 zeolite framework types are recognized, and the rate of discovery of new framework types remains slow, with only a few new types discovered annually [1]. Much computational effort has been and is being expended in the generation of new hypothetical structures. These are 'plausible' arrangements of atoms in a network of corner-linked tetrahedra. Such a hypothetical structure is intended to be a candidate for synthesis as a novel zeolite—a target at which synthetic chemists can take aim. The libraries of such hypothetical structures now contain millions of candidates. This wealth of possibilities has not, however, led to an explosion in the rate of synthesis of new zeolite frameworks.

Instead we are confronted with a new puzzle. Why is it that, of these millions of candidates, so very few actually exist in reality? Clearly a large number of the hypothetical structures are not, in fact, plausible candidates for synthesis. Indeed, we are faced with not one but two bottlenecks. First, we require criteria that will identify the right candidate structures for attempted synthesis. Second, we need more insight into the process of zeolite structure formation, which will improve our ability to generate the desired structure by control of the conditions of synthesis and use of structure-directing agents.

It is already clear from the process of structure generation that excessive distortion of the tetrahedral geometry in the framework is a sign of an implausible structure. In the crystal structures of natural and synthetic zeolites, the geometry of a TO_4 (SiO_4 or AlO_4) unit is typically quite regular, with only small (a few degrees) variation in the

O–T–O angle from the tetrahedral ideal of 109.47°, and with minimal variation in the T–O bond lengths. Larger distortions come at an excessively high energy cost. This has been identified as a rejection criterion in surveys of hypothetical structures [2]. However, this still leaves a large number of apparently plausible candidates which are predicted, using empirical interatomic potentials, to have low framework distortion energies comparable to those calculated for known existing structures. Are there additional criteria which explain why these hypothetical structures have not yet been—or cannot be—synthesised?

Method

We have been investigating this question using the approach of 'geometric simulation' [3–6]. This is a deliberately simplified modeling approach, in which we neglect detailed energetics and long-range interactions, focussing instead on the most local interactions: interatomic bonding, and contact steric repulsion. A distinctive feature of the approach is that bonding is represented, not by the conventional use of two- and three-body potentials, but rather by a geometric template or 'ghost' describing the ideal geometry of a bonded group of atoms. In this case, the bonding within a TO_4 group is represented by a regular tetrahedron, to whose vertices atoms are tethered by harmonic constraints. This approach and its many applications in mineral physics are the subject of a recent review by the current authors [6].

Results and discussion

The advantage of geometric simulation, as a complement to conventional methods, is that it focuses attention on geometric features of the framework behavior that would otherwise be obscured. In the case of zeolites, a survey of zeolite frameworks revealed an unexpected phenomenon [3, 7]. In known frameworks (naturally occurring and synthetic), the geometry of the tetrahedral units can be made ideal using geometric simulation, i.e., the atoms in the framework can be so arranged that no steric clashes occur between framework atoms, the T–O–T angles become equal to exactly the tetrahedral angle of 109.47°, and the T–O bond lengths reach their ideal values of 1.61 Å for Si–O and 1.75 Å for Al–O, to within a small numerical tolerance.

Flexibility window

This idealisation of the geometry can be carried out over a range of densities, thus defining what we term a 'flexibility window' [3]. Strikingly, known structures under ambient conditions tend to lie at the low-density edge of the window, indicating that zeolites can be considered as maximally expanded materials. We stress that it is not obvious a priori that this idealisation should be possible at all. Indeed, we found, and subsequent research has confirmed [8, 9], that this flexibility window phenomenon is almost universal in all known zeolites, with only a single known exception (framework **GOO**, goosecreekite), but is very uncommon in hypothetical structures. This immediately suggests that the existence of a flexibility window in a hypothetical structure is one of the missing criteria for the selection of good candidates for synthesis.

Flexibility and compression

In our recent research, we focus on elucidating the links between the geometric phenomenon of the flexibility window and the physical properties of zeolite framework. In particular, we make use of data on the evolution of zeolite framework structures under compression. Geometric simulations, to determine whether a compressed structure lies within its flexibility window, can be carried out given the structure under ambient conditions and the cell parameters under pressure, even if detailed refinement of the crystal structure under pressure is not possible. We find that the flexibility window is linked to structural phase transitions under pressure, both displacive and reconstructive. Here, we review two principal results, on the analcime (**ANA**) and silicalite (**MFI**) frameworks.

ANA group zeolites

The analcime group of zeolites with the **ANA** framework type [10] includes minerals with a wide variety of compositions and has four aluminosilicate end members—analcime, leucite, pollucite, and wairakite. Using geometric simulation to analyse compression data on analcime, we found that the range of densities over which the ambient (cubic) crystal structure persists matches the range of the flexibility window. The transition to a much lower symmetry (triclinic) form, which occurs at a relatively low pressure of around 1 GPa, takes place once the limit of the flexibility window is reached [11]. This intriguing result led us to investigate other members of the analcime group. Although different members of the group display a wide variety of crystal structures, in each case we observed the same phenomenon whereby a displacive phase transition under compression occurs around the edge of the flexibility window [12]. Remarkably, our results are even sensitive to Al/Si ordering in the framework. Geometric simulations for wairakite, which displays Al/Si ordering, match the experimental data when ordering is included but not when a uniform average T–O distance is used [13].

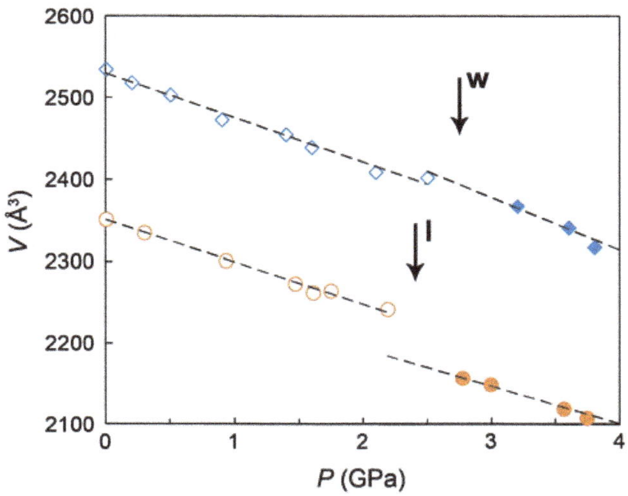

Fig. 1 P–V data for wairakite (*blue diamonds*) and leucite (*brown circles*) annotated with results of geometric simulation. Points are labeled with *open symbols* if the framework is found to be within the flexibility window and with *closed symbols* if the framework is intrinsically distorted, and thus is outside the flexibility window. For both zeolites, the ambient/low-P structure is perfectible, while the high-P structure is not. Locations of the phase transitions of wairakite (W) and leucite (L) are indicated by *arrows*. Modified from [13]

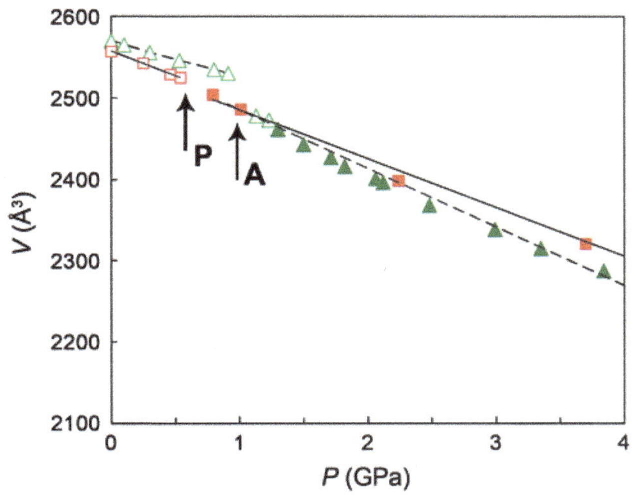

Fig. 2 P–V data for analcime (*green triangles*) and pollucite (*red squares*) annotated with results of geometric simulation. Points are labeled with *open symbols* if the framework is found to be perfectible (within the flexibility window) and with *closed symbols* if the framework is intrinsically distorted (outside the flexibility window). For both, the ambient/low-P structure is perfectible, while the high-P structure is generally not. Locations of the phase transitions of analcime (A) and pollucite (P) are indicated by *arrows*. Modified from [13]

In Figs. 1 and 2 we summarise data on analcime-group minerals under compression. Each experimental P–V data point is annotated with a open symbol when the structure lies within its flexibility window and with a closed symbol otherwise. The locations of phase transitions under compression are marked.

Silicalite

In a recent study on silicalite (**MFI** framework), we have obtained an intriguing result connecting the flexibility window to pressure-induced amorphisation of the framework [14]. This structure displays PIA when compressed using non-penetrating pressure-transmitting media (silicone oil), i.e., the empty channels within the structure gradually collapse under compression, with rebonding of the framework in a denser, more 'glassy' form. When penetrating media are used—CO_2 or argon—PIA is prevented and the structure remains crystalline up to pressures in excess of 20 GPa [15, 16].

Geometric simulations revealed some unusual features of these behaviors. During PIA, the crystalline portions of the structure appear to remain within their flexibility window. In contrast, we find that the effect of penetrating pressure-transmitting media entering the channels of the structure is to push the structure out of its flexibility window, inducing strain in the framework. Counterintuitively, it is flexibility that is necessary for PIA to proceed, whereas the introduction of strain in the framework acts to prevent its collapse and rebonding. These results are summarised in Fig. 3.

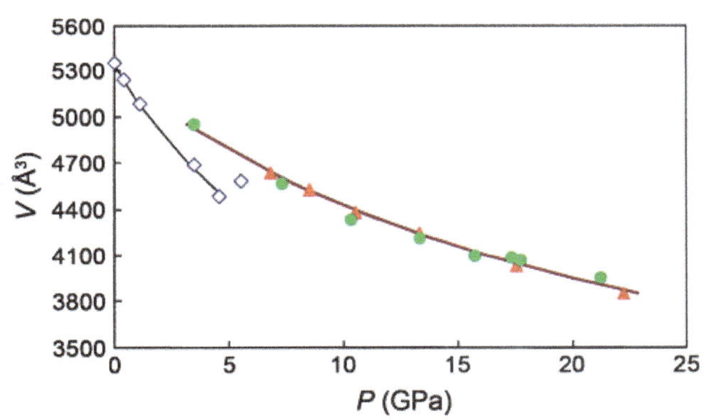

Fig. 3 P–V data for silicalite with (*green circles*—CO_2, *red triangles*—Ar), without (*open diamonds*) penetrating pressure medium. *Open symbols* indicate data within the flexibility window and *filled symbols* indicate data outside the window. The *lines* fit to a second-order Birch–Murnaghan EoS. Modified from [14]

We anticipate that these and further ongoing investigations will shed more light on the links between the geometry and physics of zeolite frameworks. Such insights may be instrumental in permitting the systematic design, synthesis and application of novel zeolite framework structures to meet the catalytic challenges of the future.

Acknowledgments AS thanks the Royal Society (University Research Fellowship) for funding, SAW thanks the Leverhulme foundation for funding.

References

1. http://www.iza-structure.org/databases/. Accessed 29 Aug 2012
2. Zwijnenburg M, Simperler A, Wells S, Bell R (2005) Tetrahedral distortion and energetic packing penalty in "Zeolite" frameworks: linked phenomena? J Phys Chem B 109(31):14783–14785
3. Sartbaeva A, Wells S, Treacy M, Thorpe M (2006) The flexibility window in Zeolites. Nat Mater 5:962–965
4. Wells S, Dove M, Tucker J (2002) Finding best-fit polyhedral rotations with geometric algebra. J Phys Cond Matter 14:4567–4584
5. Wells S, Dove M, Tucker J (2004) Reverse Monte Carlo with geometric analysis—RMC + GA. J App Cryst 37:536–544
6. Wells SA, Sartbaeva A (2012) Template-based geometric simulation of flexible frameworks. Materials 5:415–431.
7. Rivin I (2006) Geometric simulations: a lesson from virtual zeolites. Nat Mater 5:931–932
8. Kapko V, Dawson C, Rivin I, Treacy M (2012) Flexibility of ideal zeolite frameworks. Phys Rev Lett 107:164304
9. Kapko V, Dawson C, Treacy M, Thorpe M (2010) Density of mechanisms within the flexibility window of Zeolites. Phys Chem Chem Phys 12:8531-8541
10. Gatta G, Nestola F, Ballaran TB (2006) Elastic behavior, phase transition, and pressure induced structural evolution of analcime. Am Min 91:568–578
11. Sartbaeva A, Gatta G, Wells SA (2008) Flexibility window controls pressure-induced phase transition in analcime. EuroPhys Lett 83:26002.
12. Gatta GD, Sartbaeva A, Wells SA (2009) Compression behaviour and flexibility window of the analcime-like feldspathoids: experimental and theoretical findings. Eur J Miner 21:571–580.
13. Wells SA, Sartbaeva A, Gatta G (2011) Flexibility windows and phase transitions of ordered and disordered ANA framework zeolites. EuroPhys Lett 94:56001.
14. Sartbaeva A, Haines J, Cambon O, Santoro M, Gorelli F, Levelut C, Garbarino G, Wells SA (2012) Flexibility windows and compression of monoclinic and orthorhombic silicalites. PRB 85:064109.
15. Haines J, Isambert LCA, Hebert P, Kohana S, Keen D, Hammouda T, Andrault D (2009) Topologically ordered amorphous silica obtained from the collapsed siliceous zeolite, silicalite-1-F: a step toward"Perfect" glasses. J Am Chem Soc 131:12333–12338
16. Haines J, Cambon O, Santoro LCM, Gorelli F, Garbarino G (2010) Deactivation of pressure-induced amorphization in silicalite SiO_2 by insertion of guest species. J Am Chem Soc 132:8860–8861

19

Studies of amino acids for inhibition of aldol condensation and dissolution of polymeric product of aldehyde in alkaline media

Hiren M. Bhajiwala · Harshad R. Patil · Virendrakumar Gupta

Abstract Compounds with acidic and basic functional groups (aspartic acid, arginine, and cystine) are evaluated for inhibition of aldol condensation and dissolution of product in alkaline media. Inhibition efficiency of amino acids were studied at varied mole ratios of acetaldehyde to antipolymerant (2, 5, and 10), reaction temperature, and caustic concentration as a function of time (0.25, 1.0, 2.0 and 24.0 h) using UV spectrophotometry. Results obtained have been compared with efficiency of known amine derivatives. Arginine showed better efficiency when compared with aspartic acid and cysteine. Furthermore, data indicate that presence of carboxylic group in addition to amine group facilitates inhibition as well as dissolution of aldol polymer in alkaline media, while compounds with amine group acts as inhibitor for aldol condensation only.

Keywords Aldol condensation · Inhibition · Dissolution · UV spectroscopy · FTIR

Introduction

Olefins are produced by naphtha or gas (ethane/propane/other alkane) cracking in petrochemical industry. Olefins are then used as raw material to produce commodity or specialty chemicals and polymers such polyethylene, polypropylene, polyvinyl chloride, mono-ethylene glycol, etc. These chemicals and polymers are used in different sectors such as agriculture, automobile, packaging, life appliances, apparels, etc.

H. M. Bhajiwala · H. R. Patil · V. Gupta (✉)
Reliance Technology Group, Reliance Industries Limited, Hazira Complex, Surat 394510, Gujarat, India
e-mail: virendrakumar.gupta@ril.com

Cracking of naphtha to produce olefins is a highly complex chemical phenomena because it involves multiple reactions of intermediates, reactants, and products. Acid gases such as H_2S, CS_2, –RSH, etc., are also produced along with olefins. The removal of acid gases is essential for better performance of downstream catalytic purifications units and production of high-purity chemicals and monomers for further conversion to the required chemicals or polymers. The process for removal of acid gases in naphtha cracker involves treatment of cracked product with caustic solution in absorption column. This facilitates the removal of acid gases from other chemicals in the form of Na_2S. The acetaldehyde formed during cracking undergoes aldol condensation in a caustic absorption column, resulting in the formation of polymeric materials or red oil chocking the absorption column. Different types of antipolymerants are therefore added in the absorption column to prevent the formation of polymeric material for increasing the efficiency of acid removal process in naphtha cracking. In view of this, understanding the aldol condensation process of aldehyde and its prevention in alkaline media is critical in naphtha cracking process.

Aldol condensation involving aldehyde group preset in naturally occurring or synthetic chemical [9, 11] is one of the most thoroughly studied reaction in organic chemistry. Aldol condensation is catalyzed by acids as well as bases. Literature studies indicate that the nature of aldehyde compounds and catalyst type has been investigated in detail for in-depth understanding of condensation process [1, 10, 14, 17, 18, 21]. However, no studies are reported in published literature except patents for prevention of aldol condensation in alkaline media which is very critical in naphtha cracking process for efficiency improvement.

Patent literature on prevention of aldol condensation in alkaline media indicates amine containing chemicals

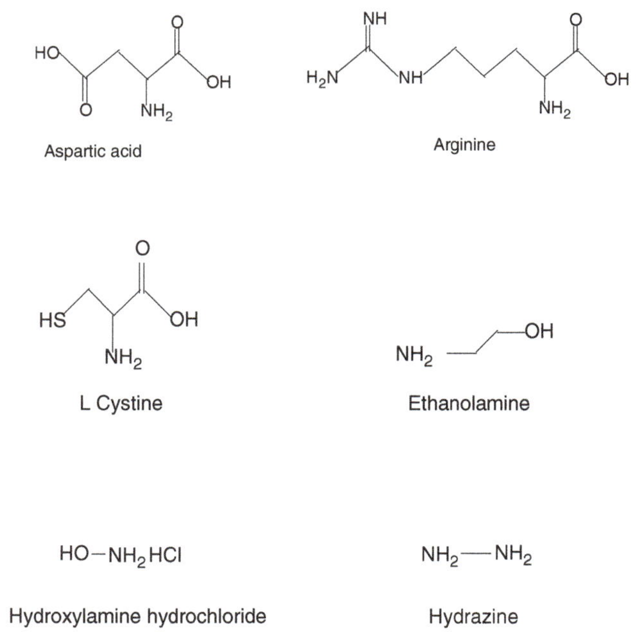

Fig. 1 Structures of amino acids and amine derivatives (used in literature) used as antipolymerant

(primary, secondary amines) and borohydrates are used as antipolymerants. The amine compounds react with aldehyde to produce imines that prevent aldol polymerization condensation reaction. The electron donating characteristics of antipolymerant [6–8] is an important parameter for determining its efficiency. The examples include hydroxylamine and its salts [15], hydrazine and its derivative [8], carbohydrazine and its derivatives [6], amino phenol, amino benzoic acid and its salts [7], urea [20], ethylenediamine [2], acetoacetate ester compounds [16] and borohydride [4]. In addition to the action of antipolymerant, the mixing phenomena and design of caustic absorption column in naphtha cracker plant are critical in preventing the formation of polymeric material. Subramaniyam et al. [19] explored the concept of using chemicals which not only prevent the formation of polymer, but also help in dissolving the polymeric material if any formed. Examples of such compounds containing both amine and carboxylic acid groups are 2-aminoethanesulfonic acid, 6-amino hexanoic acid, etc.

We have initiated a systematic work on developing high performing chemicals for prevention of aldol condensation in alkaline media and understanding the chemical structure–performance relationship [12, 13]. In the present communication, we report our studies on amino acids (aspartic acid, arginine, and cysteine, Fig. 1) as antipolymerants for the prevention of aldol condensation and dissolution of polymeric material in alkaline media. The results on amino acids are also compared with known compounds such as ethanolamine, hydroxylamine hydrochloride, hydrazine (Fig. 1).

Experimental

Chemicals

Acetaldehyde, aspartic, arginine, and cystine (AR grade, Sigma Aldrich, India.), ethanolamine, hydroxylamine hydrochloride, hydrazine (Labort, India), caustic (AR grade, SD Fine Chemical Product, India) were used for synthesis and performance reactions. Both arginine and aspartic acid are levorotary type isomers. These are naturally occurring amino acids derived from biological or animal sources.

Performance test of antipolymerants for inhibition of aldol reaction

Amino acid (0.018 mol) was mixed with 40 ml (2.5 M) aqueous sodium hydroxide in a flat-bottomed flask and placed on a hot plate at 350 rpm. Acetaldehyde was added under stirring at 350 rpm. The reaction was monitored by measuring transmittance of solution by UV spectrophotometer (PE Lambda35) at 800 nm at different time intervals for inhibition efficiency measurement [20]. Comparative experiment was conducted and UV transmittance was measured by same procedure without addition of antipolymerant and for amine derivatives reported in literature. Effect of amino acid concentration at 2, 5, and 10 mole ratio of acetaldehyde to antipolymerant and effect of temperature at 25, 40, and 55 °C were also investigated.

Performance test of antipolymerants for dissolution of aldol condensation polymer

Acetaldehyde (0.036 mol) was added to 40 ml (2.5 M) caustic solution in 100-ml conical flask under stirring. Color of the solution changes to hazy yellow after few minutes due to aldol condensation of acetaldehyde under alkaline condition. Yellow precipitate of aldol polymer started appearing after 15 min. Amino acid compound (0.018 mol) was added to this solution and transmittance of solution was measured by UV spectrophotometer at time intervals of 0.25, 1, and 2 h at 800 nm to measure dissolution efficiency of amino acid. Performance of the reported amine derivatives was also studied for dissolution of aldol condensation polymer for comparison.

Product characterization of aldol condensation polymer

IR spectra of dried acetaldehyde–amino acid complex were recorded on a Perkin Elmer make Fourier transform infrared spectrometer (FTIR) with 2 cm^{-1} resolution and 32 scans by mixing the sample with KBr.

Results and discussion

Aldol condensation reactions involving aldehyde groups are discussed in both academia and industry. Mechanism of liquid red oil formation, which is a product of aldol condensation of few numbers of aldehyde monomer, is shown in Scheme 1. Further polymerization leads to the formation of solid polymer.

In base-catalyzed aldol condensation reaction, acidic hydrogen or β-hydrogen of acetaldehyde is eliminated in the presence of hydroxyl ion. Reactive enolate of aldehyde is formed during the reaction which acts as nucleophile and attacks the electrophilic carbon of another carbonyl (aldehyde) molecule forming 2-butenal [3]. It further gives 2,4-hexenal with double bond called red oil (Scheme 1a) which is partially soluble in caustic solution

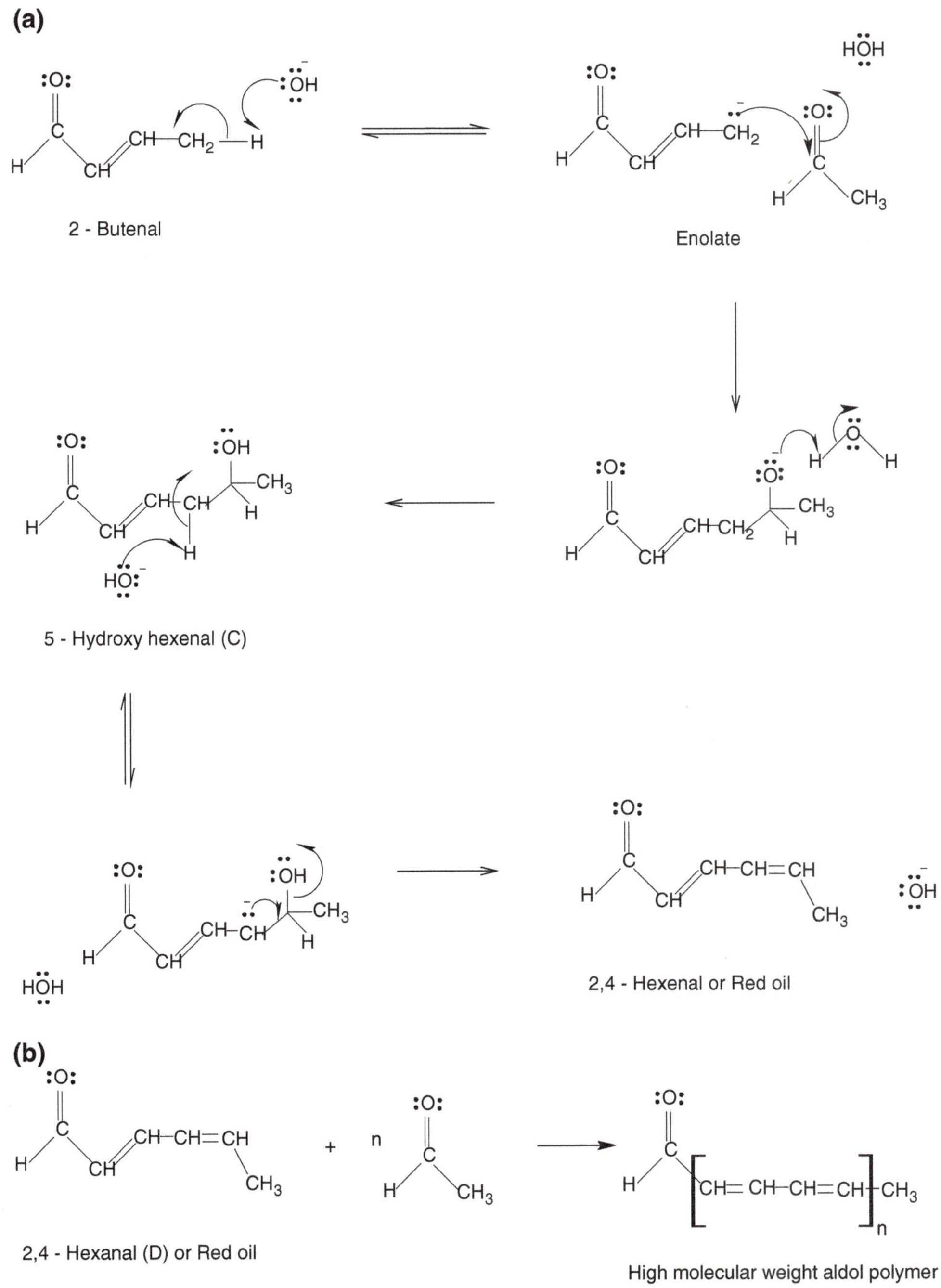

Scheme 1 Probable mechanism for formation of red oil (**a**) and polymeric product (**b**)

and insoluble in water. Aldol condensation of dissolved oxygenated hydrocarbons (carbonyls, such as aldehyde and ketones) produces polymeric products that are commonly referred to as red oil in caustic tower [5] which further polymerize to give high molecular weight red polymer (Scheme 1b).

Table 1 Studies on effect of type of compounds and quantity for inhibition efficiency of aldol condensation reaction

Entry	Compounds[a]	Mole ratio[b]	%*T* at 800 nm on UV			
			0.25 h	1 h	2 h	24 h
1	No compound		1.2	0.2	<0.1	<0.1
2	Arginine	2	99	99	99	98
3		5	99	98	94	85
4		10	91	82	74	61
5	Cystine	2	97	92	79	63
6		5	95	80	67	42
7		10	90	68	40	14
8	Aspartic acid	2	99	73	44	11
9		5	94	54	23	3
10		10	90	27	5	0.1
11	Ethanolamine	2	99	96	90	82
12		5	95	90	86	79
13		10	83	81	72	61
14	Hydroxyl amine hydrochloride	2	86	72	61	34
15		5	80	52	36	12
16		10	72	42	20	4
17	Hydrazine	2	95	88	73	46
18		5	84	72	60	34
19		15	73	61	46	21

Reaction temperature = 25 °C

[a] Compound in 40 ml 2.5 M caustic solution in DM water

[b] Mole ratio of aldehyde to inhibiting compound

Studies on efficiency of antipolymerant for inhibiting aldol condensation reaction

Effect of reaction time

Effect of amino acid compounds (aspartic acid, arginine, and cystine) on inhibition of aldol condensation reaction was investigated at various mole ratios (acetaldehyde/antipolymerant mole ratio of 2.0, 5.0, and 10.0) at 25 °C in 2.5 M caustic solutions as function of time (0.25, 1, 2, and 24 h) as shown in Table 1.

Addition reaction of aldehyde takes place at faster rate when amino acid compounds are not used. The solution turned hazy light yellow to reddish yellow due to the formation of polymer precipitates. Transmittance of the solution was reduced to below 0.1 % after 2 h of reaction (Fig. 2a). The solution turned to a thick slurry in 24 h which can affect the caustic tower efficiency. (Fig. 2b). Arginine indicated 98, 85, and 61 % transmittance at aldehyde to arginine mole ratio of 2, 5, and 10 after 24 h. Transmittance values for cystine were 63, 42, and 14 % and 11, 3 and 0.1 % transmittance for aspartic acid at aldehyde to inhibiting compound mole ratio of 2, 5, and 10 after 24 h.

Presence of amino acid compounds reduced the reaction rate of acetaldehyde to varied efficiency levels depending on their chemical structures. This study as function of time and quantity of such compounds indicated that arginine (among three compounds) inhibits reactions most efficiently followed by cystine and aspartic acid due to presence of two basic group ($-NH_2$) and one acidic group (>COOH). Visually solution remains transparent even after 1–2 h of reaction time (Fig. 2c).

Aspartic acid has shown lowest efficiency probably due to presence of two acid groups (>COOH) at chain end and one basic ($-NH_2$) group making such compound less efficient for inhibiting the reaction. Cystine exhibited better inhibition effect than aspartic acid and less than arginine

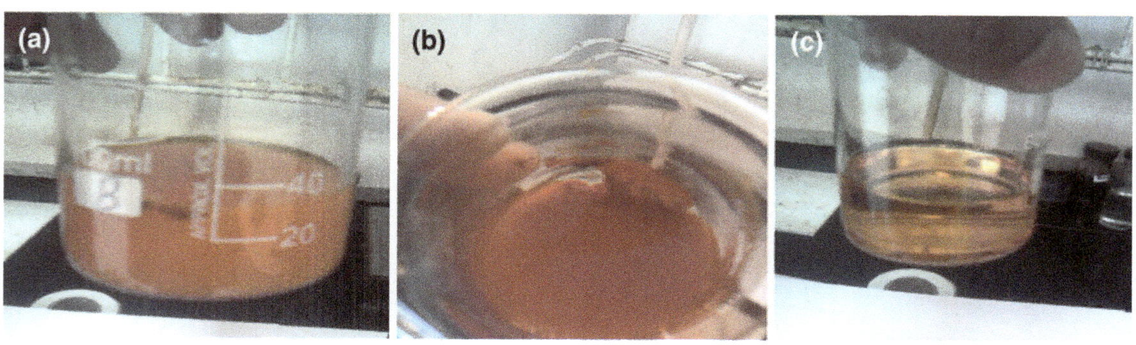

Fig. 2 **a** Acetaldehyde in caustic media without addition of compound in 1 h. **b** Acetaldehyde in caustic media without addition of compound in 24 h. **c** Acetaldehyde in caustic media with addition of arginine

Scheme 2 Probable mechanism for inhibition of aldol condensation

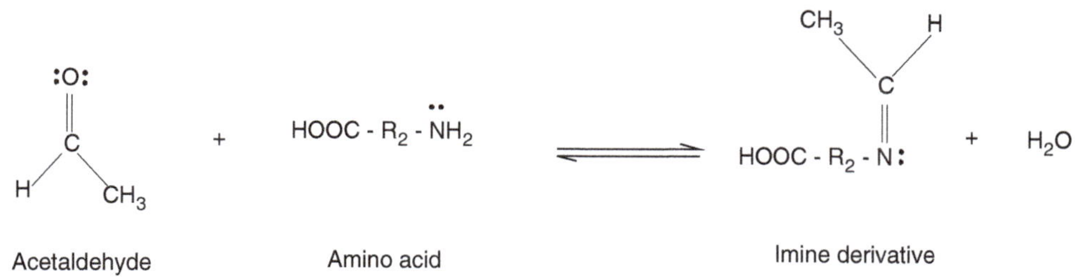

R_1= CH_3
R_2= N, S containing alkanes

Table 2 Effect of temperature on aldol condensation reaction for different chemical compounds

Entry	Compounds[a]	%*T* at 800 nm on UV after 0.25 h			%*T* at 800 nm on UV after 2 h min		
		25 °C	40 °C	55 °C	25 °C	40 °C	55 °C
1	No compound	1.2	>0.1	>0.1	>0.1	>0.1	>0.1
2	Arginine	99	99	99	99	99	99
3	Cystine	97	90	79	79	62	10
4	Aspartic acid	99	71	52	44	2	0.1

[a] Experimental condition: acetaldehyde:amino acid ratio = 2.0 added in 40 ml 2.5 M caustic solution

Table 3 Effect of catalyst (caustic) concentration on inhibition of aldol condensation reaction for different inhibition compounds

Entry	Compounds[a]	Catalysts[b]	%*T* at 800 nm on UV		
			1 h	2 h	24 h
1	No compound	–	0.2	0.1	0.1
2	Arginine	2.50	99	99	98
3		1.25	99	99	98
4		0.63	99	99	99
5	Cystine	2.50	92	79	63
6		1.25	94	85	77
7		0.63	99	89	90
8	Aspartic acid	2.50	73	44	11
9		1.25	98	98	72
10		0.63	99	99	80

Reaction temperature = 25 °C

[a] Experimental condition: acetaldehyde: amino acid ratio = 2.0

[b] Catalyst concentration molar solution (40 ml)

due to the presence of two basic (–SH and –NH_2) functional groups. Perhaps proton associated with –SH has less donating effect compared to –NH_2 resulting in lower efficiency than arginine which has two –NH_2 groups.

Three reported amine derivatives (ethanolamine, hydroxylamine hydrochloride, hydrazine) were also evaluated for inhibition efficiency. Ethanolamine showed transmittance of 82, 79, and 61 at 2, 5, and 10 mole ratio of acetaldehyde to ethanolamine. Hydroxylamine hydrochloride and hydrazine were also studied under similar condition. Performance of all evaluated compounds showed performance sequence of arginine > ethanolamine > cystine > hydroxylamine hydrochloride > hydrazine > aspartic acid.

It was evident that as ratio of acetaldehyde to inhibitor increased, transmittance of the reaction mixture decreased. Transmittance decreased rapidly for all compounds at molar ratio of 10 which indicated that sufficient inhibitor concentration is required for inhibition of aldol condensation reactions.

During aldol inhibition reaction mechanism, amine group attacks the carbonyl group of acetaldehyde to form carbinolamine intermediate which undergoes loss of water molecule to produce imine [22, 23]. Inhibition reaction is explained in Scheme 2. Imine formation in the case of aspartic acid is quite difficult due to two acidic groups at both the side ends. While in the case of arginine and cystine, electro-donating group present on both ends may be the reason for higher inhibition activity.

Effect of reaction temperature

Effect of reaction temperature on inhibition reaction was investigated at 25, 40, and 55 °C (Table 2) at acetaldehyde to inhibitor mole ratio of 2. Higher temperature increased the rate of reaction as indicated by the decrease in transmittance of solution. This explained that acetaldehydes were converted to polymeric product at faster rate at 55 and 40 °C compared to 25 °C. Results indicated that arginine was most effective inhibitor compared to cystine and aspartic acid. Arginine shows 99 %*T* at all temperatures, whereas aspartic and cystine show 0.1 and 10 %*T*, respectively, at 55 °C. Efficiency of inhibitor was in order of arginine > cystine > aspartic acid.

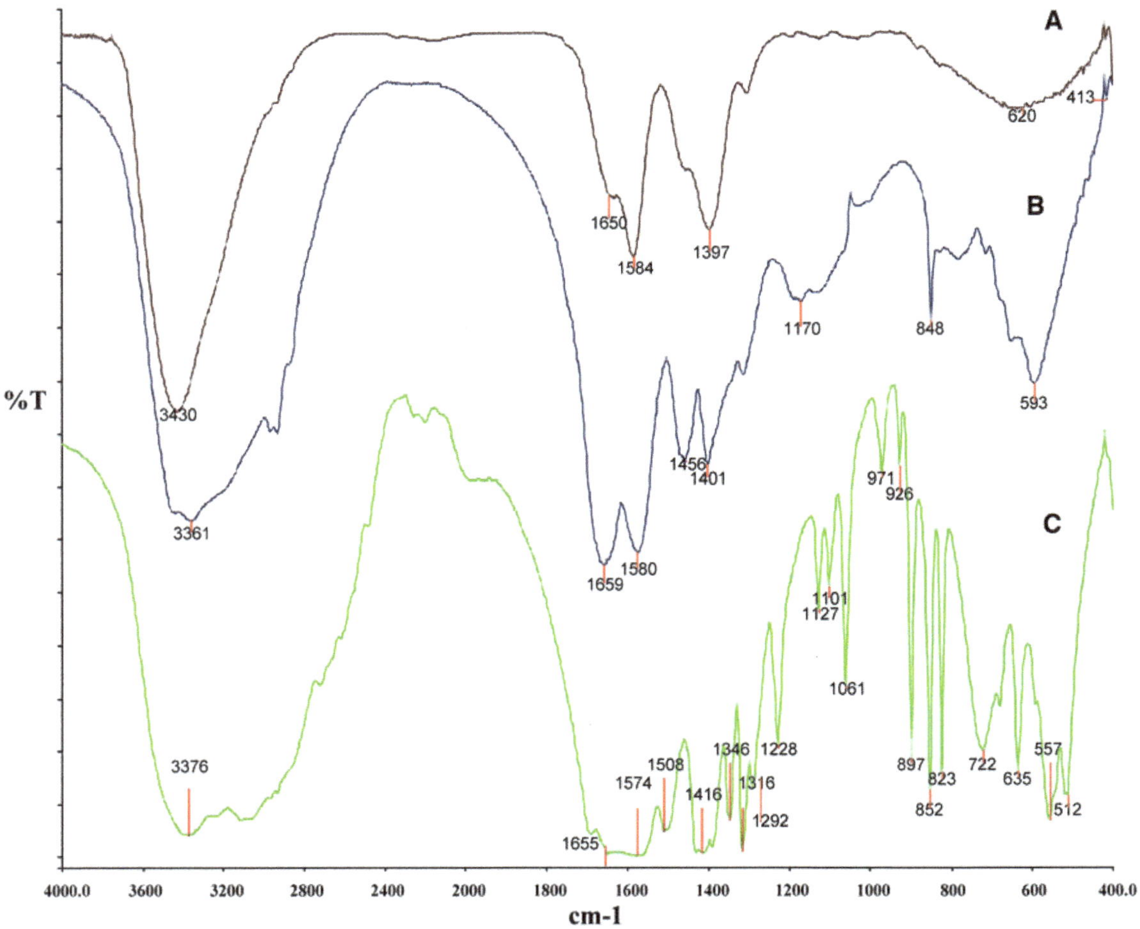

Fig. 3 IR spectra of complexes *A* acetaldehyde/cystine, *B* acetaldehyde/aspartic acid, *C* acetaldehyde/arginine

Effect of catalyst concentration

Rate of aldol condensation reaction depends on the presence of catalyst (caustic) concentration; as catalyst concentration increases reaction rate increases. Reducing concentration of the catalyst from 2.5 to 0.625 M exhibited distinct effects on transmittance (Table 3). Reduction in catalyst concentration showed increased transmittance indicating reduction in the formation of aldol polymer. Arginine shows 98–99 % transmittance at all catalyst concentrations after 24 h indicating most effective inhibition efficiency. Cystine is less effective compared to arginine with 63 and 77 % transmittance at 2.5–1.25 M concentrations. Aspartic acid showed lowest efficiency at 11 and 72 % transmittance at 2.5 and 1.25 M concentrations.

FTIR spectroscopy

Complexes of acetaldehyde with amino acids were studied through FTIR. Complexes were obtained after evaporation of water on mixing these chemicals in caustic media. Stretching band at 1,650, 1,655, and 1,659 cm^{-1} in cystine, arginine, and aspartic acid indicated the presence of >C=O of amino acid. Slightly shifted band in all three complexes at 1,584, 1,580, and 1,574 cm^{-1} (Fig. 3 a, b, c for cystine, aspartic acid, and arginine complexes with aldehydes) indicated >C=N–R (imine) formation between amino acid

Table 4 Studies on the effect of type of compounds for dissolution efficiency of product of aldol condensation reaction

Entry	Compounds[a]	%*T* at 800 nm on UV		
		0.25 h	1 h	2 h
1	No compound	0.5	<0.1	<0.1
2	Arginine	21	10	7
3	Cystine	12	8	0.2
4	Aspartic acid	8	4	0.1
5	Ethanol amine	0.8	<0.1	<0.1
6	Hydroxylamine hydrochloride	0.4	<0.1	<0.1
7	Hydrazine	0.4	<0.1	<0.1

[a] Experimental condition: acetaldehyde:amino acid ratio = 2.0, reaction temperature = 25 °C, total reactant volume = 40 ml 2.5 M caustic solution in DM water

and acetaldehyde. Shifting of the band revealed formation of strong bond between aldehyde and primary amine group. In all three products, broad band at 3,430, 3,361, and 3,376 cm^{-1} indicate –OH group of amino acid (Fig. 3a–c). Findings substantiated the theory that acetaldehyde is converted to imine due to the addition of amine-containing antipolymerant in the caustic tower.

Studies on efficiency of compound for dissolution behavior

Effect of amino acid compounds for dissolution of aldol condensation polymer was studied as function of time at aldehyde to amino acid mole ratio of 2 using the methodology described in "Experimental". Transmittance of solution which did not have amino acid compounds showed <0.1 %*T* indicating the formation of aldol polymer. When these compounds are added after formation of aldol polymer in 15 min, results indicated that arginine has better efficiency for dissolution of polymer (7–21 %*T*) compared to aspartic and cystine due to the presence of acidic and basic functional groups (Table 4). Ethanolamine, hydroxylamine hydrochloride, and hydrazine were studied for dissolution of aldol condensation polymer. These compounds showed poor performance compared to arginine, cysteine, and aspartic acid. This suggested that amine group in the compound was not effective for dissolution of aldol condensation polymer. Combination of acidic group (>COOH) and amine group is essential for dissolution of aldol condensation polymer.

Conclusions

Different amino acids were investigated as antipolymerants for inhibition and dissolution reactions of aldehyde formed during olefin production in industrial naphtha cracking process. Effects of antipolymerant concentration, catalyst base concentration, and reaction temperature were investigated as a function of time. Performance of antipolymerants was measured by UV spectrometer. Among the studied compounds, arginine was found to be very effective for inhibition and dissolution reactions compared to aspartic acid and cystine. This study revealed that compounds having higher basicity (electron donating groups) are efficient for inhibition of aldol reaction, but less effective for dissolution purpose. Similarly, more acidic functionality of compound is less effective for inhibition. Study of only amine-containing derivatives for inhibition and dissolution performance indicated the need for both acidic and basic groups in the compound used for the dissolution of aldol condensation polymer.

References

1. Annunziata R, Cinquini M, Cozzi F, Cozzi PG, Consolandi E (1992) Stereocontrol in the Mukaiyama aldol addition to α-chiral and β-thio-substituted aldehydes. J Org Chem 57:456
2. Awbrey SS (1990) Method of inhibiting fouling in caustic scrubber systems. US Patent 4952301
3. Carey FA, Chapter 18: Enols and enolates, http://www.mhhe.com/physscichemistry/carey5e/Ch18/ch18-3-4.html, 30 March 2013
4. Gary P (1996), Borohydrides to inhibit polymer formation in petrochemical caustic scrubbers. US Patent 5582808
5. Kurukchi SA, Gondolfe JM (1999) Spent caustic (pre)treatment process. US Patent 5885422
6. Lewis VE (1992) Method of inhibiting formation of fouling materials during basic washing of hydrocarbons contaminated with oxygen compounds. US Patent 5160425
7. Lewis VE, Patel NR (1998) Caustic tower trap for acetaldehyde. US Patent 5714055
8. Lewis VE, Rowe CT (1994) Process for the prevention of polymer formation in compressor systems. US Patent 5288394
9. Mukaiyama T (1982) The directed aldol reaction. Org React 28:203
10. Papa AJ (1985) Aldol condensation of enolizable aldehydes using a metal carboxylate catalyst. US Patent 4528405
11. Paterson I (1988) New symmetric aldol methodology using boron enolate. Chem Ind 12:390
12. Patil H, Bhajiwala H, Gupta V (2011) A method of cracking petrochemicals. Indian Patent 1706/MUM/2011
13. Patil H, Bhajiwala H, Gupta V (2011) Energy efficient process for synthesis of salt of aliphatic amino acid. Indian Patent 293/MUM/2012
14. Radha KP, Lakshmi PY, Munagala A (2011) Benzoic acid catalyzed aldol-type condensation of aldehydes with ethyl diazoacetate: highly diastereoselective. Eur J Org Chem 2011(26):5089–5095
15. Roling PV (1987) Method for prevention of fouling in a basic solution by addition of specific nitrogen compounds. US Patent 4673489
16. Roling PV (1993) Method for inhibiting fouling in caustic scrubber systems. US Patent 5194143
17. Saeed AM, Mojtahedi MM, Pasha GF, Ghasem F, Akbarzadeh E, Shockravi A, Mesbah AW, Massa W (2011) Switching the reactivity of dihydrothiopyran-4-one with aldehydes by aqueous organocatalysis: Baylis–Hillman, aldol, or aldol condensation reactions. Org Lett 13:5282
18. Silvana P, Mauro DN, Domenico M, Daniele N, Ada N, Romualdo R (2011) Diastereo- and enantioselective direct aldol reactions in aqueous medium: a new highly efficient proline-sugar chimeric catalyst. Adv Syn Cat 353:1443
19. Subramaniyam M, Bankar P (2006) Method for prevention of fouling in basic solution by inhibiting polymerization and solubilizing deposits using amino acids. US Patent 697824
20. Tabler DC (1970) Removal of carbonyls from polymerizable monomers. US Patent 3535399
21. Tomoyuki M, Kouichi F, Yuuji K, Masaki T (1996) Aldehyde aldol condensation process. German Patent DE 19605078 A1 19960814
22. Wade LG (2005) Organic chemistry, 6th edn, Chap 18. Prentice Hall, Upper Saddle River, NJ, p 804
23. Zhu P, Chen X, Roberts CG (2006) Method of use for aldehyde removal. US Patent 7041220

20

Microwave treatment in oil refining

Adrian Porch · Dan Slocombe · Jan Beutler · Peter Edwards · Afrah Aldawsari · Tiancun Xiao · Vladimir Kuznetsov · Hamid Almegren · Saud Aldrees · Naif Almaqati

Abstract In this paper, we discuss the potential of using microwave techniques in the refinement of heavy fraction of petroleums such as bunker oil. After discussing the fundamental issues associated with conversion of microwave energy into heat, we present measurements of the dielectric properties of heavy oils at 2.45 GHz using a highly sensitive resonant cavity method, and also over a broader frequency range (100 MHz to 8 GHz) using a coaxial probe technique. We find that the dielectric loss is very small even in these heavy oils, but still may be sufficiently large to provide efficient conversion of microwave energy into heat on untreated samples, and could be massively enhanced by means of a microwave-absorbing additive (e.g., carbon black). We conclude by discussing the design of a suitable microwave actuator for heavy oil cracking within a flow process.

Keywords Microwave · Dielectric · Petroleum

A. Porch (✉) · D. Slocombe · J. Beutler
School of Engineering,
Cardiff University, Cardiff CF24 3AA, UK
e-mail: PorchA@cf.ac.uk

P. Edwards · A. Aldawsari · T. Xiao · V. Kuznetsov · S. Aldrees · N. Almaqati
Inorganic Chemistry Laboratory, Oxford University,
Oxford OX1 3QR, UK

H. Almegren · S. Aldrees · N. Almaqati
Petrochemical Research Institute, King Abdulaziz City for Science and Technology, P.O. Box 6086, Riyadh 11442, Kingdom of Saudi Arabia

Introduction

There is an ever-increasing global demand for high value, light olefins such as ethylene, propylene and butylene, which form the essential building blocks for the petrochemical industry. However, these demands, coupled with more stringent environmental regulations for automotive fuel, cannot be met by conventional petrochemical technologies.

Here, we investigate the use of microwave technologies to assist in the high-efficiency conversion of crude oil to light olefins, aromatics and other high-value chemicals. Almost all of the targeted chemical reactions involving the breaking of C–C bonds are highly endothermic in nature, requiring considerable inputs of energy. It might be possible to heat the crude oil directly with very high power microwaves to attain the required reaction temperatures. However, this can be aided by the addition of high surface area, microwave-absorbing catalysts or additives (such as methanol or water, or solids such as carbon black), which provide a means for the selective heating of the additive within the crude oil emulsion. Either way, we expect the application of microwaves to assist in the breaking of C–C bonds from fractions of the whole crude oil to generate valuable chemicals and premier fuel products.

In this paper, we will first discuss the main issues affecting the efficient heating of materials using microwaves, then present microwave characterization of crude oil samples, and finally discuss the design of a suitable actuator for direct or assisted heating of these samples.

Literature survey

Microwaves are electromagnetic waves whose frequencies lie in the range between about 300 and 30 GHz, corresponding to free space wavelengths from about 1 m down to 1 cm, respectively. Beyond 30 GHz, the microwave spectrum gives way to the millimeter (mm) wave spectrum, and below 300 MHz the radio frequency (RF) spectrum. Microwave heating is now a standard processing tool for the industrial chemist [1–4].

In the microwave part of the electromagnetic spectrum, heating is normally a result of the conversion of the electric field energy part of the microwave signal, generally termed "dielectric heating". The efficiency of this process is determined by the dielectric loss (i.e. loss tangent) of the material. This, in turn, is related to the polar nature of the material and high conversion efficiency is found in highly polar materials (e.g., water) or those of moderate electrical conductivity (e.g., carbon black, of conductivities around 0.1–1 S/m around 900 MHz). It should be noted that metal particles do not heat in a microwave electric field, since the electric field within the particles is screened to almost zero and so is unable to drive the heating process.

Conversely, in the RF part of the spectrum, heating is a result of the conversion of magnetic energy, generally termed "induction heating", and can be initiated even in dielectric samples by the addition of small conducting or magnetic particles. Induction heating is driven by the eddy currents induced in the particles when conducting, or the magnetization currents when magnetic. The heat is then conduced by conventional means into the surrounding host material.

The advantages offered by microwaves are the rapid, uniform, and selective heating involving no direct contact between the microwave source and the heating material. The material to be heated can be thought of as simply being "bathed" in microwaves, with the resulting heat being generated volumetrically and almost uniformly throughout the material. Water-containing materials are most effectively heated around 20 GHz, but field penetration drops rapidly with increasing frequencies so that 2.45 GHz is used (as a compromise) for domestic ovens, where penetration depths are then many centimeters.

High power microwaves (HPMW) are traditionally generated by valve-type sources (e.g., magnetrons), which are very efficient at converting electrical input power into microwave power (typically around 80 %). The allocated ISM (i.e. industrial, scientific and medical) frequencies for HPMW in the UK are 896 and 2,450 MHz. Continuous microwave powers of over 100 kW can be delivered at 896 MHz, increasing to over 1 MW if the power is pulsed (even >1 GW for extremely short bursts) [5]. Generally, output power decreases with increasing frequency, but very high power gyrotron sources are now being reported even at frequencies of 30 GHz and above [6].

Microwaves are currently utilized in many industrial heating applications such as food drying and processing [7], chemical synthesis [8, 9], digestion [10, 11] and extraction [12–15]. Successful microwave pyrolysis studies have been conducted with a variety of feedstocks and resulting products (e.g., [16–20]). Two useful books that cover the application of microwaves in industrial heating applications are [21] and [22].

Microwave and RF heating methods have been used to recover oil and gas from oil shale and oil sands [15, 23, 24], but there have been only limited reports of microwave-assisted cracking of heavy oil hydrocarbons in the scientific literature. Two general reviews of the use of microwaves in the petrochemical industry are [25] and [26]. Early studies by Wan et al. used microwave-absorbing additives in reaction mixtures owing to the naturally low microwave losses of the associated organic materials, and described the microwave-assisted catalytic hydrogenation of alkenes [27, 28],. Fanson et al. [29] described the microwave-assisted cracking of hydrocarbons in the presence of a precious metal impregnated, metal oxide catalyst, with high conversion efficiency in the presence of carbon at atmospheric pressure. Tooley [30] describes an apparatus for the extreme microwave heating of liquids, with application for the cracking of waste oil. Recent studies [31] have demonstrated that even the moderate power provided by a domestic microwave oven is sufficient to improve the olefin yields from small volumes of crude oil without the addition of catalyst or absorbing additive. From this overview of the literature, we conclude that there are major opportunities for the use of microwave technologies in petroleum refinement.

Background theory for efficient heating

If an electric field is set up within a dielectric material, the interaction between the material and field is described by the material's relative permittivity ε. This links the polarization per unit volume $\underline{P}$ in terms of the internal electric field $\underline{E}_{in}$ via the relationship $\underline{P} = (\varepsilon - 1)\varepsilon_0 \underline{E}_{in}$. We consider losses by writing ε as the complex quantity $\varepsilon = \varepsilon_1 - j\varepsilon_2$, where ε_1 [or, more correctly the quantity $(\varepsilon_1 - 1)$] is proportional to the material's polarization and ε_2 the power dissipation, which we will consider in more detail below.

A resonant microwave cavity is an efficient structure for converting microwave energy into heat. These are robust metal structures that can withstand the rigors of an industrial environment that are variants of the common domestic oven, except that they are often designed for single mode operation. Very high conversion efficiency of microwave energy into heat is possible since the sample load is usually well known and characterized (unlike that for domestic ovens). Figure 1 shows schematically the simplified principles of a microwave actuator based on a cavity resonator

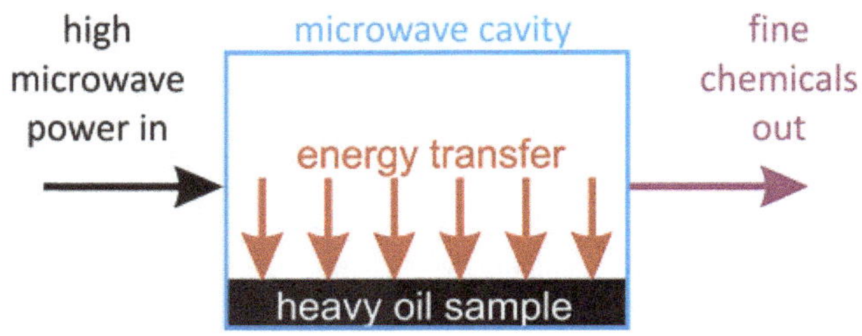

Fig. 1 A schematic of a HPMW heating system to process crude oil, based on a high power microwave cavity to enable efficient power conversion between the microwave electric field and heat in the sample. Diagnostic testing of the output (e.g., by mass spectrometry) can be used to detect for olefin generation

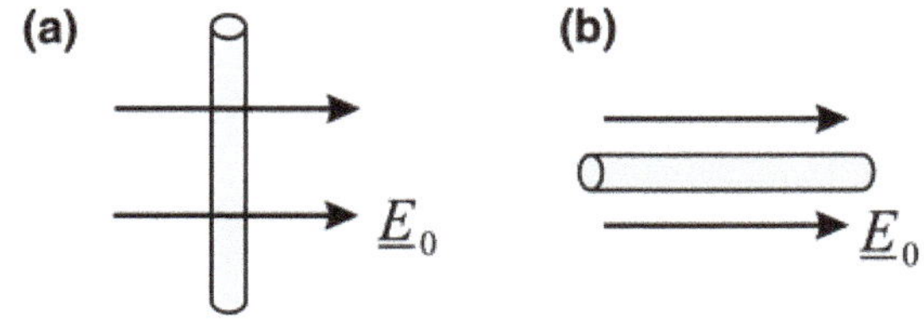

Fig. 2 Depolarization effects in dielectric cylinders. In case **a**, the magnitude of the internal electric field is reduced by a factor $2/(\varepsilon + 2)$ compared to the applied field E_0, resulting in a very small internal field when the relative permittivity $\varepsilon >> 1$. In case **b**, the continuity of the parallel component of E at a boundary between two media ensures that the magnitude of the internal electric field is approximately equal to the applied field E_0

used to process crude oil. We will now look at the design considerations based on the dielectric properties of the material to be heated.

The unloaded quality factor of a resonant cavity is given by [32]

$$Q = 2\pi f_0 \frac{\text{average stored energy}}{\text{average power dissipated}} = 2\pi f_0 \frac{U}{W} \quad (1)$$

Where f_0 is the resonant frequency. The coupling structure between the microwave source and the cavity can usually be adjusted on a bespoke system so that no power is reflected and the cavity is impedance matched to the source. The cavity is then said to be critically coupled, there is no reflected power, and the average power dissipated within the cavity equals the input power W_{in} from the source. This is clearly desirable for microwave heating applications, resulting in maximum conversion efficiency of electromagnetic energy into heat.

How much of this microwave input power ends up as heat in the sample? To answer this, we assume that the presence of the sample does not reconfigure the electromagnetic fields in the cavity to a major extent (i.e. it presents a small perturbation) and we use Eq. (1) to write

$$W_{in} = W_s + W_0 = 2\pi f_0 \frac{U}{Q} \approx W_0 \frac{Q_0}{Q} \quad (2)$$

where W_s is the average power dissipated in the sample, W_0 is that dissipated in the cavity walls, Q is the quality factor of the cavity when containing the sample and Q_0 is that when the cavity is empty. Rearranging Eq. (2), then gives us a very simple formula for the efficiency (in %) of microwave energy conversion into sample heat

$$\frac{W_s}{W_{in}} \approx \left(1 - \frac{Q}{Q_0}\right) \times 100\% \quad (3)$$

Hence, to design an efficient microwave cavity (i.e. actuator) for heating, we need to ensure that the introduction of the sample into the cavity reduces its Q factor as much as possible. To illustrate the importance and usefulness of Eq. (3), if the sample reduces the Q factor from 5,000 to 4,000 then the expected conversion efficiency when the cavity is critically coupled will be around 20 %, but if Q is reduced from 5,000 to less than 500 then the conversion efficiency will be greater than 90 %.

To ensure the greatest impact of the sample on the Q factor, we must design the actuator so that depolarization effects are kept to a minimum, thus maintaining a high electric field within the sample. The average power dissipated within a lossy dielectric is

$$W = \frac{1}{2}\omega_0 \varepsilon_0 \varepsilon_2 E_{in}^2 V \quad (4)$$

where V is the sample volume, and E_{in} is the magnitude of the internal electric field (here assumed to be uniform, which is a good general approximation). The development of polarization charges on dielectric surfaces perpendicular to the applied electric field E_0 means that $E_{in} < E_0$, which can give rise to a dramatic reduction in the heating rate in materials with large values of ε_1. This is shown schematically in Fig. 2, where we assume that a cylindrical liquid sample is introduced into a microwave cavity with an applied electric field E_0 either parallel or perpendicular to the axis of the cylinder. The ratio of internal electrical fields and power dissipation in the two cases (assuming the same applied field) are

$$\frac{E_{in}^{\parallel}}{E_{in}^{\perp}} \approx \frac{1}{2}(\varepsilon + 1) \approx \frac{1}{2}((\varepsilon_1 + 1) - j\varepsilon_2), \qquad \frac{W_{\parallel}}{W_{\perp}} \approx \frac{1}{4}\left((\varepsilon_1 + 1)^2 + \varepsilon_2^2\right) \quad (5)$$

respectively. For example, for water at 900 MHz where $\varepsilon = 80 - j3.4$, the power dissipation ratio estimated from Eq. (5) is greater than 1,600. This means that it is hugely beneficial to design an actuator so that the electric field is parallel to the sample, which for liquids is most easily accomplished by drawing them into a thin cylinder. The same is true for oil samples but the effects of depolarization are not as dramatic as for water, since for oil ε is much smaller.

We next present our measurements of the dielectric properties of crude oil, from which we can infer the heating conversion rates within a given actuator structure.

Experimental results

Two heavy oil samples were measured, one Saudi heavy crude oil and one Saudi bunker oil, both supplied by Saudi Aramco. The physical properties of the oil samples are shown in Table 1. It is seen that the bunker oil has higher sulfur and metal content than the crude oil sample, while there are less aliphatics in the bunker oil.

The first method of microwave characterization was the broadband coaxial probe, which is a standard tool for measuring the complex permittivity of liquids [33]. Our probe is a miniaturized, bespoke probe based around an Anritsu K-connector (shown schematically in Fig. 3), which has to be carefully calibrated using the procedure outlined in [34]. Briefly, the 1.6 mm diameter aperture presented to the probe is very small for frequencies below 10 GHz (where the free space wavelength is greater than 30 mm). This allows us to use a simple capacitive model for extracting the complex permittivity ε of the sample, so that the admittance of the aperture with and without the sample is $Y_s = j\omega(\varepsilon C_0 + C_f)$ and $Y_0 = j\omega(C_0 + C_f)$, where C_0 is the probe's air-spaced capacitance, C_f is the probe's fringing capacitance and ω is the (angular) frequency.

We extract experimental values of the probe admittance by measuring the voltage reflection coefficient S_{11} with and without the sample (allowing us to calibrate out to the electrical length of any cables). The capacitance terms are found from measurements of the standard polar liquids methanol and ethanol, which can be obtained in very high purity and which have well-known values of complex permittivity (both well-described by the Debye model for polar liquids). A drop of each oil sample is placed over the aperture, fully covering it to a depth of about 2 mm. All measurements were taken using an Agilent E5071B vector network analyzer (VNA) over the frequency range

Table 1 The physical properties of the oil samples used in this work

Index	Crude oil	Bunker oil
Density (kg/m^3)	860	913
API gravity	27.0	23.4
Sulfur (wt %)	2.2	3.4
Total nitrogen content (wppm)	1,300	1,600
Total metal content (wppm) (V, Ni, Fe)	120	187
Aliphatics (wt %)	37	27
Pour point (°C)	33.5	38.9
Carbon residue, RAMS (wt %)	7.1	8.5

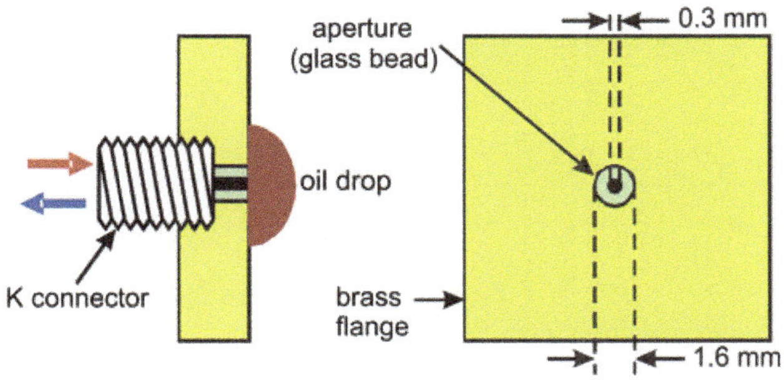

Fig. 3 A schematic diagram of the coaxial reflectance probe used to determine the broadband (0.1–8 GHz) complex permittivity ε of a sample that covers the aperture

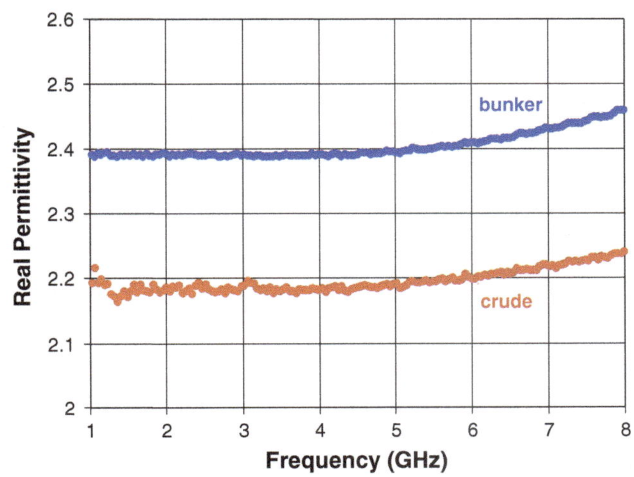

Fig. 4 The real part ε_1 of the complex permittivity of two heavy oil samples measured with the broadband coaxial probe

10–8.5 GHz and the data are read into a Mathcad program that performs the necessary computations to extract ε.

Data for the real part of complex permittivity ε_1 are shown in Fig. 4 for both heavy oil samples. The values vary little with frequency and remain in the range of 2.2–2.5 over the full bandwidth of the measurement, indicating low inherent electrical polarization. The imaginary (i.e. lossy) part of the complex permittivity ε_2 is shown in Fig. 5. Again, the values are small, consistent with the low polarization, and increase roughly quadratically with increasing frequency. At industrially relevant frequencies, the complex permittivities are found to be for crude oil $\varepsilon = 2.19 - j0.003$ at 900 MHz and $\varepsilon = 2.19 - j0.007$ at 2.45 GHz; for bunker oil $\varepsilon = 2.39 - j0.002$ at 900 MHz and $\varepsilon = 2.39 - j0.005$ at 2.45 GHz.

Referring to the physical data of Table 1, the fact that bunker oil has a slightly higher permittivity (i.e. ε_1) than crude oil over the full frequency range could be due to its higher mass density (linked also to its lower aliphatic content). The lower loss ε_2 of bunker oil (by around a factor of 2 compared to crude oil) is most likely due to a

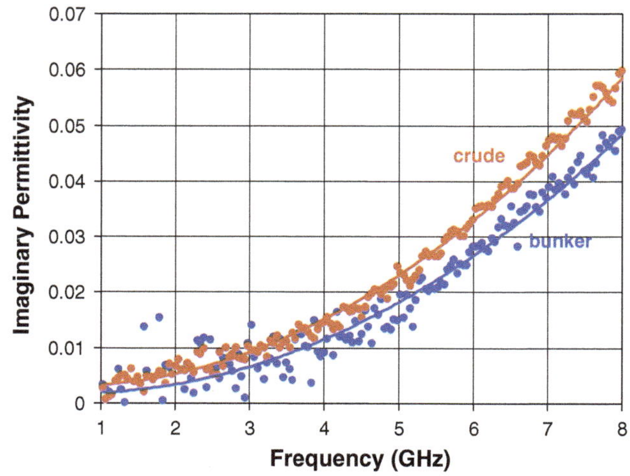

Fig. 5 The imaginary part ε_2 of the complex permittivity of two heavy oil samples again measured using the broadband coaxial probe

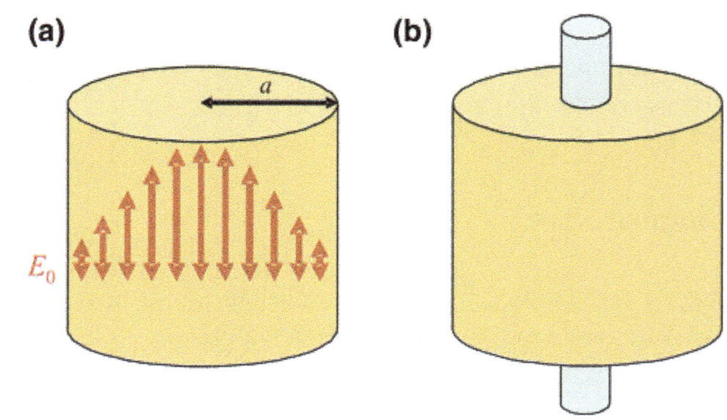

Fig. 6 **a** A schematic diagram of the TM_{010} cylindrical cavity and its electric field distribution, used to measure the complex permittivity of oil samples around 2.45 GHz. In terms of the internal radius a (in cm), the resonant frequency (in GHz) is given by $f_0 = 11.5/a$ (e.g., an internal radius of 4.6 cm gives a resonant frequency of 2.5 GHz). **b** A liquid sample (e.g., oil) is inserted along the cavity's axis, contained within a low-loss tube. This parallel field configuration minimizes the effects of sample depolarization and provides the strongest coupling between the cavity and sample. Here, the cavity is coupled via its magnetic field (not shown) by a pair of loop-terminated coaxial cables

lower concentration of polar species, but this is not linked inherently to any of the properties of Table 1.

To obtain more accurate experimental values for the loss ε_2, and hence to assess the heating conversion efficiency of these samples when placed in a suitably designed microwave cavity, the measurements were repeated in a TM_{010} cylindrical cavity operating at 2.45 GHz. The cavity is made from two D-shaped cups machined from solid aluminum, with inner cavity radius of 4.6 cm and inner cavity length of 4.0 cm. The samples were placed in polythene cylindrical tubes of inner diameter 4.5 mm, of negligible dielectric loss, running the full internal length of the cavity. A schematic of the cavity set-up is shown in Fig. 6, where it can be seen that the applied electric field is parallel to the sample, giving the maximum possible reduction in Q factor (i.e. the maximum possible electric field coupling to the sample). The cavity's microwave magnetic field is excited with a pair of identical coaxial SMA launchers terminated in short-circuit loops and the transmitted power is measured with the same VNA as for the coaxial probe measurements. Typical measurement results are shown in Fig. 7.

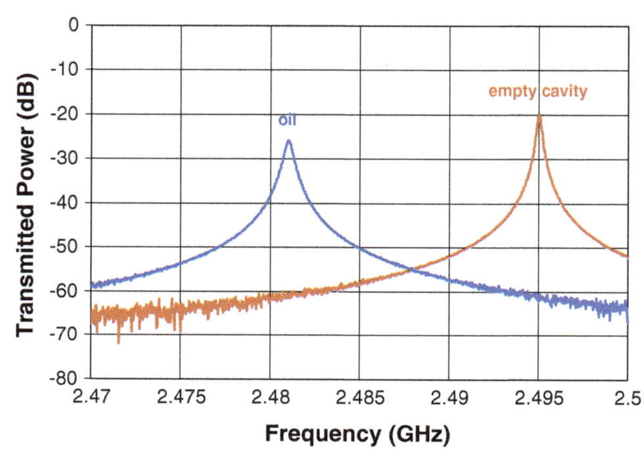

Fig. 7 The TM_{010} cavity response with and without a crude oil sample. There is very little change in the resonant bandwidth, indicative of low dielectric loss

The complex permittivity of the two heavy oil samples were calculated using first-order cavity perturbation theory [35] from the experimentally observed changes in resonant frequency f_0 and Q factor, using

$$\frac{\Delta f_0}{f_0} \approx -(\varepsilon_1 - 1)\frac{V_s}{2V_{eff}} \tag{6}$$

$$\Delta\left(\frac{1}{2Q}\right) \approx \varepsilon_2 \frac{V_s}{2V_{eff}} \tag{7}$$

where V_s is the volume of sample within the cavity (i.e. 0.64 cm^3) and V_{eff} is the effective volume of the cavity, which is found by calibration measurements of a range of metal spheres to be $69.4 \pm 0.1 cm^3$ (this compares well with the theoretically predicated value of 71.7 cm^3). At the resonant frequency of 2.48 GHz, the complex permittivity values of crude and bunker oil are $\varepsilon = 2.20 - j0.0071$ and $\varepsilon = 2.38 - j0.0048$, respectively, with ± 2 % measurement error in both real and imaginary parts of ε. These cavity data are in excellent agreement with the coaxial probe data at the same frequency, and give us confidence to predict the conversion efficiency if these samples were to be placed in a suitable microwave actuator.

Discussion

In this paper, we have reviewed the use of high power microwaves and presented some of our preliminary results

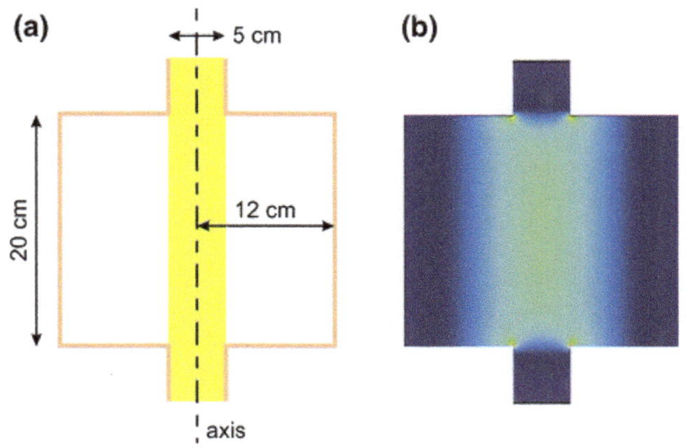

Fig. 8 **a** Schematic diagram of a microwave heater based on a copper TM_{010} cavity operating around 900 MHz (inner cavity radius of 12 cm and a length of 20 cm), with the crude oil column shown. **b** The simulated distribution of the resulting electric field magnitude, which drives the heating. The entry and exit holes for the oil each have radii of 2.5 cm, so form cylindrical waveguides which are well below their cut-off frequency. This means that the microwave fields are contained within the cylindrical cavity space and are concentrated along the axial region, where the electric field is large, approximately uniform, and directed parallel to the axis

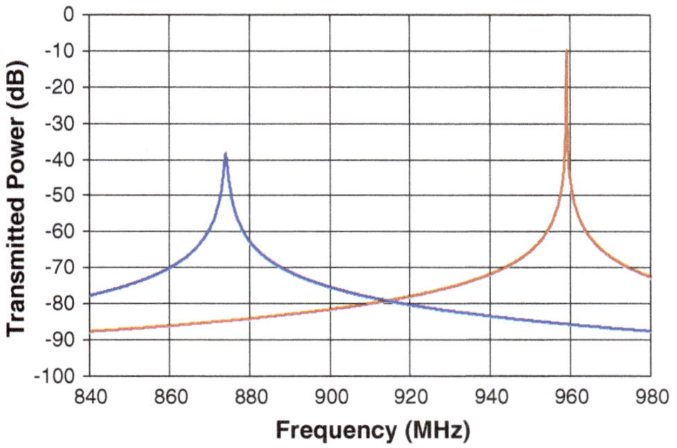

Fig. 9 The simulated response of the cavity shown in Fig. 8. Based on the measured microwave properties of crude oil at 900 MHz, if the oil forms an axial liquid column of radius 2.5 cm, the cavity Q factor drops from 32,000 when empty (if made from copper), to 1,200 when bunker oil is present (left hand trace). If the cavity is critically coupled to its high power microwave source, this means that 96 % of the input microwave power is converted to heat in the oil, with the remaining 4 % appearing as heat in the cavity walls

for the microwave characterization of crude and bunker oil samples, using both a broadband coaxial probe and highly sensitive TM_{010} cavity working around 2.45 GHz. Finally, we now discuss how we can use the measured values of complex permittivity ε to calculate the heating efficiency in the raw heavy oils, initially without additives, in a suitable microwave actuator.

We will work around 900 MHz since then a single mode (TM_{010}) cavity actuator has an internal radius of around 12 cm and so can accommodate larger liquid volumes. A suitable design is shown in Fig. 8a, where the length of the cavity (20 cm) is chosen to be long enough to give it a high Q factor when empty (around 32,000 if made from copper, or when copper coated), but short enough to ensure that higher order cavity modes (e.g., TE_{111}) do not interfere with the fundamental TM_{010} mode. Here we envisage a flow of oil through a low-loss tube, with the cylindrical inlet and outlet holes shown. We have assumed these to be of internal diameter of 5 cm, meaning that they act as waveguides well below their lowest cut-off frequency so do not affect the distribution of electromagnetic field within the cavity, or cause radiative losses which may otherwise be a safety hazard when using very high power microwave sources.

We have simulated the fields in the cavity using Poisson Superfish [36], which can used for calculating the electromagnetic fields in cylindrical cavities for axially symmetric modes (as is the case for TM_{010}). We use the complex permittivity $\varepsilon = 2.19 - j0.005$ for crude oil at 900 MHz taken from the coaxial probe data of Figs. 4 and 5. The resulting electric field magnitude is shown in Fig. 8b, which has the required property of being concentrated in the region where the oil sample resides. In addition, as with the earlier cavity perturbation experiments, the direction of $\underline{E}$ is parallel to the axis of the oil cylinder, thus giving the largest possible value of the internal electric field within the oil to drive the heating.

From these fields, we calculate that the Q factor of the cavity is reduced from 32,000 to 1,200 due to the presence of the raw crude oil, and from Eq. (3) we estimate a heating conversion efficiency for about 96 %. The simulated spectral response of the cavity with and without the crude oil sample is shown in Fig. 9. Assuming the cavity is critically coupled, this means that 100 kW of incident microwave power leads to about 96 kW being generated as heat within the sample, with the remainder being dissipated as heat in the cavity walls (which may have to be air-cooled). On the one hand, this conversion efficiency is encouragingly high, but on the other hand the volume of oil needs to be large (here $\pi \times 2.5$ cm $\times$ 2.5 cm $\times$ 20 cm $\approx$ 0.4 L) since its microwave loss is so small. Clearly, the high resulting heat capacity due to the large oil mass will limit the heating rate of the oil. To reduce the oil volume whilst preserving similar (>95 %) conversion efficiencies, the value of ε_2 can be increased dramatically by the addition of a microwave absorbing phase.

To compare the heating efficiency of different samples in a TM_{010} cavity, a useful figure of merit is the product $\varepsilon_2^{\mathrm{eff}} V_s$, which we aim to get as high as possible. Here V_s is the sample volume and $\varepsilon_2^{\mathrm{eff}}$ is the effective loss of the oil emulsion including the absorbing phase, which depends mostly on ε_2 of the lossy phase and its volume fraction. Carbon blacks (and related materials) are ideal lossy additives, with well-known high values of microwave absorption (e.g., ε_2 in the range of 1–10 for carbon black composites at microwave frequencies [37]). Hence, we can get >95 % conversion efficiency for a carbon black inclusion volume of only about 0.2 mL, meaning that the oil volume can be scaled down to 10s of mL, thus reducing its heat capacity by an equivalent ratio to the volume reduction.

Conclusions

To conclude, we have determined the complex permittivity of heavy oil samples at microwave frequencies. We have then used these results to model the efficiency of heat generation in a flow system suitable for microwave-assisted cracking. Whilst the microwave loss of even heavy oils is found to be small, it is possible to have very high microwave conversion efficiencies for moderate volumes of liquid. If the thermal mass is too large, high conversion efficiency can be maintained for smaller volumes by addition of a highly absorbing additive. In the future, we will perform experiments on such oil emulsions to characterize their microwave properties, and build an efficient desktop system to evaluate the fundamental principles associated with the microwave-assisted cracking of heavy oils.

Acknowledgments We thank Merck KGaA for the sponsorship of Dan Slocombe.

References

1. Mingos DMP, Baghurst DR (1991) Applications of microwave dielectric heating effects to synthetic problems in chemistry. Chem Soc Rev 20:1–47
2. Zhang X, Hayard DO, Mingos DMP (2003) Effects of microwave dielectric heating on heterogeneous catalysis. Catal Lett 88:33–38
3. Jones DA, Lelyveld TP, Mavrofidis SD, Kingman SW, Miles NJ (2002) Microwave heating applications in environmental engineering—a review. Resour Conserv Recycl 34(2002):75–90
4. Thostenson ET, Chou TW (1999) Microwave processing: fundamentals and applications. Compos A 30:1055–1071
5. Gold SH, Nusinovich GS (1997) Review of high-power microwave source research. Rev Sci Instrum 68:3945–3974
6. Kumar N, Singh U, Kumar A, Khatun H, Singh TP, Sinha AK (2011) Design of 35 GHz gyrotron for material processing applications. Prog Electromagn Res B 27:273–288
7. Venkatesh MS, Raghavan GSV (2004) An overview of microwave processing and dielectric properties of agri-food materials. Biosyst Eng 88:1–18
8. Kappe CO (2004) Controlled microwave heating in modern organic synthesis. Angew Chem Int Ed 43:6250–6284
9. Tian ZQ, Jiang SP, Liang YM, Shen PK (2006) Synthesis and characterization of platinum catalysts on multi-walled carbon nanotubes by intermittent microwave irradiation for fuel cell applications. J Phys Chem B 110:5343–5350
10. Chen CM, Chen M, Leu FC, Hsu SY, Wang SC, Shi SC, Chen CF (2004) Purification of multi-walled carbon nanotubes by microwave digestion method. Diam Relat Mater 13:1182–1186
11. Afridi HI, Kazi TG, Jamali MK, Kazi GH, Arain MB, Jalbani N, Shar GQ (2006) Analysis of heavy metals in scalp hair samples of hypertensive patients by conventional and microwave digestion methods. Spectrosc Lett. 39:203–214
12. Al-Harahsheh M, Kingman S, Somerfield C, Ababneh F (2009) Microwave-assisted total digestion of sulphide ores for multi-element analysis. Anal Chim Acta 638:101–105
13. El Harfi K, Mokhlisse A, Chanaa MB, Outzourhit A (2000) Pyrolysis of the Moroccan (Tarfaya) oil shales under microwave irradiation. Fuel 79:733–742
14. Lucchesi ME, Chemat F, Smadja J (1043) Solvent-free microwave extraction of essential oil from aromatic herbs: comparison with conventional hydrodistillation. J Chromatogr A 2004: 323–327
15. Robinson JP, Kingman SW, Onobrakpeya O (2008) Microwave-assisted stripping of oil contaminated drill cuttings. J Environ Manag 88:211–218
16. Ludlow-Palafox C, Chase HA (2001) Microwave-induced pyrolysis of plastic wastes. Ind Eng Chem Res 40:4749–4756
17. Menendez JA, Inguanzo M, Pis JJ (2002) Microwave-induced pyrolysis of sewage sludge. Water Res 36:3261–3264
18. Huang YF, Kuan WH, Lo SL, Lin CF (2008) Total recovery of resources and energy from rice straw using microwave-induced pyrolysis. Bioresour Technol 99:8252–8258
19. Yagmur E, Ozmak M, Aktas Z (2008) A novel method for production of activated carbon from waste tea by chemical activation with microwave energy. Fuel 87:3278–3285
20. Lei HW, Ren SJ, Julson J (2009) The effects of reaction temperature and time and particle size of corn stover on microwave pyrolysis. Energy Fuels 23:3254–3261
21. Metaxas AC, Meredith RJ (1981) IEE Power Engineering Series 4. Peter Peregrinus Ltd., London
22. Roussy G, Pearce JA (1996) Foundations and industrial applications of microwave and radio frequency fields. Wiley, New York
23. Pierre EP (1992) Microwave separation of bituminous material from tar sands. Canadian Patent 1293943
24. Bosisio RG, Cambon JL, Chavarie C, Klvana D (1977) Experimental results on the heating of Athabasca tar sand samples with microwave power. J Microw Power 12:301–307
25. Mutyala S, Fairbridge C, Paré JRJ, Bélanger JMR, Ng S, Hawkins R (2010) Microwave applications to oil sands and petroleum: a review. Fuel Process Technol 91:127–135
26. Kustov LM, Sinev IM (2010) Microwave activation of catalysts and catalytic processes. Russ J Phys Chem A 84:1676–1694
27. Wan JKS, Kriz JF (1985) Hydrodesulphurization of hydro cracked pitch. US patent 4545879
28. Wan JKS, Wolf K, Heyding RD (1984) Some chemical aspects of microwave assisted catalytic hydro-assisted processes. In: Kaliaguine S, Mahay A (eds) Catalysis on the energy scene. Elsevier, Amsterdam
29. Fanson PT, Hirato H, Masaya IBE, Suib SL, Son YC (2006) Process using microwave energy and a catalyst to crack hydrocarbons. PCT International Patent, WO/2006/083829
30. Tooley J (2008) Process for cracking of waste oil by microwave. US Patent 2008/0202982 A1
31. Mohammed AD, Mohammed IA, Ajinomoh CS (2011) Upgrading heavy crude oil potentials through microwave assisted distillation. J Innov Res Eng Sci 2:137–147

32. Pozar DM (2011) Microwave engineering. Wiley, New York
33. McLaughlin BL, Robertson PA (2007) Miniature open-ended coaxial probes for dielectric spectroscopy applications. J Phys D Appl Phys 40:45–53
34. Sulaimalebbe A, Porch A, Vidal-Iglesias FJ, Attard G (2008) Microwave properties of platinum nanoparticle films. In: IEEE MTT-S International Microwave Symposium Digest, pp 1585–1588
35. Shi LH, Akyel C, Bosisio RG (1981) Precise calculations and measurements on the complex dielectric constant of lossy materials using TM010 cavity perturbation techniques. IEEE Trans Microw Theory Tech. 29:1041–1048
36. Poisson Superfish, Los Alamos National Laboratory. http://laacg1.lanl.gov/laacg/services
37. Hotta M, Hayashi M, Lanagan MT, Agrawal DK, Nagata K (2011) Complex permittivity of graphite, carbon black and coal powders in the ranges of X-band frequencies (8.2 to 12.4 GHz) and between 1 and 10 GHz. ISIJ Int 51:1766–1772

21

Nitrides as ammonia synthesis catalysts and as potential nitrogen transfer reagents

J. S. J. Hargreaves

Abstract In this article, an overview of the application of selected metal nitrides as ammonia synthesis catalysts is presented. The potential development of some systems into nitrogen transfer reagents is also described.

Keywords Nitrogen · Hydrogen · Catalysis · Ammonia · Nitrides

Introduction

The successful development of the Haber–Bosch Process for the production of ammonia was a paramount achievement of the twentieth Century. Through the provision of an accessible route to synthetic fertiliser, it has been pivotal in sustaining a large proportion of the global population [1]. The process involves the direct combination of pure H_2 and N_2 reactants over promoted iron catalyst at moderate temperature (typically ca 400 °C) and high pressure (typically 100–200 atmospheres):

$$0.5N_{2(g)} + 1.5H_{2(g)} \leftrightarrow NH_{3(g)} \Delta H^{\circ} = -46 \text{ kJ mol}^{-1}$$

The reaction is equilibrium limited and, being exothermic, is thermodynamically favoured at lower reaction temperatures. However, the temperature of operation is dictated by the requirement to achieve acceptable process kinetics. Whilst the Haber–Bosch Process can, to some extent, be viewed as a mature technology, the development of more active (or even more poison tolerant) catalysts which would allow a reduction in the stringent conditions of operation is of significant interest. In this context, it is salient to note that, taken in its entirety, including the provision and purification of reactant feedstreams, on a global scale, the operation of the Haber Process currently consumes 1–2 % of the world's energy demand, e.g. [2, 3]. Therefore, any improvements can potentially yield massive rewards both in terms of economic and environmental benefits. The identification and application of more active catalysts could lead to lowering of the process temperatures and/or pressures of operation necessary to achieve a desired space time yield of product. Indeed, lowered reaction temperatures result in improved obtainable equilibrium yields offering the tantalising prospect of a possible win–win situation, providing sufficiently favourable kinetics can be achieved. This is, however, a very significant challenge.

Iron is cheap and abundant and, when promoted, very effective under current Haber–Bosch Process conditions. Therefore, any replacement catalyst must afford the opportunity of significantly higher activity leading to lower temperature and/or pressure of application. Indeed, a number alternative systems based upon ruthenium have been identified in the literature, e.g. [4–7]. One such system involving promoted ruthenium supported on carbon is 10–20 times more active than conventional iron systems while operating at lower reaction pressure and forms the basis of an alternative ammonia synthesis process, the Kellogg Advanced Ammonia Process [8, 9].

The focus of this overview is a description of some of the relatively recent literature relating to the activity of metal nitrides as ammonia synthesis catalysts. The potential of nitrides as possible nitrogen transfer reagents will also be considered. In some respects, this relates to historical work on the development of the Haber–Bosch Process where some early studies by Mittasch and co-workers [10] at BASF considered routes to nitrogen

J. S. J. Hargreaves (✉)
WestCHEM, School of Chemistry, University of Glasgow, Joseph Black Building, Glasgow G12 8QQ, UK
e-mail: Justin.Hargreaves@glasgow.ac.uk

fixation involving the formation of nitrides and also demonstrated some nitrides, such as molybdenum nitride, to be intermediates in the stepwise synthesis of ammonia via nitridation of metals with N_2 and subsequent reduction by H_2.

Metal nitrides as ammonia synthesis catalysts

Binary metal nitrides

In the early studies, Mittasch [10] drew attention to molybdenum metal which was stated to catalyse ammonia synthesis differently from iron in that it formed easily detectable amounts of nitride during catalysis. It was said to resemble tungsten, manganese and uranium in this respect. It is therefore to be expected that nitrides of such metals may be active ammonia synthesis catalysts. However, mention should also be made that under industrial conditions bulk iron nitridation occurs resulting in the distorted structure associated with catalytic activity [11].

A number of binary nitrides have been identified as active ammonia synthesis catalysts in the literature, including those of molybdenum [10, 12–14], uranium [15, 16], vanadium [17, 18], rhenium [19, 20] and cerium [21].

In the case of molybdenum, arguably the most well-studied binary nitride, structure sensitivity has been reported to be operative over γ-Mo_2N in studies involving systematic variation of particle size [12]. In these studies, it was proposed that the fraction of active sites decreased with decreasing particle size. However, the effect of morphology was found to be limited in a study comparing plate like γ-Mo_2N and nanorod form of γ-Mo_2N prepared by the temperature programmed ammonolysis of plate and nanorod MoO_3, respectively [14]. Such ammonolysis is known to yield γ-Mo_2N which is pseudomorphic with respect to the MoO_3 precursor [22]. Comparison of different phases of molybdenum nitride showed β-$Mo_2N_{0.78}$ to possess similar activity to its γ-Mo_2N counterpart, whereas δ-MoN was found to be inactive (although potential deleterious effects resulting from residual sulphur left over from the preparation of this phase by ammonolysis of MoS_2 could not be excluded) [14]. The β nitride phase can be prepared by reaction of MoO_3 with N_2/H_2 mixtures [23] and the δ phase by ammonolysis of either MoS_2 or $MoCl_5$ [24]. Applying ^{15}N labelling to study the mechanism of ammonia synthesis over molybdenum nitride, Aika and Ozaki [25] deduced rate determining N_2 activation leading to intermediate $N_{(a)}$. It was argued that a proportion of the ^{15}N was lost into bulk nitride, with the following scheme being presented:

$$
\begin{array}{l}
N_2 \rightarrow N_{(a)} \leftrightarrow NH_3 \\
\qquad \updownarrow \\
\quad Mo_2N
\end{array}
$$

and similarities with iron nitride, previously investigated by Logan et al. [26], were drawn. Contrary to this, Hillis et al. [27] had reported the hydrogenation of NH_x species to be rate determining and the rate of hydrogenation of bulk nitride to be around 20–50 times slower than ammonia synthesis in this system. Some studies have been undertaken on other binary nitride systems. For example, based upon unusual transient behaviour, the hydrogenation of bulk nitrogen in uranium nitride has been proposed with the suggestion being made of a possible dynamic process between nitrogen species not immediately hydrogenated and loosely bonded subsurface nitrogen species which subsequently diffuse to the surface where they are hydrogenated to form ammonia [15]. This is akin to some oxidation processes which occur by the Mars–van Krevelen process where lattice oxygen species are those active in catalytic oxidation reactions [28]. Phase stability may be a consideration in other systems, for example, in rhenium nitride, the active form of the catalyst is proposed to comprise a mixture of Re_3N and Re [19].

Ternary metal nitrides

In recent years, there have been reports that Co_3Mo_3N-related catalysts have high activity that have attracted interest [29–35]. In particular, it has been reported that when suitably promoted, the activity of this phase can exceed that of conventional iron based Haber–Bosch catalysts. Table 1 presents some activity data taken from a study by Aika and Kojima [29]. Cs^+ has been found to be a promoter up to a certain limit, beyond which phase decomposition occurs [29]. Preparation of the phase generally involves ammonolysis of hydrated cobalt molybdate precursor—perhaps surprisingly use of anhydrous cobalt

Table 1 Ammonia synthesis activities of various catalysts (adapted from [29])

Catalyst	Rate (μmol h^{-1} g^{-1})	Surface area (m^2 g^{-1})	Specific activity (μmol h^{-1} m^{-2})
Fe-K_2O-Al_2O_3	330	14	24
Co_3Mo_3N	652	21	31
Co_3Mo_3N-K5	869	17	51
Co_3Mo_3N-K30	364	8	46
Co_3Mo_3N-Cs2	986	16	62
Co_3Mo_3N-Cs10	586	10	59

Reaction conditions 400 °C and 0.1 MPa

molybdate is less effective. Ammonolysis may present a problem in large-scale application since, for example, heat transfer may be an issue as outlined in the case of the binary molybdenum nitride system [36, 37]. However, straight application of the hydrated molybdate precursor followed by in situ nitriding with 3/1 H_2/N_2 at 700 °C for 2 h produced a less active catalyst which, although containing some Co_3Mo_3N, comprised additional phases [38].

The activity of the Co_3Mo_3N phase has been rationalised in terms of the N_2 binding energy by Norskov and co-workers [32] as shown in Fig. 1 which presents a comparison of the calculated turnover frequency for a range of materials as a function of N_2 binding energy. The significance of N_2 binding energy as an important parameter for a reaction process in which N_2 dissociation is generally considered rate limiting has been discussed and justified [39]. In terms of the plot presented in Fig. 1, the high activity of Ru and Fe, the basis for the two major industrial processes, as discussed in the introduction, can be seen. Osmium can also be seen to be a highly active catalyst—a fact recognised in the early work as outlined by Mittasch [10] in which uranium was also highlighted as having high activity. The high activity of Fe, Ru and Os for ammonia synthesis represents an interesting group trend within the d block of the periodic table, although it is seldom, if ever, highlighted as so. In addition, the initial discovery of the activity of Os represents an early example of a currently popular aspect for consideration—scarcity of elements. At that stage, the global reserves of Os were only 100 kg and Os was also susceptible to oxidation to yield OsO_4 and so its replacement in the process was of paramount importance, which led to the discovery of the promoted Fe system. In Norskov's work, the high activity of the Co_3Mo_3N phase has been explained as an average of the N_2 binding energies for Co (which is too weakly binding) and Mo (which is too strongly binding) arriving at an optimum in the Sabatier curve shown in Fig. 1, with the Co_3Mo_3N catalyst reportedly being developed along this principle. It was argued that both components have to be in close proximity in order to be effective and that the (111) surface plane exhibits mixed Co and Mo sites. N was stated to not participate in the reaction but rather induce the required ordering. Support for this rationalisation was provided in previous microkinetic studies by the authors for which the activation energy was found to be intermediate between Co and Mo [32]. The volcano curve can also be used to rationalise the relatively high ammonia synthesis activity of the Ni_2Mo_3N phase [29, 33]. Fe_3Mo_3N, which is isostructural with Co_3Mo_3N, is also a high activity ternary nitride for the reaction [29, 33]. A perhaps somewhat related approach to the Sabatier curve are the experimental surface science studies by Attard and King [40] aimed towards the production of hybrid materials combining a

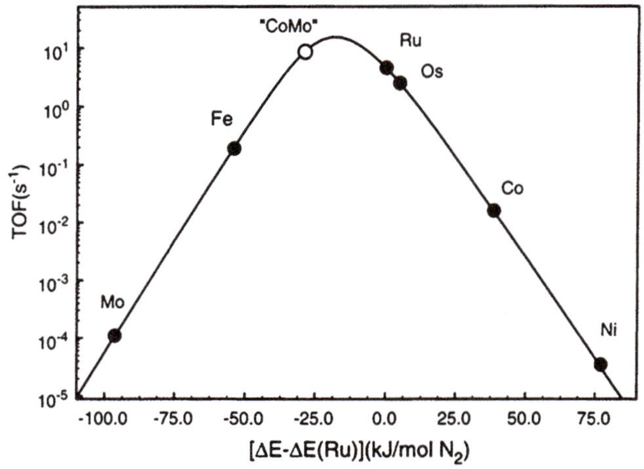

Fig. 1 Calculated turnover frequencies for ammonia synthesis as a function of the adsorption energy of nitrogen. The synthesis conditions are 400 °C, 50 bar, gas composition H_2/N_2 3/1 containing 5 % NH_3. Reproduced with permission from [32]. Copyright (2001) American Chemical Society

component with a marked ability to dissociate N_2 with a high sticking probability with a component which bonds weakly to nitrogen in order to produce a nitrogen–metal bond of similar strength to Fe–N. The former component was W, whilst Cu, Ag and Au were examples of the latter component. It was found that N_2 would not adsorb at sites consisting of both W and Cu atoms and hence it was predicted that bimetallics of W with Group IB metals would act as poor ammonia synthesis catalysts. However, if N_2 could be dissociated by another route, the possibility of copper–tungsten bimetallics functioning as effective ammonia synthesis catalysts was raised.

In the context of high ammonia synthesis activity, it is interesting to draw attention to the reported high activity of cobalt–rhenium phases [19, 41]. Whilst these do not seem to have achieved the prominence of the Co_3Mo_3N related systems, literature studies have indicated that the activity exceeds that of them. A Co/Re material ratio of 1/4 prepared by ammonolysis at 700 °C was found to be particularly effective and it was proposed that catalytic activity may relate to the presence of some form of rhenium nitride phase. However, the scarcity of rhenium and its relative cost may be a consideration in large-scale application, although it should be acknowledged that Pt–Re reforming catalysts are applied on an industrial scale and also that Co–Re catalysts for Fischer–Tropsch synthesis have been a focus of interest, e.g. [42]. Whilst many formulations were screened by Mittasch and colleagues [10] in the early development of ammonia synthesis catalysts, it is an interesting thought to note that rhenium as such would not have formed part of the matrix, since its discovery was not documented until 1925. Indeed, in these studies Co was even documented as a promoter of Mo which is fascinating

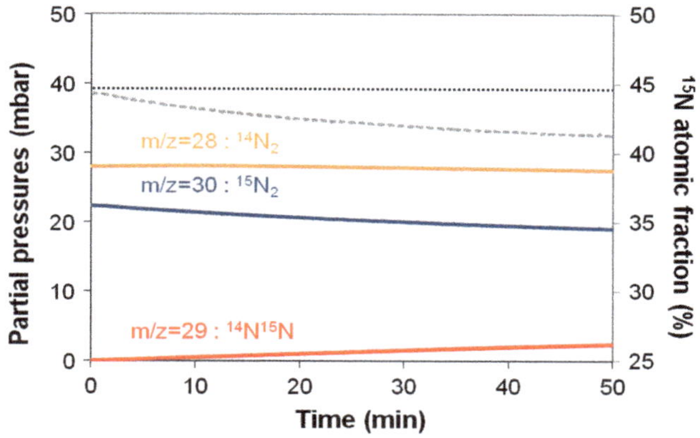

Fig. 2 Evolution of the nitrogen isotopomer partial pressures and the gas-phase ^{15}N atomic fraction during homomolecular isotopic nitrogen exchange at 450 °C over Co_3Mo_3N. Reproduced with permission from [43]. Copyright (2013) American Chemical Society

Table 2 Nitrogen isotopic exchange activity of Co_3Mo_3N as a function of pre-treatment (adapted from [43])

Pre-treatment	H_2/N_2 (3/1) 700 °C/Ar 700 °C (30 min)	H_2/N_2 (3/1) 700 °C/N_2 700 °C (30 min)	H_2/N_2 (3/1) 700 °C/ Ar 700 °C (30 min)/ N_2 700 °C (30 min)
Temperature of exchange (°C)	600	600	600
$^{15}N_2$ pressure introduced (mbar)	55.6	55.1	55.0
Total pressure at 40 min	51.4	48.5	52.3
$^{15}N_{exchanged}/N_{total331}$ at 40 min	6	28	25

in the context of the development of the Co_3Mo_3N catalytic system at a very much later stage and illustrative of the wide ranging nature of the early work conducted at BASF.

Whilst it was stated that the lattice N of the Co_3Mo_3N catalyst did not participate in the NH_3 synthesis reaction, a nitrogen isotopic exchange reaction undertaken by Hunter et al. has shown it to be exchangeable [43]. Figure 2 presents the homomolecular isotopic exchange profile in which a mixture of $^{14}N_2$ and $^{15}N_2$ is scrambled over Co_3Mo_3N at 450 °C. Most pertinent is the observation that the atomic fraction of ^{15}N in the gas phase decreases over the course of the reaction which is a clear indication of the operation of the heterolytic exchange pathway in which exchange of gas-phase species with the lattice N (which is, of course, predominantly ^{14}N reflecting the natural distribution of nitrogen isotopes) occurs and that the dissociative activation of N_2 does not occur. Isotope exchange studies of the materials with $^{15}N_{2(g)}$ only were undertaken at higher reaction temperature and the results and the reaction and pre-treatment conditions are presented in Table 2. From this data, a clear influence of the nature of pre-treatment upon the exchangeability of lattice nitrogen is observed—in these studies, the influence of adsorbed N species was discounted and it was found that ca. 25 % of lattice N species could be exchanged at 600 °C following N_2 pre-treatment at 700 °C as compared to 6 % for the corresponding pre-treatment in Ar. Whilst these reaction temperatures are significantly above that for ammonia synthesis, the presence of H_2 in that case could accentuate the lattice N reactivity. These results were taken to demonstrate the possibility that the lattice nitrogen may indeed be reactive under reaction conditions and hence that a Mars–van Krevelen type mechanism, in which NH_3 formation occurs by direct hydrogenation of lattice species which are then replenished from gas-phase N_2, could be operative and the significant effects of pre-treatment upon the extent of exchange demonstrate the sensitivity of the system, possibly involving surface restructuring effects.

Potential of nitrides as nitrogen transfer reagents

The possible existence of Mars–van Krevelen type reaction mechanisms in both the binary and ternary nitride systems raises the interesting possibility of the development of nitrides as reagents for novel nitrogen transfer pathways. The prospect of such a "mechanism" leads to the idea of a "process".

The mechanism implies two concurrent general stages to ammonia synthesis:

(i) the hydrogenation of lattice N species to yield NH_3, and
(ii) the replenishment of lattice N vacancies from gas-phase N_2

In terms of a "process" stages (i) and (ii) could be conducted under different conditions (e.g. at different temperatures) leading to potential kinetic and/or thermodynamic advantages. In oxidation catalysis there is a precedent for the distinction between "mechanism" and "process" as illustrated by the example of butane oxidation catalysed by vanadium phosphate. On the basis that this reaction occurred by a Mars–van Krevelen mechanism it has proved possible to separate the butane oxidation and catalyst re-oxidation steps into separate stages [44, 45]. In this way, the desired partial oxidation product, maleic anhydride, does not come into direct contact with O_2 in the presence of the catalyst.

Applying this general concept, it may prove possible to develop new nitrogen transfer reagents by-passing the use of NH_3 as a nitrogen transfer reagent. The potential value of this idea can be illustrated by consideration of the large-scale production of aniline via the selective reduction of nitrobenzene. Considering this process from the viewpoint of N_2 as the initial source of nitrogen, the process steps can be considered as:

$$N_2 + 3H_2 \leftrightarrow 2NH_3$$
$$2NH_3 + 4O_2 \rightarrow 2HNO_3 + 2H_2O$$
$$HNO_3 + 2H_2SO_4 \rightarrow NO_2^+ + H_3O^+ + 2HSO_4^-$$
$$C_6H_6 + NO_2^+ \rightarrow C_6H_5NO_2 + H^+$$
$$C_6H_5NO_2^+ + 3H_2 \rightarrow C_6H_5NH_2 + 2H_2O$$

As detailed above, the production of ammonia by Haber–Bosch Process is extremely energy intensive. Having applied this energy to form intermediate ammonia, the second equation (a simplified representation of the Ostwald Process) involves oxidation of some of the hydrogen atoms bonded to each ammonia molecule added in the previous process step. Taken together, these initial two stages represent an indirect way of oxidising hydrogen to yield water! The nitric acid product is then used to form the nitronium cation in corrosive mixture of concentrated acids. Having obtained the nitronium cation benzene is nitrated and the nitrobenzene product selectively reduced to yield aniline—in another stage indirectly involving the reaction between oxygen and hydrogen to yield water!

It might be imagined that the direct amination of benzene:

$$C_6H_6 + NH_3 \leftrightarrow C_6H_5NH_2 + H_2$$

is a potential alternative route, but it is thermodynamically limited and approaches to overcome this limitation by use of a cataloreactant [46] or coupled hydrogen consuming reactions [47] have proved unsuccessful.

If the lattice nitrogen of a nitride could be used in an alternative aniline forming two-stage process such as:

$$C_6H_6 + MN + 0.5H_2 \rightarrow C_6H_5NH_2 + M$$

and

$$M + \text{``N''} \rightarrow MN$$

(where MN represents a metal nitride, M the corresponding metal and "N" a nitridation agent such as NH_3 or, ideally, N_2) although a formidable challenge, the potential interest in the development of such a route is readily apparent. Indeed the nitrogen transfer and renitridation stages could be conducted under different reaction conditions. However, non-selective reduction could be a major potential issue. Aniline production is just one of a number of major "nitrogen incorporation" processes for which a similar potential approach may be imagined.

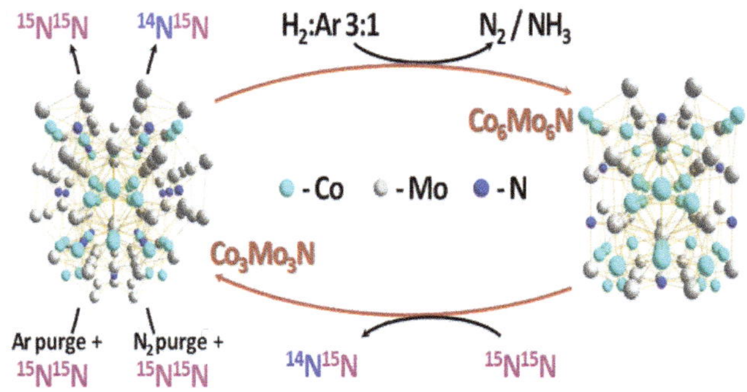

Fig. 3 Schematic of the phase transformations and isotopic exchange pathways over the Co_3Mo_3N–Co_6Mo_6N system. Reproduced with permission from [43]. Copyright (2013) American Chemical Society

Whilst not widespread, reports of the reactivity of lattice N are not without precedent, some of which were already mentioned in the section relating to binary nitride catalysts applied to ammonia synthesis. In 1948, Goodeve and Jack [48] published an investigation relating to the denitridation of iron nitride. In this study, hydrogen was shown to enhance the loss of lattice nitrogen leading to its complete conversion to ammonia in the 250–450 °C range. Interestingly, when CO was used, 25 % of the nitrogen lost formed cyanogen with isomorphous substitution of lattice N by C. Lattice nitrogen reactivity has also been implicated in NH_3 decomposition over zirconium oxynitride [49] and the ammoxidation of propane over VAlON [50].

With the Mars–van Krevelen process concept in mind, the reactivity of a number of metal nitrides towards H_2 has been investigated [14, 38, 51, 52]. Co_3Mo_3N proved a particularly interesting case in this regard—at 700 °C the previously unprecedented Co_6Mo_6N phase was formed through loss of 50 % of the lattice nitrogen [51, 52]. Despite the fact that the majority of N eliminated was in the form of N_2, with NH_3 being a minor co-product, H_2 was necessary to drive the process. In Co_3Mo_3N all the lattice nitrogen is crystallographically equivalent and it might be imagined that it would be totally lost upon reaction with H_2. However, the residual 50 % relocates to a different crystallographic site (migration from the 16c to the 8a Wyckoff site) as confirmed by powder neutron diffraction [52]. Restoration of the Co_3Mo_3N phase, which would be a necessary step for its application as a nitrogen transfer reagent, was found to be relatively facile when using a N_2/H_2 mix and achievable at higher temperature when using N_2 alone [53]. It was demonstrated that the Co_3Mo_3N/Co_6Mo_6N system can be cycled back and forth. It has tentatively been postulated that the H_2 necessary to accomplish the Co_3Mo_3N to Co_6Mo_6N phase change may function by inducing a phase change through reduction of surface stoichiometry—in terms of diffraction analysis, the transformation occurs between the two end members with no intermediate stoichiometries being formed. Isotopic nitrogen exchange over the Co_6Mo_6N phase has demonstrated that simultaneously, along with renitridation occurring leading to regeneration of the Co_3Mo_3N phase, a degree of exchange occurs with both processes ceasing simultaneously [43]. A schematic of the entire Co_3Mo_3N–Co_6Mo_6N interconversion and isotopic exchange processes is presented in Fig. 3.

Whilst Fe_3Mo_3N is iso-structural with Co_3Mo_3N, its lattice nitrogen is much less reactive. In terms of Co_3Mo_3N, the elevated temperature necessary to conduct the loss of 50 % of the lattice nitrogen would lead to carbidation upon contact with hydrocarbon reactants and so it is necessary to either find a means for enhancement of the lattice nitrogen reactivity or to develop different systems of enhanced reactivity. This could involve experimental based studies or, perhaps more excitingly, computational modelling techniques could be applied to design systems with desired lattice nitrogen reactivity. In the Co_3Mo_3N–Co_6Mo_6N system, it is probable that the renitridation step is enhanced by the maintenance of 50 % of the original nitrogen in the system. In other nitrides, such as Cu_3N, where N loss occurs at temperatures of greater potential interest for nitrogen transfer avoiding carbidation, the total loss of nitrogen which is difficult to suppress and the concomitant structural transformation have rendered renitridation extremely difficult [54].

In principle, routes to nitrogen-containing organic products such as amines could be envisaged based upon the application of carbonitrides as reagents involving hydrogenation. This could open up the exciting prospect of indirect routes from N_2 and CH_4 in three stage procedures involving nitridation, carbidation and reduction separately. In practice, each step could be conducted under different reaction conditions and the sequence of nitridation and carbidation could be reversed. This prospect is a significant challenge but arguably any demonstration of the indirect reaction of methane and nitrogen to yield any organic nitrogen-containing product, and even HCN, would be of major interest. To this end, preliminary investigations of the reduction molybdenum carbonitride systems prepared from single source routes have been undertaken [55]. In these studies, the preferential reactivity of lattice N over lattice C has been observed, indicating that it would be necessary to develop materials with more closely matched reactivity of the lattice N and C components. Again, computational modelling in which target phases are identified could play a significant role in this regard.

Related alternative strategies for ammonia synthesis

The observations outlined above indicate that it could be possible to apply Co_3Mo_3N as a material for the two-stage synthesis of ammonia in which hydrogenation of the lattice nitrogen and its subsequent replenishment from N_2 are conducted in separate stages. However, the temperatures involved in this would be high. Also, the mass fraction of N in the system is very low—an issue compounded when it is considered that only 50 % of the lattice nitrogen is "active"—and only a relatively fraction of it is hydrogenated to yield NH_3 with the major proportion being lost in the form of N_2.

The design of new, improved systems with greater efficiency of N utilisation and enhanced reactivity of N based along these lines represents an exciting opportunity for combined computational modelling and experimental studies. Cyclic operation of ammonia synthesis has been proposed previously in the literature by Amariglio and co-workers as a means of overcoming the inhibition of reaction rate by H_2 for Ru [56] and Os [57] systems. It was suggested that reaction rates of possible commercial interest could be achieved at atmospheric pressure using such an approach in these systems. For example, by applying the cyclic two-step procedure—saturation under pure N_2, followed by hydrogenation under pure H_2—extrapolation indicated that rates comparable to the industrial performance of iron catalysts could be obtained over Ru at 300 °C and atmospheric pressure.

Routes applying solar energy towards nitrogen fixation and the production of ammonia have been described recently by Pfromm and co-workers [58–60]. In these routes, metal nitrides are prepared by various means from N_2 and NH_3 is generated in a subsequent hydrolysis step. The necessary high temperatures are attained via a solar furnace. The use of a solar furnace as an environmentally friendly means to attain high temperatures for application to nitrogen fixation is an interesting concept. Historically, in the cyanamide process which immediately pre-dated the Haber–Bosch Process, nitrogen fixation was achieved by direct reaction of N_2 with calcium dicarbide producing calcium cyanamide ($CaCN_2$). Although this process required high temperatures, it was exothermic and self-sustaining once initiated [61]. However, the production of calcium carbide from CaO and C necessary for the process did involve the application of very high temperatures. $CaCN_2$ could be directly applied as a fertiliser itself or hydrolysed to liberate ammonia.

Conclusion

In this overview, selected examples of the applications of metal nitrides as catalysts for the synthesis of ammonia have been described. The potential of this class of material for the development of novel nitrogen transfer reagents has also been outlined. It can be anticipated that advances in nitride synthesis coupled with the application of more complex materials, such as quaternary systems, may result in exciting developments in these areas. It is clear that nitride materials represent an interesting opportunity. They are of increasing interest in other areas of heterogeneous catalysis.

References

1. Smil V (2002) Nitrogen and food production: proteins for human diets. AMBIO J Hum Env 31:126–131
2. Tanabe Y, Nishibayashi Y (2012) Developing more sustainable processes for ammonia synthesis. Coord Chem Rev 257:2551–2564
3. Bielawa H, Hinrichsen O, Birkner A, Muhler M (2001) The ammonia synthesis catalyst of the next generation: barium promoted oxide supported ruthenium. Angew Chemie Int Edn 40:1061
4. Jacobsen CJH (2001) A novel support for ruthenium-based ammonia synthesis catalysts. J Catal 200:1–3
5. Urabe K, Aika K, Ozaki A (1976) Activation of nitrogen by alkali metal promoted transition metal. 6. Hydrogen effect on isotopic equilibration of nitrogen and rate-determining step of ammonia-synthesis on potassium promoted ruthenium catalysts. J Catal 42:197–204
6. Rao KSR, Masthan SK, Prasad PSS, Rao PK (1991) Effect of barium addition on the ammonia synthesis activity of a cesium promoted ruthenium catalyst supported on carbon covered alumina (CAA). Appl Catal 73:L1–L5
7. Honkala K, Hellman A, Remediakis IN, Logadottir A, Carlosson A, Dahl S, Christensen CH, Norskov JK (2005) Ammonia synthesis from first principles calculations. Science 307:555–558
8. Rhodes AK (1996) New ammonia process, catalyst proven in Canadian plant. Oil Gas J 94(47):37–41
9. Liu H (2012) Ammonia synthesis catalysts—innovation and practice. World Scientific Press. ISBN: 978-981-4355-77-3
10. Mittasch A (1950) Early studies of multicomponent catalysts. In: Frankenburg WG, Rideal EK, Komorewsky VI (eds) Advances in catalysis, vol II. Academic Press, New York, pp 81–104
11. Schlogl R (2003) Catalytic synthesis of ammonia—a "never ending story?" Angew Chemie Int Edn 42:2004–2008
12. Volpe L, Boudart M (1986) Ammonia synthesis on molybdenum nitride. J Phys Chem 90:4874–4877
13. Kojima R, Aika K (2001) Molybdenum nitride and carbide catalysts for ammonia synthesis. Appl Catal A Gen 219:141–147
14. Mckay D, Hargreaves JSJ, Rico JL, Rivera JL, Sun XL (2008) The influence of phase and morphology of molybdenum nitrides on ammonia synthesis activity and reduction characteristics. J Solid State Chem 161:325–333
15. Segal N, Sebba F (1967) Ammonia synthesis catalyzed by uranium nitride. 1. Reaction mechanism. J Catal 8:105–112
16. Segal N, Sebba F (1967) Ammonia synthesis catalyzed by uranium nitride. 2. Transient behaviour. J Catal 8:113–119
17. King DA, Sebba F (1965) Catalytic synthesis of ammonia over vanadium nitride containing oxygen. 1. Reaction mechanism. J Catal 4:253–259
18. King DA, Sebba F (1965) Catalytic synthesis of ammonia over vanadium nitride containing oxygen. 2. Order-disorder transition revealed by catalytic behavior. J Catal 4:430–439
19. Kojima R, Aika K (2001) Rhenium containing binary catalysts for ammonia synthesis. Appl Catal A Gen 209:317–325
20. Kojima R, Enomoto H, Muhler M, Aika K (2003) Cesium promoted rhenium catalysts supported on alumina for ammonia synthesis. Appl Catal A Gen 245:311–322
21. Panov GI, Kharitonov AS (1985) Catalytic properties of nitrides in ammonia synthesis. React Kinet Catal Lett 29:267–274
22. Volpe L, Boudart M (1985) Compounds of molybdenum and tungsten with high surface area. 1. Nitrides. J Solid State Chem 59:332–347
23. Gong SW, Chen HK, Li W, Li BQ (2005) Synthesis of beta-$Mo_2N_{0.78}$ hydrodesulfurization catalyst in mixtures of nitrogen and hydrogen. Appl Catal A Gen 279:257–261
24. Ganin AY, Kienle L, Vajenine GV (2006) Synthesis and characterisation of hexagonal molybdenum nitrides. J Solid State Chem 179:2339–2348
25. Aika KI, Ozaki A (1969) Mechanism and isotope effect in ammonia synthesis over molybdenum nitride. J Catal 14:311–321
26. Logan SR, Moss RL, Kemball KC (1958) The catalytic decomposition of ammonia over evaporated iron films. Trans Faraday Soc 54:922–930
27. Hillis MR, Kemball C, Roberts MW (1966) Synthesis of ammonia and related processes on reduced molybdenum dioxide. Trans Faraday Soc 62:3570–3585
28. Doornkamp C, Ponec V (2000) The universal character of the Mars and van Krevelen mechanism. J Mol Chem A Chem 162:19–32
29. Kojima R, Aika K (2001) Cobalt molybdenum bimetallic nitride catalysts for ammonia synthesis—Part 1. Preparation and characterization. Appl Catal A Gen 215:149–160
30. Kojima R, Aika K (2001) Cobalt molybdenum bimetallic nitride catalysts for ammonia synthesis—Part 2. Kinetic study. Appl Catal A Gen 218:121–128
31. Kojima R, Aika K (2001) Cobalt molybdenum bimetallic catalysts for ammonia synthesis—Part 3. Reactant gas treatment. Appl Catal A Gen 219:157–170
32. Jacobsen CJH, Dahl S, Clausen BS, Bahn S, Logadottir A, Norskov JK (2001) Catalyst design by interpolation in the periodic table: bimetallic ammonia synthesis catalysts. J Am Chem Soc 123:8404–8405
33. Jacobsen CJH (2000) Novel class of ammonia synthesis catalysts. Chem Commun 12:1057–1058
34. Boisen A, Dahl S, Jacobsen CJH (2002) Promotion of binary nitride catalysts: isothermal N_2 adsorption, microkinetic model, and catalytic ammonia synthesis activity. J Catal 208:180–186
35. Kojima R, Aika K (2000) Cobalt molybdenum bimetallic catalysts for ammonia synthesis. Chem Lett 514–515
36. Wise RS, Markel EJ (1994) Catalytic NH_3 decomposition by topotactic molybdenum oxides and nitrides—effect on temperature programmed gamma-Mo_2N synthesis. J Catal 145:335–343
37. Wise RS, Markel EJ (1994) Synthesis of high surface area molybdenum nitride in mixtures of nitrogen and hydrogen. J Catal 145:344–355
38. Hargreaves JSJ, Mckay D (2009) A comparison of the reactivity of lattice nitrogen in Co_3Mo_3N and Ni_2Mo_3N catalysts. J Mol Catal A Chem 305:125–129
39. Dahl S, Logadottir A, Jacobsen CJH, Norskov J (2001) Electronic factors in catalysis: the volcano curve and the effect of promotion in catalytic ammonia synthesis. Appl Catal A Gen 222:19–29
40. Attard GA, King DA (1990) Co-adsorption of nitrogen with Cu, Ag and Au on W(100)—the role of metal adatoms in controlling surface reactivity. J Chem Soc Faraday Trans 86:2735–2741

41. Alexander A-M, Hargreaves JSJ, Mitchell C (2013) The denitridation of nitrides of iron, cobalt and rhenium under hydrogen. Top Catal 56:1963–1969
42. Iida H, Sakamoto K, Takeuchi M, Igarashi A (2013) Fischer–Tropsch synthesis over Co/SiO_2 and Co-M (M:Ru, Re)/SiO_2 catalysts prepared by a high temperature supercritical drying method. Appl Catal A Gen 466:256–263
43. Hunter SM, Gregory DH, Hargreaves JSJ, Richard M, Duprez D, Bion N (2013) A study of $^{15}N/^{14}N$ isotopic exchange over cobalt molybdenum nitrides. ACS Catal 3:1719–1725
44. Contractor RM (1999) Dupont's CBT technology for maleic anhydride. Chem Eng Sci 54:5627–5632
45. Bordes E, Contractor RM (1996) Adaptation of the microscopic properties of redox catalysts to the type of gas-solid reactor. Top Catal 3:365–375
46. Desrosiers P, Guan SH, Hagemeyer A, Lowe DM, Lugmair C, Poojary DM, Turner H, Weinberg H, Zhou XP, Armbrust R, Fengler G, Notheis U (2003) Application of combinatorial catalysis for the direct amination of benzene to aniline. Catal Today 81:319–328
47. Becker J, Holderich WF (1998) Amination of benzene in the presence of ammonia using a Group VIII metal supported on a carrier as catalyst. Catal Lett 54:125–128
48. Goodeve C, Jack KH (1948) Kinetics of nitrogen evolution from an iron-nitrogen interstitial alloy. Disc Faraday Soc 4:82–91
49. Soerijanto H, Rodel C, Wild U, Lerch M, Schomaker R, Schlogly R, Ressler T (2007) The impact of nitrogen mobility of zirconium oxynitride catalysts for ammonia decomposition. J Catal 250:19–24
50. Olea M, Florea M, Sack I, Silvy PR, Gaigneaux EM, Marin GB, Grange P (2005) Evidence for the participation of lattice nitrogen from vanadium aluminium oxynitrides in propane ammoxidation. J Catal 232:152–160
51. Mckay D, Gregory DH, Hargreaves JSJ, Hunter SM, Sun XL (2007) Towards nitrogen transfer catalysis: reactive lattice nitrogen in cobalt molybdenum nitride. Chem Commun 29:3051–3053
52. Hunter SM, Mckay D, Smith RI, Hargreaves JSJ, Gregory DH (2010) Topotactic nitrogen transfer: structural transformation in cobalt molybdenum nitrides. Chem Mater 22:2898–2907
53. Gregory DH, Hargreaves JSJ, Hunter SM (2011) On the regeneration of Co_3Mo_3N from Co_6Mo_6N and N_2. Catal Lett 141:22–26
54. Alexander A-M, Hargreaves JSJ, Mitchell C (2012) The reduction of various nitrides under hydrogen: Ni_3N, Cu_3N, Zn_3N_2 and Ta_3N_5. Top Catal 55:1046–1053
55. AlShalwi M, Hargreaves JSJ, Liggat JJ, Todd D (2012) The reactivity of lattice nitrogen species in molybdenum (oxy)carbonitrides prepared by single-source routes. Mater Res Bull 47:1251–1256
56. Rambeau G, Amariglio H (1981) Improvement of the catalytic performance of a ruthenium powder in ammonia synthesis by use of a cyclic procedure. Appl Catal 1:291–302
57. Rambeau G, Jorti A, Amariglio H (1982) Improvement of the catalytic performance of an osmium powder in ammonia synthesis by the use of a cyclic procedure. Appl Catal 3:273–282
58. Michalsky R, Pfromm PH (2011) Chromium as reactant for solar thermochemical synthesis of ammonia from steam, nitrogen and biomass at atmospheric pressure. Sol Energy 85:2642–2654
59. Michalksy R, Parman BJ, Amanor-Boadu V, Pfromm PH (2012) Solar thermochemical production of ammonia from water, air and sunlight: thermodynamic and economic analyses. Energy 42:251–260
60. Michalsky R, Pfromm PH (2012) An ionicity rationale to design solid phase metal nitride reactants for solar ammonia production. J Phys Chem C 116:23243–23251
61. Topham SA (1985) The history of the catalytic synthesis of ammonia. In: Anderson JR, Boudart M (eds) Catalysis science and technology, vol. 7, Chapter 1. Springer, Berlin

Preparation of novel mesoporous Ce-Ni_2P/SBA-15 catalysts and their catalytic performance for hydrodesulfurization of dibenzothiophene

Qingyuan Li · Pingyi Wu · Ling Lan · Ni Wei · Shengfu Ji

Abstract Ni_2P/SBA-15 precursors with Ni_2P loadings of 25 wt% and initial P/Ni of 0.8 were prepared using nickel nitride as nickel source, diammonium hydrogen phosphide as phosphorus and mesopore molecular sieve SBA-15 as support. Then Ce was introduced into the Ni_2P/SBA-15 precursor. The novel mesoporous Ce-Ni_2P/SBA-15 catalysts were prepared after temperature-programmed reduction in flowing H_2. The structure was characterized by X-ray diffraction, N_2 adsorption–desorption isotherms, NH_3 temperature-programmed desorption and X-ray photoelectron spectroscopy. The catalytic activities for the hydrodesulfurization (HDS) of dibenzothiophene (DBT) were evaluated. The results showed that only Ni_2P phase was formed in Ce-Ni_2P/SAB-15 catalysts with Ce loadings of 0–5 wt%. Ni_2P and $Ni_{12}P_5$ phases were existed in 7 wt% Ce-Ni_2P/SBA-15 catalyst. The surface area and pore volume increased when Ce was added to Ni_2P/SBA-15 catalyst. The strength of the acid sites and total acid amount of Ce-Ni_2P/SBA-15 catalysts increased with increasing Ce loadings. Ce existed in the form of $Ce^{3+}$and Ce^{4+}, Ni existed in the form of Ni^{2+} and $Ni^{\delta+}$, and P existed in the form of $P^{\delta-}$ and P^{5+}. The addition of Ce to the Ni_2P/SBA-15 catalyst decreased $Ni^{\delta+}$ concentration in Ni_2P/SBA-15 catalyst. The activity for HDS of DBT over Ni_2P/SBA-15 catalysts was affected by the addition of Ce at 300–340 °C. The catalysts exhibited a good catalytic performance of deep hydrodesulfurization of dibenzothiophene and the conversion of DBT can reach 98.9 % at 380 °C. Biphenyl was the main product over Ce-Ni_2P/SBA-15 catalysts and cyclohexylbenzene was the main product over Ni_2P/SBA-15 catalyst at 380 °C.

Keywords Ce · Ni_2P · SBA-15 · Hydrodesulfurization · Dibenzothiophene

Q. Li · P. Wu · N. Wei · S. Ji (✉)
State Key Laboratory of Chemical Resource Engineering,
Beijing University of Chemical Technology,
Beijing 100029, China
e-mail: jisf@mail.buct.edu.cn

P. Wu · L. Lan
Petrochemical Research Institute, PetroChina,
Beijing 100195, China

Introduction

Automobile industrial developments in the worldwide have increased the environmental pollution. Thus, the stricter standard for gasoline and diesel had been implemented, including decreased S content limits [1, 2]. Alumina-supported Mo or W sulfides by Ni or Co are mostly used as commercial hydrodesulfurization (HDS) catalysts on which dibenzothiophene (DBT) and 4,6-dimethyldibenzothiophene (4,6-DMDBT) are very difficult to remove [3]. So the novel high performance HDS catalysts are urgent to develop under the stringent legislation.

In the recent research, transition phosphides has been attracting attention as the next generation catalysts in the field of HDS, such as hydrogenation [4], hydrazine decomposition [5], especially hydrodesulfurization and hydrodenitrogenation [6, 7], due to their unique physical and chemical properties. Oyama et al. [8] reported that the activity of DBT HDS over Ni_2P/SiO_2 was higher than commercial Co–Mo sulfide catalysts. Mesopore molecular sieve belongs to regular channel structure that can promote large molecular diffusion in hydrocracking reaction. Wang et al. [9] had reported that the surface area and pore diameter of Ni_2P/MCM-41 were larger than that of Ni_2P/SiO_2 and Ni_2P/Al_2O_3 catalysts. It was also reported by

Oyama that the catalytic performance for HDS of 4,6-DMDBT over Ni_2P/MCM-41 catalyst was higher than that of Ni_2P/SiO_2 catalyst [10]. Our previous work reported that Ni_2P/SBA-15 catalysts exhibited excellent deep activity of DBT HDS [11].

Hydrodesulfurization performance could be tunable by adding promoter to Ni_2P catalyst. Abu et al. [12] reported that the addition of Co to Ni_2P bulk catalyst brought an increase in 4,6-DMDBT HDS activity and selectivity. In previous paper [13], our group had reported the catalytic activity of Mo-Ni_2P/SBA-15/cordierite catalysts for DBT HDS, the addition of Mo to Ni_2P/SBA-15/Cordierite can promote the DBT activity and selectivity and the mechanism over Mo-Ni_2P/SBA-15/cordierite can be changed at 380 °C. It has been shown that cerium promoter could improve catalytic performance in some catalytic reactions. Li et al. [14] demonstrated that the influence of CeO_2 on hydrogenation activity over bulk Ni_2P. CeO_2 was an effective promoter enhancing the hydrogenation activity of 1-heptene and toluene over Ni_2P catalyst. Besides, Li et al. [15] also found that CeO_2 promoters could improve the activity of quinoline hydrodenitrogenation (HDN) over Ni_2P. Cai et al. [16] reported the effect of promoter Ce on the catalytic performance of Ni/Al_2O_3 catalyst for autothermal reforming of methane to hydrogen. The result showed that the addition of Ce to the Ni/Al_2O_3 catalyst increased the active surface that promoted the activity of the catalyst. In this study, the effect of Ce on DBT HDS over Ni_2P/SBA-15 catalyst is investigated. The relationship between the structures and catalytic activities of the catalysts are discussed.

Experimental

Preparation of catalyst

SBA-15 was synthesized using the literature method [17]. Ni_2P/SBA-15 precursors with P/Ni molar ratio of 0.8 were prepared by impregnating an aqueous solution of $(NH_4)_2HPO_4$ and $Ni(NO_3)_2{\cdot}6H_2O$ (add little HNO_3 if precipitation was formed) into SBA-15 for 12 h. Then dried at 120 °C for 24 h and subsequently calcined the sample at 550 °C for 4 h. The content of Ni_2P was about 25 wt%. Prior to H_2-TPR, the precursor was crushed to 40–60 mesh. 100 mL/min H_2 flow was used to reduce the precursor by heating from room temperature (RT) to 300 °C at 10 °C min^{-1}, then from 300 to 650 °C at 1 °C min^{-1}, and finally maintained at 650 °C for 2 h. After the reduction, the catalysts were cooled to RT and passivated by 1 % O_2/Ar (v/v) for structure characterization, or for HDS of DBT. Based on the above sample, Ce-Ni_2P/SBA-15 samples were prepared by impregnating aqueous solutions of $Ce(NO_3)_3{\cdot}6H_2O$ with different concentration into the above Ni_2P/SBA-15 precursors at RT for 12 h, followed by drying the sample at 120 °C for 24 h and eventually calcining the sample at 550 °C for 4 h. The contents of Ce were 0–7 wt%.

Catalyst characterization

X-ray diffraction (XRD) patterns of Ce-Ni_2P/SBA-15 catalysts were obtained on a Rigaku D/Max 2500 VB2 +/PC diffractometer using Cu Kα radiation operating at 200 mA and 40 kV. N_2 adsorption–desorption isotherms of Ce-Ni_2P/SBA-15 catalysts were measured using a Quadrasorb SI instrument. Before measurement, the sample was outgassed for at least 5 h at 350 °C under vacuum. BET and BJH methods were used to determine the specific surface area and pore size distribution, respectively. NH_3-TPD data were collected from Thermo Electron TPD/R/O 1100 series catalytic surfaces analyzer equipped with a TC detector. Each sample was initially pretreated through heating in 20 ml/min of nitrogen from ambient to 300 °C, with a 1 h hold. The sample was then cooled down to ambient; ammonia adsorption was carried out. Physically adsorbed ammonia was removed by purging with nitrogen at 100 °C for 1 h. NH_3-TPD of the sample was carried out by increasing the temperature linearly from 40 to 800 °C with a heating rate of 10 K/min and a helium flow rate of 20 ml/min. XPS experiments were carried out on an ESCALAB 250 (Thermo Electron Co.) instrument using Al K_α as the exciting radiation at constant pass energy of 30 eV. Binding energies were calibrated using the carbon present as a contaminant (C 1s = 285.0 eV).

Catalytic activity testing

The HDS of DBT was performed in a high-pressure fixed-bed continuous-flow stainless steel reactor (9 mm in diameter and 500 mm in length) with a central thermocouple to measure the temperature of the catalyst bed. Ce-Ni_2P/SBA-15 catalyst with different Ce loadings was loaded on the fixed-bed reactor between the pads of quartz sand. The reaction feed consisting of 1 wt% DBT in decalin was introduced into the reactor by a piston pump. The HDS reaction was operated at 3.0 MPa from 300 to 380 °C, liquid hourly space velocity = 1.9 h^{-1}, and the ratio of hydrogen-to-liquid feed was 400 (v/v). The liquid products were collected at 1 h intervals and analyzed by an off-line gas chromatography [SP2100, Beijing Beifenruili Analytic Instrument (Group) Co., Ltd], equipped with a flame ionization detector and a capillary column (HJ. PONA, 50 m × 0.20 mm × 0.50 μm).

Results and discussion

XRD of Ce-Ni_2P/SBA-15 catalysts

Figure 1 shows the XRD patterns of Ce-Ni_2P/SBA-15 catalysts with different Ce loadings. In the case of Ni_2P/SBA-15 catalyst, the diffraction peak of the Ni_2P phase located at $2\theta = 40.6°$, 44.5°, 47.3° and 54.1° (PDF-03-0953) had been observed, suggesting that Ni_2P was formed in the Ni_2P/SBA-15 catalyst. None of new diffraction peak appeared in the XRD patterns of Ce-Ni_2P/SBA-15 catalysts with Ce $\leq$ 5 wt%. For 7 % Ce-Ni_2P/SBA-15 catalyst, the diffraction peak of the $Ni_{12}P_5$ phase appeared, which was located at $2\theta = 38.4°$, 41.6° and 48.9° (PDF-22-1190). This indicated that Ni_2P and $Ni_{12}P_5$ phases existed in 7 % Ce-Ni_2P/SBA-15.

The diffraction peak of Ce species was not shown with the increasing of Ce loadings in Ce-Ni_2P/SBA-15 catalysts, indicating that Ce was well dispersed in SBA-15 or Ce concentration was too low. The result agreed with the literature [14]. Besides, No cerium phosphides ($Ce_2Ni_{12}P_7$ and $CeNi_2P_2$), cerium nickel phosphide ($Ce_6Ni_6P_{17}$) or cerium oxide species were observed when Ce was added to Ni_2P bulk catalyst. In the literature [18], the formation of $Ni_{12}P_5$ was attributed to the loss of phosphorus in temperature-programmed reduction process. It has reported by Abu et al. [12] that $Ni_{12}P_5$ and Ni_2P phases were formed due to the addition of excess Co to Ni_2P bulk catalyst. Therefore, it was thought that the addition of excess Ce to Ni_2P/SBA-15 catalyst inhibited enough phosphorus contacted with nickel to form Ni_2P phase. So in 7 % Ce-Ni_2P/SBA-15 catalyst, $Ni_{12}P_5$ phase was formed. Furthermore, the intensity of Ni_2P diffraction peaks was slightly decreased with the increasing of Ce. This also corresponded to the previous research [19].

N_2 adsorption–desorption of Ce-Ni_2P/SBA-15 catalysts

N_2 adsorption–desorption isotherms of Ni_2P/SBA-15 and Ce-Ni_2P/SBA-15 catalysts are shown Fig. 2. It can be observed that the nitrogen adsorption–desorption isotherms of all catalysts were found to be of type IV according to IUPAC classification. All catalysts exhibited a H1 hysteresis loop that was a typical mesoporous solid [17]. This indicated that all catalysts retained SBA-15 mesoporous structure.

The pore structure parameters are summarized in Table 1. For Ni_2P/SBA-15 catalyst, the values of the S_{BET}, V_p and D_{BJH} were 247.8 m^2/g, 0.30 m^3/g and 3.8 nm, respectively. In the case of Ce-Ni_2P/SBA-15 catalysts, the values of S_{BET} and V_p were higher than that of Ni_2P/SBA-15 catalyst. It suggested that favorable influence on S_{BET} and V_p was shown by adding cerium in Ni_2P/SBA-15 catalyst. Abu et al. [12] recognized that the increased BET surface area of Co_xNi_2P and $Co_{0.08}MoP$ suggested a reduction in the degree of agglomeration of the larger crystallites when the Co was added. Compared with 5 % Ce-Ni_2P/SBA-15 catalyst, the values of S_{BET} and V_p of 7 % Ce-Ni_2P/SBA-15 catalyst slightly decreased. This could be attributed to the dispersibility of active component which was weakened by adding the excess amount of cerium. However, the values of D_{BJH} were similar to that of Ni_2P/SBA-15 catalyst, which indicated that Ce-Ni_2P existed in the SBA-15 pore channel. Gutierrez et al. [20] recognized that desorption pore diameter of NiMo/Zr-SBA-15 catalysts was related with metal contents. The pore diameter

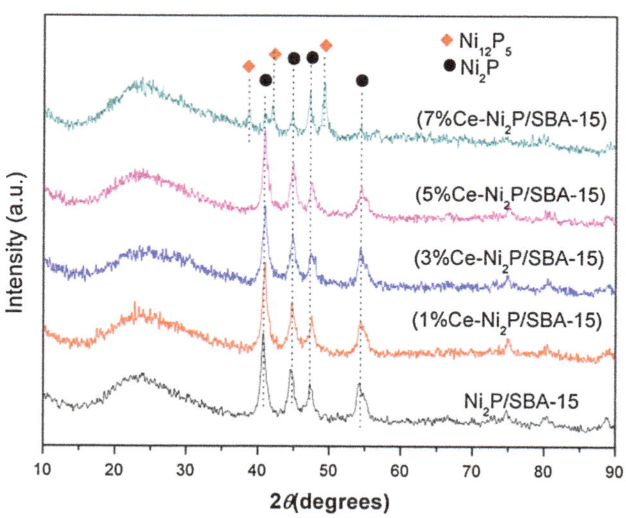

Fig. 1 XRD patterns of Ce-Ni_2P/SBA-15 catalysts with different Ce loadings

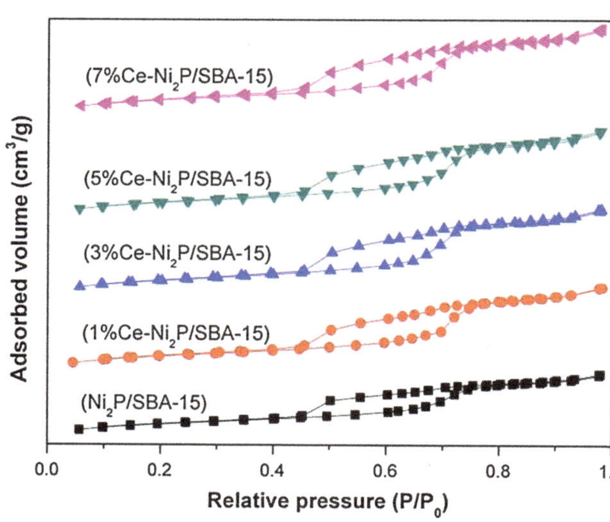

Fig. 2 N_2 adsorption–desorption isotherms of Ce-Ni_2P/SBA-15 catalysts with different Ce loadings

Table 1 Pore structure parameters of Ce-Ni_2P/SBA-15 catalyst with different Ce loadings

Catalysts	S_{BET} (m^2/g)	V_p (cm^3/g)	D_{BJH} (nm)
Ni_2P/SBA-15	247.8	0.31	3.8
1 % Ce-Ni_2P/SBA-15	248.9	0.33	3.9
3 % Ce-Ni_2P/SBA-15	254.9	0.38	3.9
5 % Ce-Ni_2P/SBA-15	265.8	0.40	3.9
7 % Ce-Ni_2P/SBA-15	260.1	0.38	3.9

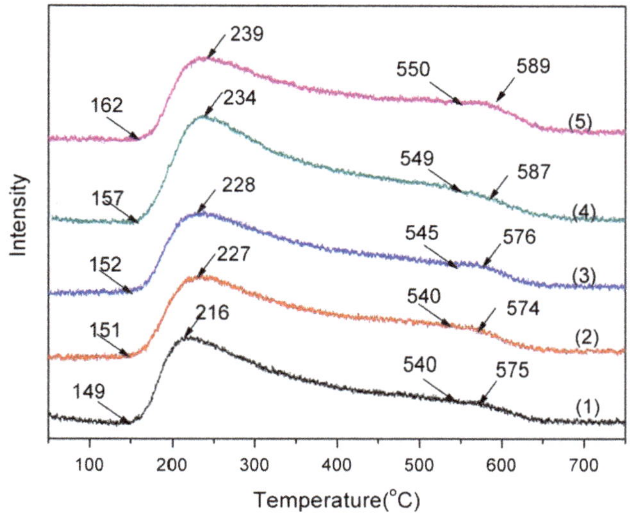

Fig. 3 NH_3-TPD profiles of Ce-Ni_2P/SBA-15 catalysts with different Ce loadings *1* Ni_2P/SBA-15, *2* 1 % Ce-Ni_2P/SBA-15, *3* 3 % Ce-Ni_2P/SBA-15, *4* 5 % Ce-Ni_2P/SBA-15, *5* 7 % Ce-Ni_2P/SBA-15

was stable due to the metal existed in the Zr-SBA-15 channel when metal loadings were low, and the pore diameter decreases because high metal loadings blocked the SBA-15 pore mouth.

NH_3-TPD of Ce-Ni_2P/SBA-15 catalysts

Hydrodesulfurization activity and selectivity were related to acidity according to some investigates. In this work, the acidity of the Ce-Ni_2P/SBA-15 catalysts was investigated by temperature-programmed desorption of ammonia (NH_3-TPD). The technique provided information on the total acidity of the solid and strength of acid sites.

The NH_3-TPD profiles of Ce-Ni_2P/SBA-15 catalysts with different Ce loadings are presented in Fig. 3. It can be observed two desorption peaks on all NH_3-TPD curves. One was the large peak at lower temperature assigned to weak acid sites. The other was small peak at higher temperature assigned to strong acid site.

Generally, the acid sites were classified into the weak (≤200 °C), medium (200–350 °C) and strong (≥350 °C) acid sites [21]. With regard to Ni_2P/SBA-15 catalyst, there were two peaks: the large desorption peak with the start desorption temperature at 145 °C and maximum desorption temperature at 216 °C, and the small desorption peak with the start desorption temperature at 540 °C and the maximum desorption temperature at 575 °C. Hu et al. [22] reported the desorption temperature at 114 °C of SBA-15 associated with Si–OH bond. Duan et al. [23] recognized that the metal atom in metal phosphide transferred into phosphorus atom resulting in the formation of acid site (252 and 620 °C). In the case of Ce-Ni_2P/SBA-15 catalysts, the start desorption temperature and the maximum desorption temperature are shifted to higher temperature with the increasing of Ce loadings. This suggested that the strength of acid sites of Ce-Ni_2P/SBA-15 catalysts was higher than that of Ni_2P/SBA-15 catalyst. Kalita et al. [24] reported that the strength of acid was related to desorption temperature. With the increasing of the desorption temperature, the strength of acid increased.

Table 2 NH_3-TPD results of Ce-Ni_2P/SBA-15 catalyst with different Ce loadings

Samples	Temperature (°C)		Acid amount (μmol/g)
	Weak acid	Strong acid	
Ni_2P/SBA-15	216	575	869
1 % Ce-Ni_2P/SBA-15	227	574	895
3 % Ce-Ni_2P/SBA-15	228	576	935
5 % Ce-Ni_2P/SBA-15	234	587	989
7 % Ce-Ni_2P/SBA-15	239	589	1,094

The total acid amount values are given in the Table 2. The value of Ni_2P/SBA-15 catalyst was 869 μmol/g, and the values of Ce-Ni_2P/SBA-15 catalysts increased with increasing Ce, which showed that Ce could improve the acid amount of Ni_2P/SBA15 catalyst. Yu et al. [25] reported that the Ce addition to the Pt–Sn/γ-Al_2O_3 catalyst can increase acid amount of Pt–Sn/γ-Al_2O_3 catalyst.

XPS of Ce-Ni_2P/SBA-15 catalysts

XPS spectra of Ce-Ni_2P/SBA-15 catalysts were shown in Fig. 4. As seen in the Fig. 4a are Ce3d5 spectra of Ce-Ni_2P/SBA-15 catalysts. The two binding energy (BE) peaks at 882.2 and 884.6 eV are assigned to Ce^{4+} and Ce^{3+} species [26]. Holgado et al. [27] reported that Ar^+ bombardment of CeO_2 produces a complete reduction to Ce_2O_3 by preferential removal of oxygen from the altered zone in the XPS measurement process. Sun et al. [28] demonstrated that the banding energy exhibited at ~882 and ~886 eV could be ascribed to CeO_2 and Ce_2O_3. In this literature, the XPS results of Ce3d5 spectra for Ce-Ni_2P/SBA-15 catalysts were consistent well with the publications.

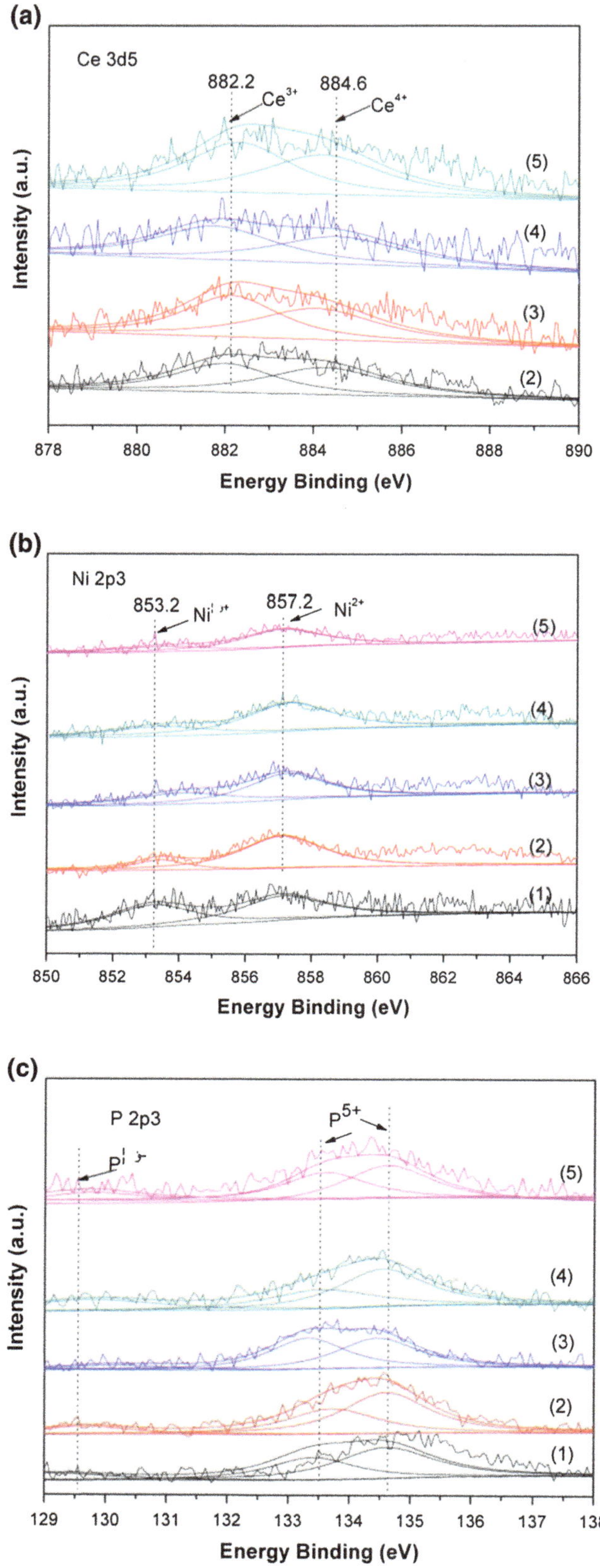

Fig. 4 XPS spectra of Ce-Ni_2P/SBA-15 catalysts with different Ce loadings **a** Ce3d5, **b** Ni2p3, **c** P2p3, *1* Ni_2P/SBA-15, *2* 1 % Ce-Ni_2P/SBA-15, *3* 3 % Ce-Ni_2P/SBA-15, *4* 5 % Ce-Ni_2P/SBA-15, *5* 7 % Ce-Ni_2P/SBA-15

Ni2p3 spectra of Ce-Ni_2P/SBA-15 catalysts are present in Fig. 4b. For Ni_2P/SBA-15 catalyst, the Ni2p3 BE peaks were at 853.5 and 857.2 eV, respectively. This could be assigned to $Ni^{\delta+}(0 < \delta < 2)$and Ni^{2+}. With regard to Ce-Ni_2P/SBA-15 catalysts, the Ni2p3 BE peaks were almost unaffected by the addition of Ce, suggesting that the addition of Ce did not affect the electronic properties of Ni_2P/SBA-15 catalyst. This was probably due to the Ce species introduced by the impregnation method which were mainly located on the surface of the catalysts. It was reported by Burns et al. [29] that the Ni 2p3/2 binding energy at 853.7 and 857.3 eV was assigned to $Ni^{\delta+}$ species and Ni^{2+} species. Ni^{2+} species was formed in Ce-Ni_2P/SBA-15 catalysts because the samples were passivated with 1 % O_2 in Ar before XPS measurement.

Figure 4c shows the P2p3 spectra of Ce-Ni_2P/SBA-15 catalysts. In the case of Ni_2P/SBA-15 catalyst, the P2p3 BE peaks located at 129.5 eV could be attributed to $P^{\delta-}$ species, and could be related to Ni_2P. The BE peaks located at 133.3 and 134.6 eV could be attributed to P^{5+} species, which could be related to phosphate ions and phosphorus pentoxide. For Ce-Ni_2P/SBA-15 catalysts, the value of the BE peaks was also located at the same position. It meant that the addition of Ce in Ni_2P/SBA-15 catalyst did not affect the valence of P. It had been reported that the formation of phosphorus pentoxide is mainly due to the superficial oxidation of nickel phosphide after the passivation by 1 % O_2/Ar (v/v) [29].

The surface atom concentration and the binding energy are shown in Table 3. As seen in the Table 3, the Ce surface atom concentration increased with the increasing of Ce loadings. $Ni^{\delta+}$ atom relative concentration evidently decreased with the increasing of Ce loadings. P atom relative concentration slightly shifted to the low value, indicating that the amount of active sites decreased when Ce was added to Ni_2P/SBA-15 catalyst. Oyama et al. had been reported that $Ni^{\delta+}$ in Ni_2P was the active site for HDS of 4,6-DMDBT [10]. So the decreasing of $Ni^{\delta+}$ species reduced the activity of DBT HDS over Ce-Ni_2P/SBA-15 catalysts.

Catalytic activity of Ce-Ni_2P/SBA-15 catalysts

The catalytic performance of Ce-Ni_2P/SBA-15 catalysts with different Ce loadings for HDS of DBT is presented in the Fig. 5a. The conversion of DBT was 40.8 % at 300 °C and 98.9 % at 380 °C over Ni_2P/SBA-15 catalyst. Compared with Ni_2P/SBA-15 catalyst, Ce-Ni_2P/SBA-15 catalysts had lower DBT conversion at 300–340 °C, and DBT conversion decreased with the increasing of Ce loadings. This was attributed to the relative concentration of $Ni^{\delta+}$ was decreased when Ce was added to Ni_2P/SBA-15 cata-

Table 3 XPS results of Ce-Ni_2P/SBA-15 catalyst with different Ce loadings

Catalysts	Ce 5d3		Ni 2p3			P 2p3			Si 2p		O 1s	
	BE (eV)	At (%)	BE (eV)	At (%)	$Ni^{\delta+}$/Ni	BE (eV)	At (%)	$P^{\delta-}$/P	BE (eV)	At (%)	BE (eV)	At (%)
Ni_2P/SBA-15	–	–	853.5; 857.2	0.66	0.30	129.5; 133.4; 134.6	2.90	0.14	103.3	27.72	533.2	68.56
1 % Ce-Ni_2P/SBA-15	882.2; 884.6	0.25	853.4; 857.2	0.58	0.25	129.4; 133.6; 134.6	4.69	0.15	103.3	27.24	533.0	66.77
3 % Ce-Ni_2P/SBA-15	882.2; 884.6	0.35	853.3; 857.1	0.56	0.20	129.6; 133.3; 134.4	4.03	0.14	103.0	25.23	532.5	69.39
5 % Ce-Ni_2P/SBA-15	882.1; 884.5	0.39	853.5; 857.2	0.60	0.19	129.4; 133.4; 134.3	3.67	0.15	103.3	28.43	533.0	66.42
7 % Ce-Ni_2P/SBA-15	882.1; 884.5	0.51	853.0; 857.2	0.62	0.17	129.5; 133.5; 134.3	4.34	0.10	103.3	25.43	532.8	68.60

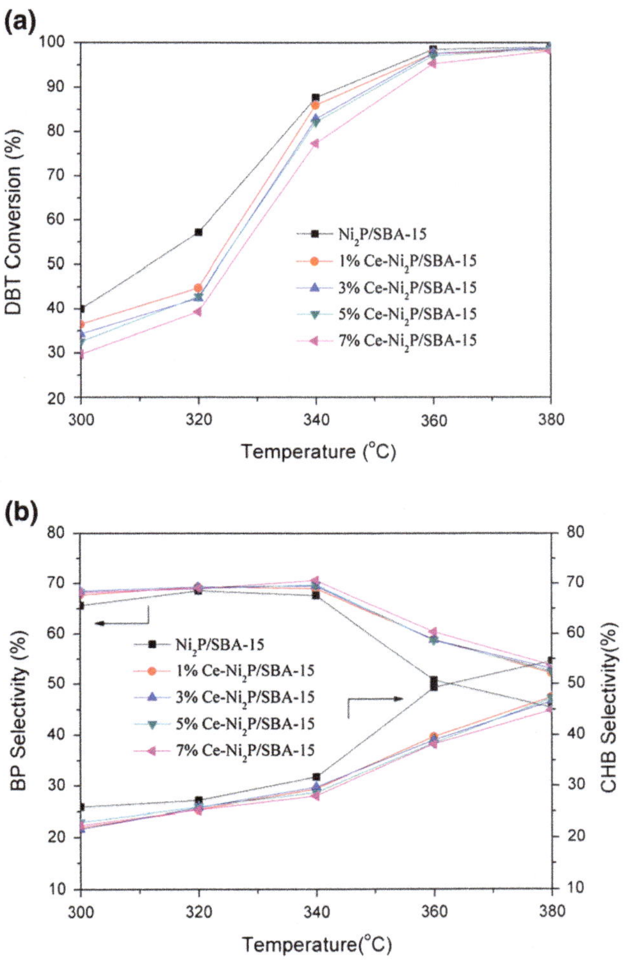

Fig. 5 HDS of DBT over Ce-Ni_2P/SBA-15 catalysts with different Ce loadings **a** activity **b** BP and CHB selectivity

lyst. In other words, the contact of the reactants and the Ni_2P active site may be restrained by adding Ce to Ni_2P/SBA-15 catalyst. 7 % Ce-Ni_2P/SBA-15 catalyst had the lowest DBT conversion at lower temperature, which can reach 29.9 % at 300 °C and 77.5 % at 340 °C.

It can be proposed that not only the low Ni_2P active sites but also the formation of $Ni_{12}P_5$ resulted in lowest activity over 7 % Ce-Ni_2P/SBA-15 catalyst. Korányi et al. [30] reported that $Ni_{12}P_5$ gave a lower active when compared with Ni_2P for HDS of DBT. At 360–380 °C, Ni_2P/SBA-15 and Ce-Ni_2P/SBA-15 catalysts exhibited similar HDS conversion. Chen et al. [31] reported that the apparent reaction rate constant increased with increasing the temperature. Higher DBT conversion was shown over Ni_2P/SBA-15 catalyst than that over Ce-Ni_2P/SBA-15 catalyst at 300–340 °C, mainly due to more active sites in Ni_2P/SBA-15 catalyst. DBT HDS reaction had high apparent reaction rate constant at 360–380 °C. So Ce-Ni_2P/SBA-15 catalyst with fewer active sites exhibited similar DBT conversion with Ni_2P/SBA-15 catalyst.

Furthermore, –OH on the surface of SBA-15 could connect Ce and form Si–O–Ce bond [32, 33], and this bond may block the mouth of the mesopore. This meant that less contents of DBT could come into the pore of Ce-Ni_2P/SBA-15 catalysts, and this could decrease the conversion of DBT. From the data of pore structure parameters and previous reference [20], we could also get this point. Then the contents of Ce reached 7 %. Some of Ce was incorporated into SBA-15 and led to the decrease of surface areas. Then the conversion of DBT could be reduced due to the decreased surface areas. This was also consistent with our previous research [34].

The Fig. 5b shows the selectivity of BP and CHB over Ce-Ni_2P/SBA-15 catalysts with different Ce loadings. BP and CHB selectivity over Ni_2P/SBA-15 catalyst was changed slightly at 300–340 °C, and BP selectivity decreased apparently at 360–380 °C. Compared with Ni_2P/SBA-15, BP and CHB selectivity over Ce-Ni_2P/SBA-15 was similar at 300–340 °C, respectively. However, BP selectivity was apparently lower than that of Ce-Ni_2P/SBA-15 at 360–380 °C. So the addition of Ce to Ni_2P/SBA-15 catalyst could regulate the selectivity of BP and CHB at 360–380 °C. Generally, BP, CHB and traces amount of H4-DBT (can be neglected) are the products of DBT HDS. The mechanism of DBT HDS could occur through two parallel reactions (seen in scheme 1), direct desulfurization (DDS) and desulfurization after hydrogenation (HYD). For DDS route, the main product is BP. For HYD route, H4-DBT is first developed and then

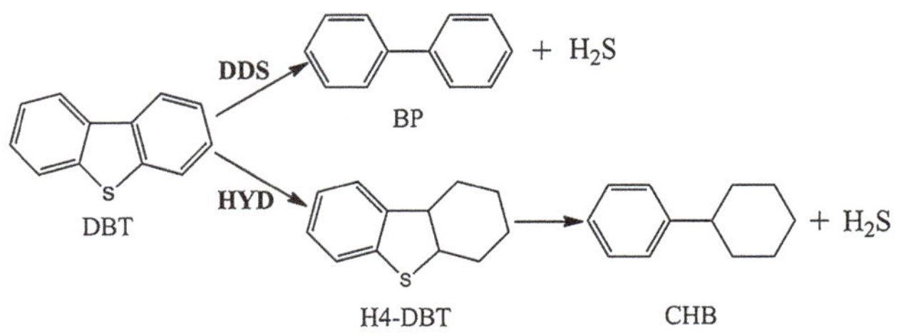

Scheme 1 The mechanism of dibenzothiophene hydrodesulfurization

quickly converted to CHB [35]. Abu et al. [12] reported that the 4,6-DMDBT selectivity can be enhanced by adding Co species to Ni_2P due to the increased acidity. In this paper, the addition of Ce to Ni_2P/SBA-15 catalyst improved the strength of acid site and the amount of acid. BP selectivity over Ce-Ni_2P/SBA-15 catalyst was higher than that of Ni_2P/SBA-15 catalyst especially at 360–380 °C, indicating that increased acid properties could enhance BP selectivity over Ce-Ni_2P/SBA-15 catalysts. On the contrary, CHB selectivity was decreased during this period. In other words, the enhanced BP selectivity followed DDS route over Ce-Ni_2P/SBA-15 catalyst when compared with Ni_2P/SBA-15 catalyst.

Conclusions

A series of Ce-Ni_2P/SBA-15 catalysts with different Ce loadings were prepared by impregnation method. The structure of as-synthesized catalysts and their HDS performance were investigated. Only Ni_2P phase existed in Ce-Ni_2P/SBA-15 when Ce $\leq$ 5 wt %, and Ni_2P and $Ni_{12}P_5$ phases existed in 7 % Ce-Ni_2P/SBA-15 catalyst. The surface area and pore volume were increased when Ce was added to Ni_2P/SBA-15 catalyst. The strength of acid site and the amount of acid over Ce-Ni_2P/SBA-15 catalyst was higher than that over Ni_2P/SBA-15 catalyst. The amount of active sites in Ni_2P/SBA-15 was higher than that of Ce-Ni_2P/SBA-15 catalyst. Compared with Ni_2P/SBA-15 catalyst, the Ce-Ni_2P/SBA-15 catalysts gave lower HDS of DBT performance at 300–340 °C. However, similar catalytic activity was showed over these catalysts for HDS of DBT at 360–380 °C. The addition of Ce to Ni_2P/SBA-15 catalyst apparently enhanced BP selectivity at 360–380 °C.

Acknowledgments The authors thank for the financial support of the National Natural Science Foundation of China (Grants 20473009) and the National Basic Research Program of China (973 Project No. 2006CB202503).

References

1. Song CS (2003) An overview of new approaches to deep desulfurization for ultra-clean gasoline, diesel fuel and jet fuel. Catal Today 86:211–263
2. Stanislaus A, Marafi A, Rana MS (2010) Recent advances in the science and technology of ultra low sulfur diesel (ULSD) production. Catal Today 153:1–68
3. Varakin AN, Nikul'shin PA, Pimerzin AA, Sal'nikov VA, Pimerzin AA (2013) Hydrogen spillover effect in the presence of CoS_x/Al_2O_3 and bulk MoS_2 in hydrodesulfurization, hydrodenitrogenation and hydrodeoxygenation. Russian J Appl Chem 86:718–726
4. Li X, Bai J, Wang A, Prins R, Wang Y (2011) Hydrodesulfurization of dibenzothiophene and its hydrogenated intermediates over bulk Ni_2P. Top Catal 54:290–298
5. Ding L, Shu Y, Wang A, Zheng M, Li L, Wang X, Zhang T (2010) Preparation and catalytic performances of ternary phosphides NiCoP for hydrazine decomposition. Appl Catal A Gen 385:232–237
6. Sal'nikov VA, Nikul'shin PA, Pimerzin (2013) The catalytic properties of transition metal sulfides synthesized from Anderson-type heteropoly compounds in hydrogenation, hydrodesulfurization, and hydrodenitrogenation reactions. Pet Chem 53:233–244
7. Biswas P, Narayanasarma P, Kotikalapudi CM, Dalai AK, Adjaye J (2011) Characterization and activity of ZrO_2 doped SBA-15 supported MiMo catalysts for HDS and HDN of bitumen derived heavy gas oil. Ind Eng Chem Res 50:7882–7895
8. Oyama ST (2003) Novel catalysts for advanced hydroprocessing: transition metal phosphides. J Catal 216:343–352
9. Wang AJ, Ruan LF, Teng Y, Li X, Lu MH, Ren J, Wang Y, Hu YK (2005) Hydrodesulfurization of dibenzothiophene over siliceous MCM-41-supported nickel phosphide catalysts. J Catal 229:314–321
10. Oyama ST, Lee YK (2008) The active site of nickel phosphide catalysts for the hydrodesulfurization of 4,6-DMDBT. J Catal 258:393–400
11. Huang X, Ji S, Wu P, Liu Q, Liu H, Zhu J, Li C (2008) Structure and hydrodesulfurization performances of Ni_2P/SBA-15 catalysts. Acta Phys Chim Sin 24:1773–1778
12. Abu II, Smith KJ (2006) The effect of cobalt addition to bulk MoP and Ni_2P catalysts for the hydrodesulfurization of 4,6-dimethyldibenzothiophene. J Catal 241:356–366
13. Guo Y, Zeng P, Ji S, Wei N, Liu H, Li C (2010) Effect of Mo promoter content on performance of Mo-Ni_2P/SBA-15/cordierite monolithic catalyst for hydrodesulfurization. Chin J Catal 31:329–334
14. Li X, Zhang YL, Wang AJ, Wang Y, Hu YK (2010) Influence of TiO_2 and CeO_2 on the hydrogenation activity of bulk Ni_2P. Catal Commun 11:1129–1132
15. Li X, Sun Z, Wang A, Wang Y (2010) Hydrodenitrogenation performance of TiO_2- and CeO_2-promoted Ni_2P catalysts. Prepr Pap Am Chem Soc Div Fuel Chem 55(1):205–206
16. Cai X, Dong X, Lin W (2008) Effect of CeO_2 on the catalytic performance of Ni/Al_2O_3 for autothermal reforming of methane. J Nat Gas Chem 17:9–102
17. Zhao D, Sun J, Li Q, Stucky GD (2000) Morphological control of highly ordered mesoporous silica SBA-15. Chem Mater 12:275–279
18. Oyama ST, Wang X, Lee YK, Bando K, Requejo FG (2002) Effect of phosphorus content in nickel phosphide catalysts studied by XAFS and other techniques. J Catal 210:207–217
19. Chen C, Liao H, Guo G, Shi Q, Chen W, Zhang R (2012) Influence of hydrodesulfurization of thiophene over Ni2P

catalysts supported on the Ce-activated bentonite. Nonmet Mines 35(4):1–3
20. Gutiérrez OY, Fuentes GA, Salcedo C, Klimova T (2006) SBA-15 supports modified by Ti and Zr grafting for NiMo hydrodesulfurization catalysts. Catal Today 116:485–497
21. Sundaramurthy V, Eswaramoorthi I, Dalai AK, Adjaye J (2008) Hydrotreating of gas oil on SBA-15 supported NiMo catalysts. Microporous Mesoporous Mater 111:560–568
22. Hu L, Ji S, Liu Q, Wu P, Zhu J, Li C (2007) Surface acidity and hydroxyl groups of SBA-15 mesoporous silica catalysts containing tungsten species. Chin J Catal 28:980–986
23. Duan X, Teng T, Wang A, Kogan VM, Li X, Wang Y (2009) Role of sulfur in hydrotreating catalysis over nickel phosphide. J Catal 261:232–240
24. Kalita P, Gupta NM, Kumar R (2007) Synergistic role of acid sites in the Ce-enhanced activity of mesoporous Ce-Al-MCM-41 catalysts in alkylation reactions: FTIR and TPD-ammonia studies. J Catal 245:338–347
25. Yu C, Ge Q, Xu H, Li W (2006) Effects of Ce addition on the Pt–Sn/γ-Al_2O_3 catalyst for propane dehydrogenation to propylene. Appl Catal A Gen 315:58–67
26. Padilla JM, Angel GD, Navarrete J (2008) Improved Pd/γ-Al_2O_3–Ce catalysts for benzene combustion. Catal Today 133–135:541–547
27. Holgado JP, Munuera G, Espinos JP, Gonzalez-Elipe AR (2000) XPS study of oxidation processes of CeO defective layers. Appl Surf Sci 158:164–171
28. Sun Z, Li X, Wang A, Wang Y, Chen Y (2012) The effect of CeO_2 on the hydrodenitrogenation performance of bulk Ni_2P. Top Catal 55:1010–1021
29. Burns AW, Gaudette AF, Bussell ME (2008) Hydrodesulfurization properties of cobalt–nickel phosphide catalysts: Ni-rich materials are highly active. J Catal 260:262–269
30. Korányi TI, Vít Z, Poduval DG, Ryoo R, Kim HS, Hensen EJM (2008) SBA-15-supported nickel phosphide hydrotreating catalysts. J Catal 253:119–131
31. Chen J, Yang H, Ring Z (2004) HDS kinetics study of dibenzothiophenic compounds in LCO. Catal Today 98:227–233
32. Subbaramaiah V, Srivastava VC, Mall ID (2013) Catalytic wet peroxidation of pyridine bearing wastewater by cerium supported SBA-15. J Hazard Mater 248–249:355–363
33. Levasseur B, Ebrahim AM, Burress J, Bandosz TJ (2011) Interactions of NO_2 at ambient temperature with cerium–zirconium mixed oxides supported on SBA-15. J Hazard Mater 197:294–303
34. Wei N, Ji SF, Wu PY, Guo YN, Liu H, Zhu JQ, Li CY (2009) Preparation of nickel phosphide/SBA-15/cordierite monolithic catalysts and catalytic activity for hydrodesulfurization of dibenzothiophene. Catal Today 1475:S66–S70
35. Chen T, Wang CM, Wang I, Tsai TC (2010) Promoter effect of vanadia on Co/Mo/Al_2O_3 catalyst for deep hydrodesulfurization via the hydrogenation reaction pathway. J Catal 272:28–36

Perspective on FCC catalyst in China

Rui Feng · Ke Qiao · You-he Wang · Zi-feng Yan

Abstract This paper provides an overview of the enormous challenge in processing heavier fluid catalytic cracking (FCC) feedstock and producing higher qualified liquid fuels. Besides optimizing the operation conditions of the FCC unit, it is crucial to design new catalysts especially for heavier and inferior feedstock. In this paper, a new concept, stepwise structure of catalyst, was postulated and a potential new catalyst based on stepwise structure design was prepared.

Keywords Fluid catalytic cracking · Heavy oil · Catalyst · Zeolite

Introduction

Fluid catalytic cracking today is one of the most important technologies for upgrading of heavy hydrocarbon fractions to light liquid fuels in most modern petroleum refineries. It is reported that the installed worldwide FCC capacity is over 2 million tons per day as of 2012, which represents 16.49 % of worldwide crude unit capacity [1, 2]. As an important secondary processing means, the typical feedstock for FCC process are gas oils derived from atmospheric and vacuum fractionators. However, because crude oil supply all over the world is becoming heavier and bastardized in quality and the demand for middle distillate is larger, the FCC process is now in rapid development.

The lighter crude oil supply is less than the demand from 2003, requiring new and enhanced production of extra heavy oils and eventually tar sands [3]. Nowadays, global residue processing is up to 0.87 billion tons every year, 23.3 % of which is processed by FCC. In China, FCC capacity was over 170 million tons in 2010 and more than 60 % of heavy oil was converted by the FCC process, providing about 75 % gasoline pool, 35 % diesel and 40 % propylene. Besides that, environmental legislation concerning aromatics and sulfur and nitrogen contents in transportation fuels are becoming more and more strict. The situation is even more severe in China due to the increasing dependence on imported crude oil and on the FCC process which produces gasoline with high olefin and high sulfur content. Therefore, it is crucial to change and adjust the FCC process conditions for improving the quality of transportation fuels and protecting the environment.

FCC reactions

FCC unit

A typical FCC unit is composed of a reactor and regenerator, distillation column and absorption stabilization system. The reactor and regenerator are considered to be the heart of the fluid catalytic cracking unit. The simplified schematic flow diagram of a typical modern FCC unit in Fig. 1 is based on the "high–low paratactic" configuration. The preheated high-boiling-point petroleum feedstock (at about 315–430 °C), consisting of long-chain hydrocarbon molecules and recycled slurry oil from the bottom of the

R. Feng · K. Qiao · Y. Wang · Z. Yan (✉)
State Key Laboratory of Heavy Oil Processing, Key Laboratory of Catalysis, China University of Petroleum, Qingdao 266580, China
e-mail: zfyancat@upc.edu.cn

R. Feng
e-mail: feng2007115@163.com

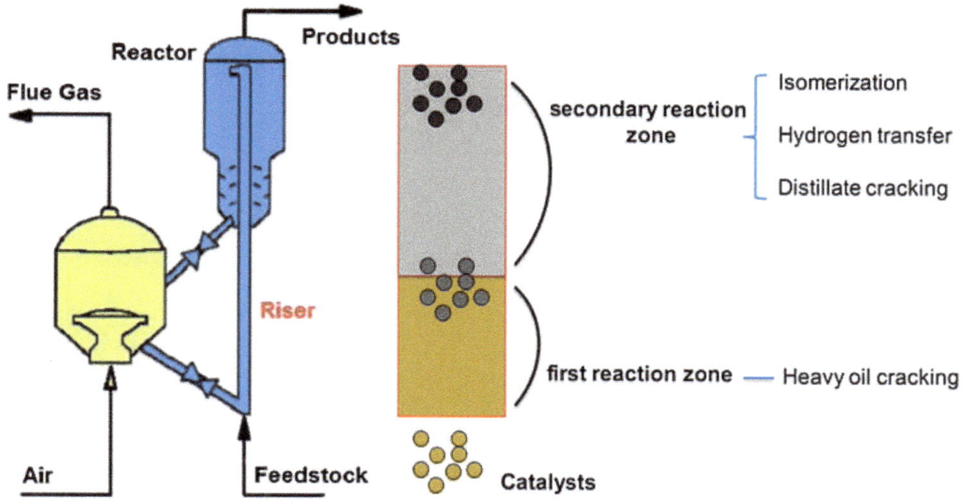

Fig. 1 A simplified diagram of fluid catalytic cracking unit as used in petroleum refineries

distillation column, is injected into the catalyst riser, where it is vaporized and cracked into smaller molecules of vapor by contact and mixing with the very hot powered catalysts from the regenerator. All of the cracking reactions take place in the riser within a period of 2–4 s or less at a temperature of about 500 °C or higher. The cracked product vapors and spent catalyst then flow upward through a set of cyclones within the reactor and they are further separated through a steam stripping section to remove any hydrocarbon vapors before spent catalysts return to the regenerator. The cracked product vapors flow from the top of the reactor to the bottom section of the distillation column, i.e., fractionators, where they are distilled into the FCC end products of cracked naphtha, fuel oil and off gas. Since the cracking reactions produce some carbonaceous material that is deposited on the catalyst and very quickly reduces the catalyst reactivity, the spent catalyst must be regenerated by burning off the deposited coke with air blown into the regenerator. The combustion of coke is exothermic and very hot (at about 700 °C). It produces a large amount of heat required for the vaporization and endothermic cracking reactions of petroleum feedstock when catalysts are recycled into the catalyst riser. The FCC unit is operated in the steady state. The operation shows no overall change with time. Therefore, the unit must be in energy balance, i.e., the energy released by coke burning in the regenerator must satisfy the heat needed for cracking reactions as well as other energy that exited.

Reaction sections

In commercial FCC operation, the main reactions of preheated feedstock in the catalyst riser are thermal cracking reactions on the surface of regenerated catalysts and catalytic cracking reactions. In these reactions, thermal cracking reactions take place in the bottom section of the riser where the temperature exceeds 600 °C via radical reaction mechanisms. Thermal cracking proceeds by the direct breaking of carbon–carbon bonds, resulting in excessive light gas yield and lower overall product values [4, 5]. Ethylene is the major product and is the chain-terminating product from the thermal cracking of paraffins. Propylene and higher olefins are also produced in decreasing yields with no branching. On the other hand, acid-catalyzed cracking reactions play a main role in the upper section of the riser at lower temperature via carbenium ion mechanism, leading to both olefins and paraffins with branching. Propylene is the chain-terminating product and a high yield of propylene over ethylene is a fingerprint for catalytic cracking compared with thermal cracking [6].

Reaction mechanisms

Catalytic cracking is catalyzed by Brønsted and Lewis acid sites of an acid solid such as SiO_2–Al_2O_3 or zeolites and involve C–C bond breakup via formation of carbocations (carbenium and carbonium ions) transition states. A study of the chemistry of catalytic cracking relies on the rich chemistry of carbocations. The basic reaction chemistries involved in catalytic cracking are discussed in very rich literature [7, 8]. A typical mechanistic scheme for catalytic cracking via carbenium ion intermediates is shown in Fig. 2 [9]; it involves initiation on either a Brønsted or Lewis acid site by protonation or H-abstraction, respectively. Propagation then occurs by hydride transfer and cracking by β-scission. Termination reactions involve reaction of a carbenium ion with either a Brønsted base to form an alkene and a Brønsted acid site, or with a Lewis base to from an alkane and reform a Lewis acid, or with a hydride ion from coke to form an alkane. Initiation and cracking can also occur by reaction of an alkane with a Brønsted acid site to form a carbonium ion, which in turn

Fig. 2 Reaction pathway network for alkane cracking

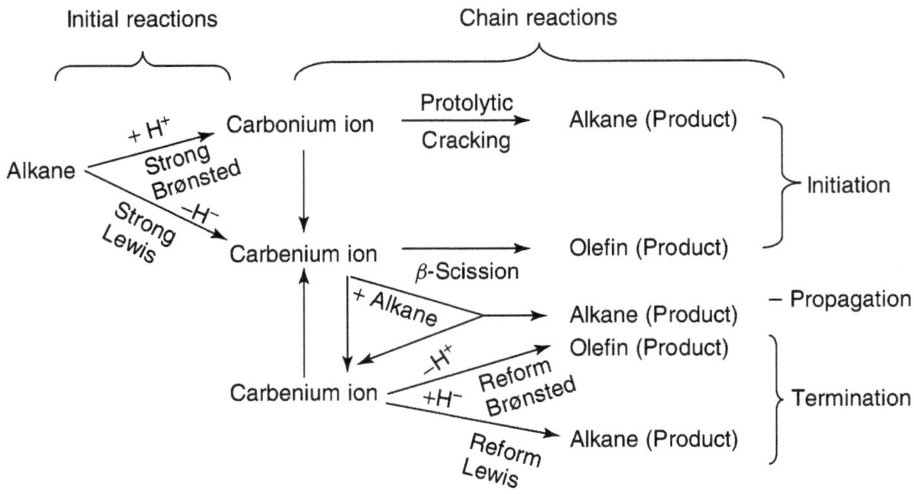

can react via β-scission to a smaller alkane and a carbenium ion or undergo H_2 elimination to a large carbenium ion.

In addition to the initial cracking reactions to form alkane and olefin, there are a number of secondary acid-catalyzed reactions that occur during hydrocarbon cracking, such as skeletal and/or double-bond position isomerizations, cyclization reactions, aromatization, alkylation, dealkylation, branching and polymerization. The secondary reactions produce a large amount of low carbon molecules such as aromatics and alkenes which are in favor of improving octane number of gasoline and propylene and butane yields. However, overcracking reactions of distillates also decrease light distillate yields and increase dry gas and coke formation, which are usually unexpected. Taking the thermal cracking into consideration, the FCC operation proceeds via parallel series reactions in the catalyst riser, converting high molecule hydrocarbons into light distillates, gases and coke.

Therefore, it is crucial to coordinate thermal cracking and secondary reactions occurring in the FCC process. One industrial successful example is the two-stage riser FCC technology (TSRFCC) proposed by the University of Petroleum (East China). The cracking reactions are reinforced by renewing the catalyst in the second stage of catalyst riser. In a similar conversion level, the TSRFCC has increased gasoline and light oil yields and lower gas yield compared with the one-stage riser (OSR).

FCC operating variables and FCC feedstock

The performance of an FCC unit is dependent on a large number of parameters. The feed composition, residence time, temperature, catalyst-to-oil ratio (CTO), hydrocarbon partial pressure, catalyst properties and riser hydrodynamics influence the conversion process in their own way.

The effects of operating parameters (reaction temperature, CTO, contact time) on thermal cracking and secondary cracking are obvious [10]. Endothermic reactions dominate the whole reactions of the FCC cracking process. Specifically, cracking and dealkylation with C–C bond breakup are endothermic and favored at high temperature, leading to high conversion of heavy hydrocarbons. In contrast, hydrogen transfer, isomerization reactions and polymerization reactions are exothermic reactions and favored at lower temperature. Generally, low reaction temperature, high CTO and short contact time of oil and catalyst may suppress the detrimental thermal cracking dramatically and benefit hydrogen transfer and isomerization reactions. In the overcracking operation caused by high temperature, thermal cracking may be enhanced, and hydrogen transfer and isomerization reactions may be decreased remarkably, while condensation reaction between olefin and aromatic hydrocarbons can be enhanced. In the overcracking operation caused by high catalyst-to-oil ratio, thermal cracking may be restrained, while hydrogen transfer and isomerization reactions in the secondary reactions can be promoted. In the overcracking

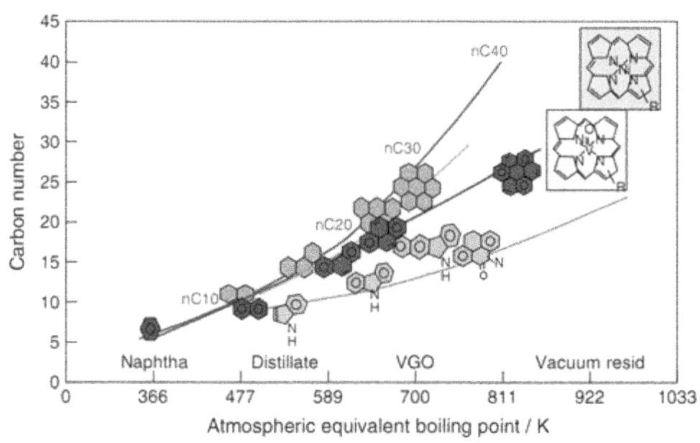

Fig. 3 Molecular distributions of petroleum

operation caused by long contact time, thermal cracking reaction increases and hydrogen transfer and isomerization reactions appear to have the same tendency as in thermal cracking reaction [6].

Generally, FCC feedstocks include paraffins, naphthenes, aromatics and polar compounds, as shown in Fig. 3 [11]. Typical feedstocks contain 50–66 % paraffins, 15–25 % naphthenes, 15–25 % aromatics and <5 % olefins. Paraffins are straight- or branched-chain hydrocarbons which are made up molecules with about 20–40 carbon atoms in the vacuum gas oil (VGO) range. Aromatics are compounds that contain at least one benzene ring, while naphthenes are saturated ring compounds. In general, FCC feedstocks contain some levels of common heteroatoms such as sulfur, nitrogen, nickel and vanadium. The composition of the feedstock influences the quantity and quality of the final products. For example, paraffinic feeds are easy to crack and produce low coke and high gasoline. However, aromatic compounds are more refractory to crack than paraffinic feedstock because they are: (i) lower in H content and (ii) associated with low diffusion coefficients. So these reactions result in lower conversion, lower gasoline and higher coke.

During cracking, the metal compounds from feedstock deposit quantitatively on the catalyst. Nickel and vanadium are particularly noticed because they are present in high concentrations and have the most detrimental effects on the cracking performance of the FCC catalyst. Nickel deposits on the catalyst surface in two primary structures of dispersed nickel oxide and nickel aluminate. The former promotes dehydrogenation reactions, consequently increasing yields of coke and hydrogen while decreasing gasoline production. Vanadium also deposits on the catalyst surface, but unlike nickel migrates through the catalyst and reacts with zeolites and rare-earth metals forming vanadate complexes, which easily destroys the zeolites. Vanadium contamination reduces conversions 3–4 times, reduces gasoline yield about 1.2 times and increases hydrogen and coke yields 0.3–0.4 times as much as nickel [12]. Also, sulfur and nitrogen in the FCC feedstock lead to SO_x and NO_x emissions in the regenerator and into distillates and must be controlled.

The traditional use of FCC converting heavy VGO, i.e., the overhead product from vacuum distillation of crude unit bottoms remains the primary duty. The conventional FCC catalysts are all effective in reducing slurry to some extent. The focus of FCC catalyst research and development is more on matrix to reduce coke and gas formation and/or how to make the active component (zeolite Y) more effective.

However, a growing number of refineries are feeding heavier and bastardized oils (residual oil or resids) along with VGO. In China, the vacuum resids added into FCC feedstock are up to 30 %; this ratio in some refineries is up to 60–70 %. It becomes more difficult to process more resids for producing gasoline and light fuel oil. The properties of resids account for the difficulties, and the reasons are as follows:

(a) Resids have increased molecular weight (MW) and higher boiling point fractions and need more energy to vaporize them to enter the catalyst riser.
(b) Resids with lower hydrogen content and increased asphalthenes, resins, polar compounds and Conradson carbon residue (CCR) are prone to form dry gas and coke.
(c) The increased sulfur and nitrogen contents in resids make it harder to meet environmental legislations.
(d) The increased contaminant metals (Ni, V, Na, Ca, Fe) in resids lead to more dry gas and coke formation, catalyst deactivation and destruction.

Significant variations in the qualities of products occur when resids are proportionally incorporated into the conventional VGO feedstock. The FCC catalysts are easier to deactivate along with higher delta coke formation and metal deposition, leading to reduced conversion of feedstock. Generally, four types of coke form in the FCC process: catalytic coke formed during the cracking reactions; contaminant coke from the activity of contaminant metals such as Ni and V; feed coke from the con-carbon in the feed; cat-to-oil coke from "unstripped" hydrocarbons that carry over into the regenerator. The severe deactivation from coke formation and metal contamination after resid incorporation is unavoidable, as a result dry gas yields including hydrogen and slurry oil yield increase while gasoline selectivity decreases.

In these cases, the spent catalysts cannot be recovered completely in the regenerator. The endothermic capacity needed for cracking reactions of FCC feedstock must be balanced with exothermic capacity when burning coke and energy lost in the flue gas and unit emission. Limited to regenerator temperature and air blasting capacity for coke burning as well as the heat removal from main column, the spent catalysts cannot be recovered to some extent. Besides optimizing FCC operating conditions, another alternative is to design and develop new FCC catalysts for FCC effective conversion.

FCC catalyst design

Composition of FCC catalysts

Modern FCC catalysts provide acid-catalyzed sites for hydrocarbon reactions. They are powered solid acids principally comprising zeolites, matrix, binder and sometimes additives.

Zeolites employed in manufacture of the FCC catalysts are synthetic versions of naturally occurring zeolites called faujasites. Virtually, all of today's catalysts contain Y zeolite or variation thereof. Zeolites possess particular characteristics of crystalline structure and acidity which are responsible for their capability to catalyze chemical reactions. The pore system of Y is characterized by a set of four tetrahedrally oriented 12-membered ring opening leading to a larger central cavity, i.e., supercage. The pore opening is 0.74 nm in diameter, while the supercage is 1.2 nm in diameter. Due to the large number of cationic sites in high surface area zeolite resulting from Si and Al atoms and possibly also due to electric field gradients set up by the cationic sites in the nanopores and cages of zeolites, a large number of acid sites with special concentration and strength are formed. Nevertheless, zeolite acid concentration and strength vary over a wide range, mainly as a function of Si/Al ratio. Generally, the acid site concentration is simply proportional to the inverse of the Si/Al ratio; in contrast, acid site strength varies inversely with Si/Al ratio. However, to modify the stability and acidity and satisfy the requirements of FCC conditions, zeolite Y is typically modified by ion exchange generating acidity, or hydrothermal treatment producing the so-called ultrastable Y (USY), or exchanging with multivalent cations such as rare-earth ions (La^{3+} and Ce^{3+}) for improving acid strength.

Generally, the matrix of the FCC catalysts is everything except active zeolites, including the binder, the clay and functional components. However, during the hydrothermal conditions of commercial FCC unit, about half of the starting zeolite is destroyed and becomes part of the matrix [13]. The overall matrix composition during use is formed by the solid-state reactions of all the non-zeolitic components. Besides the mechanical properties in fluidized state, the matrix added must to a certain extent improve bottom oil cracking, impart metals resistance or provide specialized functions such as lowering sulfur content in gasoline. Commercially, active components based on specialty aluminas are incorporated into most fluid cracking catalysts for increasing the matrix surface area and matrix acidity necessary to precrack the larger molecules into smaller ones that can then enter the pores of the zeolite. The active components also increase the pore volume of the catalyst and allow faster diffusion of resid molecules. The surface area, pore volume and thermal stability of the alumina are important for coke selectivity, bottoms cracking and metal tolerance, as well as to passivate the nickel and vanadium [14, 15]. In addition, clay may be incorporated as matrix or balance of the formulated catalysts and has certain similar functions such as precracking and mechanical strength.

New FCC catalyst design

According to thermal cracking and secondary reactions of heavy hydrocarbons occurring in the FCC process, it is believed that radical reactions for hydrocarbon precracking happen mainly on catalyst matrix at high temperature in the first reaction zone of the catalyst riser. Because chain paraffin reacts to mainly form ethylene, methane and α-olefins by radical reactions, heavy oil cracking on matrix easily forms dry gas and coke in this section. Dominant reactions in catalytic cracking take place in the secondary reaction zone of the catalyst riser. The cracking reactions in this section are based on the carbenium reaction mechanism and dependent on catalyst zeolites. The supercage of zeolite Y is critical to heavy oil cracking for offering acid sites and promoting carbenium ion formation. The unconverted slurry yield decreases remarkably with increasing zeolite Y content of the FCC catalysts.

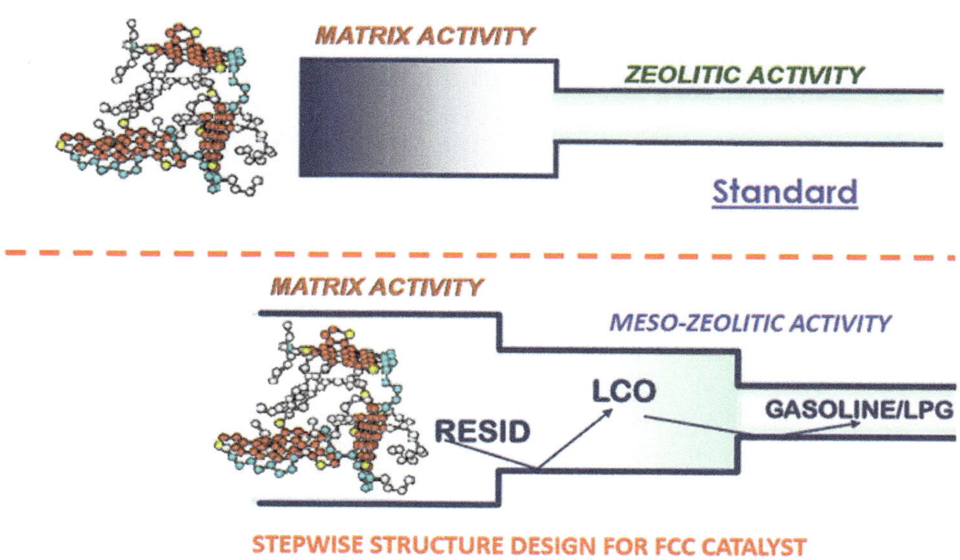

Fig. 4 Idea on new catalyst design for effective heavy oil cracking

The reaction mechanism discussed above and the philosophy of common catalyst processing FCC feedstock are generally accepted and displayed in Fig. 4. Heavy hydrocarbons first go through precracking reactions mainly on the matrix because of their accessible pores and external/internal surface active sites. These reactions are important for the conversion of FCC feedstock to heavy light oil (HCO), light cycle oil (LCO) as well as dry gases. Simultaneously, small-molecule radical ions or a small number of carbenium ions via radical ion mechanism and carbenium ion mechanism are formed which play a key role in secondary reactions. Thermal cracking is beneficial to some degree for further conversion of feedstock. Secondly, secondary cracking reactions mainly take place in the micropores of zeolites when smaller hydrocarbon is accessible to strong acid sites in the cages of zeolites. In this process, light liquid fuels with boiling point in the gasoline range and small molecule alkenes such as propylene are formed. The reactions of conventional FCC VGO feedstock are well explained by the standard scheme in Fig. 4.

However, cases are different when heavier resids are involved in the FCC feedstock. For VGO feedstock, carbon numbers are 15–30 and the dynamic diameter is smaller than 2.5 nm. The large molecule intermediates derived from precracking reactions of VGO may be accessible to active sites in zeolites (zeolite Y with 0.74 nm). However, vacuum resids have bigger carbon numbers of more than 40 and dynamic diameters of 2.5–15 nm. Most of them are too large to fit into the zeolite pores even through inadequate precracking.

Successful resid fluid cracking catalysts have become a challenge; the new catalysts require enhanced properties as follows:

(a) enhanced bottoms cracking for slurry conversion and low "feed coke";
(b) high tolerance to contaminants and low "metal coke";
(c) optimized hydrogen transfer to limit "catalytic coke" and desired balance between LPG and gasoline;
(d) open pore structure for low "stripping coke", enhanced cracking of large feed molecules and reduced overcracking of desired products.

Therefore, stepwise structure design for the FCC catalyst is introduced for heavier feedstock conversion as shown in Fig. 4. In this stage, meso-zeolites with both micro- and meso-pores are introduced into the FCC catalysts. Mesopores of zeolites are favorable for further secondary cracking reactions of large molecule intermediates. This design is beneficial for the conversion of FCC feedstock when resids are added. Simultaneously, liquid yields of gasoline and diesel increase due to limited secondary overcracking of gasoline to liquefied petroleum gas (LPG). Zeolites Y with both micropores and mesopores can be prepared in many ways, such as template method, acid or alkali post-treatment, hydrothermal treatment and so on. Modification of matrix is also used to lower matrix activity and widen pores.

The design philosophy of new catalysts is based on the idea of off-site cracking for heavy oil. A little big feed molecules precrack on matrix to produce small fraction hydrocarbon, which is feasible to enter into the supercage of zeolite Y and obtain H^+ released by Brønsted acid site to form carbeniums. Some carbeniums can go through β-scission to become smaller; however, the others diffuse out of the supercage. The carbeniums that diffuse outside encounter heavy molecules to form huge carbeniums, which leads to crack through β-scission and it is important for heavy oil cracking. Such a cracking is remarkably different form adsorption–desorption reaction that takes place on the active sites of catalysts, not resulting in coke formation, but realizing effective heavy oil cracking.

Therefore, continuous intracrystalline production of carbeniums and delivery of carbenium outside are crucial for effective cracking of heavy oil. Specifically, catalysts are designed in three ways: to advance matrix, to improve accessibility with high active center accessibility and to prepare stable zeolites. The new matrix should have lower

Table 1 Some industrial FCC catalysts for heavy oil in China

Model	R&D comp.	Active components	Applications
DVR-1 DVR-2	RIPP and catalyst complex of Qilu Petrochemical Corporation	Modified USY and REHY	Used for full VR in Daqing Petrochemical Corporation
LVR-60	RIPP	Modified USY and REHY	Used for high resids addition ratio feedstock
OREBIT-3600	RIPP and catalyst complex of Qilu Petrochemical Corporation	USY, REHY, modified Al_2O_3	Tolerance to metal performance (Ni, V)
CHV-1	RIPP and Luoyan Petrochemical Complex and Changling oil refining chemical factory	SRY (USY), REUSY	Tolerance to metal performance (Ni, V)
RAG-7	RIPP	ZSM-5	High LPG yield and high-ON gasoline

activity, reducing coke formation as much as possible. In addition, it should have highly porous structure, enhancing the thermal and hydrothermal stability and increasing the chance of contact of heavy oil molecule and carbeniums. Zeolite Y should be with high stability, high molecule accessibility (i.e., Albemarle Accessibility Index, AAI), activity maintenance and cracking selectivity.

In China, research and applications of FCC catalysts for heavy oil are popular. Table 1 gives some industrial FCC catalysts specially for processing feedstock with resid addition. All these catalysts are designed for some special requirements like increasing the conversion of heavy oil and light oil yields, decreasing coke yield, enhancing resistance to metal pollution performance and so on. However, companies differ on the development of catalysts. For instance, DVR catalysts can especially increase the cracking of Daqing full vacuum resid as well as decrease olefin content in FCC gasoline. CHV catalysts consist of zeolite with high SiO_2/Al_2O_3 ratio and have good performance in coke selectivity and isomerization reactions; at the same time these are tolerant to vanadium pollution. In addition, new FCC catalysts especially for processing of heavier FCC feedstock have been developed in China. Based on the design philosophy of the aforementioned new catalysts, the newly designed active component (i.e., zeolite) has higher thermal and hydrothermal stabilities compared with commercial zeolite. Additionally, new matrix shows weaker acidity and more open pore structures which is in favor of the diffusion of carbonium ions. The performances of this catalyst in a laboratory riser and in a pilot riser are displayed in Tables 2 and 3. Both show that new catalysts have better performance in improving conversion of feedstock and total yields, while reducing slurry and coke yields than industrial catalysts. Naphtha has high octane number but decreased olefins, beneficial for improving the quality of FCC gasoline.

Table 2 Performance of new catalyst in laboratory riser

Items	Domestic comp.	New catalyst	Foreign comp.
Dry gas, wt%	1.30	1.46	1.21
LPG, wt%	18.76	19.45	15.39
Naphtha, wt%	48.47	48.73	51.52
LCO, wt%	15.93	17.14	16.47
Slurry, wt%	7.63	4.77	6.09
Coke, wt%	7.56	8.31	8.81
Total yields, wt%	83.16	85.32	83.58
Olefins in naphtha, %	46.39	37.84	44.95
GC-RON	94.2	94.2	92.2

Table 3 Performance of new catalyst in pilot riser

Items	Blank	Final trial	Δ
Dry gas, wt%	3.64	3.14	−0.50
LPG, wt%	12.83	13.07	0.24
Naphtha, wt%	45.15	47.43	2.28
LCO, wt%	26.20	25.37	−0.83
Slurry, wt%	4.61	4.26	−0.35
Coke, wt%	7.57	6.91	−0.66
Total yields, wt%	84.18	85.69	1.51
GC-RON	88.6	90.3	1.7

Conclusions

FCC plays a critical role in China's commercial supplies of vehicular fuel and chemicals. To effectively process heavier FCC feedstock with conventional VGO and more inferior resids, a new catalyst with stepwise structure was postulated and prepared. This new catalyst is composed of matrix and zeolites with stepwise porosity and acidity that allow stepwise diffusion and cracking of heavy molecules while minimizing coke formation.

References

1. Swaty TE (2005) Global refining industry trends: the present and future. Hydrocarb Process 84:35–38
2. Bartholomew CH, Farrauto RJ (2010) Petroleum refining and processing. In: Fundamentals of industrial catalytic processes, 2nd edn. Wiley, Hoboken, pp 635–704
3. Raab D, Bayley SE (2012) A vegetation-based index of biotic integrity to assess marsh reclamation success in the Alberta oil sands, Canada. Ecol Indic 15:43–51
4. Chen L, Shi L (2005) Study on complex desulphurization additives for FCC. J Fuel Chem Tech 33:83–88
5. Shen Z, Shi W, Li H (2005) Promoting isomerization activity of USY cracking catalyst by chromium modification. J Fuel Chem Tech 33:363–366
6. Wang G, Gao J, Xu C, Feng Y (2005) Studies on thermal cracking and secondary reactions in FCC process. J Fuel Chem Tech 33:440–444
7. Tichit D, Fajula F, Figueras F et al (1985) Thermal stability and acidity of Al^{3+} crosslinked smectites. In: Imelik B, Naccache C, Coudurier G et al (eds) Catalysts by acids and bases. Elsevier, Amsterdam, pp 351–360
8. Occelli ML, O'Connor P (1994) Fluid catalytic cracking III—materials and process. ACS Symposium Series, American Chemical Society, Washington, DC, p 386
9. Wojciechowski BW, Corma A (1986) Catalytic cracking: catalysts, chemistry and kinetics. Chemical Industries, Marcel Dekker, New York, p 236
10. Karthika V, Brijet Z, Bharathi DN (2012) Design of optimal controller for fluid catalytic cracking unit. Procedia Eng 38:1150–1160

11. Altgelt KH, Boduszynski MM (1994) Composition and analysis of heavy petroleum fractions. Marcel Dekker, New York, pp 45–47
12. Park SK, Jeon HJ, Jung KS, Woo SI (2003) Regeneration of spent resid fluidized catalytic cracking catalyst by removing metal poisons such as V, Ni, and Fe. Ind Eng Chem Res 42:736–742
13. Cheng WC, Habib ET, Rajagopalan K, Roberie TG, Wormsbecher RF, Ziebarth MS (2008) Fluid catalytic cracking. In: Ertl G, Knözinger H, Schüth F, Weitkamp J (eds) Handbook of heterogeneous catalysis, 2nd edn, vol 6. Wiley-VCH Verlag GmbH, KGaA, Germany, pp 2741–2778
14. Pinto FV, Escobar AS, de Oliveira BG et al (2010) The effect of alumina on FCC catalyst in the presence of nickel and vanadium. Appl Catal A Gen 388:15–21
15. Yue MB, Xue T, Jiao WQ, Wang YM (2013) Performance of mesoporous silica-doped aluminas on nickel passivation. Mater Lett 91:115–117

24

Neural network modeling the effect of oxygenate additives on the performance of Pt–Sn/γ-Al_2O_3 catalyst in propane dehydrogenation

Younes Amini · Moslem Fattahi · Farhad Khorasheh · Saeed Sahebdelfar

Abstract The effect of oxygenate additives, water and methanol, to the feed on the performance of industrial Pt–Sn/γ-Al_2O_3 catalyst in dehydrogenation of propane was studied by neural network modeling. Because of the complex nature of the system and very low levels of oxygenate addition, neural networks were employed as an efficient and accurate tool to obtain the behavior of the system. Dehydrogenation reaction was carried out in a fixed-bed quartz reactor in the temperature range of 575–620 °C. Steady state modeling was performed in three different levels of oxygenate addition, and conversion and selectivity at different levels. The optimum amounts of water and methanol for reaction temperatures of 575, 600 and 620 °C were found to be 83.60, 125.40 and 139.34 ppm, respectively, for water and 9.98, 24.94 and 49.88 ppm for methanol by neural network method. The neural network-based optimum was compared with that obtained from experimental data. In this case, various architectures have been checked using 70 % of experimental data for training of artificial neural network (ANN). Among the various architectures multi layer perceptron network with trainlm training algorithm was found as the best architecture. Temperature and water or methanol amount for the present constituents in the feed were network input data. Output data were conversion, selectivity to propylene and yield of propylene. Comparing the obtained ANN model results with 30 % of unseen data confirms ANN excellent estimation performance. The influence of different operating conditions on the accuracy of the results was also investigated and discussed. The propylene yields, however, passed a maximum at the optimum levels of oxygenates coincided with a substantial reduction of coke formation as well. The modeling results were accurate with <0.9 % error.

Keywords Propane dehydrogenation · Pt–Sn/γ-Al_2O_3 · Oxygenates · Catalyst modifiers · Coke formation · Artificial neural network

Y. Amini
Department of Chemical Engineering, Isfahan University of Technology, Isfahan, Iran

M. Fattahi (✉) · F. Khorasheh
Department of Chemical and Petroleum Engineering, Sharif University of Technology, Azadi Avenue, P.O. Box 11365-9465, Tehran, Iran
e-mail: moslem.fattahi@gmail.com; moslemfattahi@che.sharif.edu

S. Sahebdelfar
Catalyst Research Group, Petrochemical Research and Technology Company (NPC-RT), Tehran, Iran

Introduction

Dehydrogenation of light alkanes to the corresponding alkenes is growing because of growing demand for lower alkenes for the production of polymers, polygas chemicals and oligomers as gasoline blending stocks additives [1, 2]. Propane dehydrogenation (PDH) has been considered as an alternative route for production of propylene. The reaction is a highly endothermic and equilibrium limited requiring relatively high temperatures and low pressures to achieve high propylene yields [3]:

$$C_3H_8 \Leftrightarrow C_3H_6 + H_2 \quad \Delta H^0_{298} = 124\,\text{kJ/mol} \qquad (1)$$

The reaction is generally operated at 525–625 °C near atmospheric pressures over supported platinum or chromia catalysts. Despite the simple chemistry, the reacting system is very complex due to occurrence of several side reactions and rapid catalyst deactivation. The commercial technologies differ in the catalyst formulation, reactor design, heat supply, and method of catalyst regeneration.

Pt–Sn/γ-Al_2O_3 catalyst exhibits a high activity and selectivity to propylene in PDH [4, 5]. Promoting effects of Sn on Pt/γ-Al_2O_3 have been described in terms of its role as a poison for acid sites on the alumina support as well as enhanced mobility of adsorbed hydrogen and increased dispersion of Pt on the support [6–9]. Side reactions including hydrogenolysis and cracking result in the formation of lower hydrocarbons which impact catalyst performance [10].

Deactivation of Pt–Sn/γ-Al_2O_3 by coke depositions in the dehydrogenation is an important operation problem [11, 12]. The carbonaceous deposits on catalyst surface can be removed by treatment with oxygen or steam. Some technologies use hydrogen or steam as diluents to reduce coke formation and elongate catalyst lifetime.

Steam can serve as catalyst modifier when added in low levels to the feed stream [13–15]. Recently, Barghi et al. [13] performed the kinetic modeling of PDH in presence of oxygenated additive over industrial Pt–Sn/γ-Al_2O_3 in which the effect of water addition in low level on coke formation was mathematically investigated. Pt-based catalysts are poisoned by high level of oxygenate by various mechanism. Oxygenates deplete chlorine, essential for re-dispersion of sintered platinum during catalyst regeneration in continuous catalyst regeneration (CCR) plant, from the catalyst and increase the selectivity to carbon oxides. When used in appropriate levels, however, they could exhibit beneficial effects on catalyst performance without perceptible adverse effects [16]. Nevertheless, very few investigations have reported the effect of oxygenates and its interaction mechanism in catalytic dehydrogenation of light alkanes [16, 17].

The neural network, which arose from attempts to model the functioning of the human brain, appears to be useful in the study of dehydrogenation reaction. It seems to be a promising tool to solve modeling problems in cases where the governing mechanisms cannot be formulated due to insufficient knowledge. Models based on artificial neural network (ANN) exhibit the rule-following behavior without containing any explicit representation of those rules that is the main feature of the neural network approach, which results in its wide range of applications. In recent years, the concept of neural networks (NNs) has gained wide popularity in many areas of chemical engineering such as dynamic modeling of chemical processes [18, 19], catalyst design [20], estimation of catalyst deactivation, reaction modeling [21, 22], modeling of chemical reactors [23] and modeling of complex chemical processes [24].

Applying ANN for modeling can render pre-mentioned problems and facilitate the forecasting of catalyst activity. Based on our literature survey, there was no attempt on ANN modeling of catalyst performance for dehydrogenation reactions. In this paper, we have demonstrated the use of ANN in prediction of the performance of commercial Pt–Sn/γ-Al_2O_3 catalyst in the presence of oxygenate additives, namely water and methanol, in PDH. Furthermore, the results of the modeling have been compared with the experimental data.

Materials and methods

A laboratory scale tubular flow fixed-bed quartz reactor was used for the PDH experiments. The inside diameter of the reactor was 13 mm, the length of the reactor was 90 cm, and the commercial Pt–Sn/γ-Al_2O_3 catalyst was loaded in the middle section of the reactor in between two layers of quartz particles. The catalyst loading was 1 g. All experiments were conducted at weight hourly space velocity (WHSV) of 2/h and hydrogen to propane ratio, H_2/HC, of 0.8 that are typical of industrial conditions. PDH was performed at 575, 600 and 620 °C near atmospheric pressure (0.8 bar) with different amounts of water in the feed. The product samples were analyzed 3 h after the start of the run when stable conditions were achieved. Product samples were then analyzed hourly up to 7 h after the start of the run. Details of the experimental procedures and analyses are reported elsewhere [25].

Water and methanol were introduced to the reactor feed stream using a syringe pump. The exit gasses were analyzed for light hydrocarbons and CO and CO_2 using an online Agilent 6890 gas chromatograph equipped with TCD and FID detectors. Propane conversions, product selectivities and propylene yield were defined as follows [26]:

$$\text{Propane Conversion}\,(\%) = \frac{\sum_i \left(\frac{n_i}{3}\right)[F_i]_{out} - [F_{C_3H_8}]_{out}}{\sum_i \left(\frac{n_i}{3}\right)[F_i]_{out}} \times 100 \quad (2)$$

$$\text{Selectivity for component } i\,(\%) = \frac{\left(\frac{n_i}{3}\right)[F_i]_{out}}{\sum_i \left(\frac{n_i}{3}\right)[F_i]_{out} - [F_{C_3H_8}]_{out}} \times 100 \quad (3)$$

$$\text{Propylene Yield}\,(\%) = \frac{\text{Propane Conversion} \times \text{Selectivity of Propylene}}{100} \quad (4)$$

where i includes all the components with carbon atoms in the exit gas stream, and n_i and F_i are the number of carbon atoms and molar flow of component i, respectively.

Spent catalysts were characterized by thermal gravimetric analysis (TGA). TGA tests were performed using a PL-STA 1500 model by PL Thermal Science (England) where weight measurements were registered at a rate of 50 measurements/min using a temperature ramp of 5 °C/min from the base temperature of 25 °C up to 950 °C.

Artificial neural networks modeling

NNs are inductive models inspired by the structure and the functions of biological neurons. A network is composed of units or nodes that represent the neuron bodies. The units are in the shape of conjunct loop structure which, in fact, functions like axons and dendrites [27].

Information passes among these units along interconnections. Any incoming connection has two associated values, an input value and a weight. The output of the unit is a function of the summed value. ANNs are not programmed to perform specific tasks; rather, they are trained with respect to data sets until they learn patterns used as inputs. Once they are trained, new patterns may be presented to them for prediction or classification. ANNs can automatically learn to recognize patterns in data from real systems or from physical models. An ANN can handle many inputs and produce answers that are in a form suitable for the designers.

One of the well-known topologies of NN is the multilayer perception (MLP) which is used for classification and estimation problems. One example of the layered networks is depicted in Fig. 1. In the figure, ANN input, hidden and output layers are shown. ANN training is an optimization process in which an error function is minimized by adjusting the ANN weights. When an input training pattern is introduced to ANN, output is calculated. Output is compared with the real output provided by the user [28, 29].

Output signals from each of the nodes are triggered by the signals emanating from the input nodes which can be modeled as a sigmoid relation:

$$f(\theta) = \frac{1}{1 + e^{-\theta}} \tag{5}$$

where $f(\theta)$ is the output signal from each node and θ is the activity of each node. The activity of each node is obtained as the sum of the signals received from the preceding nodes and defined as:

$$\theta = \sum \omega_i x_i \tag{6}$$

where ω_i are the weights connecting input node i to hidden node j, x_i is the input value (normalized to unity) to input node i. Signals from the hidden nodes then propagate to the output layer and generate output signal similarly as the input did.

Training of the ANN is an improvement process by which error functions can be minimized according to the network weights. When an input training pattern is introduced to ANN, output is calculated. Output is compared with the real output provided by the user. These differences can be used by improvement technique to teach neural network. The error function used here was mean square error (MSE) which is given by the following formula:

$$\mathrm{MSE} = \frac{1}{n}\sum_{i=1}^{n} \left(y_i - y_{i,\mathrm{real}}\right)^2 \tag{7}$$

Fig. 1 Structure of a neural network

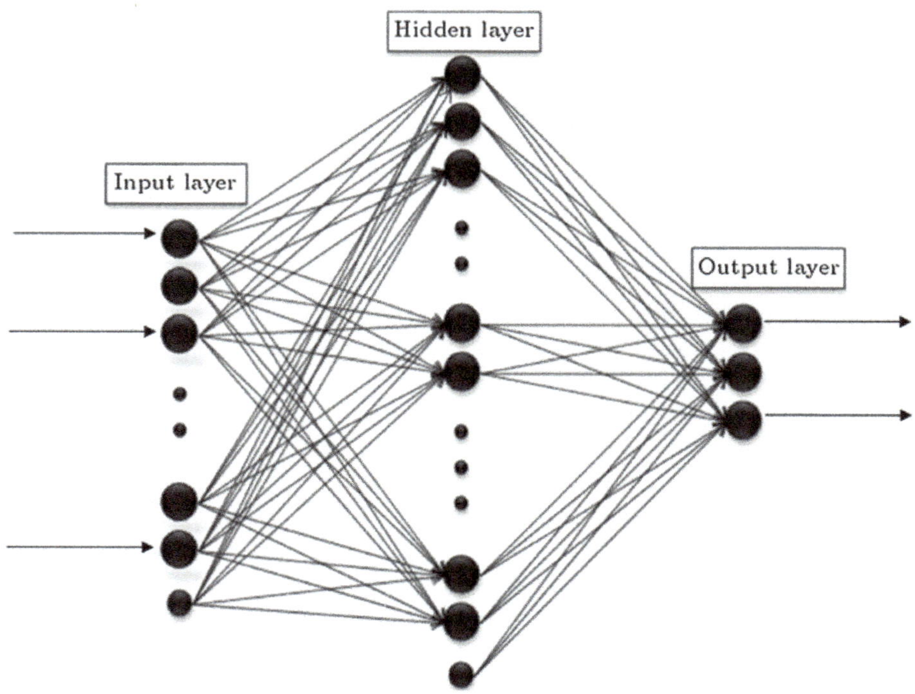

Table 1 Neural network variables and domain

Variables	Domain
Oxygenate (water or methanol) (ml/h)	0–0.6
Temperature (°C)	575–620
Conversion (mol.%)	10–50
Selectivity (mol.%)	20–100
Yield (mol.%)	10–50

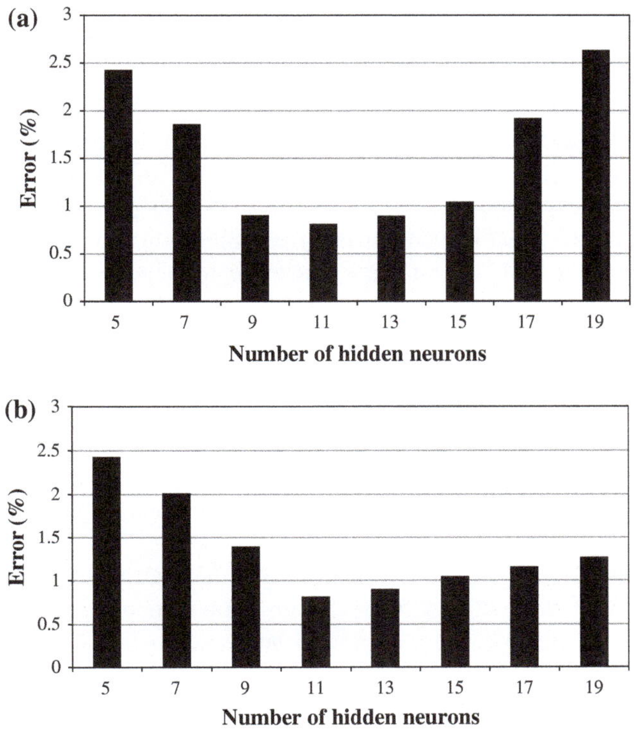

Fig. 2 Error versus number of hidden layer for **a** water and **b** methanol through the best obtained MLP network

where $y_{i,\mathrm{real}}$ is the *i*th real target and y_i is the network output corresponding to input. Thus, training process is a path from input layer to output layer to calculate an output, obtaining the error and a backward path to update the weights [30].

In this work, the neural network was trained to learn the 'function' using training set of the input and output data, in which both input and output data were experimental. Then, conversion and selectivity to propylene over the catalyst was calculated by introducing the input data of the left-out catalyst into thus trained network. This procedure was repeated for all the experimental data and conditions, and the calculated values were compared with the experimental values.

The simulation of processes with NN requires a large number of experimental data. The data were shuffled and split in two subsets: training set and test set. The splitting of samples plays an important role in evaluation of an ANN performance. The training set is used to estimate model parameter, and the test set is used to check the generalization ability of the model. The training set must be representative of the whole population of input samples. In this investigation data were obtained under various experimental conditions including different reaction temperatures and different amounts of water or methanol as additives. The use of more than 60 % of data for training and the remaining for testing is the most common approach for ANN [31]. The network was taught by 70 % of data and the remaining data were used to test the generalization capacity of the network. ANN variables and their domains are illustrated in Table 1.

Table 2 Experimental and ANN modeling conversion results for catalytic PDH in presence of water

T (°C)	Water amount (ml/h)	Conversion (mol.%)			
		Experimental		ANN Modeling	
		Start[a]	End[b]	Start	End
575	0.20	28.56	26.48	27.59	26.39
	0.25	29.74	27.88	28.65	27.54
	0.30	30.89	28.17	29.04	27.88
	0.35	24.96	22.87	–	–
	Without	26.79	24.18	26.53	24.86
600	0.40	38.76	36.54	36.73	35.41
	0.45	39.95	37.13	38.34	36.08
	Without	36.12	33.43	33.88	31.88
620	0.30	46.18	43.22	46.15	42.75
	0.40	47.57	45.02	47.91	44.61
	0.50	48.67	46.29	48.49	45.60
	0.60	45.17	39.85	44.14	40.63
	Without	45.34	42.98	44.51	41.84

[a] Start, reaction time after 3 h

[b] End, after 7 h

Using MLP network according to Fig. 2, in which MSE is plotted in conformity with the number of neurons in hidden layers for water and methanol, the optimal number for hidden layers is seven corresponding to the least square error.

Results and discussion

The developed ANN model was employed for studying variables affecting the catalyst performance. Experimental and neural network modeling results of propane conversions and selectivity and yield to propylene in the presence of small amounts of water are summarized in Tables 2, 3 and 4 and those for methanol in Tables 5, 6 and 7. Figure 3 presents the ANN modeling and experimental data on propane conversions, product selectivity and yield of

Table 3 Experimental and ANN modeling selectivity to propylene results for catalytic PDH in presence of water

T (°C)	Water amount (ml/h)	Selectivity to C_3H_6 (mol.%)			
		Experimental		ANN modeling	
		Start[a]	End[b]	Start	End
575	0.20	75.64	78.30	78.47	78.29
	0.25	77.87	79.18	79.91	79.84
	0.30	77.92	81.65	79.57	81.06
	0.35	67.53	70.25	–	–
	Without	79.26	82.64	77.76	80.38
600	0.40	84.18	87.15	87.71	89.13
	0.45	85.74	90.03	87.82	90.61
	Without	90.59	94.84	94.78	97.81
620	0.30	77.26	79.76	77.13	80.37
	0.40	75.44	79.65	74.47	79.64
	0.50	77.93	82.76	76.97	83.44
	0.60	75.73	78.32	78.98	78.22
	Without	82.78	87.98	84.03	88.93

[a] Start, reaction time after 3 h

[b] End, after 7 h

Table 4 Experimental and ANN modeling yield of propylene results for catalytic PDH in presence of water

T (°C)	Water amount (ml/h)	Yield of C_3H_6 (mol.%)			
		Experimental		ANN modeling	
		Start[a]	End[b]	Start	End
575	0.20	21.60	20.73	21.65	20.66
	0.25	24.05	22.30	22.89	21.99
	0.30	23.17	22.76	23.15	22.61
	0.35	16.85	16.07	–	–
	Without	21.23	19.98	20.64	19.97
600	0.40	32.63	31.84	32.23	31.56
	0.45	34.25	33.83	33.67	32.73
	Without	32.72	31.70	32.12	31.17
620	0.30	35.68	34.47	35.59	34.36
	0.40	35.89	35.86	35.69	35.52
	0.50	37.93	38.31	37.38	38.05
	0.60	34.21	31.21	34.83	31.79
	Without	37.53	37.81	37.41	37.22

[a] Start, reaction time after 3 h

[b] End, after 7 h

propylene versus time-on-stream at 575 °C with the optimum level 0.3 ml/h water added to the feed [25]. The agreement between the model predictions and the experimental data is good. In all runs the catalyst activity declined with time-on-stream as coke was accumulated on the catalyst surface while the selectivity to propylene increased and it can be seen that the ANN model correctly predicts the trend. This model, approximately, can be generalized to the entire range of data for water since the differences between predicted and experimental value are small indicating the capability of ANN to predict unobserved data correctly.

The indicated graphs demonstrate the good performance of the created network. The ANN simulation is a useful tool for many situations including when information for the catalytic system, such as catalyst structure and material, are unavailable. The ANN simulation is solely based on the experimental results and as such here is no need for detailed information of catalyst structure, oxygenated additive and mechanism of reaction. This is an important advantage for simulation of catalytic processes by ANNs.

Similar observations can be made with respect to the effects of water on PDH at other reaction temperatures. The optimum amount of water was found to increase with increasing temperature. As indicated by Tables 2, 3 and 4, the optimum amounts of water at reaction temperatures of 575, 600, and 620 °C were 0.3, 0.45, and 0.5 ml/h, respectively, corresponding to a feed water composition of 83.6, 125.4, and 139.34 ppm. In all cases any further increase in the feed water content beyond the optimum value would result in a substantial loss in catalyst that could also be predicted by ANN.

Figure 4 presents the propane conversions and product selectivities for PDH at 620 °C with various amounts of methanol added to the feed. The trends were qualitatively similar to those obtained when water was added to the feed. The optimum amounts of methanol at reaction temperatures of 575, 600, and 620 °C were 0.02, 0.05, and 0.1 ml/h, respectively, corresponding to a feed methanol composition

Table 5 Experimental and ANN modeling conversion results for catalytic PDH in presence of methanol

T (°C)	Methanol amount (ml/h)	Conversion (mol.%)			
		Experimental		ANN modeling	
		Start[a]	End[b]	Start	End
575	0.02	29.13	26.97	28.95	26.29
	0.05	27.13	25.14	28.64	25.51
	Without	26.79	24.18	28.21	25.62
600	0.05	38.07	35.79	38.57	36.37
	0.10	34.18	34.72	37.02	34.47
	Without	36.12	33.43	36.95	33.59
620	0.10	47.96	45.78	46.40	45.37
	0.20	41.72	36.84	40.99	36.84
	Without	45.34	42.98	45.76	43.63

[a] Start, reaction time after 3 h

[b] End, after 7 h

Table 6 Experimental and ANN modeling selectivity to propylene results for catalytic PDH in presence of methanol

T (°C)	Methanol amount (ml/h)	Selectivity to C_3H_6 (mol.%)			
		Experimental		ANN modeling	
		Start[a]	End[b]	Start	End
575	0.02	79.22	81.94	85.74	84.74
	0.05	75.87	77.41	82.98	81.14
	Without	79.26	82.64	87.37	86.70
600	0.05	86.79	90.16	91.43	91.82
	0.10	82.09	88.07	87.79	88.39
	Without	90.59	94.84	93.52	93.55
620	0.10	78.65	82.63	83.30	86.38
	0.20	64.95	69.46	68.68	74.50
	Without	82.78	87.98	87.51	88.51

[a] Start, reaction time after 3 h

[b] End, after 7 h

Table 7 Experimental and ANN modeling yield of propylene results for catalytic PDH in presence of methanol

T (°C)	Methanol amount (ml/h)	Yield of C_3H_6 (mol.%)			
		Experimental		ANN modeling	
		Start[a]	End[b]	Start	End
575	0.02	23.08	22.10	24.82	22.28
	0.05	20.58	19.46	23.77	20.69
	Without	21.23	19.98	24.65	22.21
600	0.05	33.04	32.27	35.27	33.39
	0.10	28.06	30.58	32.50	30.47
	Without	32.72	31.70	34.55	31.41
620	0.10	37.72	37.83	38.65	39.18
	0.20	27.10	25.59	28.15	27.44
	Without	37.53	37.81	40.04	38.62

[a] Start, reaction time after 3 h

[b] End, after 7 h

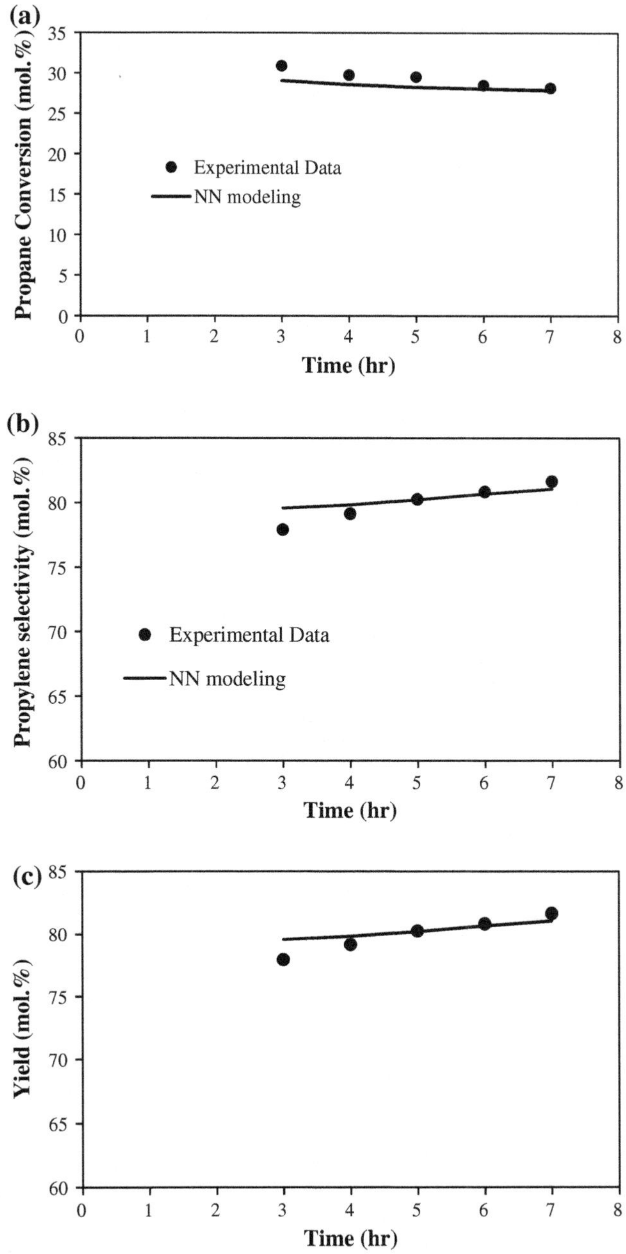

Fig. 3 ANN modeling and experimental comparison effects of water for **a** propane conversion, **b** propylene selectivity and **c** propylene yield on PDH over industrial Pt–Sn/γ-Al_2O_3 catalyst in optimum amount of water addition at 575 °C

of 9.98, 24.94, and 49.88 ppm. The agreements between the model predictions and the experimental data were slightly better for water compared with methanol as the additive. In all runs the catalyst conversion declined with time-on-stream that was correctly predicted by ANN modeling. The selectivity to propylene increased with increasing time-on-stream. This trend was also predicted by the ANN model although agreements between model predictions and experimental results are not satisfactory for the initial period. In case of the propylene yield, the predictions for methanol were not satisfactory in comparison with water. As indicated by the ANN results, the optimum amounts of methanol were lower compared with those for optimum water addition. The enhanced propane conversions and propylene yields were slightly higher when optimum amounts of water were used compared with those for optimum methanol addition. This implies that for equivalent amount of oxygen, methanol is a more efficient modifier than water.

At high temperatures carbonaceous deposits, collectively termed coke, are rapidly formed and as a consequence, catalyst deactivation and regeneration are important considerations for commercial processes. With increasing temperature, the activity of catalyst and the reaction rate increased which also resulted in an increase in

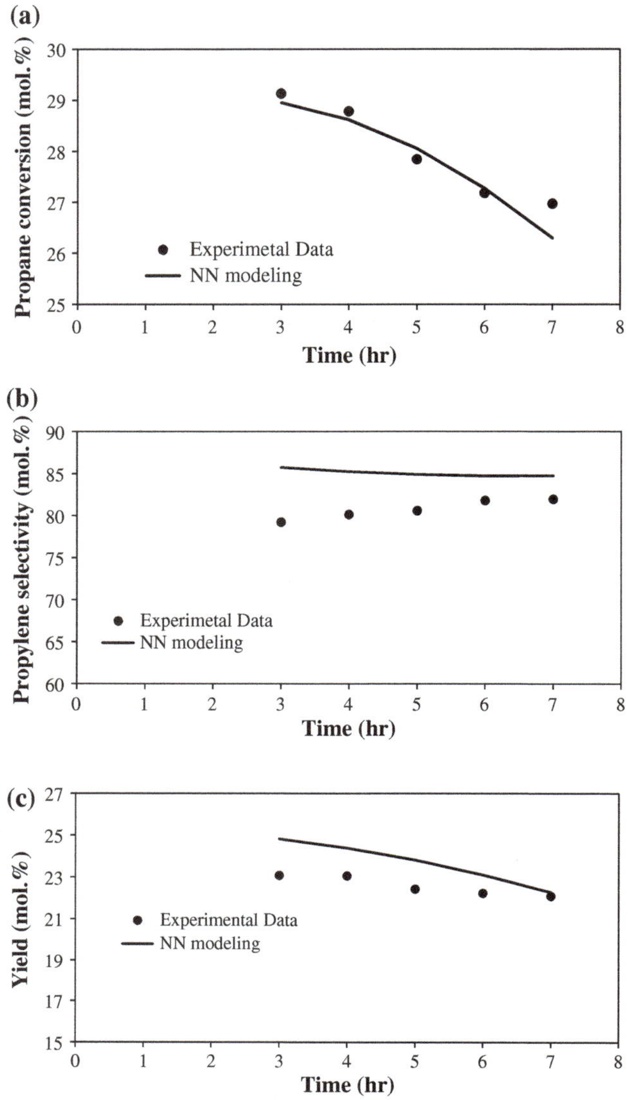

Fig. 4 ANN modeling and experimental comparison effects of methanol for **a** propane conversion, **b** propylene selectivity and **c** propylene yield on PDH over industrial Pt–Sn/γ-Al_2O_3 catalyst in optimum amount of methanol addition at 575 °C

Table 8 Coke content of spent catalysts from TGA

T (°C)	Oxygenated additive	Wt% coke of catalyst
620	0.1 ml/h methanol	4.2
620	0.5 ml/h water	3.8
620	Non	8.0
600	0.05 ml/h methanol	3.2
600	Non	5.9
575	0.3 ml/h water	2.8
575	Non	4.6

coke deposition. Water and light oxygenates can also serve as catalyst modifier when used in low levels in the feed stream. They act as catalyst depressant. Pt-based catalysts are poisoned by high level of oxygenate by various mechanisms. They also deplete chlorine, essential for re-dispersion of sintered platinum, from the catalyst and increase the selectivity to carbon oxides. When used in appropriate levels, however, they could exhibit beneficial effects on catalyst performance without other adverse effects. In general, one can conclude that with increasing reaction temperature, the optimum amounts of oxygenated additives should be increased.

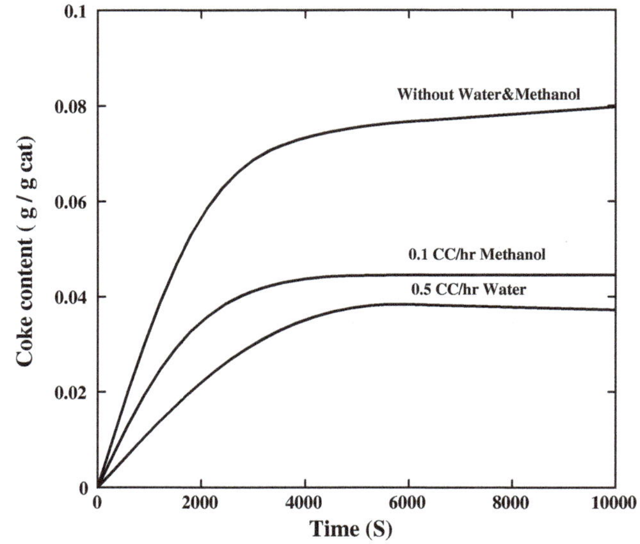

Fig. 5 Coke content versus time-on-stream(s) at 620 °C

The results from TGA for selected spent catalyst samples are presented in Table 8 in terms of the coke content (g/g cat) of spent catalysts, indicating that without any added water or methanol, the coke content of the spent catalysts increased with increasing reaction temperature. Addition of water or methanol in optimum amounts significantly reduced the coke content of spent catalysts. Comparison of the coke content at 620 °C indicated that when water was added in the optimum amount, the reduction in the coke content of the spent catalyst was slightly more than when methanol was added in the optimum amount (Fig. 5).

Conclusions

PDH was carried out over an industrial Pt–Sn/γ-Al_2O_3 catalyst with small amounts of water or methanol added to the feed. These oxygenated additives would maximize both the conversion of propane and the yield of propylene if they are added in the optimum amounts. Application of ANNs model can avoid difficulties like type of catalyst, uncertainty in porosity and mechanism of reaction, etc. in simulation of a catalyst activity. Based on different training algorithms, MLP has been found as the best network with minimum training error for modeling of catalyst conversion and selectivity. The obtained results incorporate excellent capability of ANN in modeling. The estimated values of ANN agreed very well with the experimental values, even in the case where the experimental data seemed to contain some error and diversity. The optimum amounts of oxygenated additives increased with increasing reaction temperature. For a given temperature, the optimum amount of water was more than that for methanol. These trends were well predicted by the NN model and the results lead to the conclusion that the neural network is a powerful tool for the estimation of catalyst activity, selectivity and propylene yield for sufficient amounts of oxygenated additives.

References

1. Arpe H-J (1993) Ullmann's encyclopedia of industrial chemistry, 5th edn. Wiley-VCH, New York
2. Zhang Y, Zhou Y, Qiu A, Wang Y, Xu Y, Wu P (2006) Effect of alumina binder on catalytic performance of PtSnNa/ZSM-5 catalyst for propane dehydrogenation. Ind Eng Chem Res 45:2213–2219
3. Bhasin MM, McCain JH, Vora BV, Imai T, Pujado PR (2001) Dehydrogenation and oxydehydrogenation of paraffins to olefins. Appl Catal A 221:397–419
4. Stagg SM, Querini CA, Alvarez WE, Resasco DE (1997) Isobutane dehydrogenation on Pt–Sn/SiO_2 catalysts: effect of preparation variables and regeneration treatments. J Catal 168:75–94
5. Llorca J, Homs N, Leon J, Sales J, Fierro JLG, Ramirez de la Piscina P (1999) Supported Pt-Sn catalysts highly selective for isobutane dehydrogenation: preparation, characterization and catalytic behavior. Appl Catal A 189:77–86
6. Barias OA, Holmen A, Blekkan EA (1996) Propane dehydrogenation over supported Pt and Pt–Sn catalysts: catalyst preparation, characterization, and activity measurements. J Catal 158:1–12
7. Cortright RD, Dumesic JA (1994) Microcalorimetric, spectroscopic, and kinetic studies of silica supported Pt and Pt/Sn catalysts for isobutane dehydrogenation. J Catal 148:771–778
8. Bednarova L, Lyman CE, Rytter E, Holmen A (2002) Effect of support on the size and composition of highly dispersed Pt–Sn particles. J Catal 211:335–346
9. Barias OA, Holmen A, Blekkan EA (1995) Propane dehydrogenation over supported platinum catalysts: effect of tin as a promoter. Catal Today 24:361–364
10. Gascon J, Tellez C, Herguido J, Menendez M (2005) A two-zone fluidized bed reactor for catalytic propane dehydrogenation. Chem Eng J 106:91–96
11. Gascon J, Tellez C, Herguido J, Menendez M (2003) Propane dehydrogenation over a Cr_2O_3/Al_2O_3 catalyst: transient kinetic modeling of propene and coke formation. Appl Catal A Gen 248:105–116
12. Martin N, Viniegra M, Zarate R, Espinosa G, Batina N (2005) Coke characterization for an industrial Pt–Sn/γ-Al_2O_3 reforming catalyst. Catal Today 107–108:719–725
13. Barghi B, Fattahi M, Khorasheh F (2012) Kinetic modeling of propane dehydrogenation over an industrial catalyst in presence of oxygenated compounds. React Kinet Mech Catal 107:141–155
14. Kogan SB, Herskowitz M (2001) Selective propane dehydrogenation to propylene on novel bimetallic catalysts. Catal Commun 2:179–185
15. Late L, Rundereim J-I, Blekkan EA (2004) Selective combustion of hydrogen in the presence of hydrocarbons: 1. Pt-based catalysts. Appl Catal A Gen 262:53–61
16. Kogan SB, Schramm H, Herskowitz M (2001) Dehydrogenation of propane on modified Pt/θ-alumina performance in hydrogen and steam environment. Appl Catal A 208:185–191
17. Duprez D, Hadj-Aissa M, Barbier J (1989) Effect of steam on the coking of platinum catalysts: I. Inhibiting effect of steam at low partial pressure for the dehydrogenation of cyclopentane and the coking reaction. Appl Catal 49:67–74
18. Yu DL, Gomm JB (2002) Enhanced neural network modelling for a real multi-variable chemical process. Neural Comput Appl 10:289–299
19. Zahedi G, Elkamel A, Lohi A, Jahanmiri A, Rahimpor MR (2005) Hybrid artificial neural network-first principle model formulation for the unsteady state simulation and analysis of a packed bed reactor for CO_2 hydrogenation to methanol. Chem Eng J 115:113–120
20. Omata K, Nukui N, Yamada M (2005) Artificial neural network aided design of a stable Co–MgO catalyst of high-pressure dry reforming of methane. Ind Eng Chem Res 44:296–301
21. Kito S, Satsuma A, Ishikura T, Niwa M, Murakami Y, Hattori T (2004) Application of neural network to estimation of catalyst deactivation in methanol conversion. Catal Today 97:41–47
22. Nabavi R, Niaei A, Salari D, Towfighi J (2007) Modeling of thermal cracking of LPG: application of artificial neural network in prediction of the main product yields. J Anal Appl Pyrol 80:175–181
23. Molga EJ (2003) Neural network approach to support modelling of chemical reactors: problems, resolutions, criteria of application. Chem Eng Process 42:675–695
24. Papadokonstantakis S, Machefer S, Schnitzlein K, Lygeros AI (2005) Variable selection and data pre-processing in NN modelling of complex chemical processes. Comput Chem Eng 29:1647–1659
25. Fattahi M, Khorasheh F, Sahebdelfar S, Tahriri Zangeneh F, Ganji K, Saeedizad M (2011) The effect of oxygenate additives on the performance of Pt–Sn/γ-Al_2O_3 catalyst in the propane dehydrogenation process. Scientia Iranica 18(6):1377–1383
26. Lobera MP, Tellez C, Herguido J, Menedndez M (2008) Transient kinetic modelling of propane dehydrogenation over a Pt–Sn–K/Al_2O_3 catalyst. Appl Catal A 349:156–164
27. Khataee A, Khani A (2009) Modeling of nitrate adsorption on granular activated carbon (GAC) using artificial neural network (ANN). Int J Chem React Eng 7:A5
28. Arcotumapathy V, Siahvashi A, Adesina AA (2012) A new weighted optimal combination of ANNs for catalyst design and reactor operation: methane steam reforming studies. AIChE J 58:2412–2427
29. Havel J, Lubal P, Farkova M (2002) Evaluation of chemical equilibria with the use of artificial neural networks. Polyhedron 21:1375–1384
30. Erdem Gunay M, Yildirim R (2011) Neural network analysis of selective CO oxidation over copper-based catalysts for knowledge extraction from published data in the literature. Ind Eng Chem Res 50:12488–12500
31. Zahedi G, Karami Z, Yaghoobi H (2009) Prediction of hydrate formation temperature by both statistical models and artificial neural network approaches. Energy Convers Manag 50:2052–2059

25

Carbon dioxide capture and utilization in petrochemical industry: potentials and challenges

Maryam Takht Ravanchi · Saeed Sahebdelfar

Abstract The reduction of greenhouse gas emissions is an ever-increasing challenge for production units and power plants in view of the global warming concerns. Carbon dioxide capturing from petrochemical process streams and flares has been recognized as one of the several strategies necessary for mitigating the emission of greenhouse gases into the atmosphere. To keep greenhouse gases at manageable levels, large decrease in CO_2 emissions through separation and utilization will be required. Furthermore, carbon dioxide emission potential will become an important factor in technology selection when cost of carbon dioxide emissions is set to be included in the processing cost. This article reviews the potential CO_2 separation technologies and its recycling via chemical fixation as bulk chemical products in petrochemical industry. Various separation techniques, such as absorption, adsorption, membrane separation and cryogenic, and utilization processes, such as conversion to carbon monoxide, oxidative dehydrogenation, hydrogenation and polymerization are thoroughly discussed. The technological challenges and recent developments and achievements are also presented.

Keywords Carbon dioxide · Greenhouse gas (GHG) · Carbon capture and utilization (CCU) · Catalytic conversion · C_1 chemistry · Petrochemical industry

M. Takht Ravanchi (✉) · S. Sahebdelfar
Catalyst Research Group, Petrochemical Research and Technology Company, National Petrochemical Company, No. 27, Sarv Alley, Shirazi-south, Mollasadra, P.O. Box 14358-84711, Tehran, Iran
e-mail: m.ravanchi@npc-rt.ir; maryamravanchi@gmail.com

Introduction

Environmental issues due to emissions of greenhouse gases (GHGs) have become worldwide problems. Studies have shown that increased GHG levels in atmosphere cause global warming. Carbon dioxide (CO_2), methane (CH_4), nitrous oxide (N_2O) and fluorinated gases (HFC, PFC, SF_6) are greenhouse gases among which carbon dioxide makes up a high proportion with respect to its amount present in the atmosphere (Fig. 1). Going by the prediction of Intergovernmental Panel on Climate Change (IPCC), by year 2100, the atmosphere may contain up to 570 ppm of CO_2, causing a rise in mean global temperature of around 1.9 °C and an increase in mean sea level of 3.8 m [1].

The proportion of greenhouse gases has increased significantly since the Industrial Revolution (about 1750), and global carbon dioxide (CO_2) concentration has risen by 37 %.

At present, it is accepted that the increase of GHG concentration in atmosphere is attributed to human activities. Greenhouse gas emissions are produced as the by-products of various non-energy-related industrial activities. Industrial processes can chemically transform raw materials, which often release waste gases such as CO_2. The processes include production of iron, steel, metallurgical coke, cement, lime, carbonate consuming processes (e.g., flux stone, flue gas desulfurization, and glass manufacturing), ammonia and urea, petrochemical, aluminum, soda ash, titanium dioxide, ferroalloy and glass. The production of some petrochemicals results in the release of small amounts of CO_2 emissions. Petrochemicals are chemicals isolated or derived from petroleum or natural gas. The chemical and petrochemical industry accounts for 30 % of global industrial energy use and 16 % of direct CO_2

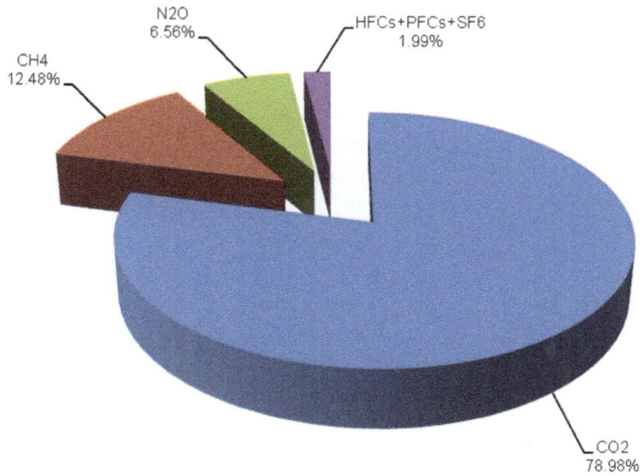

Fig. 1 Greenhouse gas emission in 2011 [1]

emissions. More than half of the energy demand is for feedstock use, which cannot be reduced through energy efficiency measures [2, 3].

The public concern about the climate change due to the greenhouse gas emissions to atmosphere led to the United Nation Framework Convention on Climate Change (UNFCCC) in 1992. The main objective of this meeting was to discuss methods for stabilization of GHG concentrations in the atmosphere at a level that the anthropogenic emissions could not interfere with the climate system. The stabilization of GHG concentrations in atmosphere can be achieved by three approaches:

- Reducing energy intensity
- Reducing carbon intensity (by use of carbon-free fuels)
- Enhancing the sequestration of CO_2

Carbon dioxide capture and sequestration (CCS) can significantly reduce emissions from CO_2 sources as well as certain industry types such as ethanol and natural gas processing plants. CCS methodologies for industrial and energy-related sources comprise three steps: CO_2 capture, CO_2 transportation and CO_2 utilization.

There are various petrochemical units that produce carbon dioxide with high purity as a by-product such as ethylene oxide and ammonia processes. Flaring is another main emission source of this greenhouse gas in petrochemical industry. In petrochemical industry, hydrocarbon wastes can be used as fuel and recycled. Therefore, the main waste products are water and CO_2. Carbon dioxide is produced in petrochemical process streams mostly by reactions with oxygenates (mainly oxygen as in oxidation of ethylene and water as in water–gas shift reaction). In fact, ethylene oxide and ammonia processes produce carbon dioxide with high purity as a by-product. However, most of the carbon dioxide in petrochemical industry is emitted in the flue gas as a result of burning fuel oil and fuel gas [4].

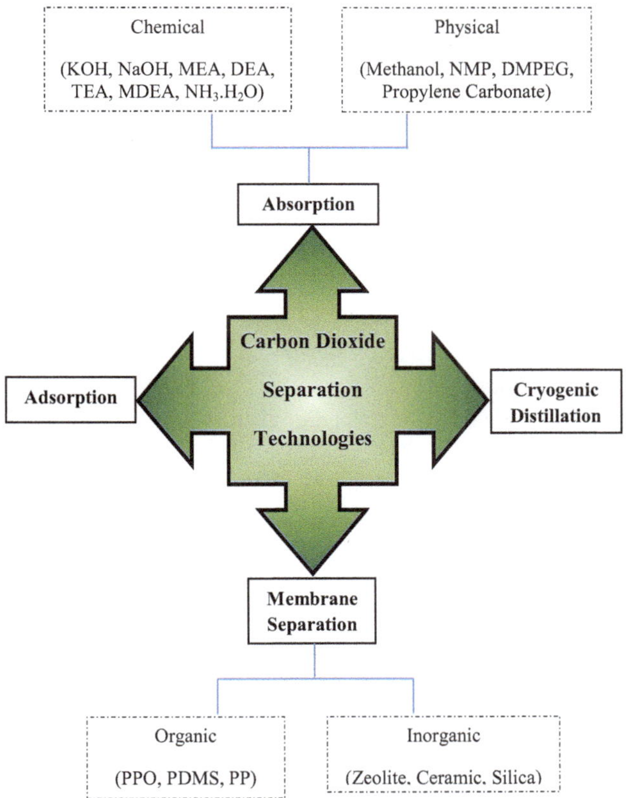

Fig. 2 Available technologies for CO_2 separation

In literature, there are some valuable reviews published in different aspects of CCS. For instance, in the field of CO_2 capture Granite and O'Brien [5]; Stewart and Hessami [6]; Olajire [7]; Pires et al. [8]; Takht Ravanchi et al. [9]; Wang et al. [10]; Kargari and Takht Ravanchi [11]; Zhao et al. [12]; Songolzadeh et al. [13]; and in the field of CO_2 utilization Wang and Zhu [14]; Song [15]; He et al. [16]; Havran et al. [17] are worth mentioning. Due to the absence of a systematic and detailed review on CO_2 separation and fixation in petrochemical industry, the present research focused on this goal.

CO_2 separation techniques

Capturing and separation of CO_2 from petrochemical gas streams are essential parameters for the carbon management for CO_2 sequestration from environment.

Generally, CO_2 separation contributes to 70–80 % of the overall CCS cost [4]. Hence, reducing separation cost is the most important issue for CCS process. In this section, common technologies of CO_2 separation from a flue gas are focused.

A wide range of technologies based on physical and chemical processes currently exist for CO_2 separation from a gas stream categorized as absorption, adsorption, membrane separation and cryogenics (Fig. 2). The choice of a suitable technology depends on the characteristics of gas stream.

Biological and hybrid processes and chemical looping combustion (CLC) are other CO_2 separation techniques which are not considered in this review. In biological processes, trees or microalgae can be used for CO_2 capture [6, 18]. Chemical looping combustion, proposed by Richter and Knoche [19], divides combustion into intermediate oxidation and reduction reactions that are performed separately with a solid oxygen carrier circulating between the separated sections. Hybrid systems can be based on membrane associated with cryogenic distillation or solvent absorption.

Absorption

Chemical absorption

The basic principle of chemical absorption is neutralization reaction in which absorbent react with CO_2 in the flue gas. The ideal chemical absorbent must have below specifications [20]:

- High reactivity with respect to CO_2—which would reduce height requirements for the absorber and/or reduce solvent circulation flow rates.
- Low regeneration cost requirements—based on a low heat of reaction with CO_2.
- High absorption capacity—which directly influences solvent circulation flow rate requirements.
- High thermal stability and reduced solvent degradation—reduced solvent waste due to thermal and chemical degradation.
- Low environmental impact.
- Low solvent costs—should be easy and cheap to produce.

Strong and weak alkaline solutions are commonly used absorbents. KOH and NaOH are examples of strong alkaline solutions, and alcohol amines (such as MEA, DEA, TEA and MDEA) and aqueous ammonia ($NH_3.H_2O$) are examples of weak alkaline solutions.

The applicability of strong alkaline solution is limited due to its strong corrosion to equipment. The alcohol amines usually suffer from the disadvantage of high energy requirement for regeneration, easy oxidation/degradation by SO_2 and O_2 in flue gas and system corrosion [21, 22]. Aqueous ammonia has the advantage of high-absorption efficiency, high-absorption capacity, low-energy requirement for absorbent regeneration and wide distribution of resources. In Table 1, advantages and disadvantages of chemical absorption by different alkaline solutions are categorized.

Table 1 Advantages and disadvantages of alkaline approaches for CO_2 capture

Process	Advantage	Disadvantage
Strong alkaline/alkaline metal solution (KOH, NaOH solution)	Fast reaction rate Large absorption capability High absorption efficiency	Cost of absorbent Strong corrosion to equipment Product treatment and disposal
Weak alkaline ($NH_3.H_2O$)	Large absorption capability and high loading capacity Low-energy requirement for absorbent regeneration Utilization of products as fertilizer Wide distribution of absorbent	Easy to volatilize and leak Thermal instability of products Corrosion to equipment
Aqueous ammines (MEA, DEA, TEA, MDEA)	Less volatility Good stability of absorbent Enhancement role used as additive	Absorption efficiency High-energy consumption for regeneration Easy degradation by SO_2 and O_2 in flue gas Resulting in system corrosion

Between several chemical absorption processes, MEA would be the first to be available for immediate industrial application. Development of new solvents has achieved some progress such as ammonia [23], piperazine promoted K_2CO_3 [24], and concentrated aqueous piperazine [25].

Physical absorption

Physical absorption takes place at high CO_2 partial pressures. This process is not economical for streams with CO_2 partial pressures lower than 15 vol%.

In physical absorption, organic solvent is used to physically, rather than chemically, absorb CO_2. In this process, CO_2 removal is based on its solubility within the solvents which depends on the partial pressure and temperature of the feed gas. Higher CO_2 partial pressure and lower temperature favor the solubility of CO_2 in absorbent. The solvents are regenerated by increasing temperature or decreasing pressure. In comparison with chemical solvents, the interaction between CO_2 and absorbent is weak.

Physical absorption process and related solvents for CO_2 removal are commercially available. Selexol process is based on dimethyl ethers of polyethylene glycol (DMPEG), a Union Carbide liquid glycol-based solvent, and has been used since 1969. Glycol is effective for CO_2 capturing at high concentration. Rectisol process is based on low temperature methanol (cold methanol). Glycol carbonate is another interesting solvent because of its high selectivity for CO_2 but it has relatively low capacity. Propylene carbonate (FLUOR process) and *N*-methyl-2-pyrrolidone (NMP-Purisol) are other physical solvents [26].

Some CO_2 capture applications benefit from a mixture of physical and chemical solvents. The most commonly used examples are Sulfinol, a mixture of the physical solvent sulfolane and the amines such as diisopropyl amine (DIPA) or methyl diethanolamine (MDEA); and Amisol, a mixture of methanol and secondary amines. These hybrid solvents attempt to exploit the positive qualities of each constituent under special conditions.

Membrane separation

Membrane separation technology is a low cost process when high purity gas streams are not fundamental [27]. Membranes can separate carbon dioxide from a gas stream by size exclusion or by chemical affinity. Chemical affinity membranes are often impregnated with a scrubbing solution or chemical functional group (e.g., amine) selective for carbon dioxide. The scrubbing solutions, as well as functional groups on the surface of sorbents, are discussed in the critical review papers [28, 29]. A large body of research has been conducted on the properties of carbon dioxide selective membranes based on inorganic materials such as zeolites, alumina, carbon, and silica. Researchers proposed the separation of carbon dioxide from flue gases by multi-walled carbon nano-tube membranes [5]. Substantial advances in the selectivity, permeability, and chemical stability of inorganic membranes are needed for successful application to gas mixtures.

Membrane process has several advantages [30]:

- Higher separation energy efficiency relative to equilibrium-based processes.
- No waste streams production.
- Existence of commercial references (e.g., CO_2 separation from natural gas).
- No regeneration energy required.
- High packing density which requires small installations.
- Increasing mass transfer area in a given volume.

Its major drawbacks can be categorized as below:

- The treatment of an enormous flow rate of CO_2 containing gas requires a very large membrane area and increases the cost of this technology.
- The need of very large, expensive and energy consuming compression equipment.

Organic (polymeric) and inorganic (zeolite, ceramic and silica) membranes are of particular interest in CO_2 separation. Polymeric membranes separate gases based on solution-diffusion mechanism. Low cost, high separation performance, ease of synthesis and mechanical stability are their advantages [31, 32]. An important limitation of polymeric membrane is its application for a high-temperature gas stream as it cannot be applied without cooling of the gas stream in order not to destroy the membrane.

It was recently demonstrated that a zeolitic imidazolate framework with high thermal (up to 390 °C) and chemical stability under refluxing organic and aqueous media shows exceptional selectivity for CO_2 capture from CO_2/CO mixtures with an extraordinary capacity for storing CO_2.

Adsorption

Adsorption is also considered a feasible process for CO_2 capture at an industrial scale. It is a selective process in which molecules contained in liquid or gaseous mixtures adhere on a solid surface, the adsorbent. These molecules, even in small concentrations in the streams, can be captured by these selective materials. The properties of the adsorbed particles (molecular size, molecular weight and polarity) and the adsorbent surface (polarity, pore size and spacing) determine the adsorption quality. As adsorption is an exothermic process, the regeneration of the adsorbents through desorption can be performed by rising the temperature. However, adsorption presents lower energy requirements and avoids the shortcomings when compared to absorption [33]. Adsorption can be either physical (physisorption) or chemical (chemisorption).

Several adsorbents have been proposed for CO_2 capture, including carbon fiber monolithic adsorbents [34], activated carbon fiber-phenolic resin composites [35], melamine-formaldehyde highly porous adsorbents [36], amine immobilized adsorbents [37], red mud [38], steam-activated anthracite [39], activated carbon [40], lithium zirconate [41], lithium silicate [42, 43], alumina, metallic oxides and zeolites [44] among others [45, 46]. Sjostrom and Krutka [47] compared several adsorbents and concluded that carbon-based materials present excellent stability, while zeolites work well under dry conditions.

In the industrial context, a continuous CO_2 capture process is preferred while the technologies often studied are the cyclic processes. These adsorption processes can be grouped into two types, according to the way in which the adsorbent is exchanged between the adsorption and desorption steps: by changing the pressure or the temperature. In pressure swing adsorption (PSA), the adsorption is performed at pressures higher than atmospheric. Desorption occurs at atmospheric pressure. However, a vacuum desorption pressure can be applied and, for commercial adsorbent 13X, the costs of CO_2 capture are comparable with the cost of MEA absorption [48]. The operation with an adsorbent with higher adsorption capacity and higher selectivity to CO_2 would increase the profitability of PSA. The vacuum swing adsorption (VSA) operates at near-ambient temperature and pressures and desorption is performed at lower pressure. It is a promising technology due to the relatively low power consumption and ease of operation.

CO_2 containing gas streams to be treated by adsorption must have high CO_2 concentrations because of the low selectivity of the most available adsorbents.

Cryogenic distillation

CO_2 cryogenic distillation involves separation of gas mixture by fractional condensation and distillation at low temperature. Cryogenic temperatures are obtained by a closed-cycle operated refrigeration system consisting of a compressor, a Joule-Thompson valve (JTV), multi-stage heat exchangers and expanders.

Cryogenic distillation is a commercial process commonly used to liquefy and purify CO_2 from high purity sources (>90 %). It involves cooling the gas to a very low temperature (lower than −73.3 °C) so that CO_2 can freeze out/liquefied and separated.

The CryoCell® technology developed by Cool Energy Ltd (Australia) uses a cryogenic process for CO_2 removal. This technology eliminates water consumption, uses of chemicals and corrosion-related issues.

Advantages and constraints

In the field of carbon dioxide capture and utilization for petrochemical industry, the separation process has high importance as it should provide CO_2 within the specified purity required and with reasonable costs. Each of the above explained separation techniques has advantages and constraints. CO_2 concentration, temperature and pressure of flue gas streams are main operating parameters based on which suitable separation process can be selected.

Chemical absorption method has many obvious advantages such as high efficiency, high absorption capacity, low cost and mature technology. These solvents are usually preferred to physical ones. Physical solvents required low regeneration energy, but they have low absorption capacity and selectivity for CO_2 separation. Physical absorption is not economical for gas streams with CO_2 concentration lower than 15 vol%. Research efforts should be directed toward developing better solvents. Future progress in solvents should be:

- to reduce energy consumption;
- to avoid damage to environment and human being (when it is vented with treated gas);
- acceptable solvent degradation and corrosion to the packed column.

More important is to combine all these in one solvent. For gas streams with high temperature (about 373 K), low pressure and low CO_2 concentration (1 atm and 10–15 % mol), bulk absorption process may be the best suitable process for CO_2 separation.

Adsorption is suitable for separating CO_2 from low flow rate stream. Although, chemical adsorbents have high capacity and selectivity, but their regeneration is difficult. Physical adsorption is the most suitable for CO_2 capture at high pressures and low temperatures.

Since the pressure drop is driving force for membrane process, the gas stream must be compressed. Since compressing is very difficult and expensive, membrane separation is suitable for high pressure stream with high concentration (>10 vol%). Inorganic membranes have high thermal and chemical stability but their selectivity is lower than polymeric membranes. Polymeric membranes are very selective for CO_2 separation but they have low thermal stability. Glassy polymeric membranes have higher selectivity, while the rubbery polymeric membranes have higher thermal stability.

Membrane processes and cryogenic distillation are efficient for gas streams with high CO_2 concentration. Major disadvantage of cryogenic distillation is the large amount of energy required to provide the refrigeration.

Consequently, for capture from flue gases of petrochemical plants, adsorption and absorption appear to be the choice at least in short- and medium-term perspective.

CO_2 utilization

The disposal or utilization of the captured CO_2 is another problem for adequate reduction of CO_2 emissions. A potential and attractive route to use the captured CO_2 is its fixation as valuable petrochemical commodities or fuels. Figure 3 presents some of the current and potential utilization routes for CO_2.

Chemical fixation of carbon dioxide

Carbon dioxide, a cheap and abundant C-containing raw material, is interesting as substrate, solvent, reactant, and extracting agent. CO_2 is a potential C_1 building block for C–C chains or as competitive source of C in chemical

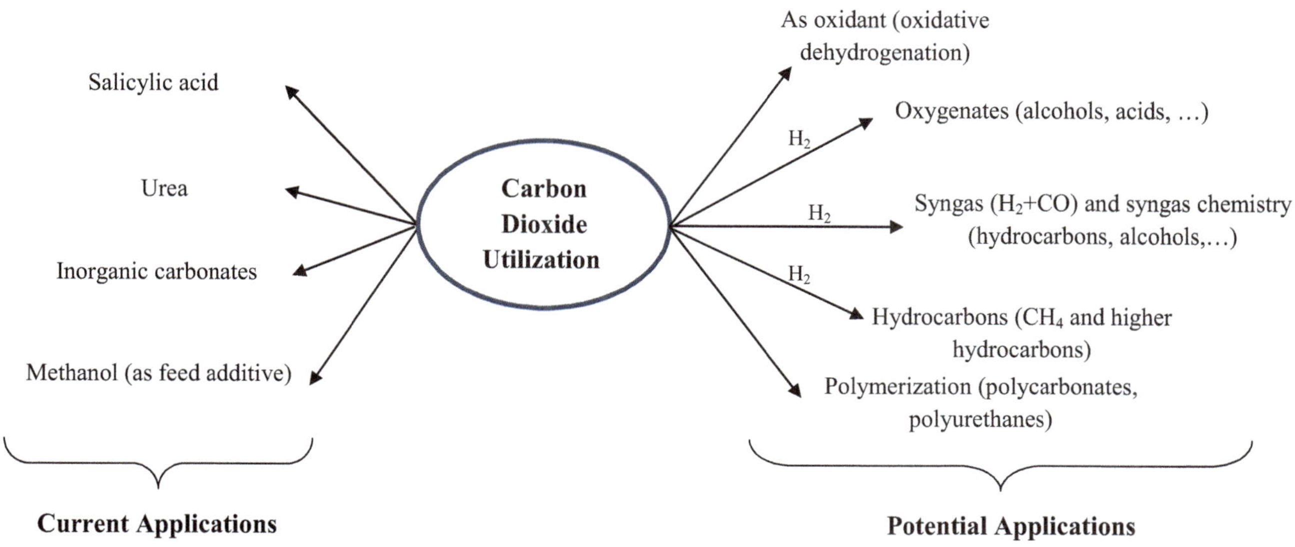

Fig. 3 Current and potential applications of CO_2 utilization

Table 2 Main uses of carbon dioxide in the chemical industry in Mtpy [55]

Process	Present	2016 forecast
Urea	114	132
Inorganic carbonates	ca. 50	70
Methanol	8	10
Total	172	217

industry and appears very attractive, although the molecule is rather inert and its reactions are energetically highly unfavorable [49]. There are several good recent review articles on CO_2 chemical fixation [17, 50–54].

The total amount of carbon dioxide used in industry is approximately 200×10^6 metric tons per year. Major industrial processes that utilize CO_2 as a raw material are listed in Table 2. Its utilization as a technological fluid, where carbon dioxide is not chemically converted and thus can be recovered at the end, is estimated at 30×10^6 metric tons per year [55]. The generation of energy used for CO_2 transformation, if based on hydrocarbon raw materials, still produces large amounts of CO_2. In addition, the resulting organic chemicals from CO_2 release carbon dioxide at the end of their use. However, despite the fact that, currently, the usage of CO_2 in the chemical industry cannot reduce significantly the global CO_2 levels, it is believed that the full potential of the fixation of CO_2 into value-added products has not yet been completely explored. The direct use of non-hydrocarbon based energy sources for reduction of CO_2 would change the above perspective and allow recycling of carbon dioxide via chemicals and liquid fuels. Thus, the further development of industrial processes, that are utilizing CO_2 for high-demand products, is of importance.

The scale of fixation for a specific product is an important parameter. Many fine chemicals or those of pharmaceutical interest produced by incorporation of CO_2 have only a limited demand worldwide (a few to a hundred ton annually). Regarding the order of magnitude of carbon dioxide emission problem (over 35 Gt annually), many of the claimed CO_2 reduction processes can be described as only a small step forward to resolve the problem and will not have a significant impact on the overall carbon emission problem. Those CO_2-fixing reactions that are feasible and show a significant bearing on overall worldwide carbon emissions are required on a very large scale. Typical examples include the conversion of CO_2 into fuels and polymeric/construction materials, which are in significant demand [54].

To date, only a few compounds have been successfully synthesized from CO_2 using heterogeneous catalytic routes, since most industrial processes utilize the more reactive but highly toxic carbon monoxide (CO) as a building block to produce higher chain hydrocarbons and liquid fuels. Processes for potential future utilization of CO_2 that have not yet been scaled up for commercialization are dependent on the development of new catalytic systems [17].

The source of energy supply should come from either renewable (solar, wind, hydro, ground heat, biomass, etc.) or integrated heat-management systems (nuclear plants or waste heat from industrial processes). Equally, the source of chemical reagents to couple with carbon dioxide should be carefully considered without any net enhancement in overall carbon emissions (to avoid a scenario that the CO_2 emission for the production of the coupling reagent is more than it captures). By taking thermodynamic limitations and the total carbon emissions of an overall process into account, many previously reported methods for carbon dioxide conversion may not be able to provide a practical way to globally reduce CO_2 emissions, although some of these reactions are of academic interest [54].

Chemistry and thermodynamic aspects

Being the most oxidized state of carbon, CO_2 as a raw material is in its lowest energy level, thus posing a major hindrance in establishing industrial processes. In other words, a large energy input is required to transform CO_2. There are four main methodologies to transform CO_2 into useful chemicals [52]:

- To use high-energy starting materials such as hydrogen, unsaturated compounds, small-membered ring compounds, and organometallics.
- To choose oxidized low-energy synthetic targets such as organic carbonates.
- To shift the equilibrium to the product side by removing a particular compound.
- To supply physical energy such as light or electricity.

Selecting appropriate reactions can lead to a negative Gibbs free energy of the reaction.

A widely accepted idea is that CO_2 is so thermodynamically and kinetically stable that it is rarely used to its fullest potential. However, due to the electron deficiency of the carbonyl carbons, CO_2 has a strong affinity toward nucleophiles and electron-donating reagents. In other words, CO_2 is an "anhydrous carbonic acid", which rapidly reacts with basic compounds [52].

Figure 4 shows the equilibrium conversion and product composition for methanation of carbon dioxide with hydrogen, obtained by constrained Gibbs free energy minimization method. This reaction is among thermodynamically favored reactions of CO_2. The equilibrium conversion is fairly high in the range of interest (300–400 °C) with a high selectivity to methane. However, as the temperature increases, the selectivity decreases and shifts to CO, indicating that the reverse reaction (dry reforming of methane) is favored. As higher organic compounds (hydrocarbons and oxygenates) are thermally unstable, lower temperatures favor higher selectivities to these compounds at the expense of lower reaction rates and CO_2 conversions.

Catalysis

Catalysts play a crucial role in CO_2 conversions, and suitable catalysts can realize interactions between CO_2 and a substrate on a transition metal center M. CO_2 has demonstrated surprising versatility by exhibiting a great variety of coordination and reaction modes in its homo- and polynuclear metal complexes. Most catalytic reactions involving CO_2 activation proceed via formal insertion of CO_2 into highly reactive M–X bonds with the formation of new C–X bonds. Coordinated CO_2 as in stable complexes is not necessarily required. Reactions are generally initiated by nucleophilic attack of X at the Lewis acidic C atom of CO_2 [49].

The conversion of carbon dioxide can be accomplished by homogeneous, heterogeneous and biological catalysts. For large-scale implementation, heterogeneous catalysis is preferred.

Homogeneous catalysis

Homogeneous catalytic processes for conversions of CO_2 are, by a large, reactions involving the incorporation of the entire CO_2 molecule into the products. These processes can be divided into two main categories: reductions of CO_2 that involve formal insertion of CO_2 into a bond in another molecule, and those that involve coupling of CO_2 with another unsaturated molecule [56].

The mild reaction conditions used in homogeneous hydrogenation of CO_2 catalyzed by transition metal complexes allow the partial hydrogenation of CO_2 to yield formic acid and derivatives:

$$CO_2 + H_2 \Leftrightarrow HCOOH \qquad \Delta H^0_{298} = -32\ \text{kJ/mol} \qquad (1)$$

Further reduction of formic acid under homogeneous conditions is more difficult and only very few examples are known reporting formation of methanol and methane catalyzed by Pd and Ru complexes in very low yields [57].

Heterogeneous catalysis

The heterogeneous catalytic conversion of CO_2 involves only hydrogenation reactions, with the exception of CO_2 reforming of CH_4, and occurs at high temperatures. Thus, the product portfolio is much more limited than for homogeneous catalysts:

$$CO_2 + H_2 \Leftrightarrow CO + H_2O \qquad \Delta H^0_{298} = +41\ \text{kJ/mol} \qquad (2)$$

$$CO_2 + CH_4 \Leftrightarrow 2CO + 2H_2 \qquad \Delta H^0_{298} = +247\ \text{kJ/mol} \qquad (3)$$

$$CO_2 + 3H_2 \Leftrightarrow CH_3OH + H_2O \qquad \Delta H^0_{298} = -49.9\ \text{kJ/mol} \qquad (4)$$

$$CO_2 + 4H_2 \Leftrightarrow CH_4 + 2H_2O \qquad \Delta H^0_{298} = -113\ \text{kJ/mol} \qquad (5)$$

Despite the limited diversity, these reactions could be potentially important for chemical utilization of CO_2 in the petrochemical industry. As these equations are equilibria, the product yields seldom reach 100 %, and, furthermore, the reactions are complicated by several equilibria occurring simultaneously at reaction temperature (see

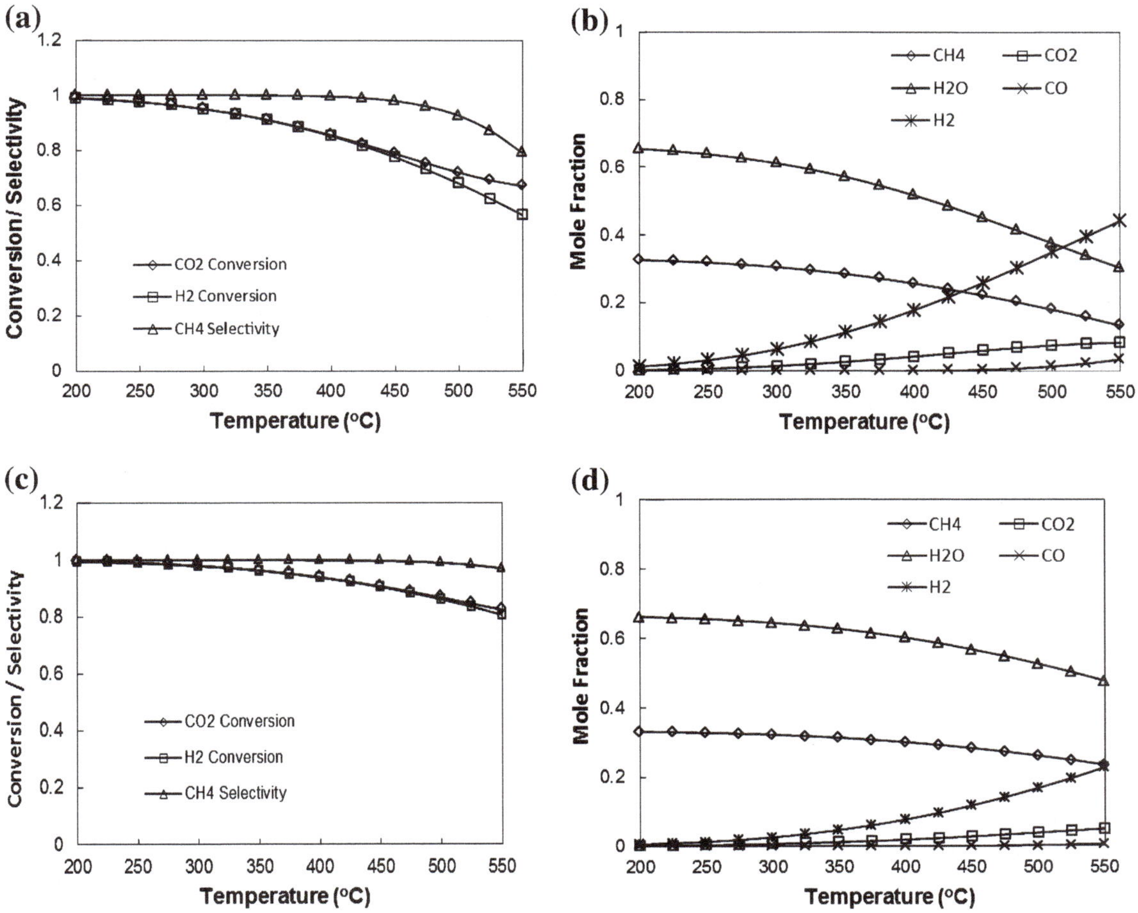

Fig. 4 The effect of temperature on conversions, selectivity and equilibrium product composition at 1 atm (**a**, **b**) and 10 atm (**c**, **d**) for stoichiometric ratio of CO_2/H_2 for methanation

Fig. 4). The art of heterogeneous catalysis is then the ability to minimize these other readily accessible reactions and steer the product formation to that desired.

Electro- and photochemical catalysis

The electro- and photochemical catalysis is identical in many respects; the major difference is the source of reducing electrons [56]. Electro- and photochemical reductions of CO_2 can also be catalyzed by metal complexes.

Methanol or related oxygenates (e.g., formic acid) can be made from CO_2 via aqueous electrocatalytic reduction in a "reversed fuel cell" without prior electrolysis of water to hydrogen [57]. The reversed fuel cell accomplishes the electrocatalytic reduction of CO_2 outside the potential range of the conventional electrolysis of water. In aqueous solutions, the selectivity depends on the electrode type. The different electrode behavior among metals can be related to their electronic configuration and can be grouped as *sp* and *d* metals. Selectivity depends on surface coverage of different reduced species. Higher surface hydrogen species concentration would lead to methane and hydrogen, while more carbon species would lead to CO and HCOOH. Copper has received much attention because it produces appreciable amounts of hydrocarbons and alcohols from CO_2 reduction, although it tends to give mixed product including hydrogen [51].

Photocatalytic reduction of CO_2 can be accomplished by suspending photosensitive semiconductor powder in aqueous solutions under irradiation, usually using UV light. Photoreduction of CO_2, however, is a completion with H_2 formation due to water decomposition, and leads to mixtures of reduced carbon products. Selectivity, therefore, is one major problem of these processes [57].

Potential processes

For effective reduction of CO_2 emissions, only reactions of high scale are of interest. Fortunately, some of heterogeneous catalytic conversions have this potential. Another need is renewable energy (hydrogen) source.

Conversion to carbon monoxide

An obvious way to use CO_2 in chemical synthesis is its reduction to CO followed by syngas (a mixture of H_2, CO and CO_2) chemistry. This can be performed by hydrogen or methane known as reverse water–gas shift (RWGS) reaction [Eq. (2)] and dry reforming of methane [Eq. (3)], respectively. Both reactions are endothermic and favored at high temperatures and susceptible to coke forming side reactions.

The commercial Fe_2O_3/Cr_2O_3, high-temperature shift catalyst might appear a logical candidate for RWGS reaction. However, despite a high initial activity, a low stability for this catalyst has been reported which was attributed to the reduction of Fe_2O_3 to Fe metal and coke formation [58]. ZnO/Cr_2O_3 catalyst, on the other hand, showed both high activity and stability with no coke formation.

The ZnO showed good specific reaction activity in the temperature range from 400 to 700 °C but this catalyst was deactivated. All the catalysts except the ZnO/Al_2O_3 (850 °C) showed low stability for the RWGS reaction and was deactivated at the reaction temperature of 600 °C. The ZnO/Al_2O_3 catalyst calcined at 850 °C was stable during 210 h under the reaction conditions of 600 °C and GHSV of 150,000 h^{-1}; showing CO selectivity of 100 % even at the pressure of 5 atm. The high stability of the ZnO/Al_2O_3 catalyst (850 °C) was attributed to the prevention of ZnO reduction by the formation of $ZnAl_2O_4$ spinel structure. The spinel structure of $ZnAl_2O_4$ phase in the ZnO/Al_2O_3 catalyst calcined at 850 °C was confirmed by XRD and electron diffraction [59].

Because of low equilibrium conversions, temperatures above 400 °C should be used. Excess hydrogen can also shift the reaction to completion. Therefore, there is need for the development of novel catalysts (e.g., Cu/ZnO, Pt/CeO_2) and reaction engineering approaches such as removal of water.

Carbon dioxide or dry reforming of methane has received much attention since 1980s. CO_2 reforming of CH_4 has great potential economic and environmental benefits. This reaction has the following important advantages:

- The formation of a suitable H_2/CO ratio for use in valuable oxygenates synthesis.
- The reduction of CO_2 and methane emissions; as both gases cause heavy greenhouse effect.
- Better use in chemical energy transmission.

The reaction is more endothermic than steam reforming. Whereas the steam reforming of methane yields syngases with a H_2/CO ratio of 3, CO_2 reforming of CH_4 results in a ratio of 1, which may be of interest for some syngas consuming reactions. The equilibrium composition is influenced by the water–gas shift (WGS) reaction. Carbon dioxide reforming needs higher temperatures (with the danger of coke formation); advantageously catalysts may be used (Ni on supports such as alumina or MgO).

Dry reforming reaction is catalyzed by Group VIII metals, such as Ni, Pt, Pd, Ru, Rh and Ir. However, there is a serious problem of catalyst deactivation due to rapid coke formation. In addition to carbon deposition, sintering, and/or metal oxidation could be the causes of catalyst deactivation. Different deactivation mechanisms are associated with different catalyst system. For example, in CO_2 reforming of CH_4 that carbon formation rather than sintering is the main cause of platinum-containing catalyst deactivation.

Precious metals are more coke resistant; however, considering the high cost and limited availability of noble metals, it is more practical in industrial standpoint to develop Ni-based catalysts with high performance and high resistance to carbon deposition. Research on the nickel catalysts used for this reaction has mainly focused on the intrinsic activity of the metal phase, stability toward carbon formation, the type of the support most suitable for improving the efficiency of the catalyst, and the reaction mechanism.

The nature of the support has also significant effects on the performances and properties of catalysts. Supports have many functions but the most important is to maintain a stable physical surface, especially at high reaction temperatures. Therefore, the support itself must resist to thermal transformation at high temperatures, which means catalyst support must have high melting points. It has been proved that type of the support and presence of modifiers greatly affect the coking tendency. The most commonly used support for dry reforming of methane is γ-Al_2O_3. Other supports such as MgO, TiO_2, SiO_2, and La_2O_3 were also used [60] and references therein). Recently, there has been interest in mixed oxides, such as $Ce_{1-x}Zr_xO_2$ [61], $LaBO_3$ [62, 63] (B = Co, Ni, Fe, Cr) and La_2NiO_4 [64, 65], either as catalysts or catalyst support. Al_2O_3 and SiO_2 are two of the most often investigated catalyst supports with high specific surface area. MgO and CaO are studied due to their high melting points which favor stable catalyst surfaces at high reaction temperatures. Also, their basicity is believed to suppress carbon formation by promoting the activation of CO_2. TiO_2 and ZrO_2 are reducible metal oxides on which metals are electron-rich and always show strong metal-support interaction (SMSI). SMSI is believed to be critical to resisting carbon formation.

Mesoporous non-crystalline zirconia has shown promising performance as a support for Ni-based catalysts. Long-term stable performance has been achieved even under conditions which favor coke formation and catalysts deactivation [66–69].

Oxidative dehydrogenation of hydrocarbons

This reaction is basically similar to the reactions discussed above and similarly carbon dioxide is not fixed. The product is an olefin and CO. Olefins are highly valued intermediates in petrochemical industry for the production of polymers and fuel additives. The simple dehydrogenation reaction is thermodynamically limited and accompanied by extensive coke formation. CO_2 is a coke removing agent as well. Bartholomew [70] reported that oxidizability in the gasification of coke was in the following order: $O_2(105) > H_2O(3) > CO_2(1) > H_2(0.003)$, where numbers in parenthesis indicated the relative ratios of the gasification with oxidants. CO_2 has been called as nontraditional, unusual or mild oxidant and oxygen transfer agent. In this sense, carbon dioxide could be proposed as an oxidant in the oxidative conversions of hydrocarbons.

The promoting effect of CO_2 is explained either by hydrogen consumption in the reverse water–gas shift reaction [Eq. (2)], which shifts the equilibrium of, for example, propane dehydrogenation [Eq. (6)] (two-step path) or its participation in direct oxidation of propane to propylene [Eq. (7)] (single-step path). Many studies considered both pathways, especially above 848 K (575 °C).

$$C_3H_8 \Leftrightarrow C_3H_6 + H_2 \qquad \Delta H^0_{298} = 124 \text{ kJ/mol} \tag{6}$$

$$C_3H_8 + CO_2 \Leftrightarrow C_3H_6 + CO + H_2O \quad \Delta H^0_{298} = 164 \text{ kJ/mol} \tag{7}$$

The oxide catalysts which are active in the non-oxidative dehydrogenation of propane are also active in the process with CO_2. Many bulk and supported oxide-based materials containing active species, such as Cr, Ga, Zn, V, In, Fe and Mn have been studied as catalysts for dehydrogenation of propane in the presence of CO_2 [71] and references therein). Among the investigated materials, those containing Cr or Ga were found to be the most active and selective one.

The single-step mechanism [direct oxidation of paraffin to olefin, Eq. (7)] has been proposed for redox oxides (Cr_2O_3, Cr_2O_5, ZnO and V_2O_5), where CO_2 can participate as an oxidizing agent in the following redox cycle [72]:

$$C_3H_8 + MeO_x \Leftrightarrow C_3H_6 + MeO_{x-1} + H_2O \tag{8}$$

$$CO_2 + MeO_{x-1} \Leftrightarrow CO + MeO_x \tag{9}$$

The two-step has been proposed for non-redox oxides such as Ga_2O_3 [73]. Production of H_2 could be an indication of two-step pathway.

The oxidative dehydrogenation of ethylbenzene to styrene over Cr or Ce catalysts is one of the most relevant of CO_2-assisted reactions [49].

Hydrogenation

The heterogeneous metal-catalyzed hydrogenation of CO_2 goes directly to methanol or methane which is the result of drastic reaction conditions applied in hydrogenations in the presence of heterogeneous catalysts. Formate, formaldehyde, and formyl species, however, have been shown by spectroscopic techniques to exist as surface intermediates during hydrogenation of carbon oxides [57].

Methanol is a versatile multi-source multi-purpose chemical. It could be used as a potentially carbon neutral fuel for automotive and fuel cell application and a feed stock for production of a variety of hydrocarbons including olefins and aromatics [74]. The methanol economy has been a suggested future economy in which methanol replaces fossil fuels as means of energy storage, fuel and raw material for synthetic hydrocarbons and their products [75].

The hydrogenation of CO_2 to methanol [Eq. (4)] is interrelated to the hydrogenation of CO [Eq. (10)] through water–gas shift reaction [Eq. (2)]. A significant disadvantage of CO_2 hydrogenation process as compared to the hydrogenation of CO is that more hydrogen is consumed because of the formation of water.

$$CO_2 + 2H_2 \Leftrightarrow CH_3OH \qquad \Delta H^0_{298} = -90.8 \text{ kJ/mol} \tag{10}$$

It is known through isotope labeling studies that in the hydrogenation of CO the primary source of methanol is CO_2 [57]. It is also known that the traditional three-component Cu catalysts (Cu–ZnO–Al_2O_3 and Cu–ZnO–Cr_2O_3) exhibiting high activity and selectivity in hydrogenation of CO to methanol show the low activity and selectivity in hydrogenation of pure CO_2. Therefore, efforts have been made to find new and more effective catalysts for direct hydrogenation. The problem is to develop catalysts with high selectivity to methanol formation but low selectivity to reverse water–gas shift reaction.

It has been widely reported that Cu–ZnO-based catalysts with additional oxides are among the most useful systems for catalytic hydrogenation of CO_2 to methanol. It has been well documented that the Cu/ZnO/ZrO_2 catalyst is active in the methanol synthesis from CO_2 and H_2 [76]. To increase its activity and stability, this basic catalyst was often modified by different oxide additives. Besides ZrO_2, other irreducible oxides such as Al_2O_3, TiO_2, and Ga_2O_3 have been investigated. An example is the high performance multicomponent Cu–ZnO–ZrO_2–Al_2O_3–Ga_2O_3 catalyst prepared by coprecipitation [77]. The addition of a small amount (0.6 wt%) of silica greatly improved the long-term stability of the catalyst [78].

As these findings indicate, ZnO is not an ideal component for CO_2 hydrogenation catalysts. As an alternative, a Cu–Ga_2O_3 catalyst was prepared using ZnO as support, and as support and a silica-supported Cu–ZnO–Ga_2O_3 sample was also investigated [79]. The latter catalyst exhibited high selectivity to methanol and the conversion to CO was very low.

Catalysts with metals other than Cu have also been prepared and used. Ru on zeolite NaY prepared by ion exchange and promoted with Co showed high activity for methanol production [80].

Methanation, that is, selective hydrogenation of CO_2 to methane [the Sabatier reaction, Eq. (4)], is potentially important transformation, although at present there is no practical reason for producing methane from carbon dioxide and hydrogen. Studies, nevertheless, have been carried out to develop suitable catalysts and find effective reaction conditions. The reaction is highly exothermic and heat removal from the reactor is a major issue.

The full reduction of CO_2 into CH_4 requires the transfer of eight electrons which leads to significant kinetic limitation. Among all catalysts studied, Ru, Rh and Ni supported on metal oxides exhibit the highest activity, and Ru has been reported to be the most selective. The supports exert a marked influence of both the adsorption and specific activity of the metals [57]. In most cases, TiO_2 is the best support material to achieve good catalyst performance, which attributed to a strong interaction between the metal and the support.

Carbon dioxide usually adsorbs dissociatively on noble metals, and hydrogen facilitates adsorption. On supported Ru, Rh and Pd catalysts, chemisorptions of CO_2 in the presence of hydrogen lead to the formation of surface carbonyl hydride [57]. Dissociation of CO_2 promoted by hydrogen, the dissociation of surface carbonyl hydride into reactive surface carbon, and hydrogenation of the latter are proposed as important steps in the hydrogenation process.

A related process is the production of ethanol and higher alcohols which have potential as gasoline additives. The reactions are thermodynamically more favorable than hydrogenation of CO_2 to methanol. The most successful catalysts are Rh-based, which is prohibitively expensive for large-scale production [81].

Polymerization

Regarding possible CO_2 mitigation by chemical fixation on a large scale, the conversion of carbon dioxide into polymeric and construction materials could be made commercially feasible.

The chemistry of polymerization catalysis is identical, often, to some of the reactions in homogeneous catalysis. Generally, however, the chemistry of copolymerization of CO_2 with organic compounds is not nearly as well developed as homogeneous catalysis. By far, the most common copolymers with CO_2 are polycarbonates, which is one of the type of biodegradable polymer [56, 82].

A promising methodology in the utilization of CO_2 is its transformation into valuable polycarbonates and/or cyclic carbonates. The synthesis of cyclic carbonates from epoxides and CO_2 (Fig. 5) is one of the major ways to transform CO_2 to valuable chemicals [51]. The first successful synthesis of cyclic carbonate using room temperature ion in liquids (ILs) of imidazolium and pyridinium salts was reported by Peng and Deng [83]. They found that 1-*n*-butyl-3-methylimidazolium tetraflouroburate was the most active catalyst with nearly 100 % selectivity, and a suitable CO_2-propylene oxide ratio was required for high conversion. The ionic liquid used as the catalyst was recyclable.

Copolymerization of monomers such as azetidines and aziridines with CO_2 for poly(urethane)s may lead to new polymers and construction materials for important uses [54]. Bayer announced to plan to market a process to produce polyurethane-based plastics using CO_2 as a raw material by 2015 [81] and references therein).

Recent developments and applications

Processes based on CO_2 are known for synthesis of hydrocarbons (via Fischer–Tropsch synthesis), energy-rich C_1 molecules (CH_4, CH_3OH, HCOOH, etc.) and fine chemicals containing functionalities such as –COO– (acids, esters, lactones), –N–COO– (carbamates), or –N–CO– (urea, amides) (Cornilis et al. 2003).

As was shown before the major drawback of dry CO_2 reforming is the high thermodynamic driving force to produce coke. The SPARG process and the CALCOR process have overcome the coking problem during reforming [81].

The sulfur passivated reforming (SPARG) process was developed to decrease the syngas H_2/CO ratio from 2.7 obtained from steam reforming to 1.8 without changing the steam reforming facility. Consequently, the SPARG process is suitable as a retrofit to the existing steam reforming plants. Because the use of CO_2 instead of steam lowers H/C ratios in the product, part of the steam was replaced by CO_2. The increased probability of carbon formation and the associated catalyst deactivation is minimized by introducing a partially sulfur-poisoned nickel catalyst. By partially sulfiding of the Ni catalyst, the sites for coke formation are blocked because of ensemble effect while sufficient sites for reforming reaction are maintained [84].

Due to lowering catalyst activity, higher operating temperatures are necessary. The process operates at 1,188–1,218 K which decreases the CH_4 slip but the

Fig. 5 Coupling of CO_2 and epoxides to afford polycarbonates and/or cyclic carbonates

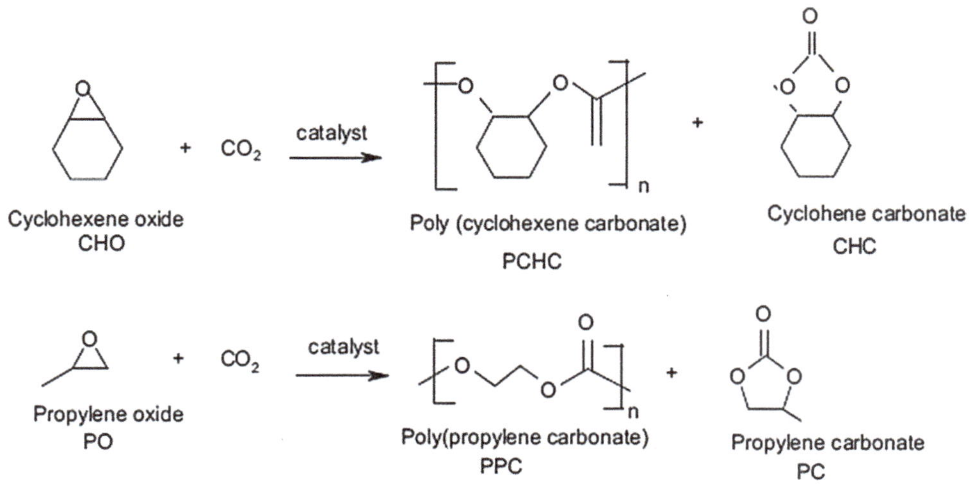

product syngas still contains about 2.7 mol % methane. When CO_2/CH_4 reforming is used to produce CO for polycarbonate synthesis, the impurities in the CO (mainly H_2 and CH_4) cause inferior mechanical properties of polycarbonates made via the phosgene route.

Under the reaction conditions applied in the SPARG process, higher hydrocarbons (C_{2+}) in the feed will be thermally cracked to form pyrolytic carbon. The C_{2+} species in the feed should be removed by a pre-reforming step. Thus, traces of sulfur might be present in the syngas produced. When a pure syngas is required, an additional cleaning step will be necessary which involves an additional investment. Therefore, a catalyst which operates carbon free without the addition of sulfur is preferred.

The CALCOR process aiming at production of high purity CO from natural gas ($H_2/CO = 0.42$ in produced syngas) or LPG was developed to overcome problems involved in transportation of the toxic CO and the need for high product quality. These factors favor the production of CO on-site even for small capacities.

The CALCOR process operates under dry reforming conditions in an excess of CO_2. Details of catalysts used are proprietary. The prevention of carbon formation is claimed to be achieved using catalysts with different activities and shape and by their specific arrangement throughout the reformer tube. The syngas produced contains below 500 ppm of methane. Two options for purification of CO produced are available: membrane purification leads to CO purity of about 98.9 vol%, while the low temperature pressure swing adsorption purification can yield CO with purity higher than 99.4 vol%. The specific CO_2 emission per ton of CO product is 15 kg CO_2 for Calcor Standard and 360 kg CO_2 for Calcor Economy (emitted CO_2 minus imported CO_2) process, compared to ca. 1,800 kg steam reforming with CO_2 recycle [85].

The new Tri-reforming process uses flue gases as CO_2 source. Tri-reforming refers to simultaneous oxidation, CO_2 and steam reforming from natural gas. It is a synergic combination of endothermic CO_2 and steam reforming and exothermic partial oxidation of CH_4. CO_2 from flue gas does not need to be separated. In fact, H_2, O_2 and CO_2 in waste flue gas from fossil fuel power plants will be used to tri-reforming natural gas and produce syngas.

Tri-reforming does have a different requirement on the catalyst composition to dry and steam reforming. It can be performed with stable operation and no carbon formation and no appreciable deactivation of tri-forming conditions. Pt/ZrO_2 and Ni/ZrO_2 showed high activity for steam and dry reforming and partial oxidation of methane [86]. Ni on ZrO_2 and ZrO_2 doped with yttrium and a lanthanide and alkaline earth metal has been patented for this process [87].

In the CAMERE (carbon dioxide hydrogenation to form methanol via a reverse water–gas-shift reaction) of Korean Institute of Science and Technology, CO_2 is converted to methanol [58]. In CAMERE process, carbon dioxide and hydrogen are converted to CO and H_2O by RWGS reaction over a $ZnAl_2O_4$ catalyst in a separate reactor and then the gas mixture is fed into a methanol synthesis reactor after water has been removed.

A commercial plant constructed by Carbon Recycling International (CRI) company for CO_2 hydrogenation to methanol is currently operative in Iceland using geothermal energy source [81, 88]. Recently in Japan, the construction of a 100 tpy demonstration plant for single-step production of methanol from CO_2 was announced [89]. The CO_2 feed comes as a byproduct of Mitsui Chemical Complex at Osaka and H_2 from water using renewable (solar) energy [90].

Many petrochemical and licensing companies are working on carbon dioxide capture and use to achieve sustainability in the petrochemical industry. As reported by Reuters on Jan 14, 2014, SABIC and KACST are examples in the Middle East. Several projects at KACST contribute to the sustainability development in the petrochemical sector. Examples include power generation with available fuels at well heads with CO_2 capture and use for enhance oil recovery (EOR), CO_2 utilization to produce syngas (dry reforming project) and utilizing carbon dioxide as regenerative agent to increase the commercial potential of methane dry reforming.

Saudi Basic Industries Corp (SABIC) has hired Germany's Linde Group to build the world's largest plant for capturing and using climate-warming carbon dioxide. The United Jubail Petrochemical Company, an affiliate of SABIC, plans to capture around 1,500 tons a day of carbon dioxide from ethylene plants and purify it for use in SABIC-owned petrochemical plants in the industrial city of Jubail. The high cost of catching gas emitted by factories and power plants and storing it safely underground have deterred commercial CCS projects.

Mitsubishi Heavy Industries, Ltd. (MHI) has signed a license agreement for carbon dioxide (CO_2) recovery technology with Gulf Petrochemical Industries Company (GPIC), a manufacturer of fertilizers and petrochemicals in Bahrain. GPIC will use the technology to recover CO_2 from flue gas emitted at its existing petrochemical plant and utilize the captured CO_2 to increase urea and methanol production. The recovery units can capture 450 metric tons of CO_2 per day, one of the world's largest capacities for the chemical application. In addition to urea production, CO_2 recovery technology can be used in chemical applications such as production of methanol and dimethyl ether (DME) and, in the food and beverage industries, production of carbonated beverages and dry ice. Another important possible application is enhanced oil recovery (EOR) enabling increased crude oil production; in this case, CO_2 is injected into an oil reservoir suffering from low productivity [91].

Conclusions

There are many potential options for carbon dioxide capture from petrochemical industry and its utilization as a C_1 feedstock for the production of petrochemical intermediates and/or products. However, as the capture and the transformation of CO_2 by itself requires energy input and CO_2 gives no enthalpy contribution to the final product, the contribution to net CO_2 emission could not be significant unless more effective techniques and renewable energy (or hydrogen) source are available. Consequently, the product demand, the source of energy and overall net carbon emissions have to be taken into consideration. Large impact of carbon dioxide capture and fixation on global CO_2 emissions could be realized only if bulk chemicals are the target of conversion. Several potential utilization routes are possible for the petrochemical industry. Based on the above discussions on CO_2 separation methods, it can be concluded that flue gases properties (mainly concentration of CO_2, temperature and pressure) are the most effective factors for selection of suitable process for CO_2 separation. Catalytic CO_2 hydrogenation seems to have the highest potential for implementation in petrochemical industry. Further R&D is needed on catalysis and hydrogen (energy) source. Electrochemical or photo-electrochemical CO_2 conversion is promising, but technology is still in its infancy.

References

1. United Nations Framework Convention on Climate Change (UNFCCC). http://unfccc.int/ghg_data/ghg_data_unfccc/ghg_profiles/items/4625.php. Accessed January 2014
2. Sahebdelfar S, Tahriri Zangeneh F, Takht Ravanchi M (2009) Chemical recycling of carbon dioxide to valuable petrochemicals. In: The 7th APCSEET conference, Qingdao, China
3. Takht Ravanchi M, Sahebdelfar S, Tahriri Zangeneh F (2009a) Carbon dioxide sequestration in petrochemical industries with the aim of reduction in greenhouse gas emissions. In: The 7th APCSEET conference, Qingdao, China
4. Seddon D (2010) Petrochemical economics technology selection in a carbon constrained world. Imperial College, London
5. Granite EJ, O'Brien T (2005) Review of novel methods for carbon dioxide separation from flue and fuel gases. Fuel Proc Technol 86:1423–1434
6. Stewart C, Hessami MA (2005) A study of methods of carbon dioxide capture and sequestration–the sustainability of a photosynthetic bioreactor approach. Energ Convers Manage 46:403–420
7. Olajire AA (2010) CO_2 capture and separation technologies for end-of-pipe applications-—a review. Energy 35:2610–2628
8. Pires JCM, Martins FG, Alvim-Ferraz MCM, Simes M (2011) Recent developments on carbon capture and storage: an overview. Chem Eng Res Des 89:1446–1460
9. Takht Ravanchi M, Sahebdelfar S, Tahriri Zangeneh F (2011) Carbon dioxide sequestration in petrochemical industries with the aim of reduction in greenhouse gas emissions. Front Chem Sci Eng 5:173–178
10. Wang M, Lawal A, Stephenson P, Sidders J, Ramshaw C (2011) Post-combustion CO_2 capture with chemical absorption: a state-of-the-art review. Chem Eng Res Des 89:1609–1624
11. Kargari A, Takht Ravanchi M (2012) Carbon dioxide: capturing and utilization. In: Liu G (ed) Greenhouse Gases INTECH, pp 1–30
12. Zhao B, Su Y, Tao W, Li L, Peng Y (2012) Post-combustion CO_2 capture by aqueous ammonia: a state-of-the-art review. Int J Greenh Gas Control 9:355–371
13. Songolzadeh M, Soleimani M, Takht Ravanchi M, Songolzadeh R (2014) Carbon dioxide separation from flue gases; a technological review emphasizing on reduction in greenhouse gas emissions. Sci World J.
14. Wang Sh, Zhu ZH (2004) Catalytic conversion of alkanes to olefins by carbon dioxide oxidative dehydrogenation—a review. Energ Fuel 18:1126–1139
15. Song Ch (2006) Global challenges and strategies for control, conversion and utilization of CO_2 for sustainable development

involving energy, catalysis, adsorption and chemical processing. Catal Today 115:2–32
16. He LN, Wang JQ, Wang JL (2009) Carbon dioxide chemistry: examples and challenges in chemical utilization of carbon dioxide. Pure Appl Chem 81:2069–2080
17. Havran V, Dudukovic MP, Lo CS (2011) Conversion of methane and carbon dioxide to higher value products. Ind Eng Chem Res 50:7089–7100
18. Skjånes K, Lindblad P, Muller J (2007) BioCO_2—a multidisciplinary, biological approach using solar energy to capture CO_2 while producing H_2 and high value products. Biomol Eng 24:405–413
19. Richter HJ, Knoche K (1983) Reversibility of combustion processes, efficiency and costing- second law analysis of processes. ACS Symp Ser 235:71–85
20. Davidson RM (2007) Post-combustion carbon capture from coal fired plants—solvent scrubbing. IEA Clean Coal Centre, CCC/125
21. Alie C, Backham L, Croiset E, Douglas PL (2005) Simulation of CO_2 capture using MEA scrubbing: a flowsheet decomposition method. Energy Conv Manag 46:475–487
22. Mamun S, Svendsen HF, Haff KA, Juliussen O (2007) Selection of new absorbents for carbon dioxide capture. Energy Conver Manage 48:251–258
23. Darde V, Thomsen K, van Well WJM, Stenby EH (2010) Chilled ammonia process for CO_2 capture. Int J Greenh Gas Control 4:131–136
24. Cullinane JT, Rochelle GT (2005) Thermodynamics of aqueous potassium carbonate, piperazine, and carbon dioxide. Fluid Phase Equilib 227:197–213
25. Freeman SA, Dugas R, Van Wagener DH, Nguyen T, Rochelle GT (2010) Carbon dioxide capture with concentrated, aqueous piperazine. Int J Greenh Gas Control 4:119–124
26. Kovvali AS, Sirkar KK (2002) Carbon dioxide separation with novel solvents as liquid membranes. Ind Eng Chem Res 41:2287–2295
27. Takht Ravanchi M, Kaghazchi T, Kargari A (2009) Application of membrane separation processes in petrochemical industry: a review. Desal 235:199–244
28. Chow JC, Watson JG, Herzog A, Benson SM, Hidy GM, Gunter WD, Penkala SJ, White CM (2003) Critical review discussion-separation and capture of CO_2 from large stationary sources and sequestration in geological formations. J Air Waste Manage Assoc 53:1172–1182
29. White CM, Strazisar BR, Granite EJ, Hoffman JS, Pennline HW (2003) Separation and capture of CO_2 from large stationary sources and sequestration in geological formations—coal beds and deep saline aquifers. J Air Waste Manage Assoc 53:645–715
30. Bounaceur R, Lape N, Roizard D, Vallieres C, Favre E (2006) Membrane processes for post-combustion carbon dioxide capture: a parametric study. Energy 31:2556–2570
31. Scholes CA, Kentish SE, Stevens GW (2009) The effect of condensable minor components on the gas separation performance of polymeric membranes for carbon dioxide capture. Energy Procedia 1:311–317
32. Xomeritakis G, Tsai CY, Jiang YB, Brinker CJ (2009) Tubular ceramic-supported sol-gel silica-based membranes for flue gas carbon dioxide capture and sequestration. J Membr Sci 341:30–36
33. Drage TC, Smith KM, Pevida C, Arenillas A, Snape CE (2009) Development of adsorbent technologies for post-combustion CO_2 capture. Energy Procedia 1:881–884
34. Thiruvenkatachari R, Su S, An H, Yu XX (2009) Post combustion CO_2 capture by carbon fibre monolithic adsorbents. Prog Energ Combust 35:438–455
35. An H, Feng B, Su S (2009) CO_2 capture capacities of activated carbon fibre-phenolic resin composites. Carbon 47:2396–2405
36. Pevida C, Snape CE, Drage TC (2009) Templated polymeric materials as adsorbents for the postcombustion capture of CO_2. Energy Procedia 1:869–874
37. Chatti R, Bansiwal AK, Thote JA, Kumar V, Jadhav P, Lokhande SK, Biniwale RB, Labhsetwar NK, Rayalu SS (2009) Amine loaded zeolites for carbon dioxide capture: amine loading and adsorption studies. Micropor Mesopor Mater 121:84–89
38. Yadav VS, Prasad M, Khan J, Amritphale SS, Singh M, Raju CB (2010) Sequestration of carbon dioxide (CO_2) using red mud. J Hazard Mater 176:1044–1050
39. Maroto-Valer MM, Tang Z, Zhang Y (2005) CO_2 capture by activated and impregnated anthracites. Fuel Process Technol 86:1487–1502
40. Pevida C, Plaza MG, Arias B, Fermoso J, Rubiera F, Pis JJ (2008) Surface modification of activated carbons for CO_2 capture. Appl Surf Sci 254:7165–7172
41. Fauth DJ, Frommell EA, Hoffman JS, Reasbeck RP, Pennline HW (2005) Eutectic salt promoted lithium zirconate: novel high temperature sorbent for CO_2 capture. Fuel Process Technol 86:1503–1521
42. Essaki K, Nakagawa K, Kato M, Uemoto H (2004) CO_2 absorption by lithium silicate at room temperature. J Chem Eng Jpn 37:772–777
43. Kato M, Nakagawa K, Essaki K, Maezawa Y, Takeda S, Kogo R (2005) Novel CO_2 absorbents using lithium-containing oxide. Int J Appl Ceramic Technol 2:467–475
44. Zhao Z, Cui X, Ma J, Li R (2007) Adsorption of carbon dioxide on alkali-modified zeolite 13× adsorbents. Int J Greenh Gas Control 1:355–359
45. Drage TC, Arenillas A, Smith KM, Pevida C, Piippo S, Snape CE (2007) Preparation of carbon dioxide adsorbents from the chemical activation of urea-formaldehyde and melamine-formaldehyde resins. Fuel 86:22–31
46. Singh R, Reddy MKR, Wilson S, Joshi K, Diniz da Costa JC, Webley P (2009) High temperature materials for CO_2 capture. Energy Procedia 1:623–630
47. Sjostrom S, Krutka H (2010) Evaluation of solid sorbents as a retrofit technology for CO_2 capture. Fuel 89:1298–1306
48. Ho MT, Allinson GW, Wiley DE (2008) Reducing the cost of CO_2 capture from flue gases using pressure swing adsorption. Ind Eng Chem Res 47:4883–4890
49. Cornils B, Herrmann WA, Schlogl R, Wong C-H (2003) Catalysis from A to Z: A Concise Encyclopedia. Wiley, Weinheim
50. Hua B, Guild C, Suib SL (2013) Thermal, electrochemical, and photochemical conversion of CO_2 to fuels and value-added products. J CO_2 Util 1:18–27
51. Olajire AA (2013) Valorization of greenhouse carbon dioxide emissions into value-added products by catalytic processes. J CO_2 Util 3–4:74–92
52. Sakakura T, Choi J-C, Yasuda H (2007) Transformation of carbon dioxide. Chem Rev 107:2365–2387
53. Tahriri Zangeneh F, Sahebdelfar S, Takht Ravanchi M (2011) Conversion of carbon dioxide to valuable petrochemicals: an approach to clean development mechanism. J Nat Gas Chem 20:219–231
54. Yu KMK, Curcic I, Gabriel J, Tsang SCE (2008) Recent advances in CO_2 capture and utilization. ChemSusChem 1:893–899
55. Aresta M, Dibenedetto A, Angelini A (2013) The changing paradigm in CO_2 utilization. J CO_2 Util 3–4:65–73
56. Heyn RH (2003) Carbon dioxide conversion. In: Horvath IT (ed) Encyclopedia of catalysis. Wiley, New Jersey, pp 19–71
57. Olah GA, Molnar A (2003) Hydrocarbon chemistry. Wiley, New Jersey

58. Park SW, Joo OS, Jung KD, Kim H, Han SH (2000) ZrO_2/Cr_2O_3 catalyst for reverse-water-gas-shift reaction of CAMERE process. Korean J Chem Eng 17:719–722
59. Joo O-S, Jung K-D (2003) Stability of $ZnAl_2O_4$ catalyst for reverse-water-gas-shift reaction (RWGSR). Bull Korean Chem Soc 24:86–90
60. Guo J, Lou H, Zhao H, Chai D, Zheng X (2004) Dry reforming of methane over nickel catalysts supported on magnesium aluminate spinels. Appl Catal A Gen 273:75–82
61. Montoya JA, Romero-Pascual E, Gimon C, Del Angel P, Monzon A (2000) Methane reforming with CO_2 over Ni/ZrO_2-CeO_2 catalysts prepared by sol-gel. Catal Today 63:71–85
62. Guo JJ, Lou H, Zhu YH, Zheng XM (2003) CO_2 reforming of CH_4 over nickel and cobalt catalysts prepared from La-based perovskite precursors. J Nat Gas Chem 12:17–22
63. Wu YY, Kawaguchi O, Matsuda T (1998) Catalytic reforming of methane with carbon dioxide on $LaBO_3$ (B = Co, Ni, Fe, Cr) catalysts. Bull Chem Soc Jpn 71:563–572
64. Guo JJ, Lou H, Zhu YH, Zheng XM (2003) La-based perovskite preparations, properties, and its catalytic action for CO_2 reforming of CH_4. Mater Lett 57:4450–4456
65. Nam JW, Chae H, Lee SH, Jung H, Lee KY (1998) Natural gas conversion. Stud Surf Sci Catal 119:843–847
66. Rezaei M, Alavi SM, Sahebdelfar S, Yan ZF (2006) Nanocrystalline zirconia as support for nickel catalyst in methane reforming with CO_2. Energy Fuels 20:923–929
67. Rezaei M, Alavi SM, Sahebdelfar S, Liu XM, Qian L, Yan ZF (2007) CO_2-CH_4 reforming over nickel catalysts supported on mesoporous nanocrystalline zirconia with high surface area. Energy Fuels 21:581–589
68. Rezaei M, Alavi SM, Sahebdelfar S, Yan ZF (2008) Effect of CO_2 concentration on the activity and stability of nickel catalyst supported on mesoporous nanocrystalline zirconia. J Nat Gas Chem 17:278–288
69. Rezaei M, Alavi SM, Sahebdelfar S, Yan ZF (2009) A highly stable catalyst in methane reforming with carbon dioxide. Scripta Mater 61:173–176
70. Bartholomew CH (1984) Catalyst deactivation. Chem Eng 91:96–112
71. Michorczyk P, Ogonowski J, Zenczak K (2011) Activity of chromium oxide deposited on different silica supports in the dehydrogenation of propane with CO_2—a comparative study. J Mol Catal A Chem 349:1–12
72. Michorczyk P, Ogonowski J, Kustrowski P, Chmielarz L (2008) Chromium oxide supported on MCM-41 as a highly active and selective catalyst for dehydrogenation of propane with CO_2. Appl Catal A Gen 349:62–69
73. Xu B, Zheng B, Hua W, Yue Y, Gao Z (2006) Support effect in dehydrogenation of propane in the presence of CO_2 over supported gallium oxide catalysts. J Catal 239:470–477
74. Olah GA (2005) Beyond oil and gas: the methanol economy. Angew Chem Int Ed 44:2636–2639
75. Olah GA, Goeppert A, Prakash GKS (2006) Beyond oil and gas: the methanol economy. Wiley-VCH, Weinheim
76. Słoczynski J, Grabowski R, Olszewski P, Kozłowska A, Stoch J, Lachowska M, Skrzypek J (2006) Effect of metal oxide additives on the activity and stability of $Cu/ZnO/ZrO_2$ catalysts in the synthesis of methanol from CO_2 and H_2. App Catal A Gen 310:127–137
77. Saito M, Fujitani T, Takeuchi M, Watanabe T (1996) Development of copper/zinc oxide-based multicomponent catalysts for methanol synthesis from carbon dioxide and hydrogen. Appl Catal A 138:311–318
78. Saito M, Takeuchi M, Fujitani T, Toyir J, Lou S, Wu J, Mabuse H, Ushikoshi K, Mori K, Watanabe T (2000) Advances in joint research between NIRE and RITE for developing a novel technology for methanol synthesis from CO_2 and H_2. Appl Organomet Chem 14:763–772
79. Toyir J, de la Piscina RR, Fierro JLG, Homs N (2001) Highly effective conversion of CO_2 to methanol over supported and promoted copper-based catalyst: influence of support and promoter. App Catal B 29:207–215
80. Bando KK, Arakawa H, Ichikuni N (1999) CO_2 hydrogenation over micro- and mesoporous oxides supported catalysts. Catal Lett 60:125–132
81. Homs N, Toyir J, de la Piscia PR (2013) Catalytic processes for activation of CO_2. In: Suib SL (ed) New and future developments in catalysis: activation of carbon dioxide, 1st edn. Elsevier, Amsterdam, pp 1–23
82. Miao CX, Wang JQ, He LN (2008) Catalytic processes for chemical conversion of carbon dioxide into cyclic carbonates and polycarbonates. Open Organic Chem J 2:68–82
83. Peng JJ, Deng YQ (2001) Cycloaddition of carbon dioxide to propylene oxide by ionic liquids. New J Chem 25:639–641
84. Verykios XE (2003) Catalytic dry reforming of natural gas for production of chemicals and hydrogen. Int J Hydrog Energy 28:1045–1063
85. Teuner STC, Neuman P, Linde FV (2001) The calcor standard and calcor economy processes. Oil Gas Euro Mag 3:44–46
86. Hegarty HES, Ross JRH, O'Conner AM (1998) Syngas production from natural gas using ZrO_2-supported metals. Catal Today 42:225–232
87. Moon DJ, Kang JS, Nho WS, Kim DH, Lee SD, Lee BG (2008) Ni-based catalyst for tri-reforming of methane and its catalytic application for the production of syngas. US 20080260628 A
88. Shulenberger AM, Jonsson FR, Ingolfsson O, Tran KC (2007) US patent application 0244208 A1
89. Trembaly JF (2008) CO_2 as a feedstock, Mitsui will make methanol from greenhouse gas. Chem Eng News 86:13
90. Olah GA, Goeppert Prakash GKS (2009) Chemical recycling of carbon dioxide to methanol and dimethyl ether: from Greenhouse Gas to renewable environmentally carbon neutral and synthetic hydrocarbons. J Org Chem 74:487–498
91. Bahrain: Betting on Industry. http://www.bahiw.com/pdf/ecnomicmarch0109.pdf. Accessed January 2014

Acyclic diene metathesis polymerization and precision polymers

Nicolas F. Sauty · Lucas Caire da Silva ·
Michael D. Schulz · Chip S. Few · Kenneth B. Wagener

Abstract The history of and major advances in the acyclic diene metathesis (ADMET) reaction are described. Because precise branch identity and frequency can be achieved by ADMET polymerizations of symmetrical α,ω-dienes, polyethylenes with precisely spaced alkyl branches of specified length have been prepared. Investigations of their morphologies and thermal properties have provided valuable insight into the behavior of polyethylene. ADMET preparation of ethylene copolymers and telechelic oligomers, as well as the properties of these materials, is also discussed.

Keywords ADMET polymerization · Olefin metathesis · Precision polymers · Polyethylene

Introduction

The development of olefin metathesis has led to remarkable achievements in organic and polymer chemistry. From allowing facile synthesis of complex natural products to producing intricate polymer architectures, olefin metathesis has had an undeniable impact on organic and polymer synthesis. There are two distinct modes of metathesis polymerization: ring-opening metathesis polymerization (ROMP) and acyclic diene metathesis (ADMET) polymerization [1, 2] (Fig. 1). The development of the ADMET reaction is the subject of this review.

N. F. Sauty · L. C. da Silva · M. D. Schulz ·
C. S. Few · K. B. Wagener (✉)
The George and Josephine Butler Polymer Research Laboratory,
Department of Chemistry and Center for Macromolecular
Science and Engineering, University of Florida,
Gainesville, FL 32611-7200, USA
e-mail: wagener@chem.ufl.edu

Historical background

The development of well-defined, efficient olefin metathesis catalysts (Fig. 2) was paramount to the development of the ADMET reaction. Although classical olefin metathesis catalysts were effective in some cases, the exact nature of the catalytically active species was not known. Early work by Grubbs [3–7] and Schrock [8–12] produced some of the first spectroscopically identifiable olefin metathesis catalysts.

The introduction of Schrock's catalysts (e.g., **[Mo]1**) enabled significant progress in olefin metathesis and its use in organic and polymer synthesis. However, these catalysts are sensitive to air, water, and some polar functional groups, and this sensitivity has limited their application in some cases. To address this issue, late transition metal catalysts were developed both by Schrock [13] and Grubbs [14]. Of these, the Grubbs ruthenium-based catalysts have been particularly successful. The first generation of these ruthenium complexes was reported in 1992 and 1993 [15–18]. Although first-generation ruthenium catalysts such as **[Ru]1** have less activity than the Mo-based Schrock catalysts, the Grubbs' complexes are more tolerant to moisture, oxygen, and polar functional groups. A second-generation of ruthenium catalysts, principally **[Ru]2**, resulted from the replacement of a phosphine with an *N*-heterocyclic carbene (NHC) ligand. These catalysts are very reactive in olefin metathesis reactions and possess greater functional group tolerance than the first-generation ruthenium catalysts [19–21]. The activity of some second-generation ruthenium catalysts is comparable to the activity of **[Mo]1** [22]. In 1999, a new class of olefin metathesis catalysts with an isopropoxy group on the benzylidene moiety, **[Ru]3**, was reported by Hoveyda and coworkers [23]. Soon thereafter, the second-generation Hoveyda–Grubbs catalysts were

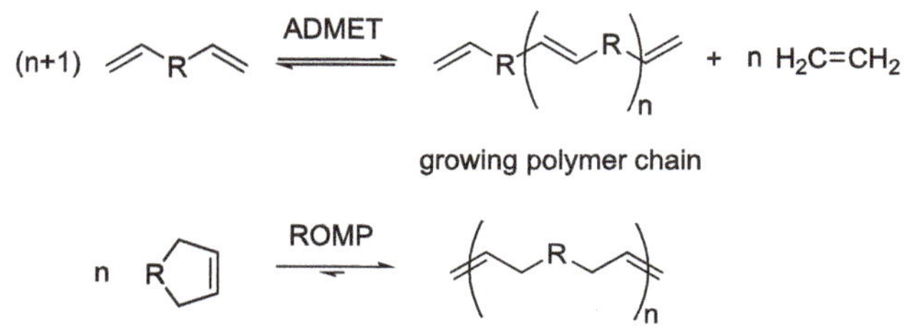

Fig. 1 The ADMET reaction

reported, all incorporating an NHC ligand (**[Ru]4** is one example). These catalysts are very effective for many different olefin metathesis reactions [22, 24].

These and other catalyst developments [25] allowed ADMET to be a useful and feasible reaction. The Wagener group began investigating unconjugated α,ω-diene polycondensation, later known as ADMET polymerization, using the $WCl_6/EtAlCl_2$ catalyst system in the late 1980s [26]. These reactions resulted in a combination of soluble oils and intractable solids. It was determined that the oil was a low molecular weight polymer—the result of olefin metathesis. It was believed that the insoluble product resulted from a vinyl addition reaction, which, in competition with metathesis, would result in a crosslinked polymer, but the solid product was not rigorously characterized. In order to test this hypothesis, Wagener and coworkers reacted $WCl_6/EtAlCl_2$ with styrene. The predicted product of olefin metathesis for this reaction would be stilbene; however, the reaction also yielded polystyrene. This result demonstrated that vinyl addition and olefin metathesis are competing reactions when these catalyst systems are used. To achieve high polymer, the competing vinyl addition reaction must be eliminated. Serendipitously, this work coincided with the report of catalyst **[Mo]1**. The concurrence of these advances enabled ADMET to be applied as a means of synthesizing high molecule weight polymer.

The ADMET reaction

Acyclic diene metathesis polymerization is an equilibrium step-growth condensation of α,ω-dienes promoted by a suitable metathesis catalyst. The reaction produces a growing unsaturated polymer chain that retains the symmetry and functionality of the monomer.

Over the years, monomer design and the development of suitable catalysts have opened the way to a variety of well-defined polymer structures which are not readily achievable by other common techniques. The ADMET polymerization proceeds via a metal-carbene mechanism, in which the metal methylidene is the true catalyst formed during the polymerization cycle (Fig. 3). This is the same metal-carbene mechanism operating in cross-metathesis [27]. Coordination of the metallic center of the catalyst with a terminal olefin from either the monomer or polymer is followed by a reversible 2 + 2 cycloaddition to produce a metallocyclobutane intermediate. This undergoes further reversion leading to either a productive or a non-productive pathway. Chain growth is promoted by the productive metathesis pathway resulting from cycloreversion, which will not restore the coordinating olefin and free catalyst. This pathway is characterized by production of ethylene, which should be removed from the reaction mixture in order to drive the equilibrium towards chain growth. ADMET polymerization affords PDIs around 2, which are typical for step-growth polymerization. Also, molecular weights up to 50,000 g/mol can be achieved, depending primarily on catalyst life time, monomer purity, and reaction conditions. It is fair to say that catalyst compatibility and monomer design/purification are the main challenges to be overcome in order to achieve high polymer with well-defined primary structure.

ADMET polymerization usually requires relatively long reaction times in order to guarantee high conversion and consequently high molecular weight polymers. Because some reactions require days to be completed, catalyst lifetime can be a limiting parameter. Catalyst lifetime depends primarily on the rate of deactivation by catalyst poisoning and can be a major problem if a compatible metathesis catalyst is not available. Grubbs' ruthenium-based and Schrock's molybdenum-based catalysts [28] are among the most used catalysts for ADMET. Grubbs' first-generation catalyst is very common due to its high tolerance towards moisture, oxygen and coordinating functional groups, even though its reactivity is not as high as Schrock's catalyst [29]. The highly reactive Schrock's catalyst

Fig. 2 Five examples of well-defined olefin metathesis catalysts

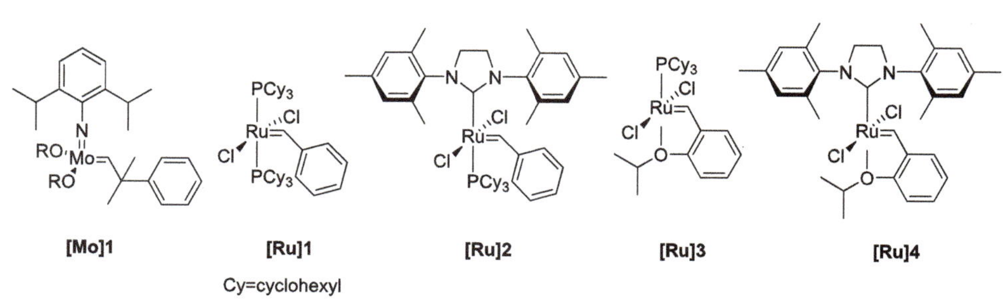

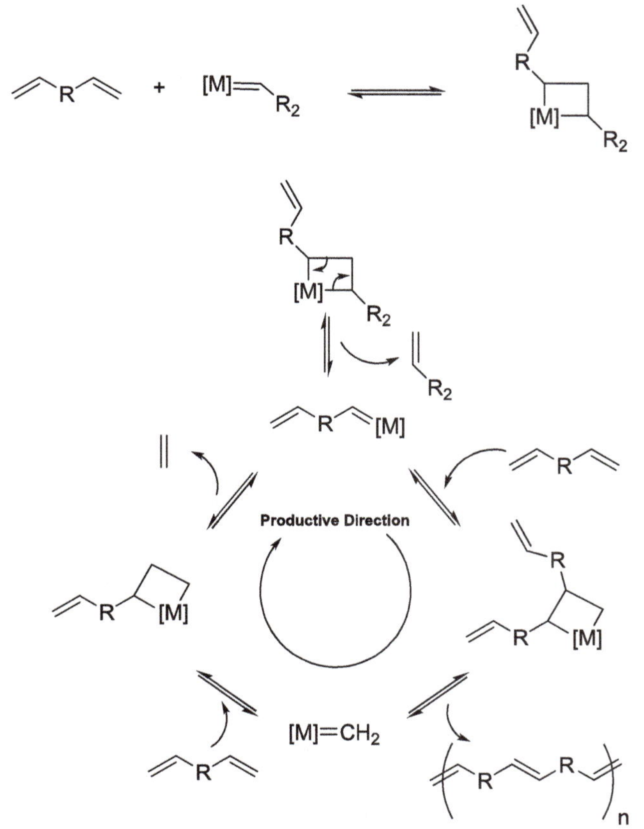

Fig. 3 The ADMET mechanism

is much more sensitive and requires careful handling and extremely dry and pure monomers. The high oxidation state of the molybdenum metal center makes the catalyst highly vulnerable to polar functional groups, water and oxygen. Traces of impurities can deactivate the catalyst rendering very poor conversion/molecular weights.

Even though there are a large number of metathesis catalysts available, many are known to promote isomerization, which ultimately results in loss of control over the polymer primary structure. Grubbs' first generation and Schrock's catalysts are known to promote minimal isomerization under regular ADMET reaction conditions, while studies on Grubbs' second generation, as well as other ruthenium-based metathesis catalysts, shows significant isomerization under similar conditions [30, 31]. Isomerization is a drawback, especially when the choice of a catalyst is limited by monomer functionality. Fortunately, several methods are available to minimize isomerization, usually by use of special additives such as benzoquinones [32, 33] or boron-containing Lewis acids [34].

Virtually any α,ω-diene can be polymerized via ADMET if a suitable catalyst is available that tolerates the desired monomer functionality. The mechanism governing metathesis ensures control over primary structure based on retention of monomer symmetry/asymmetry in the final polymer. Saturation of the internal olefins of the unsaturated ADMET polymer eliminates the *cis*/*trans* distribution, allowing for polymers with well-defined primary structures. Tacticity in ADMET polymers is not yet fully controlled and is being currently addressed by the Wagener group.

ADMET polymerization is ideally carried out in the bulk with monomer-to-catalyst ratios typically in the range of 100–500:1 under high vacuum. The increased viscosity during polymerization is overcome by performing the reaction at temperatures ranging from 40 to 75 °C under constant stirring. Where solubility is not an issue, solvents can be used to maintain low viscosity levels, as long as the catalyst is not deactivated in solution. Plenio and coworkers showed that high-boiling solvents such as 1,2-dichlorobenzene can be used in ADMET polymerization when a slight vacuum is desirable [35]. More recently, Simocko et al. [36] demonstrated that ionic liquids are excellent media to perform ADMET polymerizations. Grubbs' catalyst is usually employed when solvents need to be used, given its higher tolerance towards contaminants and polar functional groups. During the polymerization, removal of ethylene condensate can be achieved by applying high vacuum, or by purging the reaction vessel with an inert gas, in order to drive the reaction to completion. This can be achieved by applying high vacuum or by purging the reaction vessel with an inert gas in order to push the reaction to completion. Reaction progress can be easily followed by comparing the relative ^1H NMR signal intensities of the terminal olefin on the monomer and internal olefin of the resulting unsaturated polymer.

Modeling polyethylene

Polyethylene (**PE**) is the most produced polymer and still dominates the polymer market because of its low cost, versatility, and ease of manufacture [37]. Conventional industrial methods inherently afford polyethylenes with random branches of various lengths along the polymer chain, and this branching greatly affects **PE** properties [38]. Over the years, considerable effort has been devoted to controlling branching and ideally to producing linear polyethylene. In addition, copolymerization of ethylene with polar vinyl monomers often leads to a higher branching content in conjunction with the low incorporation of the polar entity in the polymer backbone. This eventually produces a gradient polymer rather than a statistical polymer.

ADMET polymerization of symmetrical diene monomers is a remarkable tool to afford flawless and impeccably regular polyolefins. Subsequent hydrogenation produces polyethylene copolymers possessing a functionality (either within the polymer backbone or side branching) at

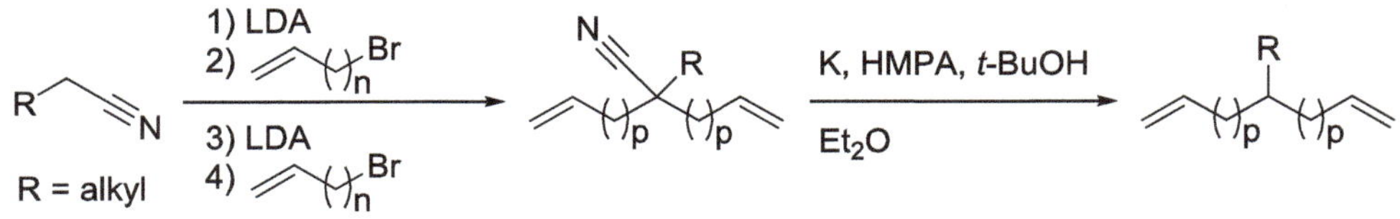

Fig. 4 Monomer synthesis

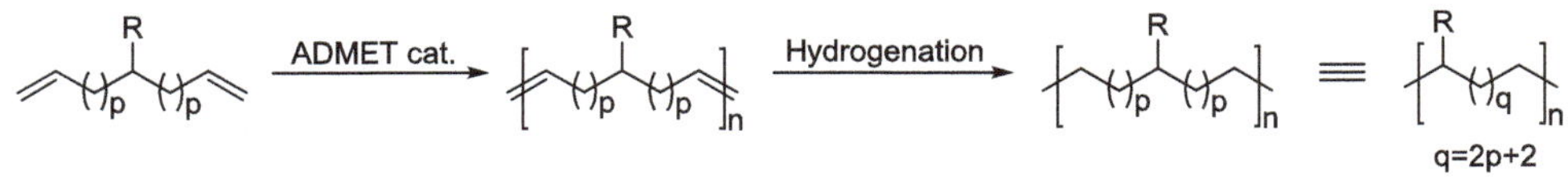

Fig. 5 Synthesis of precision alkyl-branched polyethylene

perfectly unvarying intervals (i.e., separated by a constant number of methylene spacers). Depending on the catalyst, molecular weights (M_n) of ADMET polyethylene copolymers usually range from 5,000 to 50,000 g/mol. These M_ns are sufficient to be accepted as polyethylene models, given that the entanglement molecular weight of linear polyethylene is about 1,000 g/mol [39].

ADMET polymerization offers a unique way to model branched polyethylene, as well as any copolymers of ethylene and α-alkenes (C_nH_{2n}). A two-step route involving a dialkylation of a nitrile intermediate followed a reductive decyanation [40] has been used to prepare a collection of symmetrical alkyl-branched α,ω-dienes (Fig. 4).

Subjecting these monomers to ADMET polymerization and subsequent exhaustive hydrogenation afforded polyethylenes with precise placement (every *n*th carbon) of alkyl branches on their backbones (Fig. 5).

Conversely, statistical placement of alkyl branches can be obtained by copolymerization of the aforementioned symmetrical monomer with a linear α,ω-diene (typically 1,9-decadiene) [41]. Precision alkyl-branched polyethylenes were found to possess enhanced thermal properties due to higher crystallinity, leading to sharper melting transitions at lower temperatures and greater heats of fusion. In precision polymers, two factors determine the thermal properties and morphologic parameters of these materials: the branch identity and the branch frequency. The crystalline structures have been studied using wide and small angle X-ray scattering (WAXS and SAXS), transmission electron microscopy (TEM) and infrared (IR) spectroscopy. Morphology is greatly influenced by the alkyl branch size and spacing.

Regularity in primary structure predictably translates to uniformity in the form of narrow lamellar thickness distributions in polyethylene containing precisely spaced alkyl branches [42]. The lamellar thickness corresponds to the distance between branch sites in these materials, which suggests that the chains are folding at the branch sites, with the branches residing at the interface of the lamellae [43]. As a consequence of the narrow lamellar thickness distribution, ADMET polymers have a narrower melt temperature range compared to other polyethylenes.

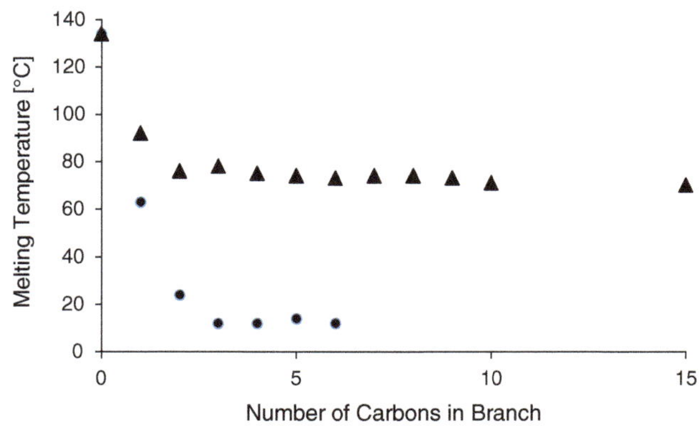

Fig. 6 The effect of alkyl branch size on T_m can be seen for ADMET polyethylenes with branches on every 21st (*filled circle*) and 39th (*filled triangle*) carbon along the polymer backbone

Multifarious branch sizes, ranging from methyl to pentadecyl, have been incorporated in a precise fashion on the polyethylene backbone. It was discovered that the resulting fusion temperatures were highly dependent on the nature of the branching (Fig. 6). As a general rule, the melting temperature dwindles with longer alkyl branches. The effect of alkyl branch size on the melting temperature can be seen for ADMET polyethylenes with branches on every 21st [44] and 39th [45] carbon. Polymers with branches excluded from the lamellae exhibit similar fusion temperatures. This expulsion of alkyl chains from the lamellae is affected by defect size, as ADMET polymers bearing smaller alkyl groups (e.g., methyl) crystallize with a greater degree of branch inclusion. This is true of precision polymers of varying branch spacing, and the maximum included branch size decreases with increasing branch interval. Indeed, when the branch is placed on every 21st carbon, methyl and ethyl branches are included in the unit cell and afford polymers displaying melting transitions at 63 and 24 °C, respectively. Conversely, polymers with branches larger than the maximum included branch size exhibit similar melting temperatures. In other words, the transition becomes independent of the branch identity with T_ms of approximately 12 °C. Similarly, when an alkyl defect is placed every 39th carbon, only a methyl branch can be included in the unit cell, and as a consequence,

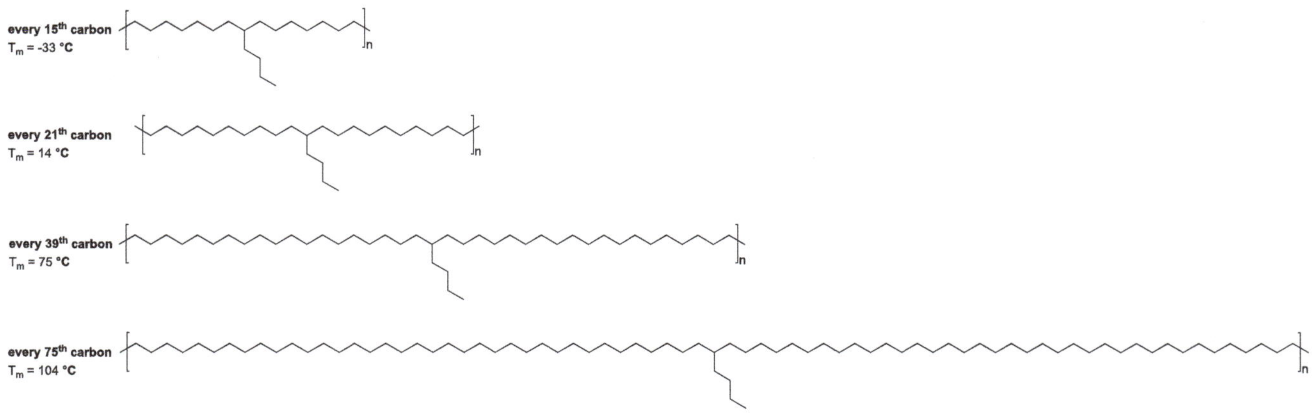

Fig. 7 Branch frequency: alkyl branch every *n*th carbon

similar melting temperatures are found for the polymers bearing longer branches.

Various branch frequencies have been investigated with the goal of diminishing the frequency (increasing the spacing) as much as possible. The general observation is that for a given branch identity, thermal transitions occur at higher temperatures with diminished frequencies. The Wagener group has synthesized butyl-branched polyethylenes in which the branch appears every 9th, 21st, 39th, and 75th [46] carbon (Fig. 7).

This has led to the discovery of a linear relationship between the branch frequency and fusion temperature of the aforementioned polymers. It also suggested a melting temperature of 136 °C for an infinitely low frequency which is quite close to that of linear ADMET polyethylene (134 °C).

The distance between alkyl branches also has an effect on the crystalline lattice, as less frequent branching leads to more ordered structure. The extreme case, with no branching, leads to orthorhombic structure. The introduction of branches leads to less-ordered crystalline structures, of which there are several examples in the literature [47, 48]. When precisely spaced alkyl groups are placed every 7th carbon or closer, the resulting polymers are amorphous.

The morphological effects described in the preceding paragraphs can be further understood by considering the crystallization behavior of representative polymers. As is the case in polyethylenes synthesized by other means, ADMET polyethylenes crystallize through the formation of a hexagonal mesophase, which expands as the chains slide past one another in the all *trans* conformation until the equilibrium lamellar thickness is reached [49]. While crystallization of longer methylene sequences occurs before shorter sequences in metallocene PE, forcing shorter methylene sequences into the amorphous phase, all methylene sequences are of the same length in ADMET polyethylene, allowing crystallites of equilibrium lamellar thickness to form from the transient hexagonal mesophase [50].

By varying the size and interval of alkyl branches on polyethylene, the resulting morphological effects are delineated. Uniform ethylene sequences allow a metastable mesophase to form, at which point the chains can reposition themselves into an energetically stable morphology, resulting in a material with narrow lamellar thickness distribution. Depending on defect size, the branches may be contained within the lamellar stem in the case of smaller branches, or excluded to the surface of the crystallite in the case of longer branches. The lamellar thickness and crystalline lattice type depend on the distance between defects, with longer intervals leading to more ordered polymorphisms.

Beyond the thermal and morphologic elucidation of precision alkyl-branched polyethylene, molecular dynamics were probed utilizing deuterium quadrupolar echo NMR in the solid state. The ^{2}H-NMR spectra of precision CD_3-branched polyethylene were acquired for a wide range of temperatures [51, 52]. A T_1-fitting methodology was applied to deconvolute the spectral contributions of the amorphous and crystalline phases, allowing isolation of two distinctive dynamic behaviors. This is the consequence of a bi-exponential behavior of the spin–lattice relaxation in the CD_3 branches. The amorphous phase exhibited a similar behavior to that of linear polyethylene [53]. Conversely, the crystalline domains displayed a harmonic librational axial rotation reorientation model with a mean jump angle up to 40° when approaching the melting transition. Various other deuterated precision polymers are being analyzed by deuterium quadrupolar echo methods with the intention of elucidating the dynamics of precision polymers.

Ethylene copolymers

The advent of ADMET polymerization not only promoted modeling of polyethylene, but also stimulated the understanding of numerous classes of polymers. Various

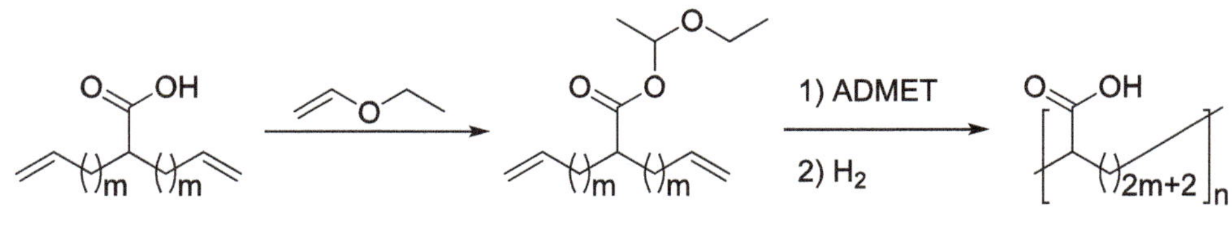

Fig. 8 Synthesis of precision poly(ethylene-*co*-acrylic acid)

copolymers of ethylene and polar vinyl intermediates were scrutinized and compared to commercial samples. All syntheses relied on the ADMET polymerization of symmetrical α,ω-diene monomers followed by exhaustive hydrogenation to afford the desired polyethylene precision copolymers. Relevant examples are briefly described in this section.

Ethylene–vinyl acetate copolymers (EVA) are widespread materials, with hot-melt adhesives and plastic wraps being common applications [54]. Although free-radical copolymerizations usually produce gradient or block polymers, perfectly random copolymers are obtained from the polymerization of ethylene and vinyl acetate due to similar reactivities [55]. However, undesired and random branching is inherently promoted via a back-biting process, significantly altering the characteristics and performance of the copolymers. However, perfectly linear ethylene–vinyl acetate copolymers can be prepared by means of ADMET polymerization [56]. A number of monomers with various methylene run lengths ranging from 8 to 12 were subjected to polymerization and subsequent hydrogenation, furnishing copolymers with different vinyl acetate contents (acetate branches separated by 18–26 carbons). Commercial samples with similar vinyl acetate ratios were compared with the ADMET copolymers. Similar to the commercial copolymers, the precision polymers displayed a linear relationship between the melting temperature and vinyl acetate content, but with dissimilar slopes. A larger slope, underlining the precision factor, evidenced the greater order in precision polymers. With a similar approach, copolymers of ethylene and vinyl alcohol were made without difficulty.

While metathesis catalysts such as **[Ru]1** can tolerate a variety of functional groups, functionalities such as primary or secondary amines, sulfides, or carboxylic acids deactivate the catalyst, thereby precluding preparation of high molecular weight materials (oligomers to about 3,000 g/mol were produced). In the case of carboxylic acids, this problem can be circumvented by protecting the acid functionality. Various agents can be employed, but ethyl vinyl ether was found to be extremely convenient [57]. The resulting 1-ethoxyethyl ester, inert to ruthenium catalysts, is quantitatively cleaved during the hydrogenation of the resulting polyolefin (Fig. 8).

The ethylene–acrylic acid copolymers were found to be amorphous when only 8 or 14 methylene units separated two consecutive branching sites. Conversely, increasing the spacing to 20 methylenes resulted in a semicrystalline material displaying dimerized acid moieties in an orthorhombic lattice. Simple treatment of the copolymer with zinc reagents, produced a zinc-neutralized material containing 22 % free acid and 66 % zinc dicarboxylates. The zinc dicarboxylate components aggregated into a self-assembled structures adopting a cubic crystalline lattice, an unprecedented observation for ionomers having a purely hydrocarbon backbone [58].

Precision ethylene–vinyl halide copolymers have also been examined [59]. Contrary to many other systems, it was discovered that regardless of the halide identity (except iodide) and appearance frequency on the polymer backbone, the halogen atoms were included in the unit cell. This yielded an orthorhombic system in the case of fluorine, and triclinic systems in the case of the larger chlorine and bromine substituents.

Motivated by the recent advances in fuel cell membranes, precision polymers incorporating sulfonic [60] and phosphonic [61, 62] acids have been investigated for potential uses as proton carriers. As with carboxylic acids, the aforementioned functionalities must be protected prior to polymerization to prevent poisoning of the metathesis catalyst. While ethyl phosphonates could straightforwardly be deprotected post-hydrogenation, the sulfonate analogues have not led to comparable results. The significant insolubility of the poly(ethylene-*co*-vinylsulfonic acid)s promotes premature precipitation of the material and prevents complete deprotection. This problem is currently being addressed by the Wagener group. With sufficient spacing (20 methylene spacers or greater), the ethylene–vinylphosphonic acid copolymers show a semicrystalline behavior. Amorphous materials are promoted by shorter methylene segments.

Finally, among others, ADMET polymerization has been utilized to produce precision materials containing ether [63], silicon [64], or amine [65] moieties.

Telechelic oligomers via ADMET

Many telechelic oligomers (oligomers with functionalized end groups) have been synthesized by ADMET. These materials are often macromonomers used to synthesize larger polymers, and they have many other uses as well. Telechelic oligomers of polybutadiene (Fig. 9) have been prepared via the reaction of 1,5-hexadiene and a mono- or

Fig. 9 Synthesis of telechelics via ADMET

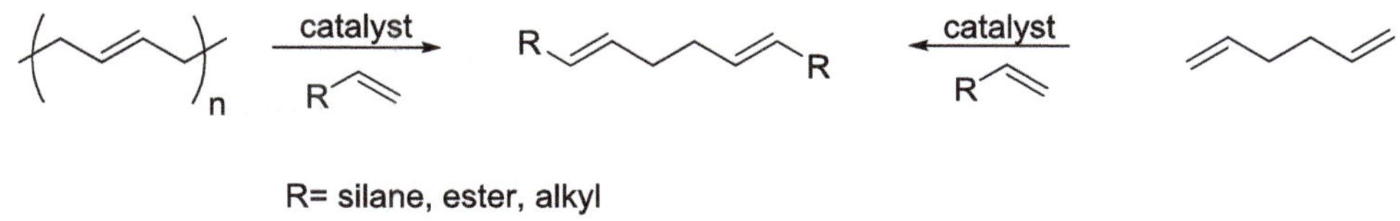

di-functional olefin [66, 67]. ADMET polybutadiene may alternatively be achieved by the polymerization of 1,5-hexadiene. This polymer can then be used to prepare telechelic oligomers by reacting with a suitable olefin. Sometimes the reaction is incomplete, producing telechelic oligomers. Additionally, polybutadiene produced by cationic polymerization can undergo ADMET depolymerization with acrylates to yield ester-terminated telechelics [68]. This approach was also used to produce a polyamide by depolymerizing polybutadiene with acryloyl chloride and subsequently reacting with 1,6-diaminohexane. Ester-end-capped polyethylene oligomers have been made by ADMET by reacting 1,9-decadiene and 9-decenyl acetate and subsequently hydrogenating the unsaturated units. Telechelic polyacetylenes have been synthesized by ADMET polymerization of 2,4-hexadiene with 1,2-disubstituted olefins [69]. Polyacetylene oligomers terminated with a variety of end groups, including silanes, alkyl groups, and esters were prepared with this method.

Hydrophobic, amorphous telechelic diols have also been synthesized by ADMET [70]. These materials could potentially be used in UV- and hydrolysis-resistant polyurethanes. The hydrocarbon backbone was produced by ADMET polymerization followed by hydrogenation of a gem-dimethyl substituted α,ω-diene, and this material was then endcapped with alcohol-containing monoenes with varied numbers of methylenes between the alcohol group and olefin. Some molecular weight control was achieved in the resulting telechelics, which had 2.0 functionality.

In another report, 1,9-decadiene was polymerized using **[Ru]1** with epoxy-containing monoenes [71].The resulting telechelic polymer was then reacted with toluene diisocyanate resulting in a copolymer containing oxazolidone linkages. ADMET has been used to synthesize ester-terminated telechelics with a similar approach [72].

Conclusion

Over the last two decades, ADMET polymerization has been extensively studied as a unique and valuable means to access and synthesize precision polymers. These materials have provided better understanding of structure–property relationships in various systems, as they serve as models for their ill-defined commercial counterparts. While significant advances have resulted from precision polymers, many more systems have yet to be explored [73].

References

1. Wagener KB, Boncella JM, Nel JG (1991) Acyclic diene metathesis (ADMET) polymerization. Macromolecules 24:2649–2657
2. Mutlu H, de Espinosa LM, Meier MAR (2011) Acyclic diene metathesis: a versatile tool for the construction of defined polymer architectures. Chem Soc Rev 40:1404–1445
3. Howard TR, Lee JB, Grubbs RH (1980) Titanium metallacarbene-metallacyclobutane reactions: stepwise metathesis. J Am Chem Soc 102:6876–6878
4. Lee JB, Gajda GJ, Schaefer WP, Howard TR, Ikariya T, Straus DA, Grubbs RH (1981) Structures of titanacyclobutanes. J Am Chem Soc 103:7358–7361
5. Lee JB, Ott KC, Grubbs RH (1982) Kinetics and stereochemistry of the titanacyclobutane–titanamethylene interconversion. Investigation of a degenerate olefin metathesis reaction. J Am Chem Soc 104:7491–7496
6. Gilliom LR, Grubbs RH (1986) A titanacyclobutane precursor to alkyl-substituted titanium-carbene complexes. Organometallics 5:721–724
7. Gilliom LR, Grubbs RH (1986) Titanacyclobutanes derived from strained cyclic olefins: the living polymerization of norbornene. J Am Chem Soc 108:733–742
8. Schaverien CJ, Dewan JC, Schrock RR (1986) A well-characterized, highly active, lewis acid free olefin metathesis catalyst. J Am Chem Soc 108:2771–2773
9. Schrock RR, DePue RT, Feldman J, Schaverien CJ, Dewan JC, Liu AH (1988) Preparation and reactivity of several alkylidene complexes of the type $W(CHR')(N\text{-}2,6\text{-}C_6H_3\text{-}i\text{-}Pr_2)(OR)_2$ and related tungstacyclobutane complexes. Controlling metathesis activity through the choice of alkoxide ligand. J Am Chem Soc 110:1423–1435
10. Schrock RR, DePue RT, Feldman J, Yap KB, Yang DC, Davis WM, Park L, DiMare M, Schofield M, Anhaus J, Walborsky E, Evitt E, Krüger C, Betz P (1990) Further studies of imido alkylidene complexes of tungsten, well-characterized olefin metathesis catalysts with controllable activity. Organometallics 9:2262–2275
11. Murdzek JS, Schrock RR (1987) Well-characterized olefin metathesis catalysts that contain molybdenum. Organometallics 6:1373–1374
12. Schrock RR, Murdzek JS, Bazan GC, Robbins J, DiMare M, O'Regan M (1990) Synthesis of molybdenum imido alkylidene complexes and some reactions involving acyclic olefins. J Am Chem Soc 112:3875–3886
13. Toreki R, Schrock RR (1990) A well-defined rhenium(VII) olefin metathesis catalyst. J Am Chem Soc 112:2448–2449

14. Trnka TM, Grubbs RH (2001) The development of $L_2X_2 = CHR$ olefin metathesis catalysts: an organometallic success story. Acc Chem Res 34:8–29
15. Nguyen ST, Johnson LK, Grubbs RH, Ziller JW (1992) Ring-opening metathesis polymerization (ROMP) of norbornene by a group VIII carbene complex in protic media. J Am Chem Soc 114:3974–3975
16. Nguyen ST, Grubbs RH, Ziller JW (1993) Synthesis and activities of new single-component ruthenium-based olefin metathesis catalysts. J Am Chem Soc 115:9858–9859
17. Schwab P, Grubbs RH, Ziller JW (1993) Synthesis and applications of $RuCl_2(=CHR')(PR_3)_2$. J Am Chem Soc 118:100–110
18. Wu Z, Nguyen ST, Grubbs RH, Ziller JW (1995) Reactions of ruthenium carbenes of the type $(PPh_3)_2(X)_2Ru=CH–CH=CPh_2$ ($X = Cl$ and CF_3COO) with strained acyclic olefins and functionalized olefins. J Am Chem Soc 117:5503–5511
19. Huang J, Stevens ED, Nolan SP, Petersen JL (1999) Olefin metathesis-active ruthenium complexes bearing a nucleophilic carbene ligand. J Am Chem Soc 121:2674–2678
20. Weskamp T, Schattenmann WC, Spiegler M, Herrmann WA (1998) A novel class of ruthenium catalysts for olefin metathesis. Angew Chem Int Ed 37:2490–2493
21. Scholl M, Trnka TM, Morgan JP, Grubbs RH (1999) Increased ring closing metathesis activity of ruthenium-based olefin metathesis catalysts coordinated with imidazolin-2-ylidene ligands. Tetrahedron Lett 40:2247–2250
22. Garber SB, Kingsbury JS, Gray BL, Hoveyda AH (2000) Efficient and recyclable monomeric and dendritic Ru-based metathesis catalysts. J Am Chem Soc 122:8168–8179
23. Kingsbury JS, Harrity JPA, Bonitatebus PJ Jr, Hoveyda AH (1999) A recyclable Ru-based metathesis catalyst. J Am Chem Soc 121:791–799
24. Gessler S, Randl S, Blechert S (2000) Synthesis and metathesis reactions of a phosphine-free dihydroimidazole carbene ruthenium complex. Tetrahedron Lett 41:9973–9976
25. Vougioukalakis GC, Grubbs RH (2010) Ruthenium-based heterocyclic carbene-coordinated olefin metathesis catalysts. Chem Rev 110:1746–1787
26. Lindmark-Hamburg M, Wagener KB (1987) Acyclic metathesis polymerization. The olefin metathesis reaction of 1,5-hexadiene and 1,9-decadiene. Macromolecules 20:2949–2951
27. Opper K, Wagener KB (2011) ADMET: metathesis polycondensation. J Polym Sci Part A Polym Chem 49:821–831
28. Oskam JH, Fox HH, Yap KB, McConville DH, O'Dell R, Lichtenstein BJ, Schrock RR (1993) Ligand variation in alkylidene complexes of the type $Mo(CHR)(NR')(OR'')_2$. J Organomet Chem 459:185–198
29. Wagener KB, Brzezinska K, Anderson JD, Younkin TR, Steppe K, DeBoer W (1997) Kinetics of acyclic diene metathesis (ADMET) polymerization. Influence of the negative neighboring group effect. Macromolecules 30:7363–7369
30. Lehman SE Jr, Schwendeman JE, O'Donnell PM, Wagener KB (2003) Olefin isomerization promoted by olefin metathesis catalysts. Inorg Chim Acta 345:190–198
31. Courchay FC, Sworen JC, Wagener KB (2003) Metathesis activity and stability of new generation ruthenium polymerization catalysts. Macromolecules 36:8231–8239
32. Fokou PA, Meier MAR (2010) Studying and suppressing olefin isomerization side reactions during ADMET polymerizations. Macromol Rapid Commun 31:368–373
33. Hong SH, Sanders DP, Lee CW, Grubbs RH (2005) Prevention of undesirable isomerization during olefin metathesis. J Am Chem Soc 127:17160–17161
34. Simocko C, Wagener KB (2013) Effects of boron-containing lewis acids on olefin metathesis. Organometallics 32:2513–2516
35. Weychardt H, Plenio H (2008) Acyclic diene metathesis polymerization of divinylarenes and divinylferrocenes with Grubbs-type olefin metathesis catalysts. Organometallics 27:1479–1485
36. Simocko C, Yang Y, Swager TM, Wagener KB (2013) Metathesis step-growth polymerization in ionic liquids. ACS Macro Lett 2:1061–1064
37. Knuuttila H, Lehtinen A, Nummila-Pakarinen A (2004) Advanced polyethylene technologies—controlled material properties, long-term properties of polyolefins. Adv Polym Sci 169:13–27
38. Fawcet EW, Gibson RQ, Perrin MH, Patton JG, Williams EG (1937) Imperial Chemical Industries Ltd. Pat 2, pp 816–883
39. Baughman TW, Wagener KB (2005) Recent advances in ADMET polymerization. In: Buchmeiser M (ed). Springer, Berlin, pp 1–42
40. Rojas G, Wagener KB (2008) Avoiding olefin isomerization during decyanation of alkylcyano α, ω-dienes: a deuterium labeling and structural study of mechanism. J Org Chem 73:4962–4970
41. Rojas G, Wagener KB (2009) Precisely and irregularly sequenced ethylene/1-hexene copolymers: a synthesis and thermal study. Macromolecules 42:1934–1947
42. Hosoda S, Nozue Y, Kawashima Y, Suita K, Seno S, Nagamatsu T, Wagener KB, Inci B, Zuluaga F, Rojas G, Leonard JK (2011) Effect of the sequence length distribution on the lamellar crystal thickness and thickness distribution of polyethylene: perfectly equisequential ADMET polyethylene vs ethylene/α-olefin copolymer. Macromolecules 44:313–319
43. Hosoda S, Nozue Y, Kawashima Y, Utsumi S, Nagamatsu T, Wagener K, Berda E, Rojas G, Baughman T, Leonard J (2009) Perfectly controlled lamella thickness and thickness distribution: a morphological study on ADMET polyolefins. Macromol Symp 282:50–64
44. Rojas G, Inci B, Wei Y, Wagener KB (2009) Precision polyethylene: changes in morphology as a function of alkyl branch size. J Am Chem Soc 131:17376–17386
45. Inci B, Lieberwirth I, Steffen W, Mezger M, Graf R, Landfester K, Wagener KB (2012) Decreasing the alkyl branch frequency in precision polyethylene: effect of alkyl branch size on nanoscale morphology. Macromolecules 45:3367–3376
46. Inci B, Wagener KB (2011) Decreasing the alkyl branch frequency in precision polyethylene: pushing the limits toward longer run lengths. J Am Chem Soc 133:11872–11875
47. Qiu W, Sworen J, Pyda M, Nowak-Pyda E, Habenschuss A, Wagener KB, Wunderlich B (2006) Effect of the precise branching of polyethylene at each 21st CH_2 group on its phase transitions, crystal structure, and morphology. Macromolecules 39:204–217
48. Lieser G, Wegner G, Smith JA, Wagener KB (2004) Morphology and packing behavior of model ethylene/propylene copolymers with precise methyl branch placement. Colloid Polym Sci 282:773–781
49. Matsui K, Seno S, Nozue Y, Shinohara Y, Amemiya Y, Berda EB, Rojas G, Wagener KB (2013) Influence of branch incorporation into the lamella crystal on the crystallization behavior of polyethylene with precisely spaced branches. Macromolecules 46:4438–4446
50. Nozue Y, Kawashima Y, Seno S, Nagamatsu T, Hosoda S, Berda EB, Rojas G, Baughman TW, Wagener KB (2011) Unusual crystallization behavior of polyethylene having precisely spaced branches. Macromolecules 44:4030–4034
51. Bowers CR, Wei Y, Aitken BS, Reeg CR, Akel CD (2012) Molecular dynamics in precision deuteriomethyl branched polyethylene from solid-state deuterium NMR. Polymer 53:2633–2642
52. Wei Y, Graf R, Sworen JC, Cheng C-Y, Bowers CR, Wagener KB, Spiess HW (2009) Local and collective motions in precise

polyolefins with alkyl branches: a combination of ^{2}H and ^{13}C solid-state NMR spectroscopy. Angew Chem Int Ed 48:4617–4620
53. Spiess WH (1983) Molecular dynamics of solid polymers as revealed by deuteron NMR. Colloid Polym Sci 261:193–209
54. Zutty NL, Faucher JA, Bonotto S, Mark HF (eds) (1967) Encyclopedia of polymers science and technology. Wiley, New York
55. Odian G (1991) Principles of polymerization, 3rd edn. Wiley, New York, p 460
56. Watson MD, Wagener KB (2000) Ethylene/vinyl acetate copolymers via acyclic diene metathesis polymerization. Examining the effect of "long" precise ethylene run lengths. Macromolecules 33:5411–5417
57. Baughman TW, Chan CD, Winey KI, Wagener KB (2007) Synthesis and morphology of well-defined poly(ethylene-co-acrylic acid) copolymers. Macromolecules 40:6564–6571
58. Seitz ME, Chan CD, Opper KL et al (2010) Nanoscale morphology in precisely sequenced poly(ethylene-co-acrylic acid) zinc ionomers. J Am Chem Soc 132:8165–8174
59. Boz E, Ghiviriga I, Nemeth AJ, Jeon K, Alamo RG, Wagener KB (2008) Random, defect-free ethylene/vinyl halide model copolymers via condensation polymerization. Macromolecules 41:25–30
60. Opper KL, Wagener KB (2009) Precision sulfonic acid ester copolymers. Macromol Rapid Commun 30:915–919
61. Opper KL, Markova D, Klapper M, Müllen K, Wagener KB (2010) Precision phosphonic acid functionalized polyolefin architectures. Macromolecules 43:3690–3698
62. Markova D, Opper KL, Wagner M, Klapper M, Wagener KB, Müllen K (2013) Synthesis of proton conducting phosphonic acid-functionalized polyolefins by the combination of ATRP and ADMET. Polym Chem 4:1351–1363
63. Sydlik SA, Delgado PA, Inomata S et al (2013) Triptycene-containing polyetherolefins via acyclic diene metathesis polymerization. J Polym Sci Part A Polym Chem 51:1695–1706
64. Matloka PP, Wagener KB (2006) The acyclic diene metathesis (ADMET) polymerization approach to silicon containing materials. J Mol Catal A Chem 257:89–98
65. Leonard JK, Wei Y, Wagener KB (2012) Synthesis and thermal characterization of precision poly(ethylene-co-vinyl amine) copolymers. Macromolecules 45:671–680
66. Marmo JC, Wagener KB (1993) Acyclic diene metathesis (ADMET) depolymerization. Synthesis of mass-exact telechelic polybutadiene oligomers. Macromolecules 26:2137–2138
67. Marmo JC, Wagener KB (1995) ADMET depolymerization. Synthesis of perfectly difunctional ($f = 2.0$) telechelic polybutadiene oligomers. Macromolecules 28:2602–2606
68. Schulz MD, Ford RR, Wagener KB (2013) Insertion metathesis depolymerization. Polym Chem 4:3656–3658
69. Tao D, Wagener KB (1993) Acyclic diene metathesis (ADMET) polymerization. Synthesis of telechelic polyacetylene. Polym Prepr 34:469–470
70. Schwendeman JE, Wagener KB (2009) Synthesis of amorphous hydrophobic telechelic hydrocarbon diols via ADMET polymerization. Macromol Chem Phys 210:1818–1833
71. Tamura H, Nakayama A (2002) Functional telechelic polymer synthesis via ADMET polymerization. J Macromol Sci Pure Appl Chem A 39:745–758
72. Tamura H, Maeda N, Matsumoto R, Nakayama A, Hayashi H, Ikushima K, Kuraya M (1999) Synthesis of ester terminated telechelic polymer via ADMET polymerization. J Macromol Sci Pure Appl Chem A 36:1153–1170
73. Atallah P, Wagener KB, Schulz MD (2013) ADMET: the future revealed. Macromolecules 46:4735–4741

Microwave-assisted conversion of ethane to ethylene

Siauw Ng · Craig Fairbridge · Sateesh Mutyala · Yan Liu · Jacqueline M. R. Bélanger · J. R. Jocelyn Paré

Abstract A novel microwave reactor has been constructed for the conversion of ethane to ethylene. The bench-top, pilot-scale facility is capable of conventional furnace heating and microwave-assisted heating, thus allowing direct comparative studies between the conventional industry-wide process and the novel microwave-assisted conversion process. The latter makes use of chemical substances called susceptors that are good microwave absorbers and can transfer heat energy to the surrounding environment. The ethane feed stream is thus heated to achieve thermal conversion to ethylene. Data validating the apparatus and preliminary data on the feasibility of conversion are presented.

Keywords Ethylene production · Microwave-assisted petrochemistry · Ethane-to-ethylene conversion · Microwave susceptors

Introduction

As a continuation of our work in further developing technologies for novel microwave-assisted processes (MAP^{TM}) [1–13], we have undertaken a research program aimed at assessing the feasibility of using microwaves as a processing tool in the industrial sector. More specifically, we have an interest in assessing and validating applications for the petrochemical and petroleum sectors with the aim to reduce energy intensity as well as greenhouse gas emissions.

Ethylene production is an important industrial activity in Canada. For example, NOVA Chemicals alone—Canada's largest producer of ethylene—has a publicly reported production capacity of over 2.5 billion kg/year. The process used currently is relatively simple [14, 15]: ethane is fed into long tubular reactors at temperatures of roughly 850–900 °C and pressures slightly over atmospheric. Once the materials leave the reactor, they must be cooled to 450 °C within 50 ms to prevent further polymerization of the ethylene or its degradation into side products. Currently, no catalyst is used in the Canadian plant as Canada still has access to significant quantities of pure ethane from its natural gas production. This situation is likely to change in the near future as excess ethane boil-off is rapidly depleting, thus raising the likelihood of the need to implement chemical catalytic processes.

Experimental setup

The experimental apparatus consisted of three main sections: supply, reaction, and sampling. A simplified schematic of the overall experimental apparatus is shown in Fig. 1. The following provides a general description for each section.

Supply section

The supply section consisted of three feed lines for hydrocarbon feedstock: ethane (in this study), nitrogen, and steam. Ethane (Praxair, 99.9 % pure) was fed to the reaction section from a regulated supply cylinder using a mass

S. Ng · C. Fairbridge · S. Mutyala · Y. Liu
Natural Resources Canada, CanmetENERGY, One Oil Patch Drive, Devon, AB T9G 1A8, Canada

J. M. R. Bélanger · J. R. J. Paré (✉)
Your World, 94 Newton Street, Moncton, NB E1E 3A4, Canada
e-mail: jrjpare@microwaveworldwide.com

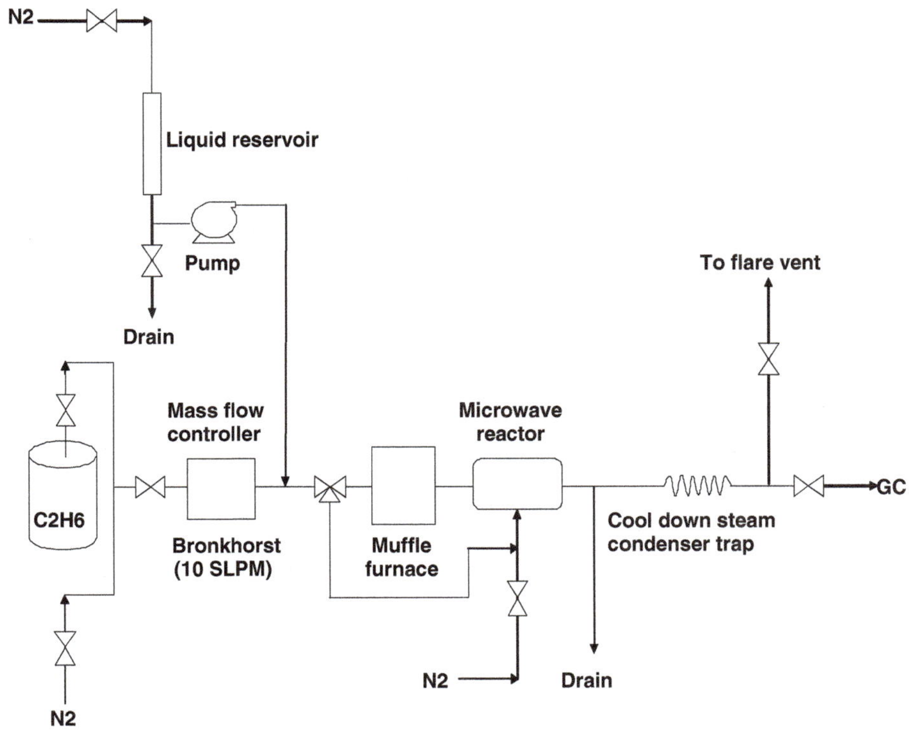

Fig. 1 Simplified schematic diagram of the overall ethylene conversion bench-top pilot plant

flow controller (Bronkhorst). The flow rate of ethane was set by the controller with an accuracy of ±0.1 standard liters per minute (SLPM). The flow rate was the volumetric flow rate corrected to a temperature of 20 °C and pressure of 101.3 kPa. When the ethane was fed to the reactor, a pressure relief valve (not shown in Fig. 1) was used to equalize the pressures upstream and downstream from the flow-regulating valve to prevent backpressure on the feedstock lines and downstream pressure fluctuations.

Reaction section

The bench-top pilot facility is capable of conventional furnace heating and microwave-assisted heating. The furnace heater section can also be used to preheat the feedstock prior to entry into the microwave reactor section if desired. The system can accommodate tubular reactors of varying diameters. Figure 2 shows the overall layout of the apparatus. The path taken by the feedstock, namely directly into the microwave reactor or via the furnace section, is controlled by a switchable multipath flow valve.

Microwave reactor system

The system has been described previously in detail from the design and modeling [16] as well as from the construction and assembly standpoints [17]. The main components of the experimental arrangement in this study are a microwave generator and a cavity (reactor system) connected via appropriate waveguides. The microwave system includes adequate protection systems to prevent any reflected power from damaging the magnetron head.

The microwave generator is a 2,450-MHz, 3-kW GMP 30 KSM (Sairem) that can be operated at nominal power ratings ranging between 300 and 3,000 W. The magnetron of the generator is protected by a high-flow water cooling system set at 4 °C. A rectangular waveguide with a standard size of WR340 is located at the power head. It is connected to a WR430 waveguide via a waveguide transition. The latter waveguide serves as a wall to the cavity and is slotted on the side facing the cavity, allowing long uniform emission of microwaves into the cavity. Five slots centered at the middle of the wall are cut, with alternate angles at 145°, 35°, 145°, 35°, and 145°. The rectangular reactor cavity has overall dimensions of 40 × 40 × 30 cm.

A three-stub impedance-matching tuner is fitted between the emission head and the slotted waveguide. The generator is equipped with emitted and reflected power measurement devices in real time. The cavity was designed for multipurpose applications. It can be operated with a single waveguide. Alternatively, it can also be operated by attaching a second slotted waveguide on the opposite face of the cavity. That second waveguide has slots cut differently so as to avoid the potential for obtaining high-field points that could lead to undesirable arcing phenomena [16]. The reactor was designed to be housed in the

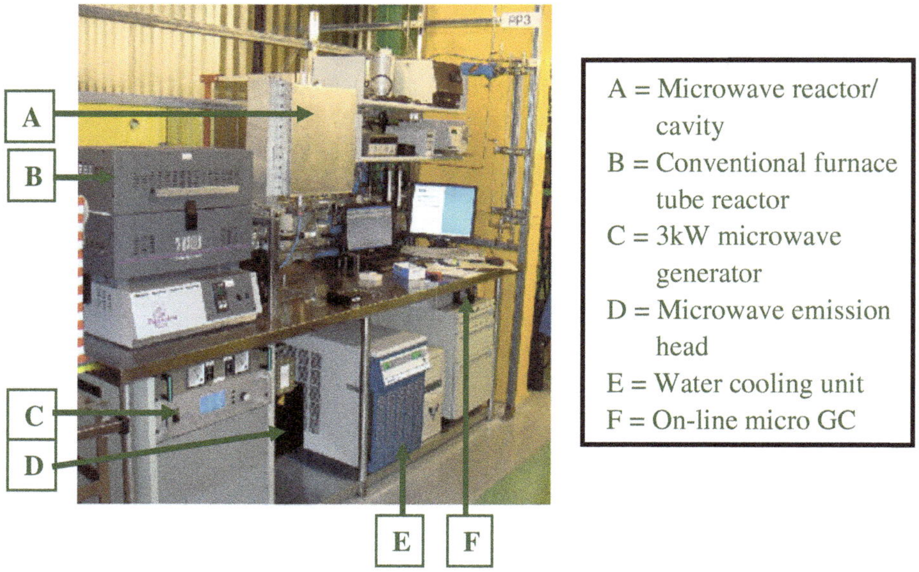

Fig. 2 Photograph of the bench-top pilot facility showing the overall layout

rectangular microwave cavity [17], which is most appropriate for applying microwave heating to continuous flows of fluids or to solids in granular form. Two cylindrical metal chimneys centered on the top and bottom walls of the cavity were fitted to allow the insertion of one tubular reactor. The open-ended tubular reactor passes vertically through these holes into the microwave cavity.

In this work, a quartz tube (2.54 cm or 1″ ID) was used as reactor. The tube was packed with the susceptor to allow for the direct transfer of microwave energy onto the susceptor. Silicon carbide (SiC), a good microwave absorber as reported by Kim et al. [18], was selected as the susceptor for this work. The ethane feed flow rate was set to 0.5 SLPM. The height of the dielectric sample was set between 5 and 30 cm. As the material in the reactor is heated, its dielectric properties change, thus causing the impedance to change with time. We used an Agilent network analyzer E5070B to insure that optimal conditions were found. The three-stub tuner was adjusted to compensate for impedance changes and a reflector plate was also used to adjust the maximum field strength in the reactor tube. Once the process steady state had been reached, the impedance-matching device no longer required much adjustment.

Temperature evolution was monitored via an optical pyrometer (Luxtron) located at the center of the packed bed. The optical pyrometer lens head operated in the range from 300 to 2,000 °C and is known not to be affected when used within electromagnetic fields.

Conventional furnace reactor system

The bench-top pilot system includes a conventional electrical muffle furnace (operable up to 1,200 °C) that can house quartz tube reactors having diameters up to around 5 cm. A Bench Link data recorder was used to record the temperature profiles during the reaction. Flow rates allowed for processing ethane in the range of liters per hour. The reactor muffle furnace temperature was adjusted when nitrogen gas was flowing until the system reached the desired temperature conditions before ethane was introduced. A condenser accessory was available for use when experiments were carried out using superheated steam to dilute the ethane (to reduce its partial pressure). Steam is normally present to reduce carbon deposition when working under conventional processing conditions.

Sampling section

Analysis of product gases was performed by an on-line four-channel micro gas analyzer (Agilent 3000A micro GC) located after the outlet of the reactor and after a cold trap used to remove solid particles, such as coke produced from secondary reactions, prior to entering into the micro gas chromatograph. The micro GC was calibrated with standard gases and gaseous mixtures prior to the experimental run.

Results and discussion

One of the main problems with the conventional heating process is the overheating of some zones within the reactor, which leads to coking, a process that usually results in significant losses in production and reduced energy efficiency. To mitigate this problem, the process makes use of injected superheated steam to reduce the partial vapor pressure of ethane and to "clean" the surface of the reactor, thus minimizing the potential for hot spots along the tubular structure. Steam may actually account for as much as 30 % of the feedstock. It is believed that thermal control can be improved using microwaves because, inherently, microwaves do not exhibit thermal inertia. In fact, a large amount of the energy used in the existing process is due to the need to generate and superheat the steam. Water has a heat capacity roughly double that of ethylene and therefore accounts for a significant portion of the energy needed in that process unit. It is believed that eliminating or reducing the amount of steam required would significantly reduce the energy requirements of the process.

The technical challenges facing the novel microwave-based process are mostly associated with high-volume flow regimes. One Canada-based plant, for example, produces 135 Mt of ethylene per hour. Despite the gigantic scale of the process, it is believed to be amenable to microwave treatment because, much like in the conventional process, the reactors required would be relatively small in diameter and, as such, do not represent a significant challenge with respect to the microwave penetration parameter. An issue that remains is that of scale-up to provide a cavity having uniform electric field quality over the necessary length, which does represent a significant technical challenge that will require innovative cavity design approaches [16–23].

One additional challenge is the introduction of novel catalysts into the reactors based on their selective microwave absorption. Energy efficiency would be increased using catalysts capable of selectively absorbing the microwaves and serving to heat the ethane flowing through the catalyst bed.

Microwave-assisted process

As a preliminary test for microwave dehydrogenation of ethane, the performance of selected dielectric materials was investigated in fixed-bed vertical reactor using pure, non-preheated, ethane. The reactor could be divided into zones 1 (gas inlet), 2 (reaction), and 3 (gas outlet) packed with quartz wool (7.6 cm or 3″ bed, as support), α-SiC particles (16 mesh, 5 cm or 2″ bed), and quartz wool again (7.6 cm bed), respectively. During reaction, product gas was analyzed every 8 min starting at 735 °C and terminated at ~760 °C after 32 min test duration. Four sets of data were collected. Ethane conversion, ethylene selectivity, and yields of major and minor gaseous products were calculated from the micro GC data. Coke formation was noticed mainly in zone 2, little in zone 3, and none in zone 1. Coke was determined from the weight difference of the reactor plus SiC before and after the combustion, a non-specific test for coke determination with greater error as expected. The results are given in Figs. 3, 4 and Table 1. Figure 3 demonstrates sharp increases in ethylene yield

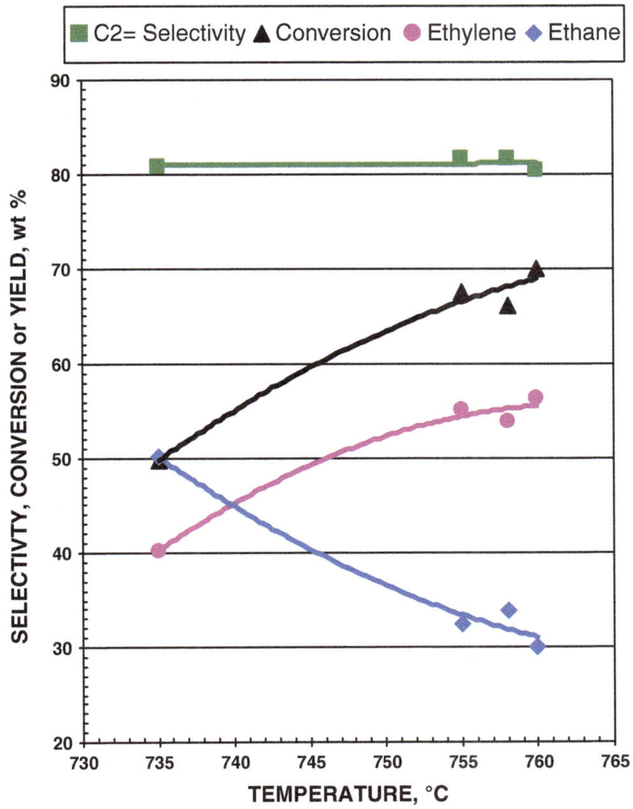

Fig. 3 Selectivity, conversion, and yields of major reaction products over a range of temperature for typical microwave-assisted processes

and conversion rendering a relatively constant selectivity over a period from 735 to 760 °C.

Conventional thermal process

Conventional pyrolysis of ethane was conducted at 800 ± 1 °C using a horizontal quartz reactor (2.54 cm or 1″ ID × 30 cm or 12″ L). Similarly, the reactor could be divided into zones 1 (gas inlet), 2 and 3 (gas outlet) packed with quartz wool (7.6 cm or 3″ bed, for gas preheating), quartz particles (16 mesh, 15.2 cm or 6″ bed), and quartz wool again (7.6 cm bed), respectively. When reaching steady state at 800 ± 1 °C, GC data were collected for every 7–15 min over a total of 180 min period producing 16 sets of data. Average data are shown in Table 1. Coke formation was observed mainly in zones 2 and 3, some in zone 1 (in area close to zone 2) as demonstrated in Fig. 5, indicating that coke could also be formed in quartz wool bed if the temperature was high enough. Note that coke formation along the reactor was mostly covered by a metering tape in Fig. 5. Total coke yield was determined at the end of the test through combustion followed by CO_2 absorption by ascarite (composed of NaOH) forming Na_2CO_3 and H_2O which was absorbed by $Mg(ClO_4)_2$. It

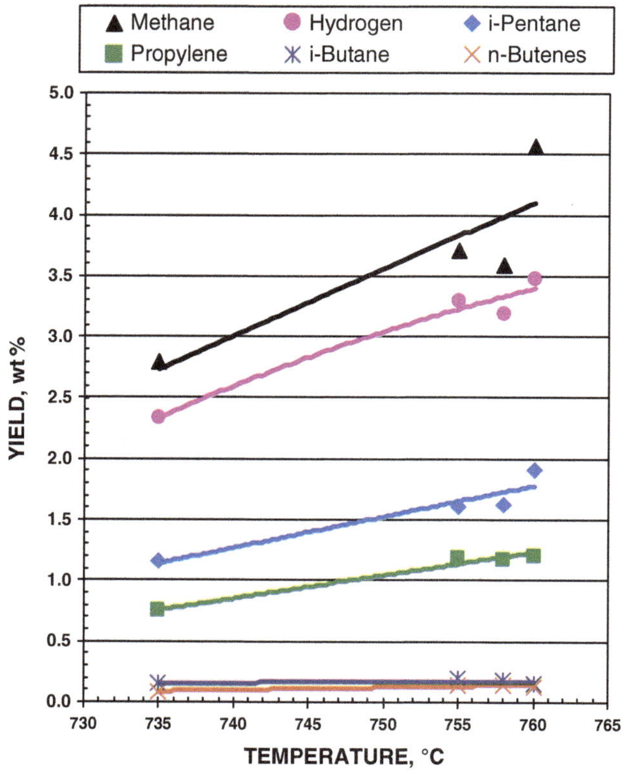

Fig. 4 Yields of minor reaction products over a range of temperature for typical microwave-assisted processes

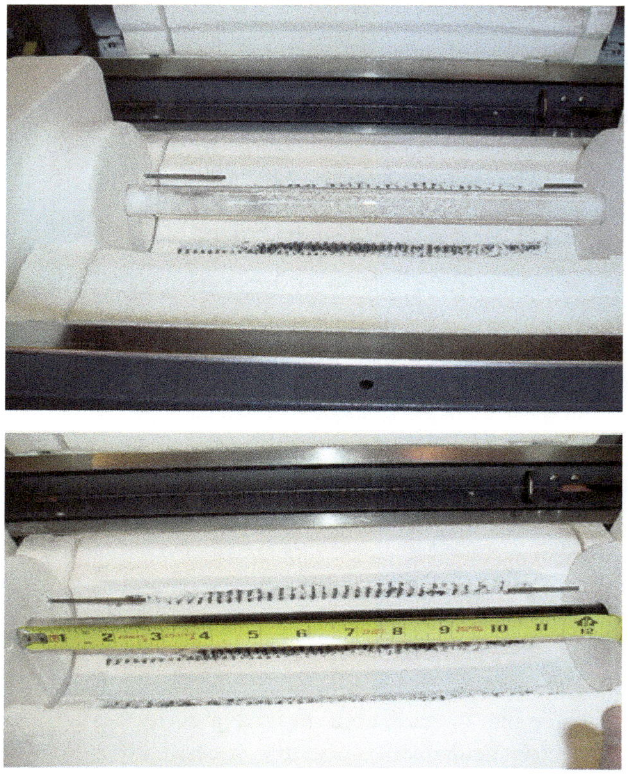

Fig. 5 Interior of the conventional electrical muffle furnace with reactor filled with quartz particles and quartz wool; *top* before reaction; *bottom* after carrying out an ethane-to-ethylene conversion process

Table 1 Comparison of conversion, selectivity, and yields between microwave-assisted and thermal cracking processes

	Microwave-assisted	Thermal cracking	Microwave/ thermal × 100 (%)
Temperature (°C)	757.5 ± 5.0	800 ± 1	
Ethane flow rate (L/min)	0.5	0.5	
Bed packing material	SiC particles	Quartz particles	
Length of bed (cm)	5	15.2	
Test duration (min)	24	164	
Number of data point	3	16	
Conversion (wt%)	67.89	71.71	94.7
Ethylene selectivity (%)	81.18	81.59	99.5
Yields (wt%)			
Ethylene	55.10	58.51	94.2
Ethane (unconverted)	32.11	28.29	–
Methane	3.96	4.70	84.2
Hydrogen	3.31	3.41	97.3
Iso-pentane	1.71	2.44	70.2
Propylene	1.19	1.75	68.1
Iso-butane	0.17	0.21	80.6
n-Butenes	0.13	0.33	39.6
Coke	2.29	0.14	–
Mass balance (wt%)	99.92	99.61	

can be seen from Table 1 that at 757.5 ± 5.0 °C, microwave results, in terms of conversion and gaseous product yields, are lower than their corresponding counterparts at 800 ± 1 °C by conventional heating. However, the ratio of the former to the latter multiplied by 100 is, in general, greater than 70 %, especially for the core product (ethylene) which is over 94 %. Thus, with lower temperatures by ~40 °C but still high ethylene yield, microwave treatment demonstrates the potential huge energy savings over conventional heating, considering the high volumes of gases to be treated.

The lower microwave results are likely due to the lower temperature of the experiment (~760 vs. 800 °C), although it can also be attributable to the shorter SiC bed (5 cm) in microwave study vs. longer bed (30 cm) filled with quartz particles (15.2 cm) and quartz wool, in the conventional test. It is anticipated that ethane will undergo more cracking when passed through a longer reaction zone (longer retention time). Thus, for all practical purposes and without attempts to further optimize the process, it is believed that, when compared with conventional ethane pyrolysis, the microwave-assisted cracking will give similar conversion, ethylene yield, and other gaseous product yields at 800 °C based on the trends observed in Figs. 3 and 4.

It is also noteworthy that the reactor outlet temperature when using the microwave reactor was only a few degrees above room temperature. This rapid self-cooling phenomenon proves to be useful in reducing or avoiding the polymerization of the products as it occurs in the commercial conventional process which requires a fast heat exchange at the outlet of the reactor.

Conclusions

A novel cavity for the microwave-assisted conversion of ethane to ethylene has been tested and validated from an operational standpoint. The results obtained in this study showed that similar ethylene yields exist between the two approaches, conventional and microwave. Furthermore, yields of components from secondary reactions are also similar in magnitude for the two processes. This suggests that the same reaction pathway is being followed in both cases.

These results are promising from a process standpoint and clearly demonstrate the feasibility of the microwave-assisted process. Further, these preliminary results warrant further work to optimize the application, specifically in terms of increased temperature and further reductions in reaction times. This experimental optimization work will also focus on increasing the electric field intensity, a parameter known to be of importance [8], by making use of a second slotted waveguide facing the one used in the present work [17].

These results would justify the development of an additional program area of designing and manufacturing novel microwave-susceptible materials that would serve as enhanced reactor components or as catalysts. The overall goal is to provide microwave-specific conditions that cannot be achieved via conventional processing, including higher process temperatures, shorter residence times, and no preheating (hence a significantly larger temperature difference between the susceptor surface and the reagent gas).

The approach is based on selective heating with the aim of further reducing energy consumption while exploring new chemical pathways. This new industrial research tool opens the door to other studies aimed at determining whether the reaction kinetics can be altered by judicious application of microwaves using complementary support and catalytic materials to reduce overall energy usage and greenhouse gas emissions.

An objective of the work was to see if maximum microwave field could be adjusted within the sample to make microwave-assisted process more energy efficient. Experimental data showed energy deposition efficiency very close to the modeling data [16].

Another objective of the work was to see if microwave-assisted ethane conversion could be energy efficient. Although no final data were obtained and further work is required to take into account all process units energy requirements, this work has demonstrated that the reaction carried out at lower temperature gave similar products without pre-heating and super-heated steam for the microwave work. The latter alone could bring energy savings of 50 %. This number will be reduced of course by the fact that, in a conventional process, some of the reactor heat is recuperated during the quenching step and used in other process units. Preliminary projections still show a minimum of 30 % overall energy savings if large-scale feasibility is demonstrated at similar conversion rates.

Acknowledgments We are indebted toward NOVA Chemicals of Canada for technical information about industrial operational requirements. The authors would like to acknowledge support from the Government of Canada's interdepartmental Program of Energy Research and Development, PERD 1.1.3. Petroleum Conversion for Cleaner Air.

References

1. Paré JRJ, Sigouin M, Lapointe J (1991) Microwave-assisted natural products extraction. U.S. Patent 5,002,784
2. Paré JRJ (1994) Microwave extraction of volatile oils and apparatus therefor. U.S. Patent 5,338,557
3. Paré JRJ (1995) Microwave-assisted generation of volatiles, of supercritical fluid, and apparatus therefor. U.S. Patent 5,337,426
4. Paré JRJ (1995) Microwave-assisted extraction from materials containing organic matter. U.S. Patent 5,458,897
5. Paré JRJ (1996) Microwave-assisted generation of volatiles, of supercritical fluid, and apparatus therefor. U.S. Patent 5,519,947
6. Paré JRJ (1997) Microwave-assisted generation of volatiles, of supercritical fluid, and apparatus therefor. U.S. Patent 5,675,909
7. Paré JRJ (1998) Microwave-assisted generation of volatiles, of supercritical fluid, and apparatus therefor. U.S. Patent 5,732,476
8. Paré JRJ, Bélanger JMR, Punt MM (2000) Controlled energy density microwave-assisted processes. U. S. Patent 6,061,926
9. Paré JRJ, Bélanger JMR, Stafford SS (1994) Microwave-assisted process (MAP™): a new tool for the analytical laboratory. Tr Anal Chem 13:176–184
10. Paré JRJ, Bélanger JMR (1997) Microwave-assisted processes (MAP™): principles and applications. In: Paré JRJ, Bélanger JMR (eds) Instrumental methods in food analysis Chap. 10. Elsevier, Amsterdam
11. Paré JRJ, Bélanger JMR (2011) Microwave-assisted extraction. In: Pawliszyn J, Lord H (eds) Handbook of sample preparation Chap. 12. John Wiley, New York
12. Bélanger JMR, Paré JRJ, Poon O, Fairbridge C, Ng S, Mutyala S, Hawkins R (2008) Remarks on various applications of microwave energy. J Microw Power Electromagn Energy 42:24–44

13. Mutyala S, Fairbridge C, Paré JRJ, Bélanger JMR, Ng S, Hawkins R (2010) Microwave applications to oil sands and petroleum: a review. Fuel Process Technol 91:127–135
14. Van Goethem MWM, Barendregt S, Grievink J, Moulijn JA, Verheijen PJT (2007) Ideal chemical conversion concept for the industrial production of ethene from hydrocarbons. Ind Eng Chem Res 46:4045–4062
15. Froment GF, Van de Steene BO, Van Damme P, Narayanan S, Goosens AG (1976) Thermal cracking of ethane and ethane-propane mixtures. Pilot plant versus industrial data. Ind Eng Chem Process Design Develop 15:495
16. Mutyala S, Fairbridge C, Paré JRJ, Ankam B, Bélanger JMR (2011) Design and numerical simulation of a high-efficiency microwave applicator for industrial processes. Hydrocarbon World 6(1):71–75
17. Liu Y, Fairbridge C, Paré JRJ, Ng S, Bélanger JMR, Mutyala S (2011) Optimization of microwave applicator for improved energy efficiency and homogeneity. Chimica Oggi/Chemistry Today 29(4):14–17
18. Kim DK, Cha CY, Lee WT, Kim JH (2001) Microwave dehydrogenation of ethane to ethylene. J Ind Eng Chem 7(6):363–374
19. Sutton WH (1992) Microwave processing of ceramics—an overview, microwave processing of materials iii. In: Proceedings of the symposium at materials research society, 269:3–20
20. Stevenson AF (1948) Theory of slots in rectangular waveguides. J Appl Phys 19:24–38
21. Stegen RJ (1951) Longitudinal shunt slot characteristics, Hughes Aircraft Company, Tech Memo No. 261
22. Elliott RS (1981) Antenna theory and design. Prentice-Hall, Englewood Cliffs, pp 407–414
23. Edgar RH, Osepchuk J (2001) Consumer, commercial, and industrial microwave ovens and heating systems. In: Datta AK, Anantheswaram RC (eds) Handbook of microwave technology for food applications. Marcel Dekker, New York, pp 215–277

28

Catalysis using colloidal-supported gold-based nanoparticles

Nikolaos Dimitratos · Ceri Hammond · Christopher J. Kiely · Graham J. Hutchings

Abstract The discovery of novel materials that can be active, selective and stable catalysts for the efficient transformation of organic molecules to useful products is of high importance. In recent years, there has been significant interest in the utilisation of supported gold-based nanoparticles that can be effective catalysts for a broad range of chemical processes. In this paper, we describe and discuss the utilisation of gold-based nanoparticles as efficient catalysts for a range of important reactions, with particular emphasis placed on our team recent research.

Keywords Gold catalysis · Nanoparticles · Alkane activation · Alcohol oxidation

Introduction

For a long period, gold was considered to be one of the least catalytically useful materials. Gold is unique among the metallic elements because of its resistance to oxidation and corrosion; its well-known history of usage in coinage and in jewellery perfectly illustrates this chemical nobility [1]. However, this perception has changed dramatically over recent years, following the seminal discoveries of Bond, Hutchings and Haruta, who identified that finely divided gold particles were particularly active catalysts for the hydrogenation of olefins [2], the hydrochlorination of ethyne to vinyl chloride [3] and the low temperature oxidation of CO [4], respectively. From these studies, it was identified that when Au is present as very small particles, with a mean diameter below 10 nm, it becomes surprisingly catalytically active, especially at low temperatures. After these important discoveries, many research groups in academia and industry have focused their research upon exploring the catalytic performance of Au catalysts for a wide range of oxidation and hydrogenation reactions [5, 6]. New discoveries have been reported since that period, and novel, designed supported gold nanoparticles have shown to be extremely effective catalysts for the oxidation of CO [7–11], the selective oxidation of alcohols and polyols [12, 13], the epoxidation of olefins [14, 15], the hydrochlorination of ethyne [16], the selective hydrogenation of unsaturated carbonyl and nitro groups [17], and the direct synthesis of hydrogen peroxide from molecular hydrogen and oxygen [18]. Whilst these monometallic Au nanoparticles have been shown to be effective redox catalysts, we have shown over the last decade that using bimetallic catalysts, such as Au–Pd catalysts, a significant improvement in terms of catalytic reactivity and stability for a range of reactions, such as the oxidation of alcohols [19] and hydrogen peroxide synthesis [20–22], could be observed. In this paper we review some of our recent work focusing on the synthesis of Au–Pd, Au–Pt and Au–Pd–Pt alloy and core–shell nanoparticles with colloidal methods, and their catalytic performance in a range of reactions, namely alkane activation (toluene and methane oxidation) and the oxidation of alcohols and polyols using molecular oxygen or hydrogen peroxide/*tert*-butyl hydroperoxide as green oxidants. These two reactions have a very broad

N. Dimitratos (✉) · C. Hammond · G. J. Hutchings
School of Chemistry, Cardiff University, Main Building, Park Place, Cardiff CF10 3AT, UK
e-mail: sacnd1@cardiff.ac.uk

G. J. Hutchings
e-mail: hutch@cardiff.ac.uk

C. J. Kiely
Center for Advanced Materials and Nanotechnology, Lehigh University, 5 East Packer Avenue, Bethlehem, PA 18015-3195, USA

interest and remain great challenges yet to be solved by the academic and industrial communities. In the case of alcohol and polyol oxidation, control of the selectivity to the desired product (aldehyde/ketone versus carboxylic acid or ester) is the key. Conversely, in the case of alkane oxidation the key challenge is to enhance activity at mild conditions, avoiding harsh reaction conditions and corrosive solvent systems, whilst concurrently controlling selectivity towards the desired product. This paper reviews some of the essential aspects of various bimetallic and trimetallic catalysts prepared using a colloidal method for these reactions.

Case I. Selective oxidation of alcohols and polyols

A variety of important industrial chemicals can be obtained by the selective oxidation of alcohols and polyols into the corresponding carbonylic or carboxylic derivatives through the use of heterogeneous catalysts [23, 24]. In our studies, we have primarily used benzyl alcohol and glycerol as model substrates for studying the catalytic performance of supported Au-based nanoparticles for alcohol/polyol oxidation. The methodologies for synthesising Au-based nanoparticles are primarily based on colloidal methods, and to a lesser extent impregnation methods. The advantages of colloidal methods rely on the fact that metal nanoparticles with a narrow particle size distribution and with a specific morphology can be synthesised. Nano-sized metal colloids are usually synthesised by an in situ reduction method from a suitable metal precursor, such as chemical reduction (with $NaBH_4$ or alcohol), photoreduction, electrochemical reduction or thermal decomposition [25]. Small molecular ligands, surfactants and polymers have been used to stabilise the so-formed metallic colloids. In some cases, metal colloids are stable over long periods even without coagulation in a solvent. However, for prolonged stability in a sol state, and for obtaining a uniform, controlled particle size, use of a stabilising ligand is essential. Typically, polymers such as polyvinyl alcohol and polyvinylpyrrolidone and various other surfactants have frequently been used. In the case of polymeric stabilisers, the typical stabilisation mechanism is based on steric stabilisation. However, electrostatic stabilisation is also possible through the utilisation of electrostatic stabilisers such as citrate. Our preferred method is based on steric stabilisation. The synthesis of colloidal metal nanoparticles typically requires (1) the use of inorganic metal precursors, such as $HAuCl_4$, $PdCl_2$ and H_2PtCl_6; (2) the use of H_2O as solvent; (3) dilute aqueous solutions of polyvinyl alcohol (PVA) as stabiliser and $NaBH_4$ as the reducing agent. For the synthesis of Au-based colloids, the first step involves the dissolution of the desired metal precursors in aqueous solution, followed by the addition of the desired amount of PVA. By adding $NaBH_4$ as the reducing agent, the formation of a colloidal solution is subsequently achieved. The final step involves the immobilisation of the Au-based colloids onto the desired heterogeneous support, such as carbon, TiO_2 or other metal oxide supports [26]. The synthesis of alloyed or core–shell structures is possible by modifying the reduction step. In the case of forming alloy nanoparticles, a simultaneous reduction process should be followed, whereas in the case of core–shell structures a sequential reduction protocol is essential.

The catalytic performance of supported Au–Pd nanoparticles with alloy and core–shell structures synthesised by colloidal methods was carried out for the aerobic oxidation of benzyl alcohol. High catalytic efficiency of Au–Pd catalysts was demonstrated with very high TOFs (turnover frequencies) (40,000 h^{-1}) at mild reaction conditions (120 °C, PO_2 = 10 bar) when random Au–Pd alloy nanoparticles were used (Table 1) [26, 27]. Extensive STEM-XDS and XPS studies were performed to characterise the synthesised materials. These studies showed the presence of random homogeneous alloys with metallic oxidation state for Au and the majority of Pd also present in metallic state. Higher activity of the supported Au–Pd catalysts synthesised by the colloidal method instead of the impregnation

Table 1 Benzyl alcohol oxidation results after 2 h reaction for sol-immobilised Au–Pd catalysts supported on C and TiO_2: the effect of the order of metal reduction in the preparation [27]

Catalyst	Conversion (%)	Selectivity (%)						TOF/h^{-1b}
		Benzene	Toluene	Benzaldehyde	Benzoic acid	Benzyl benzoate	Acetal	
1 % Au + Pd/TiO_2^a	61.2	0.5	26.7	69.2	1.7	1.9	0	15,400
1 % Pd(Au)/TiO_2^a	57.8	0.2	18.1	77.1	1.8	2.8	0	19,300
1 % Au(Pd)/TiO_2^a	48.7	0.4	23.0	72.3	1.9	2.5	0	17,400
1 % Au + Pd/C[a]	81.1	0.5	40.9	55.0	1.3	2.1	0.2	35,400
1 % Pd(Au)/C[a]	70.4	0.4	35.1	62.6	2.0	0.0	0	41,900
1 % Au(Pd)/C[a]	79.2	0.7	28.8	65.0	2.7	2.7	0	24,300

[a] Reaction conditions: benzyl alcohol, 0.1 g of catalyst, T = 120 °C, pO_2 = 150 psi, stirring rate 1,500 rpm

[b] Calculation of TOF (h^{-1}) after 0.5 h of reaction. TOF numbers were calculated on the basis of total loading of metals

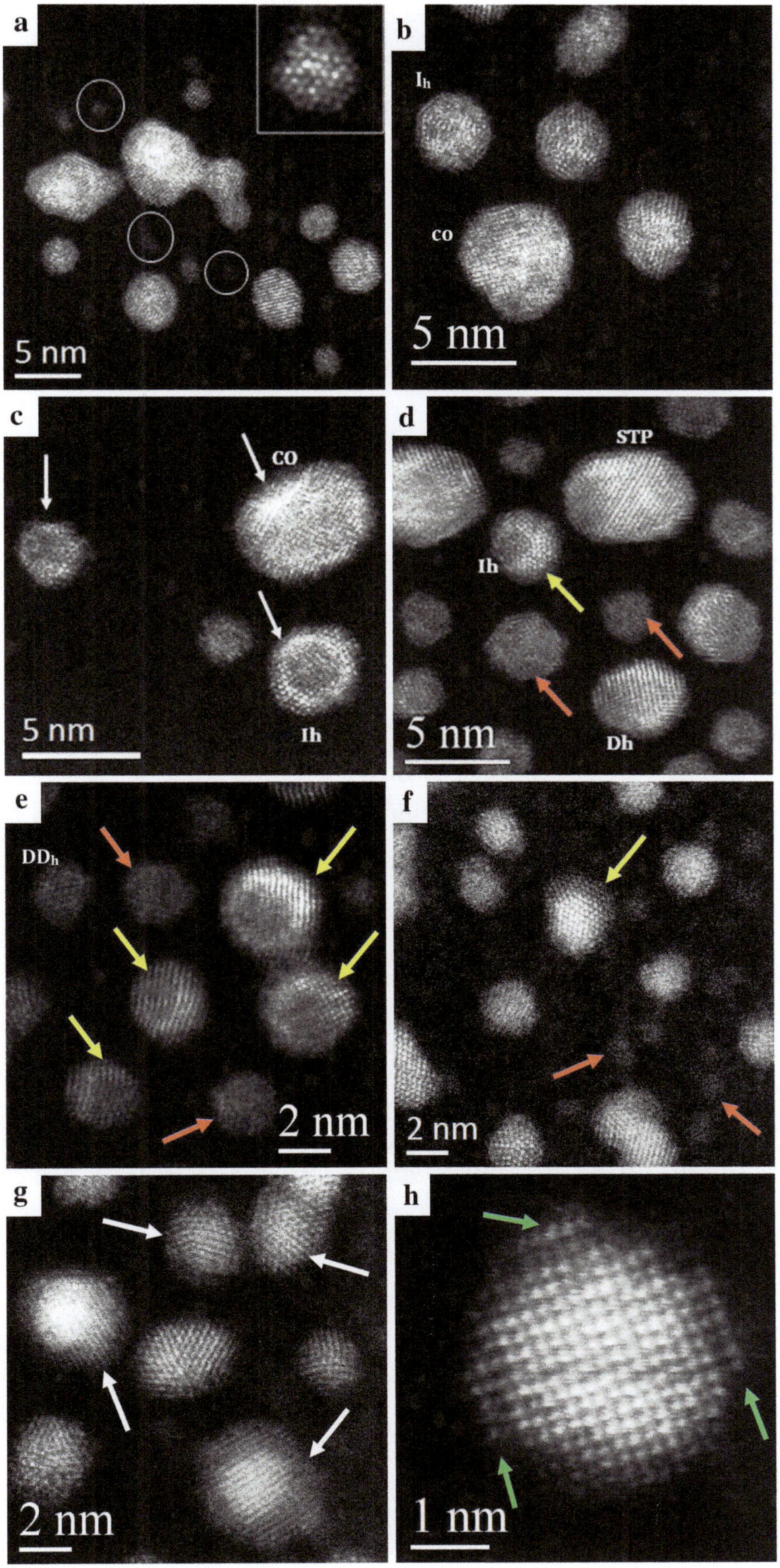

Fig. 1 STEM-HAADF images of the starting colloids dispersed onto a continuous carbon thin film: **a**, **b** Au + Pd colloids; **c**–**e** Au(Pd) colloids; and **f**–**h** Pd(Au) colloids. *Circles* indicate the presence of nanometer scale clusters of atoms; *white arrows* indicate the presence of Pd particles with complete Au coverage; *yellow arrows* indicate the presence of Pd particles with incomplete Au coverage; *red arrows* indicate the presence of Pd particles with no Au coverage; *green arrows* indicate the presence of bright atoms observed in the Pd shell indicating incorporation of Au atoms into the Pd shell region. Reprinted with permission from reference [28]. Copyright (2011) Royal Society of Chemistry

Table 2 Benzyl alcohol oxidation results at iso-conversion for sol-immobilised Au–Pd catalysts supported on C and TiO_2: the effect of the order of metal reduction in the preparation [27]

Catalyst	Conversion (%)	Selectivity (%)					
		Benzene	Toluene	Benzaldehyde	Benzoic acid	Benzyl benzoate	Acetal
1 % Au + Pd/TiO_2^a	50	0.5	26.9	69.1	1.5	2.0	0
1 % Pd(Au)/TiO_2^a	50	0.2	23.0	72.8	1.5	2.5	0
1 % Au(Pd)/TiO_2^a	50	0.4	23.0	72.2	1.9	2.5	0
1 % Au + Pd/C[a]	50	0.3	44.0	51.3	1.7	2.7	0.1
1 % Pd(Au)/C[a]	50	0.3	41.6	54.6	1.6	1.8	0.1
1 % Au(Pd)/C[a]	50	0.6	38.9	57.1	1.6	1.7	0.1

[a] Reaction conditions: benzyl alcohol, 0.1 g of catalyst, T = 120 °C, pO_2 = 150 psi

Table 3 Benzyl alcohol oxidation results after 2 h reaction with various Au–Pd–Pt catalysts supported on carbon prepared by the sol-immobilisation method [30]

Catalyst	Conversion (%)	Selectivity (%)				
		Toluene	Benzaldehyde	Benzoic acid	Benzyl benzoate	TOF/h^{-1}[b]
1 wt% (0.3 Au + 0.4 Pd + 0.3 Pt)/C[a]	35.4	0.9	83.3	7.9	7.9	16,000
1 wt% (0.4 Au + 0.4 Pd + 0.2 Pt)/C[a]	36.7	0.6	81.7	8.0	9.7	22,100
1 wt% (0.45 Au + 0.45 Pd + 0.1 Pt)/C[a]	53.9	0.1	80.2	13.1	6.6	31,900

[a] Reaction conditions: benzyl alcohol, 0.05 g of catalyst, T = 120 °C, pO_2 = 150 psi, stirring rate 1,500 rpm.

[b] Calculation of TOF (h^{-1}) value was done after 0.5 h of reaction, and was based on the total metal loading

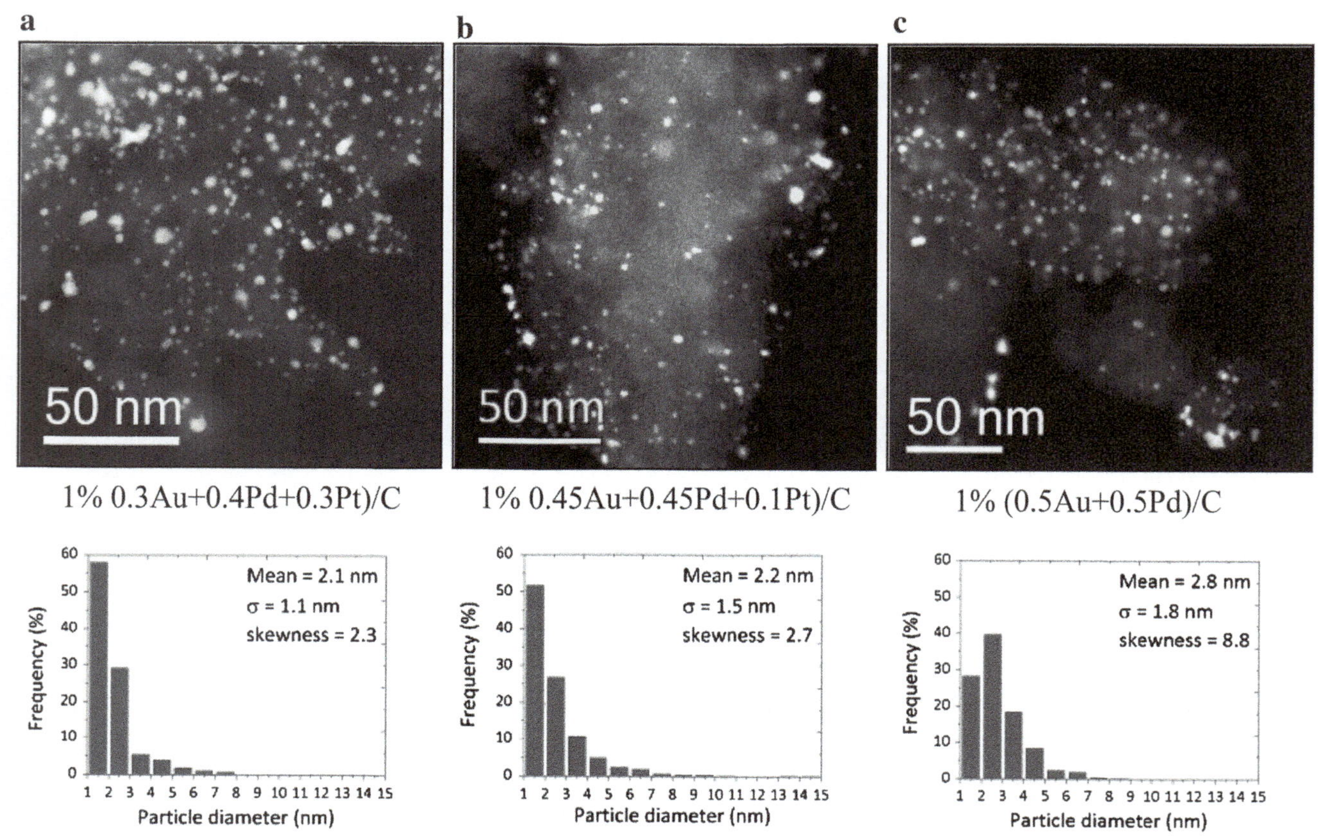

Fig. 2 Representative low-magnification STEM-HAADF images and corresponding particle size distributions of selected trimetallic, catalysts characterized in this study; namely; **a** 1 wt% (0.3 Au–0.4 Pd–0.3 Pt)/C; **b** 1 wt% (0.45 Au–0.45 Pd–0.1 Pt)/C; **c** 1 wt% (0.5 Au–0.5 Pd)/C. Reprinted with permission from reference [30]. Copyright (2013) Royal Society of Chemistry

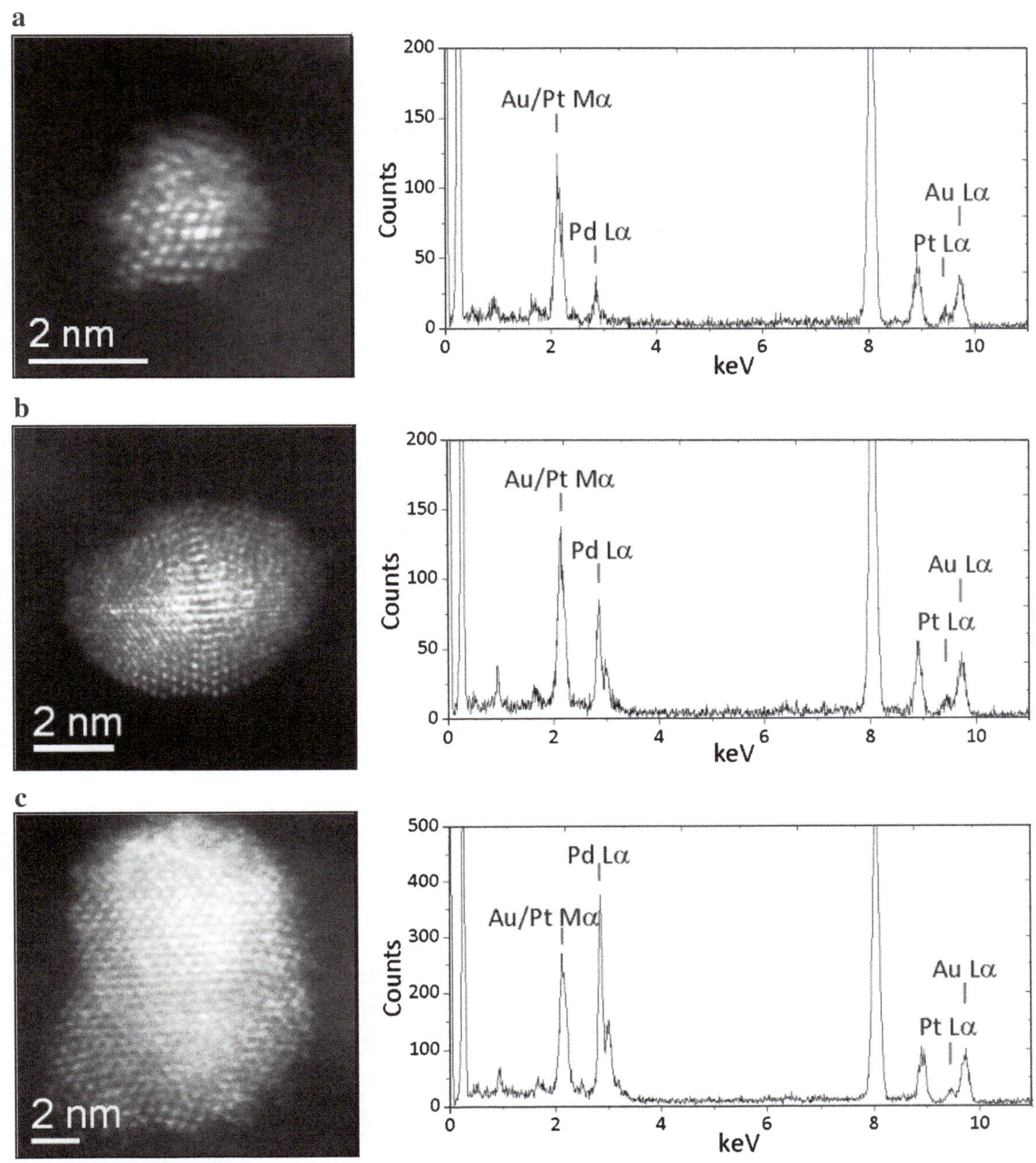

Fig. 3 High-magnification STEM-HAADF images and the corresponding X-ray energy dispersive spectra acquired from representative **a** 2 nm, **b** 5 nm and **c** 10 nm particles in the sol-immobilised 1 wt% (0.45 Au + 0.45 Pd + 0.1 Pt)/C catalyst dried at 120 °C. Reprinted with permission from reference [30]. Copyright (2013) Royal Society of Chemistry

method was observed, and it was mainly attributed to the significantly smaller average particle size, the narrower particle size distribution and the metallic oxidation state. In subsequent studies we investigated the catalytic performance of supported Au–Pd nanoparticles with alloy and core–shell structures. As we described above, for the synthesis of supported bimetallic Au–Pd catalysts with core–shell structures, the preparation strategy was based on the sequential addition and reduction of each metal. The bimetallic sols were subsequently deposited onto the chosen supports, carbon and titania. STEM-HAADF studies were performed to study the morphology of the synthesised colloids, and the formation of both alloyed and core–shell structures was shown (Fig. 1) [28]. It was found that the catalytic activity for the aerobic oxidation of benzyl alcohol could be carried out at mild conditions (120 °C, $PO_2 = 10$ bar), and the order of metal addition has a marked effect on activity as well as selectivity (Table 1). The choice of support (carbon versus titania) was also shown to significantly affect both catalytic activity and the

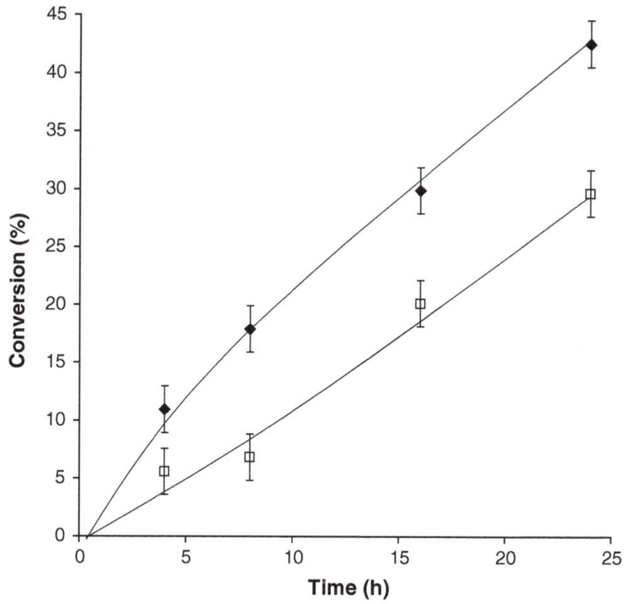

Fig. 4 Time-on-line data for selective oxidation of glycerol by Au–Me/MgO catalysts, with Me = Pt (*filled diamond*) or Pd (*unfilled circle*). Reaction conditions: 1:3 molar fraction Au:Me, total metal loading 1 % by mass, water (10 mL), 0.3 mol/L glycerol, mole fraction of glycerol/metal = 500, T = ambient, pO_2 = 300 kPa. *Points* represent the mean value obtained from ten reaction runs, and the error bars are defined by the range of these measurements. Reprinted with permission from reference [33]. Copyright (2011) Wiley–VCH

distribution of products (selectivity), with carbon-supported materials exhibiting both an increase of activity by a factor of 2, and a lower selectivity to benzaldehyde at iso-conversion compared to the analogous TiO_2-supported catalysts (Table 2). Mechanistic studies were performed for the oxidation of benzyl alcohol using an Au–Pd alloy catalyst synthesised by colloidal method, and it was found that in the absence of oxygen benzyl alcohol was transformed into benzaldehyde and toluene at initial equal rates [29]. The introduction of oxygen significantly increased the rate of benzyl alcohol consumption and benzaldehyde production at the expense of toluene formation. It was found that at low partial pressures (below 3 bar of oxygen), the catalytic rate was dependent on oxygen, suggesting that oxygen can participate in the reaction pathway as an adsorbed species, whereas above 3 bar of oxygen the reaction is zero order in oxygen. The formation of benzaldehyde was found to be dependent upon the concentration of oxygen at the surface. Despite the fact that the high catalytic efficiency of supported Au–Pd nanoparticles synthesised by colloidal method is evident, the challenge of achieving high selectivity to benzaldehyde remained still an issue. For tackling this challenge, either a low reaction temperature approach was chosen—resulting in a low turnover frequency (TOF)—or the conversion was kept at low levels (below 30 %). Another approach included the addition of a small amount of a third metal, such as Pt, and the utilisation of supported trimetallic Au–Pd-Pt nanoparticles [30]. Using a simultaneous reduction method for the synthesis of pre-formed Au–Pd-Pt colloids, and by varying Au–Pd-Pt atomic ratios the catalytic performance of the synthesised supported Au–Pd-Pt nanoparticles was investigated. Variation of Au–Pd-Pt atomic ratios led to marked improvements in

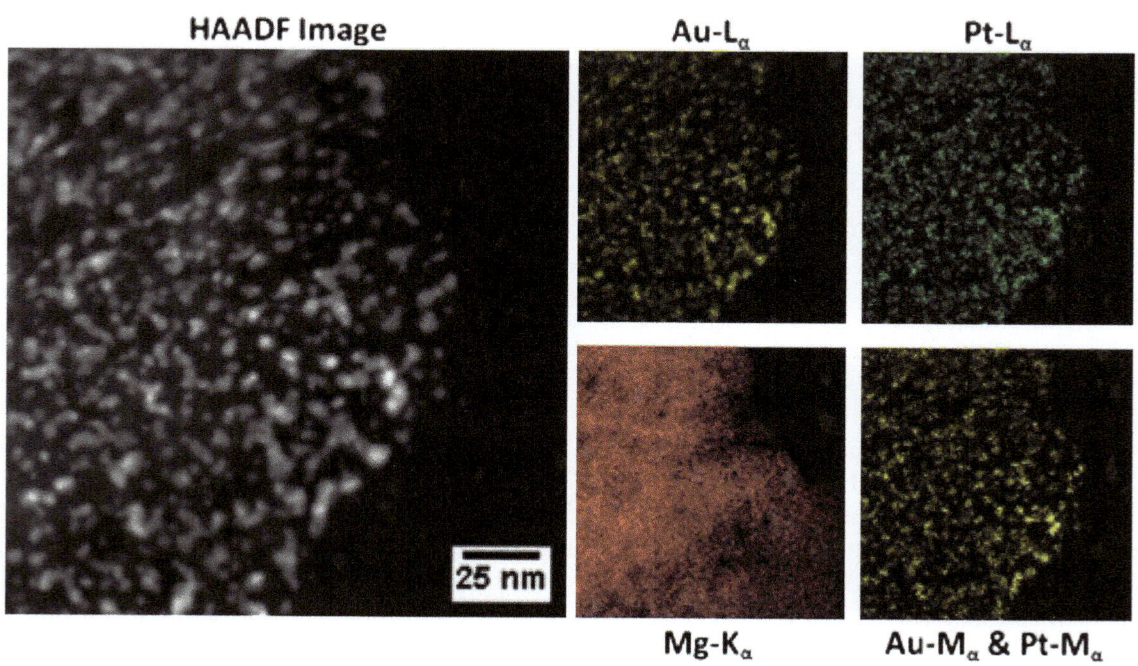

Fig. 5 HAADF image of the AuPt (1:1)/MgO catalyst (dried at 120 °C), along with the corresponding Au–L_α, Pt–L_α, Mg–K_α, and (Au–M_α + Pt–M_α) elemental maps from the same area. Reprinted with permission from reference [33]. Copyright (2011) Wiley–VCH

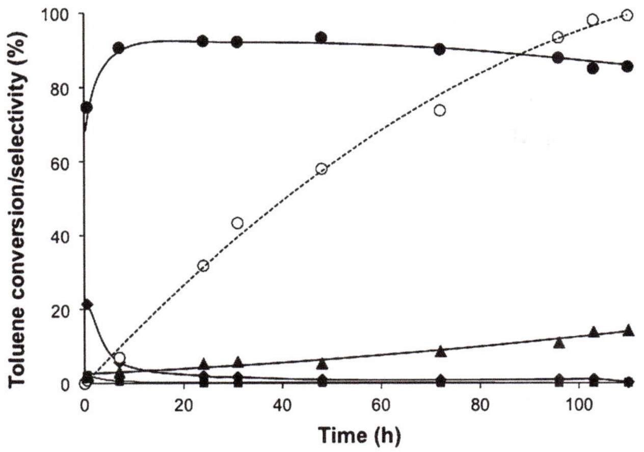

Fig. 6 Toluene conversion and selectivity to partial oxidation products. Reaction conditions: 160 °C, 0.1 MPa pO_2, 20 ml toluene, 0.8 g of catalyst (1 wt% AuPd/C prepared by sol immobilisation with a 1:1.85 Au/Pd ratio), toluene/metal molar ratio of 3,250 and reaction time: 110 h. Key: *unfilled circle* conversion *filled square* selectivity to benzyl alcohol, ·· selectivity to benzaldehyde, *closed up triangle* selectivity to benzoic acid, *filled circle* selectivity to benzyl benzoate. Reprinted with permission from reference [34]. Copyright (2011) The American Association for the Advancement of Science

benzaldehyde selectivity (95 %) and significantly decreased the formation of toluene as by-product, whilst maintaining a high level of benzyl alcohol conversion (Table 3). STEM-HAADF-XEDS studies showed that the mean particle size was around 2–3 nm, and homogeneous trimetallic alloys were obtained (Fig. 2). However, depending on the size of the individual nanoparticles a clear size-dependent composition variation was found, with the smaller (<2 nm) particles found to be Pd-deficient, and particles larger than 10 nm observed to be Pd rich (Fig. 3). The beneficial effect of adding Pt atoms seems to be related to the role of Pt as electronic (Au and Pt have higher electronegativity than Pd, therefore can be accepted of electrons from Pd) or strain modifier to the Au–Pd nanoparticles, which reduces the capability of the Pd sites to be active for disproportionation reactions and therefore improving the selectivity to benzaldehyde. Therefore, the challenge of synthesising Au–Pd-Pt nanoparticles with a controlled-size metal composition still remains.

Another organic compound that has recently attracted significant research interest is glycerol. The main reason is the following: glycerol, as a by-product from biodiesel production, is highly abundant and is readily available. As a waste material, its disposal via chemical fixation is highly favourable, and a large number of products can indeed be obtained from its chemical transformation. Nevertheless, whilst its high functionality leads to a number of potential products, particularly through oxidation, its high functionality demands exceptional control over reaction selectivity, and targeted selectivity to the desired products remains a significant challenge [31]. In previous studies, we have shown the efficient oxidative transformation of glycerol under alkaline conditions using supported Au–Pd nanoparticles prepared by colloidal and impregnation methods [32]. A key challenge in the case of upgrading glycerol is the avoidance of alkaline conditions (without the use of NaOH as base), where the desired product will be the free carboxylic acid instead of the salt of acid. In our studies, Au–Pt nanoparticles were synthesised at various Au–Pt atomic ratios, and were subsequently immobilised on MgO [33]. By performing the aerobic oxidation of glycerol at mild conditions (3 bar of O_2, 40–60 °C) and without base, the efficient transformation of glycerol to glyceric acid was achieved at a selectivity of 80 % and conversion levels of 43 % (Fig. 4). Moreover, the general applicability of the novel Au–Pt/MgO catalysts was demonstrated by the efficient oxidation of ethylene and propylene diols. Extensive STEM-XEDS studies showed that the individual metal nanoparticles in the Au–Pt/MgO samples contain both Au and Pt metals, indicating the presence of Au–Pt alloy nanoparticles (Fig. 5). STEM-HAADF analysis showed that the majority of Au–Pt particles exhibited a cuboctahedral structure, exposing a mixture of [111] and [200] facets.

Case II. Selective oxidation of hydrocarbons

The selective oxidation of hydrocarbons to useful chemical products is one of the greatest challenges in academia and

Fig. 7 Reaction scheme for the oxidation of toluene to benzyl benzoate

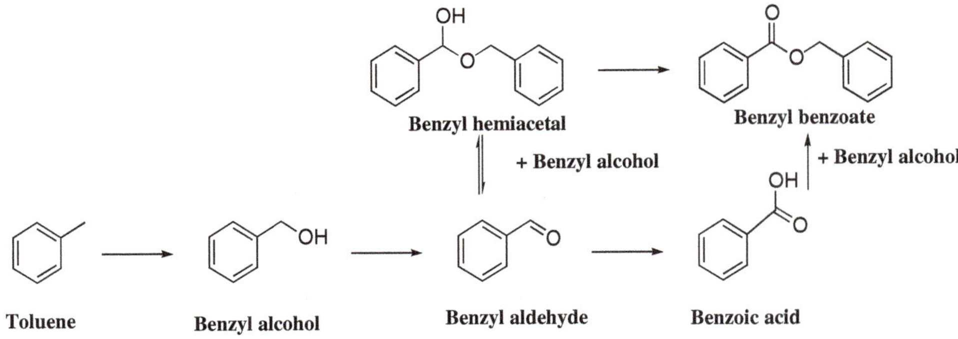

industry. However, given their low reactivity and high levels of stability, the activation of typically inert C–H bonds often requires high reaction temperature, resulting in poor selectivity to the desired oxygenated organic products. Utilisation of catalysts that can lower the reaction temperature required for alkane activation, and thereby performing the catalytic transformation of hydrocarbons efficiently at lower temperatures, could result in an improved selectivity to oxygenated products. Taking into account that Au–Pd supported nanoparticles synthesised by colloidal methods showed high catalytic efficiency for the oxidation of alcohols, polyols and hydrogen peroxide synthesis, we investigated the catalytic performance of these materials for the oxidation of toluene and methane at mild conditions. In the case of toluene, the efficient transformation of toluene to benzyl benzoate (90–95 % selectivity) was achieved at mild conditions (120–160 °C, 10 bar of O_2) using molecular oxygen or *tert*-butyl hydroperoxide as oxidant (Fig. 6) [34]. Mechanistic studies revealed that the oxidation proceeds via the initial formation of benzyl alcohol, which is rapidly oxidised to benzaldehyde. The formation of the analogous hemiacetal is followed by oxidation to the ester (Fig. 7). In subsequent studies, the effect of the oxidant was studied, and instead of molecular oxygen, *tert*-butyl hydroperoxide (TBHP) was used as the principal oxygen donor. High conversion of toluene at mild conditions (80 °C, atmospheric pressure) was achieved with benzoic acid being the major product (Fig. 8) [35]. Mechanistic studies were performed in order to elucidate the reaction mechanism, and especially the role of TBHP, and it was found for the first time that surface-bound oxygen-centred radicals play a role in activating toluene, as demonstrated by EPR studies.

The direct conversion of methane to methanol at mild conditions is one of the most important challenges that academia and industry face. Taking into account from previous studies that H_2O_2 or *tert*-butyl hydroperoxide has been used efficiently to oxidise methane, we thought that a hydroperoxide intermediate may be effective for the oxidation of methane. Therefore, the catalytic performance of supported Au–Pd nanoparticles for the liquid-phase oxidation of methane was carried out using either H_2O_2, or in situ generated H_2O_2 from H_2 and O_2 as oxidant [36]. Using supported Au–Pd nanoparticles synthesised by the impregnation method, high selectivity to oxygenated products (90 % including methyl hydroperoxide) and moderate selectivity to methanol (50 %) at mild reaction conditions (50 °C) were achieved. The efficient oxidation of methane to methanol was possible even at 2 °C, with an oxygenate selectivity of 93 % found at these sub-ambient temperatures. Isotopic labelling studies were performed and confirmed that in the presence of the catalyst, all of the observed products were derived from methane (Fig. 9). EPR studies were also performed to investigate the nature of radical species that

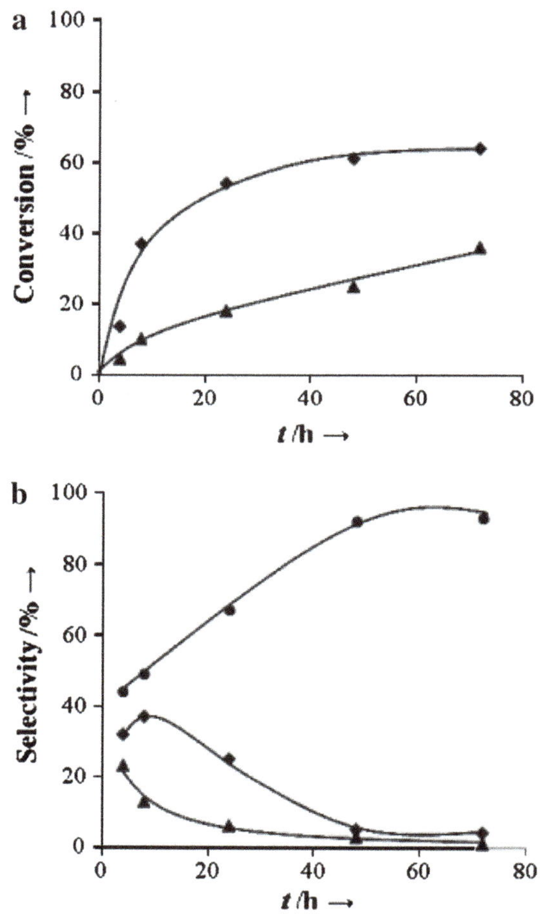

Fig. 8 Time-on-line data for the reaction of toluene. **a** Conversion of toluene (*filled triangle*) and TBHP (*filled diamond*). **b** Selectivities to benzaldehyde (*filled diamond*), benzyl alcohol (*filled triangle*) and benzoic acid (*filled circle*). Reaction conditions: molar ratio toluene/TBHP = 1, 80 8C, catalyst (0.2 g), substrate/metal molar ratio = 6,500. Reprinted with permission from reference [35]. Copyright (2012) Wiley–VCH

may produce during reaction. It was shown that both $\cdot CH_3$ and $\cdot OH$ radicals were formed during reaction, and that the methane oxidation mechanism for supported metal nanoparticles involves $\cdot CH_3$ radical that can react either with surface-bound $\cdot OOH$ to form $CH_3OO\cdot$, and eventually CH_3OH and/or H_2CO, or with $\cdot OH$ to produce CH_3OH directly. At this stage, we believe that the major reaction pathway is the one via surface-bound $\cdot OOH$ due to the observation of methyl hydroperoxide as the primary reaction product. Using the in situ generated H_2O_2 approach, a similar catalytic productivity was observed but with much higher methanol selectivity (70–80 %).

Concluding remarks

Supported bimetallic (Au–Pd, Au–Pt) and trimetallic (Au–Pd–Pt) nanoparticles are very effective catalysts for the "clean" and efficient oxidation of organic

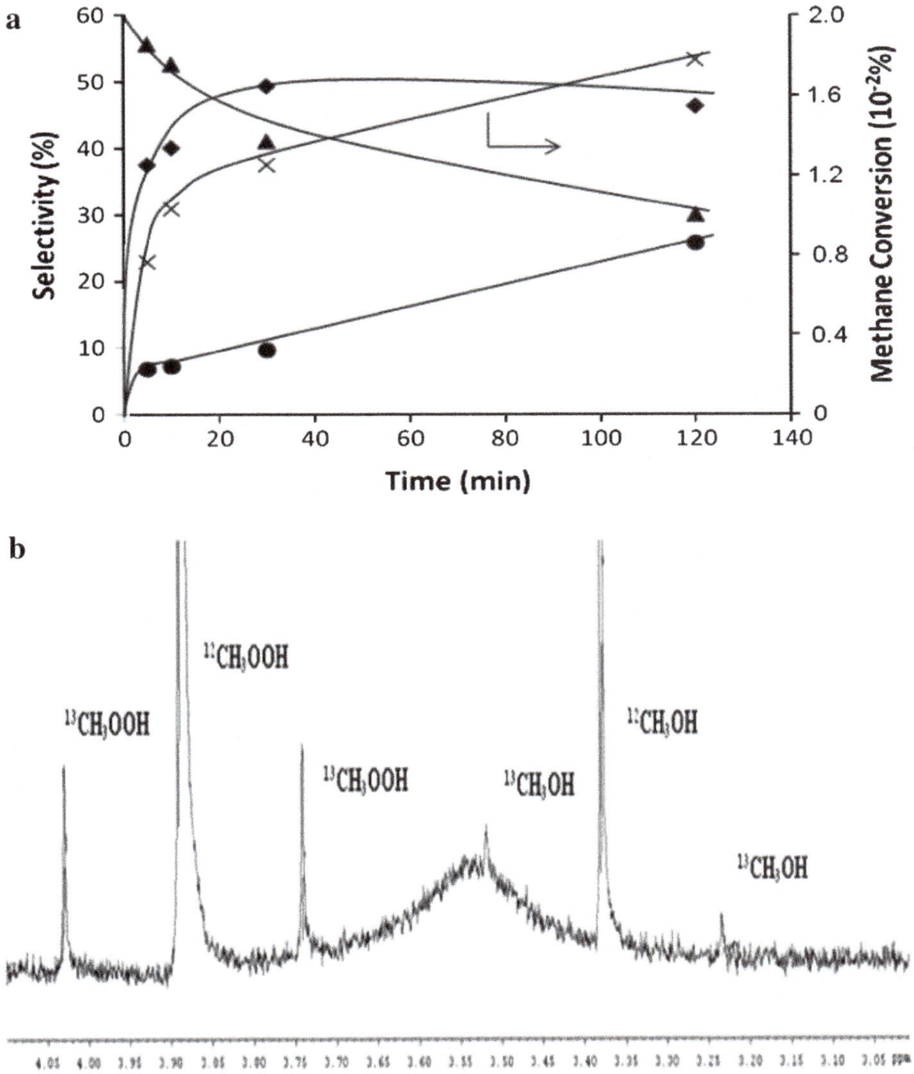

Fig. 9 **a** Time-on-line plot of methane oxidation with addition of H_2O_2 in the presence of a 1 wt%AuPd/TiO_{2IW} catalyst. Key: *closed up triangle* selectivity to methyl hydroperoxide, *filled diamond* selectivity to methanol, *filled circle* selectivity to carbon dioxide, **x** methane conversion. Reaction conditions: $P_{(CH4)}$, 30.5 bar; [H_2O_2], 0.5 M; T, 50 °C; stirring rate, 1,500 rpm; catalyst mass, 10 mg. **b** ^{1}H-NMR spectrum of the reaction filtrate collected when using a solution initially spiked with $^{13}CH_3OOH$ in a typical reaction using 5 wt%Au–Pd/TiO_{2IW} and pure $^{12}CH_4$. Reprinted with permission from reference [36]. Copyright (2013) Wiley–VCH

compounds (alcohols, polyols and alkanes) with molecular oxygen, hydrogen peroxide or alkyl hydroperoxides as "green" oxidants. A closer examination of the bimetallic/trimetallic catalysts reveals that it is important not only to control the particle size, but also to focus on the synthesis of structures consisting of an alloy or a core–shell structure, since electronic and geometric properties are affected in this manner. In the majority of the investigated reactions, it is generally accepted that the synthesis of nanoparticles of small particle size (2–6 nm) seems to be key for obtaining high catalytic activity. However, the particle size, shape and final morphology have all been shown to affect each oxidation process in a different way. In the future, control of the desired structure will play a major role in the development of new and more efficient catalysts, especially for industrial applications. Therefore, the main challenges that researchers will face in the near future are: (1) understanding the formation and the role of active sites, (2) developing more efficient, cheaper nanoparticles, particularly at lower metal loading, with prolonged catalytic life time, (3) scaling up nanoparticle synthesis, and (4) expanding the utilisation of Au-based and bimetallic-based catalysts in a larger variety of organic reactions. Finally, it is evident that the area of gold catalysis still remains an area of intense research, and is therefore a promising and growing area, with new materials and new reactions being regularly discovered and explored.

References

1. Bond GC, Thompson DT (1999) Catalysis by gold. Catal Rev Sci Eng 41:319–388
2. Bond GC, Sermon PA, Webb G, Buchanan DA, Wells PB (1973) Hydrogenation over supported gold catalysts. J Chem Soc Chem Commun (13):444-445

3. Hutchings GJ (1985) Vapor-phase hydrochlorination of acetylene—correlation of catalytic activity of supported metal chloride catalysts. J Catal 96:292–295
4. Haruta M, Kobayashi T, Sano H, Yamada N (1987) Novel gold catalysts for the oxidation of carbon-monoxide at a temperature far below 0 °C. Chem Lett 16:405–408
5. Hashmi ASK, Hutchings GJ (2006) Gold catalysis. Angew Chem Int Ed 45:7896–7936
6. Della Pina C, Falletta E, Prati L, Rossi M (2008) Selective oxidation using gold. Chem Soc Rev 37:2077–2095
7. Herzing AA, Kiely CJ, Carley AF, Landon P, Hutchings GJ (2008) Identification of active gold nanoclusters on iron oxide supports for CO oxidation. Science 321:1331–1335
8. Xie XW, Li Y, Liu ZQ, Haruta M, Shen WJ (2009) Low-temperature oxidation of CO catalysed by Co3O4 nanorods. Nature 458:746–749
9. Haruta M (2011) Spiers memorial lecture role of perimeter interfaces in catalysis by gold nanoparticles. Faraday Discuss 152:11–32
10. Yoshida H, Kuwauchi Y, Jinschek JR et al (2012) Visualizing gas molecules interacting with supported nanoparticulate catalysts at reaction conditions. Science 335:317–319
11. Akita T, Kohyama M, Haruta M (2013) Electron microscopy study of gold nanoparticles deposited on transition metal oxides. Acc Chem Res 46:1773–1782
12. Abad A, Concepcion P, Corma A, Garcia H (2005) A collaborative effect between gold and a support induces the selective oxidation of alcohols. Angew Chem Int Ed 44:4066–4069
13. Carrettin S, McMorn P, Johnston P, Griffin K, Hutchings GJ (2002) Selective oxidation of glycerol to glyceric acid using a gold catalyst in aqueous sodium hydroxide. Chem Commun 7:696–697
14. Sinha AK, Seelan S, Tsubota S, Haruta M (2004) A three-dimensional mesoporous titanosilicate support for gold nanoparticles: vapor-phase epoxidation of propene with high conversion. Angew Chem Int Ed 43:1546–1548
15. Hughes MD, Xu YJ, Jenkins P et al (2005) Tunable gold catalysts for selective hydrocarbon oxidation under mild conditions. Nature 437:1132–1135
16. Nkosi B, Adams MD, Coville NJ, Hutchings GJ (1991) Hydrochlorination of acetylene using carbon-supported gold catalysts—a study of catalyst reactivation. J Catal 128:378–386
17. Corma A, Serna P (2006) Chemoselective hydrogenation of nitro compounds with supported gold catalysts. Science 313:332–334
18. Landon P, Collier PJ, Papworth AJ, Kiely CJ, Hutchings GJ (2002) Direct formation of hydrogen peroxide from H-2/O-2 using a gold catalyst. Chem Commun (18):2058–2059
19. Enache DI, Edwards JK, Landon P et al (2006) Solvent-free oxidation of primary alcohols to aldehydes using Au-Pd/TiO2 catalysts. Science 311:362–365
20. Edwards JK, Solsona BE, Landon P et al (2005) Direct synthesis of hydrogen peroxide from H-2 and O-2 using TiO2-supported Au–Pd catalysts. J Catal 236:69–79
21. Edwards JK, Solsona B, En N et al (2009) Switching off hydrogen peroxide hydrogenation in the direct synthesis process. Science 323:1037–1041
22. Edwards JK, Ntainjua E, Carley AF, Herzing AA, Kiely CJ, Hutchings GJ (2009) Direct Synthesis of H_2O_2 from H-2 and O-2 over Gold, Palladium, and Gold-Palladium Catalysts Supported on Acid-Pretreated TiO2. Angew Chem Int Ed 48:8512–8515
23. Mallat T, Baiker A (2004) Oxidation of alcohols with molecular oxygen on solid catalysts. Chem Rev 104:3037–3058
24. Dimitratos N, Lopez-Sanchez JA, Hutchings GJ (2012) Selective liquid phase oxidation with supported metal nanoparticles. Chem Sci 3:20–44
25. Daniel MC, Astruc D (2004) Gold nanoparticles: assembly, supramolecular chemistry, quantum-size-related properties, and applications toward biology, catalysis, and nanotechnology. Chem Rev 104:293–346
26. Lopez-Sanchez JA, Dimitratos N, Miedziak P et al (2008) Au–Pd supported nanocrystals prepared by a sol immobilisation technique as catalysts for selective chemical synthesis. Phys Chem Chem Phys 10:1921–1930
27. Dimitratos N, Lopez-Sanchez JA, Morgan D et al (2009) Solvent-free oxidation of benzyl alcohol using Au–Pd catalysts prepared by sol immobilisation. Phys Chem Chem Phys 11:5142–5153
28. Tiruvalam RC, Pritchard JC, Dimitratos N et al (2011) Aberration corrected analytical electron microscopy studies of sol-immobilized Au plus Pd, Au(Pd) and Pd(Au) catalysts used for benzyl alcohol oxidation and hydrogen peroxide production. Faraday Discuss 152:63–86
29. Meenakshisundaram S, Nowicka E, Miedziak PJ et al (2010) Oxidation of alcohols using supported gold and gold–palladium nanoparticles. Faraday Discuss 145:341–356
30. He Q, Miedziak PJ, Kesavan L et al (2013) Switching-off toluene formation in the solvent-free oxidation of benzyl alcohol using supported trimetallic Au–Pd–Pt nanoparticles. Faraday Discuss 162:365–378
31. Zhou CHC, Beltramini JN, Fan YX, Lu GQM (2008) Chemoselective catalytic conversion of glycerol as a biorenewable source to valuable commodity chemicals. Chem Soc Rev 37:527–549
32. Dimitratos N, Lopez-Sanchez JA, Anthonykutty JM et al (2009) Oxidation of glycerol using gold–palladium alloy-supported nanocrystals. Phys Chem Chem Phys 11:4952–4961
33. Brett GL, He Q, Hammond C et al (2011) Selective oxidation of glycerol by highly active bimetallic catalysts at ambient temperature under base-free conditions. Angew Chem Int Ed 50:10136–10139
34. Kesavan L, Tiruvalam R, Ab Rahim MH et al (2011) Solvent-free oxidation of primary carbon–hydrogen bonds in toluene using Au–Pd alloy nanoparticles. Science 331:195–199
35. bin Saiman MI, Brett GL, Tiruvalam R et al (2012) Involvement of surface-bound radicals in the oxidation of toluene using supported Au–Pd nanoparticles. Angew Chem Int Ed 51:5981–5985
36. Ab Rahim MH, Forde MM, Jenkins RL et al (2013) Oxidation of methane to methanol with hydrogen peroxide using supported gold–palladium alloy nanoparticles. Angew Chem Int Ed 52:1280–1284

Permissions

All chapters in this book were first published in APR, by Springer; hereby published with permission under the Creative Commons Attribution License or equivalent. Every chapter published in this book has been scrutinized by our experts. Their significance has been extensively debated. The topics covered herein carry significant findings which will fuel the growth of the discipline. They may even be implemented as practical applications or may be referred to as a beginning point for another development.

The contributors of this book come from diverse backgrounds, making this book a truly international effort. This book will bring forth new frontiers with its revolutionizing research information and detailed analysis of the nascent developments around the world.

We would like to thank all the contributing authors for lending their expertise to make the book truly unique. They have played a crucial role in the development of this book. Without their invaluable contributions this book wouldn't have been possible. They have made vital efforts to compile up to date information on the varied aspects of this subject to make this book a valuable addition to the collection of many professionals and students.

This book was conceptualized with the vision of imparting up-to-date information and advanced data in this field. To ensure the same, a matchless editorial board was set up. Every individual on the board went through rigorous rounds of assessment to prove their worth. After which they invested a large part of their time researching and compiling the most relevant data for our readers.

The editorial board has been involved in producing this book since its inception. They have spent rigorous hours researching and exploring the diverse topics which have resulted in the successful publishing of this book. They have passed on their knowledge of decades through this book. To expedite this challenging task, the publisher supported the team at every step. A small team of assistant editors was also appointed to further simplify the editing procedure and attain best results for the readers.

Apart from the editorial board, the designing team has also invested a significant amount of their time in understanding the subject and creating the most relevant covers. They scrutinized every image to scout for the most suitable representation of the subject and create an appropriate cover for the book.

The publishing team has been an ardent support to the editorial, designing and production team. Their endless efforts to recruit the best for this project, has resulted in the accomplishment of this book. They are a veteran in the field of academics and their pool of knowledge is as vast as their experience in printing. Their expertise and guidance has proved useful at every step. Their uncompromising quality standards have made this book an exceptional effort. Their encouragement from time to time has been an inspiration for everyone.

The publisher and the editorial board hope that this book will prove to be a valuable piece of knowledge for researchers, students, practitioners and scholars across the globe.

List of Contributors

Zoubida Lounis
Institut de Maintenance et de Se´curite´ Industrielle, Universite´ d'Oran, 1524, El Manouer, Oran, Algeria
Laboratoire de Chimie Des, Mate´riaux Universite´ d'Oran, 1524, El Manouer, Oran, Algeria

Naouel Boumesla
Laboratoire de Chimie Des, Mate´riaux Universite´ d'Oran, 1524, El Manouer, Oran, Algeria

A. E. K. Bengueddach
Laboratoire de Chimie Des, Mate´riaux Universite´ d'Oran, 1524, El Manouer, Oran, Algeria

Wu Zong Zhou
School of Chemistry, University of St Andrews, St Andrews, Fife KY16 9ST, UK

Feng Jiao Yu
School of Chemistry, University of St Andrews, St Andrews, Fife KY16 9ST, UK

Heather F. Greer
School of Chemistry, University of St Andrews, St Andrews, Fife KY16 9ST, UK

Zheng Jiang
Inorganic Chemistry Laboratory, Department of Chemistry, University of Oxford, Oxford OX1 3QR, UK

Peter P. Edwards
Inorganic Chemistry Laboratory, Department of Chemistry, University of Oxford, Oxford OX1 3QR, UK

Martin O. Jones
ISIS Facility, STFC Rutherford Appleton Laboratory, Chilton, Didcot, Oxon OX11 0QX, UK

Andrew D. Taylor
ISIS Facility, STFC Rutherford Appleton Laboratory, Chilton, Didcot, Oxon OX11 0QX, UK

Stewart F. Parker
ISIS Facility, STFC Rutherford Appleton Laboratory, Chilton, Didcot, Oxon OX11 0QX, UK

Nikolaos Dimitratos
Department of Chemistry, University College London, 20 Gordon Street, London WC1H 0AJ, UK

Jennifer K. Edwards
School of Chemistry, Cardiff University, Main Building, Park Place, Cardiff CF10 3AT, UK

Christopher J. Kiely
Department of Materials Science and Engineering, Lehigh University, 5 East Packer Avenue, Bethlehem, PA 18015-3195, USA

Graham J. Hutchings
School of Chemistry, Cardiff University, Main Building, Park Place, Cardiff CF10 3AT, UK

Abdulhamid A. Alsaygh
Petrochemical Institute at King Abdulaziz City for and Technology, PO Box 6086, Riyadh 11442, Saudi Arabia
Hamid Al-Megren
Petrochemical Research Institute, King Abdulaziz City for Science and Technology, P. O. Box 6086, Riyadh 11442, Saudi Arabia

Tiancun Xiao
Guangzhou Boxenergytech Ltd, Guangzhou Boneng Nnengyuan Keji Youxian Gongsi, Guangzhou City, People's Republic of China
KOPRC, Inorganic Chemistry Laboratory, Oxford University, South Parks Road, Oxford, OX1 3QR, UK

Mohammad AlKinany
Petrochemical Research Institute, King Abdulaziz City for Science and Technology, P. O. Box 6086, Riyadh 11442, Saudi Arabia

Saud Aldree
Petrochemical Research Institute, King Abdulaziz City for Science and Technology, P. O. Box 6086, Riyadh 11442, Saudi Arabia

Yu Huang
Guangzhou Boxenergytech Ltd, Guangzhou Boneng Nnengyuan Keji Youxian Gongsi, Guangzhou City, People's Republic of China

Haoyi Chen
Guangzhou Boxenergytech Ltd, Guangzhou Boneng Nnengyuan Keji Youxian Gongsi, Guangzhou City, People's Republic of China

Peter P. Edwards
KOPRC, Inorganic Chemistry Laboratory, Oxford University, South Parks Road, Oxford, OX1 3QR, UK

Vladimir Kuznetsov
KOPRC, Inorganic Chemistry Laboratory, Oxford University, South Parks Road, Oxford, OX1 3QR, UK

W. Trakarnpruk
Department of Chemistry, Petrochemistry and Polymer Science, Faculty of Science, Chulalongkorn University, Bangkok 10330, Thailand

J. Jatupisarnpong
Department of Chemistry, Petrochemistry and Polymer Science, Faculty of Science, Chulalongkorn University, Bangkok 10330, Thailand

Mohammed C. Al-Kinany
Petrochemicals Research Institute, King Abdulaziz City for Science and Technology, P.O.Box 6086, Riyadh 11442, Saudi Arabia

Hamid A. Al-Megren
Petrochemicals Research Institute, King Abdulaziz City for Science and Technology, P.O.Box 6086, Riyadh 11442, Saudi Arabia

Eyad A. Al-Ghilan
Petrochemicals Research Institute, King Abdulaziz City for Science and Technology, P.O.Box 6086, Riyadh 11442, Saudi Arabia

Peter P. Edwards
Inorganic Chemistry Laboratory, Wolfson Catalysis Centre, KACST-Oxford Petrochemicals Research Centre (KOPRC), University of Oxford, South Parks Road, Oxford OX1 3QR, UK

Tiancun Xiao
Inorganic Chemistry Laboratory, Wolfson Catalysis Centre, KACST-Oxford Petrochemicals Research Centre (KOPRC), University of Oxford, South Parks Road, Oxford OX1 3QR, UK

Ahmad. S. Al-Shammari
Inorganic Chemistry Laboratory, Wolfson Catalysis Centre, KACST-Oxford Petrochemicals Research Centre (KOPRC), University of Oxford, South Parks Road, Oxford OX1 3QR, UK

Saud A. Al-Drees
Petrochemicals Research Institute, King Abdulaziz City for Science and Technology, P.O.Box 6086, Riyadh 11442, Saudi Arabia

Faisal M. Alotaibi
Petrochemicals Research Institute, King Abdulazeez City for Science and Technology (KACST), PO Box: 6086, Riyadh 11442, Saudi Arabia

Raed H. Abudawood
Research and Development Center Saudi Aramco, Dhahran 31311, Saudi Arabia

Hamid A. Al-Megren
Petrochemicals Research Institute, King Abdulazeez City for Science and Technology (KACST), PO Box: 6086, Riyadh 11442, Saudi Arabia

Mohammed C. Al-Kinany
Petrochemicals Research Institute, King Abdulazeez City for Science and Technology (KACST), PO Box: 6086, Riyadh 11442, Saudi Arabia

Arthur A. Garforth
School of Chemical Engineering and Analytical Science, The University of Manchester, Manchester M13 9PL, UK

Rajagopalan Vijayaraghavan
School of Advanced Sciences, Centre for Nano Materials, VIT University, Vellore 632 014, India

A. Alshammari
National Nanotechnology Research Centre, King Abdulaziz City for Science and Technology, King Abdullah Road, P.O. Box 6086, Riyadh 11442, Saudi Arabia

A. Köckritz
Leibniz-Institut für Katalyse e.V. an der Universität Rostock, Albert-Einstein-Str. 29a, 18059 Rostock, Germany

V. N. Kalevaru
Leibniz-Institut für Katalyse e.V. an der Universität Rostock, Albert-Einstein-Str. 29a, 18059 Rostock, Germany

A. Bagabas
National Nanotechnology Research Centre, King Abdulaziz City for Science and Technology, King Abdullah Road, P.O. Box 6086, Riyadh 11442, Saudi Arabia
Leibniz-Institut für Katalyse e.V. an der Universität Rostock, Albert-Einstein-Str. 29a, 18059 Rostock, Germany

A. Martin
Leibniz-Institut für Katalyse e.V. an der Universität Rostock, Albert-Einstein-Str. 29a, 18059 Rostock, Germany

Ataallah Soltani Goharrizi
Department of Chemical Engineering, Faculty of Engineering, Shahid Bahonar University of Kerman, Jomhoori Blvd, 76175 Kerman, Iran

Bahador Abolpour
Department of Chemical Engineering, Faculty of Engineering, Shahid Bahonar University of Kerman, Jomhoori Blvd, 76175 Kerman, Iran

Faisal M. Alotaibi
Petrochemicals Research Institute, King Abdulaziz City for Science and Technology (KACST), P.O.Box 6086, Riyadh 11442, Saudi Arabia

Raed H. Abudawood
Research and Development Center, Saudi Aramco, Dhahran 31311, Saudi Arabia

Hamid A. Al-Megren
Petrochemicals Research Institute, King Abdulaziz City for Science and Technology (KACST), P.O.Box 6086, Riyadh 11442, Saudi Arabia

Mohammed C. Al-Kinany
Petrochemicals Research Institute, King Abdulaziz City for Science and Technology (KACST), P.O.Box 6086, Riyadh 11442, Saudi Arabia

Essam H. Jamea
SABIC Technology Center, Saudi Basic Industries Corporation, Riyadh 11551, Saudi Arabia

Arthur A. Garforth
School of Chemical Engineering and Analytical Science, The University of Manchester, Manchester M13 9PL, UK

Abdulhamid A. Alsaygh
Petroleum and Petrochemicals Research Institute, King Abdulaziz City for Science and Technology (KACST), P.O. Box 6086, Riyadh 11442, Saudi Arabia

Jehan Al-hamidi
Chemistry Department, College of Science, Princess Nora Bint Abdulrahman University, Riyadh, Saudi Arabia

Fares D. Alsewailem
Petroleum and Petrochemicals Research Institute, King Abdulaziz City for Science and Technology (KACST), P.O. Box 6086, Riyadh 11442, Saudi Arabia

Ibrahim M. Al-Najjar
Petroleum and Petrochemicals Research Institute, King Abdulaziz City for Science and Technology (KACST), P.O. Box 6086, Riyadh 11442, Saudi Arabia

Vladimir L. Kuznetsov
Department of Chemistry, University of Oxford, South Parks Road, Oxford OX1 3QR, UK

Liam J. France
Department of Chemistry, University of Oxford, South Parks Rd, Oxford OX1 3QR, UKXian Du

Naif Almuqati
Department of Chemistry, University of Oxford, South Parks Rd, Oxford OX1 3QR, UKXian Du

Vladimir L. Kuznetsov
Department of Chemistry, University of Oxford, South Parks Rd, Oxford OX1 3QR, UKXian Du

Yongxiang Zhao
School of Chemistry and Chemical Engineering, Shanxi University, Taiyuan 030006, China

Jiang Zheng
Department of Chemistry, University of Oxford, South Parks Rd, Oxford OX1 3QR, UKXian Du

Tiancun Xiao
Department of Chemistry, University of Oxford, South Parks Rd, Oxford OX1 3QR, UKXian Du

Abdulaziz Bagabas
Petrochemical Research Institute, King Abdulaziz City of Science and Technology (KASCT), King Abdullah Rd, Riyadh 11442, Saudi Arabia

Hamid Almegren
Petrochemical Research Institute, King Abdulaziz City of Science and Technology (KASCT), King Abdullah Rd, Riyadh 11442, Saudi Arabia

Peter P. Edwards
Department of Chemistry, University of Oxford, South Parks Rd, Oxford OX1 3QR, UKXian Du

Th. Abdel-Moghny
Applications Department, Egyptian Petroleum Research Institute, Ahmed El-Zomer, Nasr City, Box. No. 11727, Cairo, Egypt

Ramadan S. A. Mohamed
Geology Department, Faculty of Science, El-Minia University,El-Minia, Egypt

E. El-Sayed
Geology Department, Faculty of Science, El-Minia University,El-Minia, Egypt

Shoukry Mohammed Aly
Executive General Manager Collection and Treatment of Used Oil Petrotreat Co, Cairo, Egyp

Moustafa Gamal Snousy
Geology Department, Faculty of Science, El-Minia University,El-Minia, Egypt

Devendra Pakhare
Department of Chemical Engineering, Louisiana State University, Baton Rouge, LA 70803, USA

Daniel Haynes
National Energy Technology Laboratory, U.S. Department of Energy, 3610 Collins Ferry Rd., Morgantown, WV 26507, USA

Dushyant Shekhawat
National Energy Technology Laboratory, U.S. Department of Energy, 3610 Collins Ferry Rd., Morgantown, WV 26507, USA

James Spivey
Department of Chemical Engineering, Louisiana State University, Baton Rouge, LA 70803, USA

Asel Sartbaeva
Department of Chemistry, Inorganic Chemistry Laboratory, University of Oxford, South Parks Road, Oxford OX1 3QR, UK

Stephen A. Wells
Department of Physics and Centre for Scientific Computing, University of Warwick, Gibbet Hill Road, Coventry CV4 7AL, UK

Hiren M. Bhajiwala
Reliance Technology Group, Reliance Industries Limited, Hazira Complex, Surat 394510, Gujarat, India

Harshad R. Patil
Reliance Technology Group, Reliance Industries Limited, Hazira Complex, Surat 394510, Gujarat, India

Virendrakumar Gupta
Reliance Technology Group, Reliance Industries Limited, Hazira Complex, Surat 394510, Gujarat, India

Adrian Porch
School of Engineering, Cardiff University, Cardiff CF24 3AA, UK

Dan Slocombe
School of Engineering, Cardiff University, Cardiff CF24 3AA, UK

Jan Beutler
School of Engineering, Cardiff University, Cardiff CF24 3AA, UK

Peter Edwards
Inorganic Chemistry Laboratory, Oxford University, Oxford OX1 3QR, UK Afrah Aldawsari

Tiancun Xiao
Inorganic Chemistry Laboratory, Oxford University, Oxford OX1 3QR, UK Afrah Aldawsari

Vladimir Kuznetsov
Inorganic Chemistry Laboratory, Oxford University, Oxford OX1 3QR, UK Afrah Aldawsari

Hamid Almegren
Petrochemical Research Institute, King Abdulaziz City for Science and Technology, P.O. Box 6086, Riyadh 11442, Kingdom of Saudi Arabia

Saud Aldrees
Inorganic Chemistry Laboratory, Oxford University, Oxford OX1 3QR, UK Afrah Aldawsari
Petrochemical Research Institute, King Abdulaziz City for Science and Technology, P.O. Box 6086, Riyadh 11442, Kingdom of Saudi Arabia

Naif Almaqati
Inorganic Chemistry Laboratory, Oxford University, Oxford OX1 3QR, UK Afrah Aldawsari
Petrochemical Research Institute, King Abdulaziz City for Science and Technology, P.O. Box 6086, Riyadh 11442, Kingdom of Saudi Arabia

J. S. J. Hargreaves
WestCHEM, School of Chemistry, University of Glasgow, Joseph Black Building, Glasgow G12 8QQ, UK

Qingyuan Li
State Key Laboratory of Chemical Resource Engineering, Beijing University of Chemical Technology, Beijing 100029, China

Pingyi Wu
Petrochemical Research Institute, PetroChina, Beijing 100195, China
State Key Laboratory of Chemical Resource Engineering, Beijing University of Chemical Technology, Beijing 100029, China

Ling Lan
Petrochemical Research Institute, PetroChina, Beijing 100195, China

Ni Wei
State Key Laboratory of Chemical Resource Engineering, Beijing University of Chemical Technology, Beijing 100029, China

Shengfu Ji
State Key Laboratory of Chemical Resource Engineering, Beijing University of Chemical Technology, Beijing 100029, China

Rui Feng
State Key Laboratory of Heavy Oil Processing, Key Laboratory of Catalysis, China University of Petroleum, Qingdao 266580, China

Ke Qiao
State Key Laboratory of Heavy Oil Processing, Key Laboratory of Catalysis, China University of Petroleum, Qingdao 266580, China

You-he Wang
State Key Laboratory of Heavy Oil Processing, Key Laboratory of Catalysis, China University of Petroleum, Qingdao 266580, China

Zi-feng Yan
State Key Laboratory of Heavy Oil Processing, Key Laboratory of Catalysis, China University of Petroleum, Qingdao 266580, China

Younes Amini
Department of Chemical Engineering, Isfahan University of Technology, Isfahan, Iran

Moslem Fattahi
Department of Chemical and Petroleum Engineering, Sharif University of Technology, Azadi Avenue, P.O. Box 11365-9465, Tehran, Iran

Farhad Khorasheh
Department of Chemical and Petroleum Engineering, Sharif University of Technology, Azadi Avenue, P.O. Box 11365-9465, Tehran, Iran

Saeed Sahebdelfar
Catalyst Research Group, Petrochemical Research and Technology Company (NPC-RT), Tehran, Iran

Maryam Takht Ravanchi
Catalyst Research Group, Petrochemical Research and Technology Company, National Petrochemical Company, No. 27, Sarv Alley, Shirazi-south, Mollasadra,P.O. Box 14358-84711, Tehran, Iran

Saeed Sahebdelfar
Catalyst Research Group, Petrochemical Research and Technology Company, National Petrochemical Company, No. 27, Sarv Alley, Shirazi-south, Mollasadra,P.O. Box 14358-84711, Tehran, Iran

Nicolas F. Sauty
The George and Josephine Butler Polymer Research Laboratory, Department of Chemistry and Center for Macromolecular Science and Engineering, University of Florida, Gainesville, FL 32611-7200, USA

Lucas Caire da Silva
The George and Josephine Butler Polymer Research Laboratory, Department of Chemistry and Center for Macromolecular Science and Engineering, University of Florida, Gainesville, FL 32611-7200, USA

Michael D. Schulz
The George and Josephine Butler Polymer Research Laboratory, Department of Chemistry and Center for Macromolecular Science and Engineering, University of Florida, Gainesville, FL 32611-7200, USA

Chip S. Few
The George and Josephine Butler Polymer Research Laboratory, Department of Chemistry and Center for Macromolecular Science and Engineering, University of Florida, Gainesville, FL 32611-7200, USA

Kenneth B. Wagener
The George and Josephine Butler Polymer Research Laboratory, Department of Chemistry and Center for Macromolecular Science and Engineering, University of Florida, Gainesville, FL 32611-7200, USA

Siauw Ng
Natural Resources Canada, CanmetENERGY, One Oil Patch Drive, Devon, AB T9G 1A8, Canada

Craig Fairbridge
Natural Resources Canada, CanmetENERGY, One Oil Patch Drive, Devon, AB T9G 1A8, Canada

Sateesh Mutyala
Natural Resources Canada, CanmetENERGY, One Oil Patch Drive, Devon, AB T9G 1A8, Canada

Yan Liu
Natural Resources Canada, CanmetENERGY, One Oil Patch Drive, Devon, AB T9G 1A8, Canada

Jacqueline M. R. Bélanger
Your World, 94 Newton Street, Moncton, NB E1E 3A4, Canada

J. R. Jocelyn Paré
Your World, 94 Newton Street, Moncton, NB E1E 3A4, Canada

Nikolaos Dimitratos
School of Chemistry, Cardiff University, Main Building, Park Place, Cardiff CF10 3AT, UK

Ceri Hammond
School of Chemistry, Cardiff University, Main Building, Park Place, Cardiff CF10 3AT, UK

Christopher J. Kiely
Center for Advanced Materials and Nanotechnology, Lehigh University, 5 East Packer Avenue, Bethlehem, PA 18015-3195, USA

Graham J. Hutchings
School of Chemistry, Cardiff University, Main Building, Park Place, Cardiff CF10 3AT, UK